DIFFERENTIAL EQUATIONS

A Systems Approach

DIFFERENTIAL EQUATIONS

A Systems Approach

Jack Goldberg
University of Michigan

Merle C. Potter
Michigan State University

PRENTICE-HALL, Upper Saddle River, NJ 07458

Library of Congress Cataloging-in-Publication Data
Goldberg, Jack L. (Jack Leonard)
Differential Equations: a systems approach / Jack L. Goldberg, Merle Potter.
p. cm.
Includes index.
ISBN 0-13-211319-8
1. Differential equations. II. Potter, Merle C. II Title.
QA372.G598 1997
555'.35--dc21 97-1159
CIP

Acquisitions Editor: George Lobell
Editorial Assistant: Gale Epps
Editorial Director: Tim Bozik
Editor-in-Chief: Jerome Grant
Assistant Vice President of Production and Manufactuing: David W. Riccardi
Editorial/Production Supervisor: Robert C. Walters
Managing Editor: Linda Mihatov Behrens
Executive Managing Editor: Kathleen Schiaparelli
Manufacturing Buyer: Alan Fischer
Manufacturing Manager: Trudy Pisciotti
Marketing Manager:Melody Marcus
Marketing Assistant: Jennifer Pan
Creative Director: Paula Maylahn
Art Director: Jayne Conte
Cover Desigener: Bruce Kenselaar
Cover Photo: Yasuhiro Ishimoto/KB. Team Disney Building. Architect Arata Isozaki

Printed in the United States of America
10 9 8 7 6 5 4 3 2 1

ISBN 0-13-211319-8

PRENTICE-HALL INTERNATIONAL (UK) LIMITED, LONDON
PRENTICE-HALL of AUSTRALIA PTY. LIMITED, SIDNEY
PRENTICE-HALL HISPANOAMERICANA, S.A., MEXICO
PRENTICE-HALL of INDIA PRIVATE LIMITED, NEW DELHI
PRENTICE-HALL of JAPAN INC., TOKYO
SIMON & SCHUSTER ASIA PTE LTD., SINGAPORE
EDITORA PRENTICE-HALL DO BRASIL, LTDA., RIO DE JANEIRO

To our families,
our friends
and our students

Contents

List of Figures

PREFACE

The purpose of this book is to present a thorough introduction to methods for solving differential equations. Clear and precise explanations are illustrated with numerous examples to provide both the theory and the application of that theory. Hundreds of exercises, which have been ordered in level of difficulty, then allow students to develop their understanding of the theory and their skills at problem solving.

This book is unique in that matrices are used as an integral part of the presentation of differential equations. This allows an easy extension to computer simulations of the equations and their solutioons. Many computer solutions and exercises using MATLAB are included in the text. Also, we have included numerous Projects at the end of most chapters to allow more in-depth work for the interested student. These projects can easily be adapted to team efforts. The text *A Guided Tour of Differential Equations using Computer Technology* by Alexandra Skidmore and Margie Hale provide natural and useful resource for this text.

There is sufficient material for two semesters of study. The topics can be arranged so that the first course, which is required in many curricula, fits the unique needs of the user. A second, often elective course, can then be construsted from the remaining material.

We have class tested this book in manuscript form in a first course in differential equations at the University of Michigan. The course serves hundreds of Engineering and Science maajors ever semester. Relying on this feed-back, we offer a Guide to the Instructor.

We owe a dept of thanks to the many individuals whose criticism, suggestions, hard work and thoughtfulness helped us design and complete this work. To Vijay Pant, Mark Gockenbach, John Aarsvold, Michael Bean and Petra Bonfert at the University of Michigan for invaluable assistence in using Unix and LaTeX; to our students who suffered through the embryonic stages of this work in the form of class notes; and to the reviewers who kept us on our toes with their incitefull comments: Daniel Sweet, Univeristy of Maryland, College Park; Tasso Kaper, Boston University; Frank M Cholewinski, Clemson University; Matt Insall, Univeristy of Missouri at Rolla; James L.

Handley, Montana Tech; and James W. Carlson, Creighton Univerisity. In spite of this army of helpers, we take sole responsibility for the errors and omissions.

INSTRUCTOR'S GUIDE

Accompanying this text is a *Guided Tour of Differential Equations using Computer Technology* by Alexandra Skidmore and Margie Hale. This supplementary book offers computer projects in differential equations with necessary keystroke instructions for Maple, Mathematica and Derive. This work can be combined with selected examples and exercises to constitute either a one or two semester course.

In what follows, we outline the possibility of ones semester course following either the traditional study of the nth-order equation with systems being relegated to the background, or a study of systems of linear differential equations through the medium of matrix algebra. There is sufficient material here so that the instructor can pick and choose among the latter four chapters. Indeed, one can easily see how to use this text in a two semester survey of the initial and boundary value problems. Our suggestions follow.

Chapter 0: Complex Numbers, Roots and Matrices

Section 1—3 provide a review of the theory of equations. All of this material is needed in the study of the constant coefficient equation, but particular emphasis should be given to complex roots and the exponential form of complex numbers. Sections 4—16 are best left until needed in Chapter 2.

Chapter 1: First-order Differential Equations

Here we acquaint the student with first-order linear equations. This is a good place to explain what it means to solve a differential equation, to explain the distinction between specific and general solutions and to describe and illustrate initial-value problems. Towards this end, we present a number of simple applied problems. The instructor is encouraged to select two or three examples.

This chapter also provides an introduction to change of variables in a linear equation by offering the method of variation of parameters for the first-order linear equation.

Section 5 offers a brief excursion into the most common integrable nonlinear equations, variables separable and exact equations. Some users may wish to skip this section; this can be safely done because no future work depends explicity on this material.

Chapter 2: Linear Systems

Instructors who wish to use this text for a more traditional course without relying on systems material may skip this chapter and go immediatlely to Chapter 3. For those using a systems approach, we recommend first returning to Chapter 0, Sections 4—11 to review notation, row-reduction, determinants and linear independence. In this regard, the relationship between $\det \mathbf{A} = 0$ and $\mathbf{Ax} = \mathbf{0}$ having nontrivial solutions usually needs special emphasis.

We recommend doing all sections in this chapter although one may skip Sections 2.5 and 2.8 to save time.

Chapter 3: Second-Order Equations

This is the traditional treatment of the second-order, constant coefficient linear equation. One might choose to skip Section 3.2 because it is a bit more theoretical than the remaining sections and skip Section 3.3 since sectionally continuous functions are used only in Chapter 5, The Laplace Transform (its natural home!) and Chapter 8.

As in Chapter 1, we offer the instructor a choice of the classical applications to electrical ciruits and spring-mass systems.

We have made Section 3.11, The Cauchy-Euler equation of second-order optional. Section 3.12 is the standard, and in our opinion, rather difficult approach to variation of paramters. We believe that Section 4.6, variation of parameters for systems, is a far more natural method of presentation; first, because it is so similar to variation of paramters for the scalar first-order equation and second, because there is no need to puzzle over the hard-to-motivate equations which comprise the classical approach. (One never sees the classical variation of parameters for the third-order equation, and not just because it is tedious!)

Chapter 4: Higher-Order Equations

Sections 1—3 handle the nth-order equation without reference to systems. These sections, then, provide a natural extension of the material of Chapter 3 to higher order equations. We recommend these sections because they are good practice on the ideas introduced earlier. Sections 4—6 discuss the central idea of converting an nth-order equation to a (companion) system. For instructors using a systems approach we recommend doing the entire chapter, albeit some of the more theoretical aspects can be done lightly or even skipped.

Chapter 5: The Laplace Transform

As mentioned in the preface, Chapter 5 is one of optional chapters. We begin with the traditional approach to the Laplace transform with applications to sectionally continuous forcing functions (square-wave, saw-tooth, and the like). Section 5.8 is optional and Section 5.12 treats the Laplace transform of systems.

Chapter 6: Series Methods

This optional chapter presents a fairly complete discussion of power series methods in the solution of linear differential equations with variable coefficients. Besides the expansion of solutions about an ordinary point, we present solutions about the regular singular point (the Frobenius series) and a method we call the Wronskian method. The classical Bessel, Legendre, Laguerre, Hermite and Bessel-Clifford equations are used as illustrations.

Chapter 7: Numerical Methods

This chapter introduces the methods of Euler, Heun and Runge-Kutta. We use the Euler method to explain the main ideas in numerical approximations. The Heun and Runge-Kutta methods are presented for comparison purposes and to illustrate how improved accuracy is obtained. The last section is devoted to an explaination of these methods applied to systems.

Chapter 8: Boundary-value Problems

This chapter introduces the boundary-value problem as a consequence of the separation of variables in the wave, diffusion and Laplace equations. This leads directly to the expansion of functions in Fourier series. Motivation for this material is provided by physical applications.

DIFFERENTIAL EQUATIONS

A Systems Approach

Chapter 0

Complex Numbers, Roots of Polynomial Equations, and Elements of Matrix Theory

0.1 Introduction

This chapter contains several topics needed in the study of differential equations. For some students, the chapter will serve as an introduction to these ideas; for others, it will serve as a review. In either case, the material presented here will be useful as a reference source. We begin a brief review of complex arithmetic.

A number z is complex if it is of the form $x + iy$, where x and y are real and $i^2 = -1$. The real numbers are a subset of the complex numbers obtained by setting $y = 0$. The real numbers x and y are called the *real part* and *imaginary part* of z, respectively. We write

$$\operatorname{Re} z = x \quad \text{and} \quad \operatorname{Im} z = y \tag{0.1.1}$$

and hence,

$$z = \operatorname{Re} z + i \operatorname{Im} z$$

If $z = x + iy$, then $\overline{z} = x - iy$ is called the *complex conjugate* of z. In terms of z and $\overline{z}$, we see that

$$\operatorname{Re} z = \frac{1}{2}(z + \overline{z}) \quad \text{and} \quad \operatorname{Im} z = \frac{1}{2i}(z - \overline{z}) \tag{0.1.2}$$

We define the *modulus* or *absolute value* of z, written $r = |z|$, by

$$r = |z| = \sqrt{x^2 + y^2} \tag{0.1.3}$$

and an *argument* of z, written $\theta = \arg z$, by

$$\theta = \arg z = \arctan \frac{y}{x} \tag{0.1.4}$$

Figure 0.1 shows the geometric relationships between these terms.

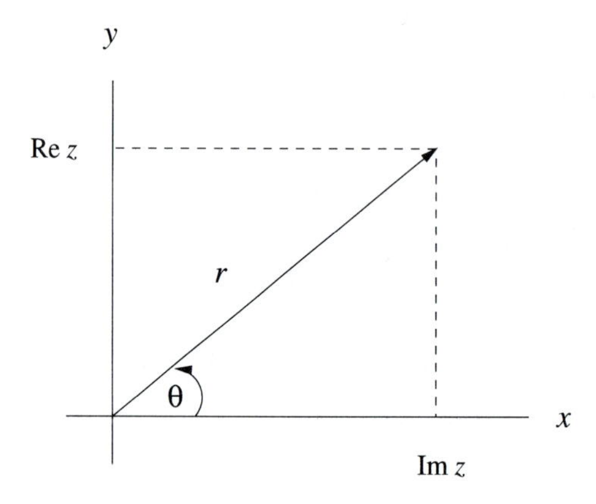

Figure 0.1: The Complex Plane

A number of elementary arithmetic results connected with complex numbers are listed in the following theorem:

Theorem 0.1.1. *For the complex numbers z and w,*

(a) $z = 0$ *if and only if* $\operatorname{Re} z = 0$ *and* $\operatorname{Im} z = 0$.

(b) $\overline{(z + w)} = \overline{z} + \overline{w}$ *and* $\overline{zw} = \overline{z}\,\overline{w}$.

(c) $z\overline{z} = |z|^2 = x^2 + y^2 \geq 0$.

(d) $\operatorname{Re}(z + w) = \operatorname{Re} z + \operatorname{Re} w, \quad \operatorname{Im}(z + w) = \operatorname{Im} z + \operatorname{Im} w$.

(e) *For each real k,* $\operatorname{Re}(kz) = k\operatorname{Re} z$ *and* $\operatorname{Im}(kz) = k \operatorname{Im} z$.

(f) $\operatorname{Re}(zw) = (\operatorname{Re} z)(\operatorname{Re} w) - (\operatorname{Im} z)(\operatorname{Im} w)$.

(g) $\operatorname{Im}(zw) = (\operatorname{Im} z)(\operatorname{Re} w) + (\operatorname{Re} z)(\operatorname{Im} w)$.

Proof: The proof for Part (c) follows; the remaining arguments are left to the student. (See Problem 6.) Suppose $z = x + iy$. Then, because $i^2 = -1$ and by the definition of $\overline{z}$,

$$z\overline{z} = (x + iy)(x - iy) = x^2 + y^2 = |z|^2$$

Finally, $x^2 + y^2 \geq 0$ because x and y are both real. □

Example 0.1.1. *For the three complex numbers, $z_1 = 5 - 2i$, $z_2 = 4i$, and $z_3 = -2$, identify their real and imaginary parts and find their complex conjugates and absolute values.*

Solution: Their real and imaginary parts and their conjugates are

$$\begin{array}{lll} \operatorname{Re} z_1 = 5, & \operatorname{Im} z_1 = -2, & \overline{z}_1 = 5 + 2i \\ \operatorname{Re} z_2 = 0, & \operatorname{Im} z_2 = 4, & \overline{z}_2 = -4i \\ \operatorname{Re} z_3 = -2, & \operatorname{Im} z_3 = 0, & \overline{z}_3 = -2 \end{array}$$

Their absolute values are

$$\begin{aligned} |\overline{z}_1| &= \sqrt{5^2 + 2^2} = \sqrt{29} \\ |\overline{z}_2| &= \sqrt{0^2 + (-4)^2} = 4 \\ |\overline{z}_3| &= \sqrt{2^2 + 0^2} = 2 \end{aligned}$$

Example 0.1.2. *Use the results in* Theorem 0.1.1 *to find the real and imaginary parts of* $1/z$.

Solution: Let $z = x + iy$ so that $\overline{z} = x - iy$. Then from Part (c) of Theorem 0.1.1,

$$\frac{1}{z} = \frac{\overline{z}}{z\overline{z}} = \frac{x - iy}{x^2 + y^2} = \frac{x}{x^2 + y^2} - \frac{iy}{x^2 + y^2} = \frac{x}{|z|^2} - i\frac{y}{|z|^2}$$

and hence,

$$\operatorname{Re}\frac{1}{z} = \frac{x}{|z|^2} \quad \text{and} \quad \operatorname{Im}\frac{1}{z} = -\frac{y}{|z|^2}$$

• EXERCISES

1. Identify the real and imaginary parts of the following numbers.
   (a) $3 - 2i$. (b) $-4 + 9i$.
   (c) 5. (d) $-4i$.
2. Write the complex conjugate of the following numbers.
   (a) $4 - 2i$. (b) -5.
   (c) $-5i$. (d) $-3 - 4i$.
3. Calculate each of the following products.
   (a) $(1 - 2i)\,3i$. (b) $(2 + i)(4 - 2i)$.
   (c) $(5 - 2i)(i - 3)$. (d) $(1 + i)(1 - i)$.

(e) $(3 + 2i)(1 - 4i)$.

4. Find the real and imaginary parts of the reciprocals of each number.
   (a) $1 - i$. (b) $2 + 3i$.
   (c) $2 + i$. (d) $3 + 4i$.
   (e) $2i$.
5. If $1/z = 5 - 12i$, what is z?
6. Prove Parts (a), (b), (d), (e), (f) and (g) of Theorem 0.1.1.
7. Show that the complex conjugate of $\overline{z}$ is z.
8. Show that $\text{Re}\,(iz) = -\text{Im}\,z$ and $\text{Im}\,(iz) = \text{Re}\,z$.

MATLAB

MATLAB uses "real(z)", "imag(z)", "conj(z)", "abs(z)", and "angle(z)" for $\text{Re}\,z$, $\text{Im}\,z$, $\overline{z}$, $|z|$, and $\arg z$, respectively. Use MATLAB to answer Problems 9—14.

9. Repeat Problem 1.
10. Repeat Problem 2.
11. Repeat Problem 3.
12. Repeat Problem 4.
13. Find the real and imaginary parts of
    (a) $\ln(i)$. (b) $\ln(-1)$. (c) i^i. (d) 2^i.
    (e) $\sqrt{i}$. (f) $e^{(1+i)}$. (g) $e^{\pi i}$. (h) $i^{1/3}$.
14. For each part in Problem 13, find $\arg(z)$ and $|z|$.

0.2 The Cartesian and Exponential Forms

Complex numbers are plotted on a rectangular grid in much the same manner as pairs of real numbers (x, y) are plotted on a rectangular Cartesian coordinate system. Indeed, simply identify the complex number $z = x + iy$ with the pair of real numbers (x, y), and plot (x, y). The y-axis is called the *imaginary axis* because the second coordinate is the imaginary part of z; the x-axis is called the *real axis*. In this geometric context, $z = x + iy$ is known as the *Cartesian form* of z. See the plot of z given in Figure 0.1.

We can solve for x and y in terms of θ and r from $x = r \cos\theta$ and $y = r \sin\theta$. (These are the familiar equations relating the polar coordinates of a point to its Cartesian coordinates.) Note that these equations and $z = x + iy$ imply

$$z = r(\sin\theta + i \sin\theta) \tag{0.2.1}$$

For example, we can find r and θ for the complex number $z = 3 + 4i$. We have $r = 5$, $\cos\theta = 0.6$ and $\sin\theta = 0.8$.

It has been known since the 18th century that the exponential function and the trigonometric functions are related. This remarkable relationship is called *Euler's formula* and is given by[1]

$$e^{i\theta} = \cos\,\theta + i\,\sin\,\theta \tag{0.2.2}$$

The *exponential form* of z follows from (0.2.1) and (0.2.2):

$$z = re^{i\theta} \tag{0.2.3}$$

which implies

$$\overline{z} = re^{-i\theta} = r\,(\cos\,\theta - i\,\sin\,\theta) \tag{0.2.4}$$

From the preceding definitions, we see that

$$z = re^{(\theta + 2\pi n)i} \tag{0.2.5}$$

provides the same value for z, for all integers n. This is the case because $\cos\,(\theta + 2\pi n) = \cos\,\theta$ and $\sin\,(\theta + 2\pi n) = \sin\,\theta$. Nonzero values of n are used to find roots and logarithms of complex numbers. See Examples 0.2.5 and 0.2.6.

Example 0.2.1. *Convert $z = 1 + i$ into exponential form.*

Solution: By definition, $x = y = 1$, z is in the first quadrant, and $r = \sqrt{2}$. From elementary trigonometry, $\tan^{-1}(y/x) = \tan^{-1}(1) = \frac{1}{4}\pi$. Using (0.2.1) we have

$$z = \sqrt{2}\left(\cos\,\tfrac{1}{4}\pi + i\,\sin\,\tfrac{1}{4}\pi\right)$$

and from (0.2.3), the exponential form is $z = \sqrt{2}e^{i\pi/4}$.

Example 0.2.2. *Convert $z = 2e^{i\pi}$ into Cartesian form.*

Solution: Here we are given $r = 2$ and $\theta = \pi$. Hence

$$z = 2\,(\cos\,\pi + i\,\sin\,\pi) = 2\,(-1 + i0) = -2$$

---

[1] Euler's formula implies the striking result $e^{i\pi} = -1$ obtained by choosing $\theta = \pi$.

It is possible to present a rigorous development of the algebra of complex numbers, but such a task would be far too great a diversion from our main goals. Instead, we shall rely upon formal manipulations to carry us through to the more useful of the scores of elementary theorems. The next theorem lists some of the results most often needed in differential equations.

Theorem 0.2.1. *For the complex number* $z = x + iy$

$$\begin{aligned} &(a) \quad \operatorname{Re} e^z = e^x \cos y \quad \textit{and} \quad \operatorname{Im} e^z = e^x \sin y \\ &(b) \quad \cos y = \tfrac{1}{2}\left(e^{iy} + e^{-iy}\right) \\ &(c) \quad \sin y = \tfrac{1}{2i}\left(e^{iy} - e^{-iy}\right) \\ &(d) \quad |e^z| = e^x > 0 \end{aligned}$$

Proof: We cannot prove these results because we have not given sufficiently careful definitions. However, the arguments we will provide are suggestive and useful in that they illustrate the type of manipulative techniques that work for complex numbers. We illustrate the argument for (a). Using simple algebra, we have

$$e^z = e^{x+iy} = e^x e^{iy} = e^x (\cos y + i \sin y)$$

from (0.2.2). The argument is completed by referring to the definitions of real and imaginary parts of a complex number. Now we show how (b) and (c) may be established. From (a) with $x = 0$, we find that

$$e^{iy} = \cos y + i \sin y \quad \text{and} \quad e^{-iy} = \cos y - i \sin y$$

Add these two equations and obtain (b). Subtract the second from the first and obtain (c). For (d), we use (a) to find

$$\begin{aligned} |e^z| &= \sqrt{(e^x \cos y)^2 + (e^x \sin y)^2} \\ &= \sqrt{e^{2x}\left(\cos^2 y + \sin^2 y\right)} = e^x \end{aligned}$$

since $\cos^2 y + \sin^2 y = 1$. Finally, $e^x > 0$ is a known fact about (real) exponential functions. □

Example 0.2.3. *Find the real and imaginary parts of* $(1 + 2i)\, e^{(1-i)t}$.

Solution: Let $z(t) = (1 + 2i)\, e^{(1-i)t}$. Then,

$$e^{(1-i)t} = e^t e^{-it} = e^t (\cos t - i \sin t)$$

so $z(t)$ can now be evaluated by elementary complex arithmetic:

$$\begin{aligned} z(t) &= (1+2i)\, e^t\, (\cos t - i \sin t) \\ &= e^t\, ((\cos t + 2\sin t) + i(2\cos t - \sin t)) \end{aligned}$$

where we have multiplied $(1+2i)$ with $(\cos t - i \sin t)$ using (f) of Theorem 0.1.1. Hence,

$$\operatorname{Re} z(t) = e^t\, (\cos t + 2\sin t)$$

and

$$\operatorname{Im} z(t) = e^t\, (2\cos t - \sin t)$$

Example 0.2.4. *Find the real and imaginary parts of*

$$z(t) = \frac{13}{2-3i} e^{2it}$$

Solution: We have

$$\begin{aligned} z(t) &= \frac{13}{2-3i} e^{2it} \\ &= 13 \frac{2+3i}{(2-3i)\,(2+3i)} e^{2it} \\ &= 13 \frac{2+3i}{13} (\cos 2t + i \sin 2t) \\ &= (2\cos 2t - 3\sin 2t) + i\, (3\cos 2t + 2\sin 2t) \end{aligned}$$

Hence,

$$\operatorname{Re} z(t) = 2\cos 2t - 3\sin 2t, \quad \operatorname{Im} z(t) = 3\cos 2t + 2\sin 2t$$

Example 0.2.5. *Find the natural logarithm of i and $3+4i$.*

Solution: We use (0.2.3) and write the complex number i as

$$i = e^{i\pi/2}$$

recognizing that $r = 1$ and choosing $\theta = \pi/2$. Then

$$\ln i = \ln e^{i\pi/2} = i\pi/2$$

Use (0.2.3) and write $3+4i=5e^{0.9273i}$. Then,

$$\begin{aligned}\ln(3+4i) &= \ln\left(5e^{0.9273i}\right)=\ln 5+\ln e^{0.9273i}\\ &= 1.609+0.9273i\end{aligned}$$

Example 0.2.6. *Use Euler's formula to find $\sqrt{i}$.*

Solution: As in the preceding example, (0.2.3) is used to write $i=re^{i\theta}=e^{i\pi/2}$. Then

$$\begin{aligned}\sqrt{i}=i^{1/2}=\left(e^{i\pi/2}\right)^{1/2} &= e^{i\pi/4}=\cos\tfrac{1}{4}\pi+i\sin\tfrac{1}{4}\pi\\ &= \tfrac{\sqrt{2}}{2}+i\tfrac{\sqrt{2}}{2}=\tfrac{\sqrt{2}}{2}(1+i)\end{aligned}$$

Letting $n=1$ in (0.2.5) leads to a second root of $\sqrt{i}$, (see also Problem 7 in the exercise set that follows):

$$\left(e^{(\pi/2+2\pi)i}\right)^{1/2}=e^{i5\pi/4}=\cos\tfrac{5\pi}{4}+i\sin\tfrac{5\pi}{4}=-\tfrac{\sqrt{2}}{2}(1+i)$$

- **EXERCISES**

1. Plot the following numbers in the complex plane, and express them in exponential form.
   (a) $3-4i$. (b) $-5+12i$.
   (c) $4+3i$. (d) $4-4i$.
   (e) $3+4i$. f) $5-12i$.
   (g) $4+4i$. (h) $-2i$.
   (i) -10.
2. Convert the following complex numbers into into Cartesian form.
   (a) $13e^{i\pi/2}$. (b) $6e^{0.4i}$.
3. Convert the following complex numbers into Cartesian form.
   (a) $e^{2+i\pi}$. (b) $e^{-1+i\pi/2}$.
   (c) $e^{(2+i\pi)t}$. (d) $3e^{(1+i\pi)t/2}$.
4. Find the real and imaginary parts of the following functions.
   (a) $2ie^{it}$. (b) $(2-31)\,e^{-it}$.
   (c) $ie^{(2+3i)t}$. (d) $(1-4i)\,e^{(3-2i)t}$.
   (e) $\ln 2i$. (f) $\ln(1+2i)$.
   (g) $\ln(2-3i)$. (h) $\ln(-2)$.
5. Express the following numbers in Cartesian form.
   (a) $(3+4i)^{1/2}$. (b) $(-5+12i)^{1/3}$.

6. Compute $\left(\pm\frac{\sqrt{2}}{2}(1+i)\right)^2$ to verify that these complex numbers are indeed two square roots of i.
7. Use the argument given in Example 0.2.6 to obtain the square root of i except begin with $i = \cos\frac{5}{2}\pi + i\sin\frac{5}{2}\pi = e^{5\pi i/2}$.

MATLAB

8. Use "angle(z)" and "abs(z)" to solve Problems 1–3.
9. Use "real(z)" and "imag(z)" to solve Problems 4–5.
10. Repeat Problem 7 by computing $(3+4i)$^ 1/2 and $(-5+12i)$^ 1/3.
11. Find the real and imaginary parts of
    (a) $e^{(1+i)}$.
    (b) $e^{\pi i}$.
    (c) $\ln(i)$.
    (d) $\ln(-1)$.
    (e) $\ln(2+3i)$.
    (f) i^i.
    (g) 2^i.

0.3 Roots of Polynomial Equations and Numbers

Let $p_n(x)$ and $q_n(x)$ be polynomials of degree n. Then, after multiplying through by the reciprocal of the coefficient of x^n, we may assume that $p_n(x)$ has the form

$$p_n(x) = x^n + a_{n-1}x^{n-1} + \cdots + a_1x + a_0 \tag{0.3.1}$$

Polynomials whose highest power has coefficient 1 are called *monic*. The coefficients a_i of p_n may not be real, although for our purposes they always are. In any case, the Fundamental Theorem of Algebra states that the equation $p_n(x) = 0$ has n solutions among the complex numbers.

Suppose that $x = r$ is a root of $p_n(x)$. This means that

$$p_n(r) = r^n + a_{n-1}r^{n-1} + \cdots + a_1r + a_0 = 0 \tag{0.3.2}$$

If $p_n(x)$ is divided by $x - r$, we obtain the identity

$$\frac{p_n(x)}{x-r} = q_{n-1}(x) + \frac{R}{x-r} \tag{0.3.3}$$

where R is a constant and $q_{n-1}(x)$ is a polynomial of degree $n-1$. Hence,

$$p_n(x) = (x-r)\,q_{n-1}(x) + R \tag{0.3.4}$$

But (0.3.4) is an identity in x, so that by setting $x = r$, we see that $R = 0$ if and only if $p_n(r) = 0$, that is, if and only if r is a root of $p_n(x)$. We combine this crucial result with the to obtain the following theorem:

Theorem 0.3.1. *For every polynomial*

$$p_n(x) = x^n + a_{n-1}x^{n-1} + \cdots + a_1 x + a_0$$

there exist n complex numbers $r_1, r_2, \ldots, r_n$ called its roots, such that

$$\begin{aligned} &(a) \quad p_n(r_i) = 0 \text{ for all } i = 1, 2, \ldots, n \\ &(b) \quad p_n(x) = (x - r_1)(x - r_2) \cdots (x - r_n) \end{aligned}$$

Moreover, if $p_n(r) = 0$, then $r = r_i$ for some i.

Proof: Parts (a) and (b) result from applying the and repeated use of (0.3.4). The final assertion follows from (b), since

$$p_n(r) = (r - r_1)(r - r_2) \cdots (r - r_n) = 0$$

if and only if one of the n factors vanish. Note that the roots are not assumed to be distinct. Indeed, if the root r_i is repeated exactly m times, the factor $(x - r_i)^m$ appears in (b), and r_i is said to have *multiplicity* m. □

A common situation arising particularly often in applications is the case in which all the coefficients of $p_n(x)$ are real. *Unless explicitly stated to the contrary, the coefficients of all polynomials appearing in the text will be real.* Now if $r = a + ib$ is a root of such a polynomial, then so is its complex conjugate $\overline{r} = a - ib$.

Theorem 0.3.2. *Suppose $r = a + ib$, $b \neq 0$, is a root of*

$$p_n(x) = x^n + a_{n-1}x^{n-1} + \cdots + a_1 x + a_0$$

Then $\overline{r} = a - ib$ is also a root.

Proof: The argument makes extensive use of three facts: (1) $\overline{ab} = \overline{a}\overline{b}$. (2) $\overline{a+b} = \overline{a} + \overline{b}$. (3) $\overline{a} = a$ if and only if a is real. Since the coefficients of $p_n(x)$ are all real,

$$\begin{aligned} \overline{p_n(x)} &= \overline{x^n + a_{n-1}x^{n-1} + \cdots + a_1 x + a_0} \\ &= \overline{x^n} + \overline{a_{n-1}x^{n-1}} + \cdots + \overline{a_1 x} + \overline{a_0} \\ &= \overline{x^n} + a_{n-1}\overline{x^{n-1}} + \cdots + a_1 \overline{x} + a_0 = p_n(\overline{x}) \end{aligned}$$

Hence, $p_n(\overline{r_i}) = \overline{p_n(r_i)} = \overline{0} = 0$. That is, $\overline{r_i}$ is a root of $p_n(x)$. □

Example 0.3.1. *Find all the solutions of* $x^3 = 1$.

Solution: By inspection we see that $r_1 = 1$ is a solution. Hence the polynomial $x^3 - 1$ has the factor $x - 1$. We divide $x^3 - 1$ by $x - 1$ to obtain the factorization

$$x^3 - 1 = (x-1)(x^2 + x + 1)$$

The quadratic formula yields the two remaining solutions,

$$r_2 = \tfrac{1}{2}\left(-1 + \sqrt{1-4}\right) = -\tfrac{1}{2} + \tfrac{1}{2}i\sqrt{3}$$

and its complex conjugate

$$r_3 = \overline{r_2} = -\tfrac{1}{2} - \tfrac{1}{2}i\sqrt{3}$$

Note that the magnitude of all three roots is one and that they are oriented 120° from each other, as shown in Figure 0.2.

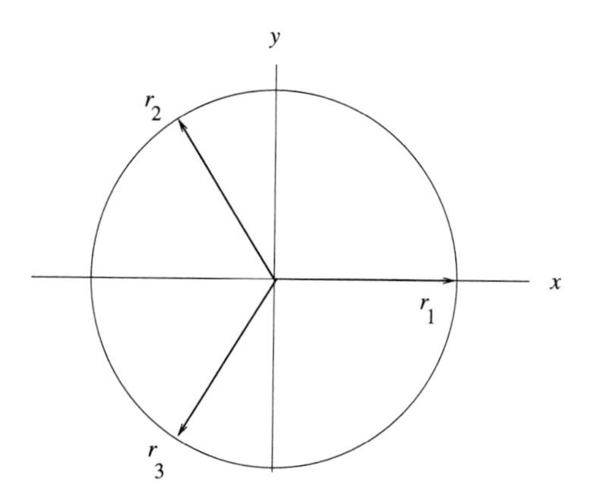

Figure 0.2: The Cube Roots of 1

Alternatively, we could express $z = 1$ in exponential form using (0.2.5):

$$z = 1 = e^{2\pi n i}, \quad n = 0, 1, 2, \ldots$$

Choose $n = 0$ to obtain $1^{1/3} = e^{0i/3} = 1$. Choose $n = 1$ to obtain

$$1^{1/3} = e^{2\pi i/3} = \cos\tfrac{2\pi}{3} + i\sin\tfrac{2\pi}{3} = -\tfrac{1}{2} + \tfrac{\sqrt{3}}{2}i$$

Finally, choose $n = 2$, and we have

$$1^{1/3} = e^{4\pi i/3} = \cos\frac{4\pi}{3} + i\sin\frac{4\pi}{3} = -\frac{1}{2} - \frac{\sqrt{3}}{2}$$

For $n = 3$ or greater, we get only repeats of these three roots.

Example 0.3.2. *Find the roots of $x^4 - 1$ and factor this polynomial.*

Solution: Since $x^4 - 1$ is the difference of two squares, we have

$$x^4 - 1 = (x^2 - 1)(x^2 + 1) = (x - 1)(x + 1)(x - i)(x + i)$$

Thus the roots are $r_1 = 1$, $r_2 = -1$, $r_3 = i$, and $r_4 = -i$.

Example 0.3.3. *Find a monic polynomial of degree 3 whose roots are $0, 1, 2$.*

Solution: Since $(x - 0), (x - 1)$ and $(x - 2)$ must be factors of any such polynomial, we have

$$p_3(x) = x\,(x - 1)\,(x - 2) = x^3 - 3x^2 + 2x$$

Example 0.3.4. *Find a monic polynomial of lowest degree with real coefficients having the roots $1, 1,$ and $1 - i$.*

Solution: Because the roots of a polynomial with real coefficients come in complex conjugate pairs, the fact that $1 - i$ is a root implies that $1 + i$ is also a root. Hence,

$$(x - (1 - i))\,(x - (1 + i)) = x^2 - 2x + 2$$

is a factor of $p(x)$. Likewise, $(x - 1)^2$ is also a factor. Therefore,

$$p(x) = (x^2 - 2x + 2)(x - 1)^2 = x^4 - 4x^3 + 7x^2 - 6x + 2$$

has the required roots. (No lower degree polynomial could have four roots, so this is the monic polynomial of least degree with these roots.)

• EXERCISES

1. Find the monic polynomial of least degree that satisfies the following conditions.
   (a) The roots are $1+i, 1-i$.
   (b) The roots are $1, 1, -1$.
   (c) The polynomial is divisible by (x^2-1) and has a root $r=2$.
   (d) The polynomial has a root $r = 2-3i$.
   (e) The polynomial has a root $r = a+ib, \quad b \neq 0$.
2. Find the solutions of the following equations and verify Theorem 0.3.1.
   (a) $x^5 = x$. (b) $x^3 - x^2 + x - 1 = 0$.
   (c) $x^4 - 3x^2 = 4$. (d) $x^6 = 1$.
   (e) $x^3 - 3x^2 + 4x = 2$. (f) $x^4 - x^3 + x^2 = x$.
3. Show that $x^n - 1 = (x-1)\left(x^{n-1} + x^{n-2} + \cdots + x + 1\right)$.
4. Show that $x^n + 1 = (x+1)\left(x^{n-1} - x^{n-2} + \cdots - x + 1\right)$, n odd.

MATLAB

5. Find the four roots of $x^4 - 4x^3 + 2x^2 + 4x + 4 = 0$.
6. Find a real monic polynomial of degree 4 whose roots are
   (a) $1, 1, 1, 1$ (b) $\pm 1, \pm i$
   (c) $\pm\sqrt{1 \pm i}$
7. At the MATLAB prompt,

```
≫ q = sort(angle(r));
s = exp(i*[q',min(q)]');
axis('square');
plot(s)
```

   for $n = 2, 3, \ldots, 9$.

Explain the output!

0.4 Matrix Notation and Terminology

A *matrix* is a rectangular array of numbers; its *size* is the number of rows by the number of columns in the array. Thus, the matrices

$$\text{(a)} \begin{bmatrix} 1 & 0 & -1 \\ 2 & 5 & 7 \end{bmatrix} \quad \text{(b)} \begin{bmatrix} -2 & 1 & 1 \\ 0 & 0 & 0 \\ 1 & 1 & 1 \end{bmatrix} \quad \text{(c)} \begin{bmatrix} x \\ y \\ z \end{bmatrix}$$

$$\text{(d)}\ [0] \quad \text{(e)}\ [1 \quad 1-i \quad 1+i]$$

have sizes 2×3, $3\times 3, 1\times 3, 1\times 1$, and 1×3, respectively. Sometimes we call the size of a matrix its dimensions. The size 2×3 is read "two by three."

In general, matrices are denoted by boldface capital letters. So we use $\mathbf{A}$ to denote the $p \times q$ matrix whose entries are a_{ij}. That is,

$$\mathbf{A} = \begin{bmatrix} a_{11} & a_{12} & \cdots & a_{1q} \\ a_{21} & a_{22} & \cdots & a_{2q} \\ \cdot & \cdot & & \cdot \\ \cdot & \cdot & & \cdot \\ a_{p1} & a_{p2} & \cdots & a_{pq} \end{bmatrix} \tag{0.4.1}$$

The first subscript of a_{ij} defines its row position, the second subscript its column position. Matrices with one column are called *vectors* and have their own notation. We use boldfaced lower-case letters to represent vectors in keeping with our convention for matrices. For example, the columns of $\mathbf{A}$ are the vectors

$$\mathbf{a_1} = \begin{bmatrix} a_{11} \\ a_{21} \\ \vdots \\ a_{p1} \end{bmatrix}, \quad \mathbf{a_2} = \begin{bmatrix} a_{12} \\ a_{22} \\ \vdots \\ a_{p2} \end{bmatrix}, \quad \cdots \quad \mathbf{a_q} = \begin{bmatrix} a_{1q} \\ a_{2q} \\ \vdots \\ a_{pq} \end{bmatrix} \tag{0.4.2}$$

For the system of simultaneous equations

$$\begin{aligned} 2x_1 - x_2 + x_3 - x_4 &= 1 \\ x_1 - x_3 &= 1 \\ x_2 + x_3 - x_4 &= -1 \end{aligned} \tag{0.4.3}$$

the *coefficient matrix* is

$$\mathbf{A} = \begin{bmatrix} 2 & -1 & 1 & -1 \\ 1 & 0 & -1 & 0 \\ 0 & 1 & 1 & -1 \end{bmatrix} \tag{0.4.4}$$

The vector of the right-hand side constants is denoted by $\mathbf{b}$ and is given by

$$\mathbf{b} = \begin{bmatrix} 1 \\ 1 \\ -1 \end{bmatrix} \tag{0.4.5}$$

while the *augmented matrix* $\mathbf{B}$ is given in terms of $\mathbf{A}$ and $\mathbf{b}$:

$$\mathbf{B} = \left[\begin{array}{cccc|c} 2 & -1 & 1 & -1 & 1 \\ 1 & 0 & -1 & 0 & 1 \\ 0 & 1 & 1 & -1 & -1 \end{array}\right] \tag{0.4.6}$$

Thus the augmented matrix is the coefficient matrix with an extra column containing the right-hand constants. The vector of unknowns is denoted by $\mathbf{x}$ and is written

$$\mathbf{x} = \begin{bmatrix} x_1 \\ x_2 \\ x_3 \\ x_4 \end{bmatrix} \tag{0.4.7}$$

Matrices with the same number of rows and columns are called *square matrices.* The *diagonal* entries of a square matrix are those with equal subscripts. For example, the diagonal entries of the matrix (b) given at the beginning of this section are $a_{11} = -2, a_{22} = 0$, and $a_{33} = 1$. The *off-diagonal* entries of $\mathbf{A}$ are those a_{ij}, with $i \neq j$. Square matrices whose off-diagonal entries are all zero are called *diagonal matrices.* The matrices

$$\mathbf{A} = \begin{bmatrix} 1 & 0 \\ 0 & 5 \end{bmatrix} \quad \mathbf{B} = \begin{bmatrix} 0 & 0 & 0 \\ 0 & 0 & 0 \\ 0 & 0 & 0 \end{bmatrix} \quad \mathbf{C} = \begin{bmatrix} 1 & 0 & 0 \\ 0 & 2 & 0 \\ 0 & 0 & i \end{bmatrix}$$

are diagonal matrices, as are all 1×1 matrices. The $\mathbf{I}_n$ matrices, where n is a positive integer, are $n \times n$ diagonal matrices all of whose diagonal entries are 1. These matrices are called *identity matrices.* So,

$$\mathbf{I}_1 = [1] \quad \mathbf{I}_2 = \begin{bmatrix} 1 & 0 \\ 0 & 1 \end{bmatrix} \quad \mathbf{I}_3 = \begin{bmatrix} 1 & 0 & 0 \\ 0 & 1 & 0 \\ 0 & 0 & 1 \end{bmatrix}$$

are identity matrices. The columns of $\mathbf{I}_n$ are assigned the special notation $\mathbf{e}_1, \mathbf{e}_2, \ldots, \mathbf{e}_n$; that is,

$$\mathbf{e}_1 = \begin{bmatrix} 1 \\ 0 \\ \vdots \\ 0 \end{bmatrix} \quad \mathbf{e}_2 = \begin{bmatrix} 0 \\ 1 \\ \vdots \\ 0 \end{bmatrix} \quad \mathbf{e}_n = \begin{bmatrix} 0 \\ 0 \\ \vdots \\ 1 \end{bmatrix} \tag{0.4.8}$$

When the context makes clear the size of $\mathbf{I}_n$ we drop the subscript n. Square matrices whose off-diagonal entries lying below the diagonal are zero are called *upper triangular matrices.* Similarly, *lower triangular matrices* are square matrices whose off-diagonal entries lying above the diagonal are zero. For example,

$$\mathbf{A} = \begin{bmatrix} 1 & 0 \\ 2 & 1 \end{bmatrix}, \quad \mathbf{B} = \begin{bmatrix} 1 & 0 \\ 0 & 1 \end{bmatrix}, \quad \mathbf{I}_n$$

are all lower triangular. Similarly,

$$\mathbf{C} = \begin{bmatrix} 1 & 1 \\ 0 & 1 \end{bmatrix}, \quad \mathbf{D} = [7], \quad \mathbf{I}_n$$

are all upper triangular matrices. Note that diagonal matrices are simultaneously upper and lower triangular. Conversely, if $\mathbf{A}$ is both upper and lower triangular, $\mathbf{A}$ is diagonal. We define the zero matrix $\mathbf{O}_n$ as the $n \times n$ matrix, all of whose entries are zero. So $\mathbf{O}_n$ is also diagonal.

Finally, we shall have occasion to augment a matrix $\mathbf{A}_{p\times q}$ with more than a single column. We write $[\mathbf{A}_{p\times q}|\mathbf{B}_{p\times n}]$ for the augmented matrix whose first q columns are the columns of $\mathbf{A}$ and whose last n columns are the columns of $\mathbf{B}$. For example, using the matrices $\mathbf{A}, \mathbf{B}$ and $\mathbf{C}$ above, we have

$$[\mathbf{A}|\mathbf{B}] = \left[\begin{array}{cc|cc} 1 & 0 & 1 & 0 \\ 2 & 1 & 0 & 1 \end{array}\right], \quad [\mathbf{A}|\mathbf{C}] = \left[\begin{array}{cc|cc} 1 & 0 & 1 & 1 \\ 2 & 1 & 0 & 1 \end{array}\right]$$

In practice we will omit the vertical separator between the coefficients in augmented matrices whenever clarity is not sacrificed by this omission.

• EXERCISES

1. What are the coefficient and augmented matrices of the following systems? State the sizes of each.
   (a) $x_1 = 0, \quad x_2 = 1, \quad x_3 = 1.$
   (b) $x_1 + x_2 + x_3 = -1.$
   (c) $x_1 = x_2 + x_3 = -1.$
   (d) $x_1 = x_2 = x_3 = 0, \quad x_1 = 1 + x_2.$
   (e) $x_1 = x_2 = x_3, \quad x_3 = 1.$

2. Suppose $\mathbf{A}$ is a 3×3 diagonal matrix and is the coefficient matrix of a system whose right-hand side constants are zero. What are the explicit equations of this system in terms of the unknowns x_1, x_2, x_3?

3. Repeat Problem 2, except that $\mathbf{A}$ is now upper triangular.

4. Suppose $\mathbf{A}$ is a coefficient matrix of a system whose right-hand side vector is $\mathbf{b}$, and $\mathbf{b}$ has only nonzero entries. What can be said about the solution of this system if $\mathbf{A}$ has a row of zeros?

5. Suppose $\mathbf{A}$ is a coefficient matrix of a system whose right-hand side vector is $\mathbf{b}$. What can be said about its solutions if two rows of $\mathbf{A}$ are identical, but the corresponding entries in $\mathbf{b}$ are not equal?

MATLAB

Refer to your manual for the MATLAB commands that accomplish the following constructions.

6. Generate the matrix $\begin{bmatrix}-2 & 0 & 2 & 4 & 6\end{bmatrix}$.

7. Generate the matrix $\mathbf{A} = [m^n]$ whose size is $m \times n = 4 \times 6$. Compare this matrix with $\mathbf{B}=[n^m]$ with size 6×4.

8. Generate the matrices $\mathbf{A}$ of (0.4.4) and $\mathbf{b}$ of (0.4.5). What is the result of $[Ab]$? of $[A; b]$?

9. Construct $\mathbf{B}$ having just the 1st and 3rd columns of $\mathbf{A}$ in (0.4.4).

10. Construct a diagonal matrix whose diagonal entries are $1, -1, 2$.

11. Construct a vector whose entries are the diagonal elements of the matrix $\mathbf{A}$ of (0.4.4).

12. Construct an upper and lower triangular matrix whose nonzero entries agree with those of $\mathbf{A}$ given in (0.4.4).

13. Generate the matrices (a) "magic(3)", (b) "magic(4)", and deduce the properties that define these matrices.

14. Generate the matrices (a) "zeros(3,4)", (b) "zeros(5)", (c) "ones(5,2)", (d) "ones(5)", (e) "ones(eye(4))", (f) "ones(zeros(2,3))".

15. Generate the matrices "hadamard(n)" for $n = 1, 2, 3$. Deduce the defining properties of these matrices.

16. Generate the matrices "hilb(n)" for $n = 1, 2, 3$. Deduce the defining properties of these matrices.

0.5 The Solution of Simultaneous Equations

Before we can effectively use matrix theory to solve systems of simultaneous linear differential equations, it is necessary to study how matrices are used to solve simultaneous linear algebraic equations.

Consider, for example, the following system:

$$\begin{aligned} x_1 + x_3 &= 1 \\ 2x_1 + x_2 + x_3 &= 0 \\ x_1 + x_2 + 2x_3 &= 1 \end{aligned} \tag{0.5.1}$$

We can solve these equations by elimination. First eliminate the first unknown x_1 from the second and third equations by adding -2 times the first equation to the second; then add -1 times the first to the third. This results in the equivalent system[2]

$$\begin{aligned} x_1 + x_3 &= 1 \\ x_2 - x_3 &= -2 \\ x_2 + x_3 &= 0 \end{aligned} \tag{0.5.2}$$

Next eliminate the second unknown x_2 from the third equation by adding -1 times the second equation to the third to obtain the equivalent system

$$\begin{aligned} x_1 + x_3 &= 1 \\ x_2 - x_3 &= -2 \\ 2x_3 &= 2 \end{aligned} \tag{0.5.3}$$

We now deduce that $x_3 = 1$ from the last equation, and from this, deduce that $x_2 = -1$ from the second equation and, lastly, $x_1 = 0$ from the first equation. This is the solution of (0.5.1).

It should be clear that this "elimination" process depends on the coefficients of the unknowns in each equation. This can be seen by studying the augmented matrix for the system in (0.5.1):

$$\mathbf{B} = \left[\begin{array}{ccc|c} 1 & 0 & 1 & 1 \\ 2 & 1 & 1 & 0 \\ 1 & 1 & 2 & 1 \end{array}\right] \tag{0.5.4}$$

The first column of $\mathbf{B}$ is the set of coefficients of x_1. The elimination of x_1 from the second equation corresponds to adding -2 times each entry in the first row of $\mathbf{B}$ to each corresponding entry in the second row. Likewise, -1 times the entries in the first row added to the corresponding entries in the third row amounts to eliminating x_1 from the third equation. When these operations are performed on $\mathbf{B}$, we obtain the augmented matrix of (0.5.2):

$$\mathbf{C} = \left[\begin{array}{ccc|c} 1 & 0 & 1 & 1 \\ 0 & 1 & -1 & -2 \\ 0 & 1 & 1 & 0 \end{array}\right] \tag{0.5.5}$$

The appearance of the zeros in column one indicates that x_1 does not appear in the second and third equations. Finally, we eliminate x_2 from the third equation by adding -1 times each entry in the second row of $\mathbf{C}$ to each corresponding entry in the third row. This results in

$$\mathbf{D} = \left[\begin{array}{ccc|c} 1 & 0 & 1 & 1 \\ 0 & 1 & -1 & -2 \\ 0 & 0 & 2 & 2 \end{array}\right] \tag{0.5.6}$$

[2]Two linear systems are equivalent if they have the same solution set.

which is the augmented matrix of (0.5.3). Note that it is upper triangular, the object of the elimination technique.

It is unnecessary to perform any further simplification of (0.5.3) or its representation as the augmented matrix **D** in order to solve the given system. However, it is sometimes useful to do so anyway. We divide row three in **D** by 2, then add the third row to the second, and add -1 times the third row to the first, resulting in

$$\mathbf{E} = \left[\begin{array}{ccc|c} 1 & 0 & 0 & 0 \\ 0 & 1 & 0 & -1 \\ 0 & 0 & 1 & 1 \end{array}\right] \tag{0.5.7}$$

The matrix **E** is the augmented matrix of the system

$$x_1 = 0, \quad x_2 = -1, \quad x_3 = 1 \tag{0.5.8}$$

which is the solution given after (0.5.3). Note that when the coefficient matrix is "reduced" to the identity matrix, as is the case in (0.5.7), the column of right-hand side coefficients is the solution vector.

The arithmetic steps used to go from **B** to **E** are called *elementary row operations.* These operations are classified into three types:

(1) *Interchange any two rows.*
(2) *Multiply a row by a nonzero scalar.*
(3) *Add a multiple of one row to another, different row.*

The crucial point here is that the elementary row operations replace one set of equations with another set, the latter having the same solution set as the former. This means, for instance, that the solutions of (0.5.8) are the solutions of (0.5.1) — a fact we have deduced earlier. This method of solution is referred to as *Gaussian elimination.*

Example 0.5.1. *Solve the following system using Gaussian elimination:*

$$\begin{aligned} x_1 + x_3 &= 1 \\ 2x_1 + x_3 &= 0 \\ x_1 + x_2 + x_3 &= 1 \end{aligned}$$

Solution: The augmented matrix of this system is:

$$\mathbf{B} = \left[\begin{array}{ccc|c} 1 & 0 & 1 & 1 \\ 2 & 0 & 1 & 0 \\ 1 & 1 & 1 & 1 \end{array}\right]$$

The elementary row operations used in the elimination process are given implicitly by exhibiting the sequence of matrices resulting from these operations. Thus,

$$\mathbf{B} = \left[\begin{array}{ccc|c} 1 & 0 & 1 & 1 \\ 2 & 0 & 1 & 0 \\ 1 & 1 & 1 & 1 \end{array}\right] \to \left[\begin{array}{ccc|c} 1 & 0 & 1 & 1 \\ 0 & 0 & -1 & -2 \\ 0 & 1 & 0 & 0 \end{array}\right] \to$$

$$\left[\begin{array}{ccc|c} 1 & 0 & 0 & -1 \\ 0 & 0 & -1 & -2 \\ 0 & 1 & 0 & 0 \end{array}\right] \to \left[\begin{array}{ccc|c} 1 & 0 & 0 & -1 \\ 0 & 1 & 0 & 0 \\ 0 & 0 & 1 & 2 \end{array}\right] = \mathbf{C}$$

describes the stages undergone by the augmented matrix **B** as elementary row operations are applied to it. The matrix immediately after the first arrow was obtained by two operations. The final matrix **C** results from one operation applied to the matrix immediately preceding it and then interchanging rows. We call this type of display an *arrow diagram.* Such diagrams imply that each matrix is derived from the one to its left by a few elementary row operations. In the preceding diagram, the rightmost matrix is the augmented matrix of the system

$$x_1 = -1, \quad x_2 = 0, \quad x_3 = 2$$

The solution of this system is $x_1 = -1, x_2 = 0$, and $x_3 = 2$, and this is the solution (and the only one) of the original system.

Example 0.5.2. *Solve the following system by applying elementary row operations to the augmented matrix :*

$$\begin{aligned} -x_1 + x_3 &= -1, \\ x_1 + x_2 &= 1 \\ x_1 + x_3 &= -1 \end{aligned}$$

Solution: Here is one arrow diagram for the augmented matrix of this system:

$$\mathbf{B} = \left[\begin{array}{ccc|c} -1 & 0 & 1 & -1 \\ 1 & 1 & 0 & 1 \\ 1 & 0 & 1 & -1 \end{array}\right] \to \left[\begin{array}{ccc|c} 1 & 0 & -1 & 1 \\ 0 & 1 & 1 & 0 \\ 0 & 0 & 2 & -2 \end{array}\right] \to \left[\begin{array}{ccc|c} 1 & 0 & 0 & 0 \\ 0 & 1 & 0 & 1 \\ 0 & 0 & 1 & -1 \end{array}\right]$$

The last matrix in this diagram is the augmented matrix of the system $x_1 = 0$, $x_2 = 1$, $x_3 = -1$. Thus, the solution of the given system is $x_1 = 0, x_2 = 1$, and $x_3 = -1$.

Example 0.5.3. *Apply elementary row operations to the following system to obtain its solution.*

$$\begin{aligned} x_1 + x_2 + x_3 &= 1 \\ x_1 - x_2 + x_3 &= 3 \\ x_1 + x_3 &= 2 \end{aligned}$$

Solution: As in the previous example,

$$\mathbf{B} = \left[\begin{array}{ccc|c} 1 & 1 & 1 & 1 \\ 1 & -1 & 1 & 3 \\ 1 & 0 & 1 & 2 \end{array}\right] \to \left[\begin{array}{ccc|c} 1 & 1 & 1 & 1 \\ 0 & -2 & 0 & 2 \\ 0 & -1 & 0 & 1 \end{array}\right] \to$$

$$\left[\begin{array}{ccc|c} 1 & 1 & 1 & 1 \\ 0 & 1 & 0 & -1 \\ 0 & 1 & 0 & 1 \end{array}\right] \to \left[\begin{array}{ccc|c} 1 & 0 & 1 & 2 \\ 0 & 1 & 0 & -1 \\ 0 & 0 & 0 & 0 \end{array}\right]$$

The last matrix in this system is the augmented matrix of the system

$$\begin{aligned} x_1 + x_3 &= 2 \\ x_2 &= -1 \end{aligned}$$

This system has infinitely many solutions. We can represent them all by setting $x_3 = t$. Then $x_1 = 2 - t$. ($x_2 = -1$ always.) Therefore, for every t, there results a solution of the given system. Indeed, we can verify this fact by substituting these values into the original system to obtain the identities

$$\begin{aligned} 2 - t - 1 + t &= 1 \\ 2 - t + 1 + t &= 3 \\ 2 - t + t &= 2 \end{aligned}$$

In the system in Example 0.5.3, either x_1 or x_3 may be chosen arbitrarily, but not both! When we have a choice, it is conventional to select the highest subscripted unknown. Also, the system in Example 0.5.3 has no solutions if any constant but 2 appears on the right-hand side of the third equation. To see why this is so, study the next closely related example.

Example 0.5.4. *For which k does the following system have a solution?*

$$\begin{aligned} x_1 + x_2 + x_3 &= 1 \\ x_1 - x_2 + x_3 &= 3 \\ x_1 + x_3 &= k \end{aligned}$$

Solution: As in the previous example, an appropriate arrow diagram is easy to construct.

$$\mathbf{B} = \left[\begin{array}{ccc|c} 1 & 1 & 1 & 1 \\ 1 & -1 & 1 & 3 \\ 1 & 0 & 1 & k \end{array}\right] \to \left[\begin{array}{ccc|c} 1 & 1 & 1 & 1 \\ 0 & -2 & 0 & 2 \\ 0 & -1 & 0 & k-1 \end{array}\right] \to$$

$$\left[\begin{array}{ccc|c} 1 & 1 & 1 & 1 \\ 0 & 1 & 0 & -1 \\ 0 & 1 & 0 & 1-k \end{array}\right] \to \left[\begin{array}{ccc|c} 1 & 0 & 1 & 2 \\ 0 & 1 & 0 & -1 \\ 0 & 0 & 0 & 2-k \end{array}\right]$$

The equation corresponding to the third row of the last matrix in the arrow diagram is $0 \cdot x_1 + 0 \cdot x_2 + 0 \cdot x_3 = 2 - k$ which is inconsistent unless $k = 2$. If $k = 2$, the system reduces to the one solved in Example 0.5.3.

• EXERCISES

Reduce each matrix to upper triangular form by repeated use of elementary row operations.

1. $\begin{bmatrix} a & b \\ c & d \end{bmatrix}, \quad a \neq 0.$

2. $\begin{bmatrix} 0 & b \\ c & d \end{bmatrix}.$

3. $\begin{bmatrix} 1 & 0 & 0 \\ -2 & 2 & 0 \\ 1 & 3 & -1 \end{bmatrix}.$

4. $\begin{bmatrix} 0 & 1 & 0 \\ 1 & 0 & 0 \\ 0 & 0 & 1 \end{bmatrix}.$

5. $\begin{bmatrix} 1 & 2 & 1 \\ 2 & 4 & -2 \\ 0 & 0 & 1 \end{bmatrix}.$

Solve the systems given in Problems 6–13 by applying elementary row operations.

6. $x_1 - x_2 = 6$
   $x_1 + x_2 = 0$

7. $2x_1 - 2x_2 = 4$
   $2x_1 + x_2 = 3$

8. $x_1 - x_2 = 2$
   $x_1 + x_2 = 11$

9. $x_1 + x_2 = 4$
   $2x_1 + 3x_2 = 8$

10. $x_1 + x_2 - x_3 = 0$
    $x_2 + x_3 = 3$

11. $x_1 - 3x_2 + x_3 = -2$
    $-3x_2 + x_3 = 0$

12. $x_1 + x_2 + x_3 = 4$
    $x_1 - x_2 - x_3 = 2$
    $x_1 - 2x_2 = 0$

13. $x_1 + 2x_2 + x_3 = -2$
    $x_1 + x_2 = 3$
    $x_1 + x_2 + x_3 = 4$

MATLAB

14. Use "rref(B)" to check the answer obtained in Example 0.5.3.
15. Use "rref(B)" to obtain the solutions for Problems 6–13.

0.6 The Algebra of Matrices

We have seen the convenience afforded by simply operating on the array of coefficients of a system of equations rather than on the equations themselves. From this point of view, matrices are convenient packages for the coefficients of systems. We now take a more abstract perspective by considering matrices as having an existence of their own, apart from their connections with simultaneous equations. Our first task is to invent an algebra of matrices in which we shall define a variety of arithmetic operations most of which are analogues of the operations in ordinary algebra. As we shall see, however, the definition of matrix multiplication is neither easily guessed nor simple to put to use.

Let the $m \times n$ matrices $\mathbf{A}$ and $\mathbf{B}$ be given by

$$\mathbf{A} = \begin{bmatrix} a_{11} & a_{12} & \dots & a_{1n} \\ a_{21} & a_{22} & \dots & a_{2n} \\ \vdots & \vdots & & \vdots \\ a_{m1} & a_{m2} & \dots & a_{mn} \end{bmatrix}, \quad \mathbf{B} = \begin{bmatrix} b_{11} & b_{12} & \dots & b_{1n} \\ b_{21} & b_{22} & \dots & b_{2n} \\ \vdots & \vdots & & \vdots \\ b_{m1} & b_{m2} & \dots & b_{mn} \end{bmatrix} \tag{0.6.1}$$

Then $\mathbf{A} = \mathbf{B}$ if $a_{ij} = b_{ij}$ for all i and j. Implicit in this definition of equality is the assumption that the sizes of $\mathbf{A}$ and $\mathbf{B}$ are the same. Thus, the equality of two matrices expresses the equality of the corresponding entries of the matrices.

We now define the addition of matrices $\mathbf{A} + \mathbf{B}$ and the multiplication of a matrix by an arbitrary scalar k:

$$\mathbf{A} + \mathbf{B} = \begin{bmatrix} a_{11} + b_{11} & a_{12} + b_{12} & \dots & a_{1n} + b_{1n} \\ a_{21} + b_{21} & a_{22} + b_{22} & \dots & a_{2n} + b_{2n} \\ \vdots & \vdots & & \vdots \\ a_{m1} + b_{m1} & a_{m2} + b_{m2} & \dots & a_{mn} + b_{mn} \end{bmatrix} \tag{0.6.2}$$

$$k\mathbf{A} = \begin{bmatrix} ka_{11} & ka_{12} & \dots & ka_{1n} \\ ka_{21} & ka_{22} & \dots & ka_{2n} \\ \vdots & \vdots & & \vdots \\ ka_{m1} & ka_{m2} & \dots & ka_{mn} \end{bmatrix} \tag{0.6.3}$$

The definitions in (0.6.2) and (0.6.3) lead easily to the following algebraic rules:

(a) $\mathbf{A} + \mathbf{B} = \mathbf{B} + \mathbf{A}$ (b) $\mathbf{A} + (\mathbf{B} + \mathbf{C}) = (\mathbf{A} + \mathbf{B}) + \mathbf{C}$
(c) $\mathbf{A} + \mathbf{O} = \mathbf{A}$ (d) $\mathbf{A} + (-1)\,\mathbf{A} = \mathbf{O}$
(e) $0\mathbf{A} = \mathbf{O}$ (f) $k\,(h\mathbf{A}) = (kh)\,\mathbf{A}$
(g) $k\,(\mathbf{A} + \mathbf{B}) = k\mathbf{A} + k\mathbf{B}$ (h) $(k + h)\,\mathbf{A} = k\mathbf{A} + h\mathbf{A}$

Note that these properties are analogous to the rules of ordinary (scalar) algebra.

We shall occasionally find it helpful to "stand a matrix on its side" by which we mean replacing the first row by the first column, the second row by the second column, and so on, until all columns have become rows. The matrix that results from this interchange is called the *transpose* indexmatrix!transpose of the original matrix, and $[a_{ij}]^T = [a_{ji}]$. We write $\mathbf{A}^T$ to represent the transpose of $\mathbf{A}$. Note that if a matrix is square, its transpose is also square; however, if $\mathbf{A}$ is $m \times n$, then $\mathbf{A}^T$ is $n \times m$. Here is an illustration:

$$\mathbf{A} = \begin{bmatrix} 2 & 0 \\ 3 & -1 \\ 1 & 1 \\ 0 & 0 \end{bmatrix}, \quad \mathbf{A}^T = \begin{bmatrix} 2 & 3 & 1 & 0 \\ 0 & -1 & 1 & 0 \end{bmatrix}$$

The transpose of a vector is a matrix with a single row, sometimes called a *row vector.* For example, $\mathbf{e}_1^T = [1, 0, \cdots, 0]$. We write row vectors with commas separating their various entries when it is necessary to avoid confusion.

Note that the definitions of transposition and addition leads to the conclusion that $\mathbf{C} = \mathbf{A} + \mathbf{B}$ implies that $\mathbf{C}^T = \mathbf{A}^T + \mathbf{B}^T$.

A matrix $\mathbf{A}$ is *symmetric* if it is equal to its own transpose, that is, $\mathbf{A}^T = \mathbf{A}$ and is *antisymmetric* (*skewsymmetric*) if $\mathbf{A}^T = -\mathbf{A}$. Symmetric and antisymmetric matrices must be square. The matrices

$$\mathbf{C} = \begin{bmatrix} 2 & 1 & 3 & 4 \\ 1 & 0 & -2 & 0 \\ 3 & -2 & 1 & -1 \\ 4 & 0 & -1 & 0 \end{bmatrix}, \quad \mathbf{D} = \begin{bmatrix} 0 & -1 & 2 \\ 1 & 0 & -3 \\ -2 & 3 & 0 \end{bmatrix}$$

are symmetric and antisymmetric, respectively. Also, $(\mathbf{A}^T)^T = \mathbf{A}$. From these facts, we see that $\mathbf{A} + \mathbf{A}^T$ is symmetric and $\mathbf{A} - \mathbf{A}^T$ is antisymmetric because

$$\begin{aligned} \left(\mathbf{A} + \mathbf{A}^T\right)^T &= \mathbf{A}^T + \left(\mathbf{A}^T\right)^T \\ &= \mathbf{A}^T + \mathbf{A} \end{aligned} \tag{0.6.4}$$

$$\begin{aligned} \left(\mathbf{A} - \mathbf{A}^T\right)^T &= \mathbf{A}^T - \left(\mathbf{A}^T\right)^T \\ &= \mathbf{A}^T - \mathbf{A} \\ &= -\left(\mathbf{A} - \mathbf{A}^T\right) \end{aligned} \tag{0.6.5}$$

Example 0.6.1. *Find* $\mathbf{A}+\mathbf{B}$ *and* $\mathbf{B}-5\mathbf{A}$ *for the matrices*

$$\mathbf{A}=\begin{bmatrix} 0 & 2 & 5 \\ 1 & -2 & 1 \\ 2 & 3 & 1 \end{bmatrix}, \qquad \mathbf{B}=\begin{bmatrix} -1 & 2 & 0 \\ 0 & 2 & 1 \\ 6 & -6 & 0 \end{bmatrix}$$

Solution: To find the sum of $\mathbf{A}$ and $\mathbf{B}$, we simply add corresponding entries:

$$\mathbf{A}+\mathbf{B}=\begin{bmatrix} -1 & 4 & 5 \\ 1 & 0 & 2 \\ 8 & -3 & 1 \end{bmatrix}$$

From (0.6.2) and (0.6.3), it follows that

$$\mathbf{A}-5\mathbf{B}=\begin{bmatrix} 5 & -8 & 5 \\ 1 & -12 & -4 \\ -28 & 33 & 1 \end{bmatrix}$$

• EXERCISES

1. Prove that $(k\mathbf{A})^T = k\mathbf{A}^T$.
2. Prove that symmetric and skew-symmetric matrices must be square.
3. What matrices are simultaneously symmetric and skew-symmetric?
4. Show that the diagonal entries of a skew-symmetric matrix are zero.
5. Prove that $\left(\mathbf{A}^T\right)^T = \mathbf{A}$.
6. Prove that $(\mathbf{A}+\mathbf{B})^T = \mathbf{A}^T + \mathbf{B}^T$.
7. Prove that $\mathbf{A}_s = \left(\mathbf{A}+\mathbf{A}^T\right)/2$ is symmetric.
8. Prove that $\mathbf{A}_a = \left(\mathbf{A}-\mathbf{A}^T\right)/2$ is antisymmetric.
9. Show that $\mathbf{A} = \mathbf{A}_s + \mathbf{A}_a$.
10. Show that antisymmetric matrices have zero diagonal elements.
11. Given these matrices

$$\mathbf{A}=\begin{bmatrix} 2 & 1 & 0 \\ 1 & -1 & -2 \\ 4 & 2 & 0 \end{bmatrix}, \quad \mathbf{B}=\begin{bmatrix} 1 & 1 & 1 \\ 0 & 0 & 0 \\ 2 & 1 & -3 \end{bmatrix} \quad \mathbf{C}=\begin{bmatrix} 2 & 3 & -1 \\ 0 & 2 & 0 \\ -1 & 2 & -1 \end{bmatrix}$$

perform the following operations:

(a) $\mathbf{A}+\mathbf{B}$ (b) $\mathbf{B}+\mathbf{A}$
(c) $\mathbf{A}+(\mathbf{B}+\mathbf{C})$ (d) $(\mathbf{A}+\mathbf{B})+\mathbf{C}$
(e) $4(\mathbf{A}+\mathbf{B})$ (f) $4\mathbf{A}+4\mathbf{B}$
(g) $\mathbf{A}^T$ (h) $\mathbf{A}_s+\mathbf{A}_a$. (See Problem 9.)

12. Verify that $(\mathbf{A}+\mathbf{B})^T = \mathbf{A}^T + \mathbf{B}^T$ for the matrices given in Problem 11.

MATLAB

14. Let **A** be the augmented matrix in Example 0.5.3. Invoke these commands.
   (a) "zeros(A)", "zeros(5)", "zeros(3,4)".
   (b) "ones(A)", "ones(5)", "ones(3,4)".
   (c) "rot90(A)".
   (d) "flipir(A)", "flipud(A)".
   (e) "diag(A)", "diag(A,1)", "diag([1 2 3])", "diag(diag(A))".
   (f) "triu(A)", "tril(A,2)".

0.7 Matrix Multiplication

If $\mathbf{A}$ is $m \times q$ with entries a_{ij} and $\mathbf{B}$ is $q \times n$ with entries b_{ij}, then the product $\mathbf{C} = \mathbf{AB}$ is $m \times n$ with entries c_{ij} defined by the equations,

$$c_{ij} = \sum_{k=1}^{q} a_{ik} b_{kj} \tag{0.7.1}$$

In order that the definition (0.7.1) make sense, it is necessary that each row of $\mathbf{A}$ have as many entries as each column of $\mathbf{B}$. This means that the number of columns of $\mathbf{A}$ (which gives the number of entries in each row) must be the same as the number of rows of $\mathbf{B}$ (which gives the number of entries in each column). Hence, if $\mathbf{A}$ is 2×3 and $\mathbf{B}$ is 3×4, then $\mathbf{AB}$ is defined but $\mathbf{BA}$ is not. So $\mathbf{A}_{p\times n}\mathbf{B}_{m\times q}$ is defined only when $n = m$. Matrix multiplication is therefore not commutative! Indeed, not only must we take it that $\mathbf{AB} \neq \mathbf{BA}$ even if both products are defined, it is most often the case that this is so. Unless we have some reason to believe the contrary, we assume that $\mathbf{AB} \neq \mathbf{BA}$.

Written out, the summation (0.7.1) is

$$c_{ij} = a_{i1}b_{1j} + a_{i2}b_{2j} + \cdots + a_{ik}b_{kj} + \cdots + a_{in}b_{nj} \tag{0.7.2}$$

Thus the entry in the ith row, j^{th} column of the product $\mathbf{AB}$ is computed from the entries taken from the ith row of $\mathbf{A}$ and the jth column of $\mathbf{B}$. It is instructive to display the definition of $\mathbf{C} = \mathbf{AB}$:

$$\mathbf{C} = \begin{bmatrix} \cdots & * & \cdots \\ & \vdots & \\ \cdots & c_{ij} & \cdots \\ & \vdots & \\ \cdots & * & \cdots \end{bmatrix} = \begin{bmatrix} * & \cdots & * & \cdots & * \\ \vdots & & \vdots & & \vdots \\ a_{i1} & \cdots & a_{ij} & \cdots & a_{in} \\ \vdots & & \vdots & & \vdots \\ * & \cdots & * & \cdots & * \end{bmatrix} \begin{bmatrix} \cdots & b_{1j} & \cdots \\ & \vdots & \\ \cdots & b_{ij} & \cdots \\ & \vdots & \\ \cdots & b_{mj} & \cdots \end{bmatrix}$$

In the product $\mathbf{AB}$, the matrix $\mathbf{A}$ is the *premultiplier* and $\mathbf{B}$ is the *postmultiplier.*

Example 0.7.1. *Compute* $\mathbf{AB}$, $\mathbf{BA}$, $\mathbf{CA}$, $\mathbf{DE}$, *and* $\mathbf{ED}$ *for the matrices*

$$\mathbf{A} = \begin{bmatrix} 2 \\ 3 \\ -4 \end{bmatrix}, \qquad \mathbf{B} = \begin{bmatrix} 2 & -1 & 0 \end{bmatrix}, \qquad \mathbf{C} = \begin{bmatrix} 2 & 3 & -1 \\ 0 & 1 & 4 \end{bmatrix}$$

$$\mathbf{D} = \begin{bmatrix} 3 & 0 & 1 \\ 2 & -2 & 1 \\ 0 & 2 & 0 \end{bmatrix}, \qquad \mathbf{E} = \begin{bmatrix} 2 & -1 & 1 \\ 1 & 0 & 0 \\ 2 & 0 & 1 \end{bmatrix}$$

Solution: First we compute the products $\mathbf{AB}$ and $\mathbf{BA}$:

$$\mathbf{AB} = \begin{bmatrix} 2 \\ 3 \\ -4 \end{bmatrix} \begin{bmatrix} 2 & -1 & 0 \end{bmatrix} = \begin{bmatrix} 4 & -2 & 0 \\ 6 & -3 & 0 \\ -8 & 4 & 0 \end{bmatrix}$$

$$\mathbf{BA} = \begin{bmatrix} 2 & -1 & 0 \end{bmatrix} \begin{bmatrix} 2 \\ 3 \\ -4 \end{bmatrix} = \begin{bmatrix} 2 \times 2 + (-1) \times 3 + 0 \times (-4) \end{bmatrix} = \begin{bmatrix} 1 \end{bmatrix}$$

Next,

$$\mathbf{CA} = \begin{bmatrix} 2 & 3 & -1 \\ 0 & 1 & 4 \end{bmatrix} \begin{bmatrix} 2 \\ 3 \\ -4 \end{bmatrix} = \begin{bmatrix} 2 \times 2 + 3 \times 3 + (-1) \times (-4) \\ 0 \times 2 + 1 \times 3 + 4 \times (-4) \end{bmatrix} = \begin{bmatrix} 17 \\ -13 \end{bmatrix}$$

The product $\mathbf{AC}$ does not exist. The products $\mathbf{DE}$ and $\mathbf{ED}$ both exist but are unequal:

$$\mathbf{DE} = \begin{bmatrix} 8 & -3 & 4 \\ 4 & -2 & 3 \\ 2 & 0 & 0 \end{bmatrix}, \qquad \mathbf{ED} = \begin{bmatrix} 4 & 4 & 1 \\ 3 & 0 & 1 \\ 6 & 2 & 2 \end{bmatrix}$$

In certain special circumstances, $\mathbf{AB} = \mathbf{BA}$. Two illustrations are

$$\mathbf{AI} = \mathbf{IA} = \mathbf{A} \tag{0.7.3}$$

$$\mathbf{AO} = \mathbf{OA} = \mathbf{O} \tag{0.7.4}$$

The following facts are true for all $\mathbf{A}, \mathbf{B}$, and $\mathbf{C}$ for which the sizes are compatible:

$$\begin{aligned}\mathbf{A}(\mathbf{BC}) &= (\mathbf{AB})\mathbf{C} \\ \mathbf{A}(\mathbf{A}+\mathbf{C}) &= \mathbf{AB}+\mathbf{AC} \\ (\mathbf{B}+\mathbf{C})\mathbf{A} &= \mathbf{BA}+\mathbf{CA}\end{aligned} \tag{0.7.5}$$

A striking example of the peculiarity of matrix multiplication is illustrated by the product

$$\begin{bmatrix} 1 & 1 \\ -1 & -1 \end{bmatrix}\begin{bmatrix} 1 & 1 \\ -1 & -1 \end{bmatrix} = \begin{bmatrix} 0 & 0 \\ 0 & 0 \end{bmatrix}$$

This shows that $\mathbf{AB} = \mathbf{O}$ is possible without either $\mathbf{A}$ or $\mathbf{B}$ having a single zero entry. This example also illustrates that $\mathbf{A}^2 = \mathbf{O}$ does not imply $\mathbf{A} = \mathbf{O}$.

Here is another example showing the differences between matrix and ordinary algebra:

$$\mathbf{AB} = \begin{bmatrix} 1 & 1 \\ -1 & -1 \end{bmatrix}\begin{bmatrix} 1 & 1 \\ 0 & 0 \end{bmatrix} = \begin{bmatrix} 1 & 1 \\ -1 & -1 \end{bmatrix}\begin{bmatrix} 0 & 0 \\ 1 & 1 \end{bmatrix} = \mathbf{AC}$$

which shows that $\mathbf{AB} = \mathbf{AC}$ does not imply that $\mathbf{B} = \mathbf{C}$.

The failure of commutativity of matrix multiplication complicates the rules of algebra. For example,

$$\begin{aligned}(\mathbf{A}+\mathbf{B})^2 &= (\mathbf{A}+\mathbf{B})(\mathbf{A}+\mathbf{B}) = (\mathbf{A}+\mathbf{B})\mathbf{A} + (\mathbf{A}+\mathbf{B})\mathbf{B} \\ &= \mathbf{A}^2 + \mathbf{BA} + \mathbf{AB} + \mathbf{B}^2 \neq \mathbf{A}^2 + 2\mathbf{BA} + \mathbf{B}^2\end{aligned}$$

since $\mathbf{AB} \neq \mathbf{BA}$. However, $(\mathbf{A}+\mathbf{I})^2 = \mathbf{A}^2 + 2\mathbf{A} + \mathbf{I}$ because $\mathbf{AI} = \mathbf{IA} = \mathbf{A}$. More generally,

$$\begin{aligned}(\mathbf{A}+\alpha\mathbf{I})(\mathbf{A}+\beta\mathbf{I}) &= (\mathbf{A}+\beta\mathbf{I})(\mathbf{A}+\alpha\mathbf{I}) \\ &= \mathbf{A}^2 + (\alpha+\beta)\mathbf{A} + \alpha\beta\mathbf{I}\end{aligned}$$

The reader is asked to provide a proof for this assertion in the exercises. See Problem 14(b).

The transpose of the product of two matrices is the product of the transposes taken in the reverse order:

$$(\mathbf{AB})^T = \mathbf{B}^T\mathbf{A}^T \tag{0.7.6}$$

The result extends in the obvious way for the product of three or more matrices $(\mathbf{ABC})^T = \mathbf{C}^T\mathbf{B}^T\mathbf{A}^T$. The proof of (0.7.6) is reserved for the exercises. See Problem 16. We illustrate this theorem with the following example.

Example 0.7.2. *Verify (0.7.6) by using the matrices*

$$\mathbf{A} = \begin{bmatrix} 3 \\ 0 \\ -1 \end{bmatrix} \qquad \mathbf{B} = \begin{bmatrix} 2 & -1 & 1 \end{bmatrix}$$

Solution: The product $\mathbf{AB}$ is found to be

$$\mathbf{AB} = \begin{bmatrix} 6 & -3 & 3 \\ 0 & 0 & 0 \\ -2 & 1 & -1 \end{bmatrix}$$

and $\mathbf{B}^T\mathbf{A}^T$ is given by

$$\mathbf{B}^T\mathbf{A}^T = \begin{bmatrix} 2 \\ -1 \\ 1 \end{bmatrix} \begin{bmatrix} 3 & 0 & -1 \end{bmatrix} = \begin{bmatrix} 6 & 0 & -2 \\ -3 & 0 & 1 \\ 3 & 0 & -1 \end{bmatrix}$$

Example 0.7.3. *Show that* $\mathbf{b}_2 = \mathbf{Be}_2$ *is the second column of* $\mathbf{B}$.

Solution: The entries of the ith row, 2nd column of the product $\mathbf{Be}_2$ are given by (0.7.2) with $j = 2$, namely;

$$c_{i2} = b_{i1}0 + b_{i2}1 + \cdots + b_{in}0 = b_{i2}$$

for all i. So $\mathbf{Be}_2 = \mathbf{b}_2$, which is the second column of $\mathbf{B}$. This same argument shows that $\mathbf{b}_j = \mathbf{Be}_j$ for $j = 1, 2, \ldots, n$.

We can also show that $\mathbf{e}_j^T\mathbf{B}$ yields the jth row of $\mathbf{B}$. The argument is simple: Transpose $\mathbf{B}^T\mathbf{e}_j$ to obtain $\mathbf{e}_j^T\mathbf{B}$, and interpret the result!

The most general system of m equations in n unknowns has the form

$$\begin{aligned} a_{11}x_1 + a_{12}x_2 + \cdots + a_{1n}x_n &= b_1 \\ a_{21}x_1 + a_{22}x_2 + \cdots + a_{2n}x_n &= b_2 \\ &\vdots \\ a_{m1}x_1 + a_{m2}x_2 + \cdots + a_{mn}x_n &= b_m \end{aligned} \tag{0.7.7}$$

Matrix multiplication provides a means for writing this system in compact form. (Indeed, this property of matrix multiplication is so important, that if there were no other reason for doing so, we would still define multiplication in the way we do.) Set

$$\mathbf{A} = \begin{bmatrix} a_{11} & a_{12} & \dots & a_{1n} \\ a_{21} & a_{22} & \dots & a_{2n} \\ \vdots & \vdots & & \vdots \\ a_{m1} & a_{m2} & \dots & a_{mn} \end{bmatrix}, \quad \mathbf{x} = \begin{bmatrix} x_1 \\ x_2 \\ \vdots \\ x_n \end{bmatrix}, \quad \mathbf{b} = \begin{bmatrix} b_1 \\ b_2 \\ \vdots \\ b_m \end{bmatrix}$$

Then (0.7.7) can be written as

$$\mathbf{Ax} = \mathbf{b} \tag{0.7.8}$$

We shall use this notation repeatedly. Without explicit reference to the size of $\mathbf{A}$, we shall assume that $\mathbf{A}$ is $m \times n$, so that $\mathbf{x}$ is a vector with n entries and $\mathbf{b}$ is a vector with m entries, although in most applications $m = n$. For example, the system in Example 0.5.4 takes the matrix-vector form

$$\begin{bmatrix} 1 & 1 & 1 \\ 1 & -1 & 1 \\ 1 & 0 & 1 \end{bmatrix} \begin{bmatrix} x_1 \\ x_2 \\ x_3 \end{bmatrix} = \begin{bmatrix} 1 \\ 3 \\ k \end{bmatrix}$$

•EXERCISES

1. Find each product.

(a) $\begin{bmatrix} 1 & 3 \\ 3 & 1 \end{bmatrix} \begin{bmatrix} 1 & 4 \\ 4 & 1 \end{bmatrix}$.

(b) $\begin{bmatrix} -6 & 7 \\ 7 & -8 \end{bmatrix} \begin{bmatrix} 8 & 7 \\ 7 & 6 \end{bmatrix}$.

(c) $\begin{bmatrix} 1 & 1 & 0 \\ 0 & 1 & 2 \\ 0 & 1 & 1 \end{bmatrix} \begin{bmatrix} 1 & -2 & 2 \\ 0 & 2 & 2 \\ 0 & 0 & 1 \end{bmatrix}$.

(d) $\begin{bmatrix} 2 & 0 & 0 \\ 0 & 1 & 0 \\ 0 & 1 & -1 \end{bmatrix} \begin{bmatrix} x \\ y \\ z \end{bmatrix}$.

(e) $\begin{bmatrix} 2 & 5 \\ 1 & 3 \end{bmatrix} \begin{bmatrix} -4 & -15 \\ 2 & 7 \end{bmatrix} \begin{bmatrix} 3 & -5 \\ -1 & 2 \end{bmatrix}$.

(f) $\begin{bmatrix} a & b & c \end{bmatrix} \begin{bmatrix} a \\ b \\ c \end{bmatrix}$.

(g) $\begin{bmatrix} 0 \end{bmatrix} \begin{bmatrix} 1 & 7 & -2 \end{bmatrix}$.

2. Given these matrices

$$\mathbf{A} = \begin{bmatrix} 2 & 0 & 0 \\ 0 & -1 & 0 \\ 0 & 0 & 3 \end{bmatrix}, \quad \mathbf{B} = \begin{bmatrix} 2 & 1 & 3 \\ 1 & -1 & 2 \\ 1 & 3 & 2 \end{bmatrix},$$

$$\mathbf{C} = \begin{bmatrix} 0 \\ 1 \\ 1 \end{bmatrix}, \quad \mathbf{D} = \begin{bmatrix} 2 & 4 & -1 \end{bmatrix}.$$

(a) Find $\mathbf{AB}$ and $\mathbf{BA}$.

(b) Find $\mathbf{DC}$ and $\mathbf{C}^T\mathbf{A}$.

(c) Find $\mathbf{DAC}$.

(d) Find $\mathbf{DBC}$.

3. Given these matrices

$$\mathbf{A} = \begin{bmatrix} 0 \\ -1 \\ 0 \end{bmatrix}, \qquad \mathbf{B} = \begin{bmatrix} 2 & 4 & -1 \end{bmatrix},$$

$$\mathbf{C} = \begin{bmatrix} 3 & 2 & 1 \\ -2 & 0 & -1 \\ 1 & 0 & 1 \end{bmatrix}, \qquad \mathbf{D} = \begin{bmatrix} -1 & 0 & 2 \\ 1 & 2 & 1 \\ 2 & -1 & -1 \end{bmatrix},$$

find the following products:

(a) $\mathbf{AB}$. (b) $\mathbf{BA}$.

(c) $\mathbf{CA}$. (d) $\mathbf{CD}$.

(e) $\mathbf{BD}$. (f) $\mathbf{DA}$.

4. Given the matrices

$$\mathbf{A} = \begin{bmatrix} 0 & 3 & 1 \\ -1 & 2 & 0 \\ 0 & 0 & 1 \end{bmatrix}, \qquad \mathbf{B} = \begin{bmatrix} 1 & 0 & 0 \\ -1 & 2 & 1 \\ 2 & -1 & -1 \end{bmatrix},$$

$$\mathbf{C} = \begin{bmatrix} 2 \\ 0 \\ -1 \end{bmatrix}, \qquad \mathbf{D} = \begin{bmatrix} 1 & 2 & 0 \end{bmatrix},$$

find the following products when they exist:

(a) $(\mathbf{A}+\mathbf{B})\,\mathbf{C}$ and $\mathbf{AC}+\mathbf{BC}$. (b) $\mathbf{A}\,(\mathbf{BC})$ and $(\mathbf{AB})\,\mathbf{C}$.

(c) $\mathbf{D}\,(\mathbf{A}+\mathbf{B})$ and $\mathbf{DA}+\mathbf{DB}$. (d) $(\mathbf{AB})^T$ and $\mathbf{B}^T\mathbf{A}^T$.

(e) $\mathbf{A}^T\mathbf{A}$ and $\mathbf{AA}^T$. (f) $\mathbf{C}^T\mathbf{C}$ and $\mathbf{CC}^T$.

(g) $\mathbf{A}^2$ and $\mathbf{A}^3$. (h) $\mathbf{C}^2$.

(i) $\mathbf{A}+\mathbf{C}$. (j) $\mathbf{A}^2 - 2\mathbf{B} + 3\mathbf{I}$.

(k) $2\mathbf{AC} + \mathbf{DB} - 4\mathbf{I}$.

5. Find $\mathbf{A}^2, \mathbf{A}^3$ and $\mathbf{A}^n$ for $\mathbf{A} = \begin{bmatrix} 0 & 1 & 1 \\ 0 & 0 & 1 \\ 0 & 0 & 0 \end{bmatrix}$.

6. Find a formula for $\mathbf{A}^n$ where $\mathbf{A} = \begin{bmatrix} 1 & 1 \\ 0 & 1 \end{bmatrix}$.

7. For $\mathbf{A} = \begin{bmatrix} 1 & 2 \\ 0 & 1 \end{bmatrix}$ compute

(a) $3\mathbf{A}^2 - 9\mathbf{A} + 6\mathbf{I}$. (b) $3\,(\mathbf{A}-\mathbf{I})\,(\mathbf{A}-2\mathbf{I})$.

(c) $3\,(\mathbf{A}-2\mathbf{I})\,(\mathbf{A}-\mathbf{I})$.

8. Verify

$$(\mathbf{A}+\mathbf{I})^2 = \mathbf{A}^2 + 2\mathbf{A} + \mathbf{I}$$

and

$$(\mathbf{A}+\mathbf{I})^3 = \mathbf{A}^3 + 3\mathbf{A}^2 + 3\mathbf{A} + \mathbf{I}$$

for each of the following $\mathbf{A}$:

(a) $\mathbf{A} = \mathbf{I}$.

(b) $\mathbf{A} = \mathbf{O}$.

(c) $\mathbf{A} = \begin{bmatrix} 1 & 1 & 1 \\ 1 & 1 & 1 \\ 1 & 1 & 1 \end{bmatrix}$.

(d) $\mathbf{A} = \begin{bmatrix} 1 & 0 & -1 \\ 1 & 2 & 2 \\ -1 & 1 & 0 \end{bmatrix}$.

9. Find the following products:

(a) $\begin{bmatrix} x & y & z \end{bmatrix} \begin{bmatrix} 1 & -1 & 0 \\ -1 & 1 & 1 \\ 0 & 1 & -1 \end{bmatrix} \begin{bmatrix} x \\ y \\ z \end{bmatrix}$.

(b) $\begin{bmatrix} x & y & z \end{bmatrix} \begin{bmatrix} 2 & 0 & 0 \\ 0 & 1 & 0 \\ 1 & 1 & 1 \end{bmatrix} \begin{bmatrix} x \\ y \\ z \end{bmatrix}$.

10. If $\mathbf{AB} = \mathbf{BA}$, show that $\mathbf{A}$ and $\mathbf{B}$ are square matrices of the same size.

11. If $\mathbf{A}$ and $\mathbf{B}$ are upper triangular, show that $\mathbf{AB}$ is also.

12. Show that $(\mathbf{AB})^T = \mathbf{B}^T\mathbf{A}^T$ implies $(\mathbf{ABC})^T = \mathbf{C}^T\mathbf{B}^T\mathbf{A}^T$.

13. Show by an example that $(\mathbf{AB})^2 \neq \mathbf{A}^2\mathbf{B}^2$.

14. Let $\mathbf{A} = \begin{bmatrix} 1 & 1 \\ 0 & 1 \end{bmatrix}$ and $\mathbf{B} = \begin{bmatrix} 1 & 0 \\ 1 & 0 \end{bmatrix}$.

(a) Show that $\mathbf{A}^T\mathbf{A}$ and $\mathbf{A}\mathbf{A}^T$ are symmetric.

(b) Show that $(\mathbf{A} + \alpha\mathbf{I})(\mathbf{A} + \beta\mathbf{I}) = \mathbf{A}^2 + (\alpha + \beta)\mathbf{A} + \alpha\beta\mathbf{I}$

15. Find an example of two 3×3 matrices which shows that $\mathbf{AB} = \mathbf{O}$ is possible with $\mathbf{A} \neq \mathbf{O}$, $\mathbf{B} \neq \mathbf{O}$.

16. Use the definition of matrix multiplication to find the (i, j) term in $(\mathbf{AB})^T$ and $\mathbf{B}^T\mathbf{A}^T$, and show that they are equal.

MATLAB

17. Use MATLAB to verify the results obtained for Problems 1–5.

18. Use MATLAB to verify the results obtained for Problems 6—9.

19. If $\mathbf{A} = (a_{ij})$ and $f(t)$ is a function, then the matrix $\mathbf{B} = f(\mathbf{A})$ is interpreted to mean $\mathbf{B} = (f(a_{ij}))$. In MATLAB these matrix functions are invoked by "f(A)". Compute exp($\mathbf{A}$) and $\sqrt{\mathbf{A}}$

$$\mathbf{A} = \begin{bmatrix} 2 & 1 & 3 \\ 1 & -1 & 2 \\ 1 & 3 & 2 \end{bmatrix}$$

0.8 The Inverse of a Matrix

Division is not a process defined for matrices. In its place, and to serve similar purposes, we introduce the notion of the inverse of a matrix. The square matrix $\mathbf{A}$ is said to be *nonsingular*, or to have an *inverse*, or is *invertible*, if there exists a square matrix $\mathbf{B}$ such that

$$\mathbf{AB} = \mathbf{I} = \mathbf{BA} \tag{0.8.1}$$

It should be immediately clear that not all matrices have inverses, since if $\mathbf{A} = \mathbf{O}$ (0.8.1) is false for every $\mathbf{B}$. However, if there exists a $\mathbf{B}$ such that (0.8.1) holds, then there is only one such $\mathbf{B}$. Here's why. Assume that $\mathbf{C}$ is an inverse of $\mathbf{A}$. Then $\mathbf{AC} = \mathbf{I}$, and by premultiplying this by $\mathbf{B}$ we obtain

$$\mathbf{B}(\mathbf{AC}) = \mathbf{BI} = \mathbf{B} \tag{0.8.2}$$

But,

$$\mathbf{B}(\mathbf{AC}) = (\mathbf{BA})\mathbf{C} = \mathbf{IC} = \mathbf{C} \tag{0.8.3}$$

Hence, $\mathbf{B} = \mathbf{C}$.

Since there is never more than one inverse of $\mathbf{A}$, we call $\mathbf{B}$ *the inverse* of $\mathbf{A}$ if (0.8.1) holds and write $\mathbf{A}^{-1}$ for $\mathbf{B}$. With this notation, (0.8.1) can be written

$$\mathbf{AA}^{-1} = \mathbf{A}^{-1}\mathbf{A} = \mathbf{I} \tag{0.8.4}$$

A square matrix that is not invertible is one for which $\mathbf{A}^{-1}$ does not exist. Such matrices are called *singular* or *noninvertible*. The following matrices are singular:

$$\mathbf{A} = \begin{bmatrix} 1 & 1 & 1 \\ 1 & 1 & 1 \\ 1 & 1 & 1 \end{bmatrix}, \quad \mathbf{B} = \begin{bmatrix} 0 & 0 \\ a & b \end{bmatrix}, \quad \mathbf{C} = \begin{bmatrix} 1 & 1 & 1 \\ 2 & 2 & 2 \\ 3 & 3 & 3 \end{bmatrix}, \quad \mathbf{D} = \begin{bmatrix} 0 \end{bmatrix}$$

Example 0.8.1. *Show that the preceding matrix* $\mathbf{C}$ *is singular.*

Solution: We attempt to find $\mathbf{C}^{-1}$ by examining $\mathbf{C}\mathbf{C}^{-1} = \mathbf{I}$. Let the entries of $\mathbf{C}^{-1}$ be denoted by α_{ij}. Then we have

$$\begin{bmatrix} 1 & 1 & 1 \\ 2 & 2 & 2 \\ 3 & 3 & 3 \end{bmatrix} \begin{bmatrix} \alpha_{11} & \alpha_{12} & \alpha_{13} \\ \alpha_{21} & \alpha_{22} & \alpha_{23} \\ \alpha_{31} & \alpha_{32} & \alpha_{33} \end{bmatrix} = \begin{bmatrix} 1 & 0 & 0 \\ 0 & 1 & 0 \\ 0 & 0 & 1 \end{bmatrix}$$

But we arrive at a contradiction by computing the first two entries in the first column:

$$\begin{aligned} \alpha_{11} + \alpha_{21} + \alpha_{31} &= 1 \\ 2\alpha_{11} + 2\alpha_{21} + 2\alpha_{31} &= 0 \end{aligned}$$

Because of this contradiction, we conclude that $\mathbf{C}$ can have no inverse. Therefore, $\mathbf{C}$ is singular.

Except in rather special circumstances, it is not a trivial task to discover whether $\mathbf{A}$ is singular, particularly if the size of $\mathbf{A}$ is large, say, 8 or more. Moreover, knowing that $\mathbf{A}$ is invertible still leaves open the question of how to find its inverse. We will offer two cases in which we can show that a matrix is invertible by displaying its inverse.

Let $\mathbf{D}$ be a diagonal matrix with nonzero diagonal entries; that is,

$$\mathbf{D} = \begin{bmatrix} d_{11} & 0 & \dots & 0 \\ 0 & d_{22} & \dots & 0 \\ \vdots & \vdots & & \vdots \\ 0 & 0 & \dots & d_{nn} \end{bmatrix}$$

Then

$$\mathbf{D}^{-1} = \begin{bmatrix} d_{11}^{-1} & 0 & \dots & 0 \\ 0 & d_{22}^{-1} & \dots & 0 \\ \vdots & \vdots & & \vdots \\ 0 & 0 & \dots & d_{nn}^{-1} \end{bmatrix}$$

since $\mathbf{D}\mathbf{D}^{-1}$ is obviously $\mathbf{I}$.

If $\mathbf{A}$ is the 2×2 matrix $\begin{bmatrix} a & b \\ c & d \end{bmatrix}$ with $ad - bc \neq 0$, then

$$\mathbf{A}^{-1} = (ad - bc)^{-1} \begin{bmatrix} d & -b \\ -c & a \end{bmatrix} \tag{0.8.5}$$

We verify that $\mathbf{A}^{-1}\,\mathbf{A} = \mathbf{I}$:

$$\begin{bmatrix} d & -b \\ -c & a \end{bmatrix} \begin{bmatrix} a & b \\ c & d \end{bmatrix} = \begin{bmatrix} ad - bc & 0 \\ 0 & ad - bc \end{bmatrix} = (ad - bc)\,\mathbf{I}.$$

Theorem 0.8.1. *Suppose* $\mathbf{A}$ *and* $\mathbf{B}$ *are both invertible. Then*

$$\left(\mathbf{A}^{-1}\right)^{-1} = \mathbf{A} \tag{0.8.6}$$

$$(\mathbf{AB})^{-1} = \mathbf{B}^{-1}\mathbf{A}^{-1} \tag{0.8.7}$$

$$\left(\mathbf{A}^{T}\right)^{-1} = \left(\mathbf{A}^{-1}\right)^{T} \tag{0.8.8}$$

Proof: The proof of (0.8.6) is most easily seen by setting $\mathbf{A}^{-1} = \mathbf{B}$. Then (0.8.4) shows that $\mathbf{A}$ is the inverse of $\mathbf{B}$, which is the assertion made by (0.8.6). We prove (0.8.7) by verifying that $(\mathbf{B}^{-1}\mathbf{A}^{-1})(\mathbf{AB}) = \mathbf{I}$ and that $(\mathbf{AB})(\mathbf{B}^{-1}\mathbf{A}^{-1}) = \mathbf{I}$. We have

$$\left(\mathbf{B}^{-1}\mathbf{A}^{-1}\right)(\mathbf{AB}) = \mathbf{B}^{-1}\left(\mathbf{A}^{-1}\mathbf{A}\right)\mathbf{B} = \mathbf{B}^{-1}\mathbf{IB} = \mathbf{B}^{-1}\mathbf{B} = \mathbf{I}$$

and then, in the other order,

$$(\mathbf{AB})\left(\mathbf{B}^{-1}\mathbf{A}^{-1}\right) = \mathbf{A}\left(\mathbf{BB}^{-1}\right)\mathbf{A}^{-1} = \mathbf{AA}^{-1} = \mathbf{I}$$

The proof of (0.8.8) follows by taking transposes of $\mathbf{A}^{-1}\mathbf{A} = \mathbf{AA}^{-1} = \mathbf{I}$ and using $(\mathbf{AB})^{T} = \mathbf{B}^{T}\mathbf{A}^{T}$. Thus,

$$\left(\mathbf{A}^{-1}\mathbf{A}\right)^{T} = \mathbf{A}^{T}\left(\mathbf{A}^{-1}\right)^{T} = \mathbf{I}^{T} = \mathbf{I}, \quad \left(\mathbf{AA}^{-1}\right)^{T} = \left(\mathbf{A}^{-1}\right)^{T}\mathbf{A}^{T} = \mathbf{I}^{T} = \mathbf{I}.$$

These two equations establish that $(\mathbf{A}^{-1})^{T}$ is the inverse of $\mathbf{A}^{T}$. □

The existence of an inverse is a remedy for the lack of a "law of cancellation." For suppose $\mathbf{AB} = \mathbf{AC}$ and $\mathbf{A}$ is invertible. If this equation is multiplied on the left by $\mathbf{A}^{-1}$, the left-hand side simplifies as follows:

$$\mathbf{A}^{-1}(\mathbf{AB}) = \left(\mathbf{A}^{-1}\mathbf{A}\right)\mathbf{B} = \mathbf{B}$$

On the other hand, the right-hand side also simplifies:

$$\mathbf{A}^{-1}(\mathbf{AC}) = \left(\mathbf{A}^{-1}\mathbf{A}\right)\mathbf{C} = \mathbf{C}$$

Hence, $\mathbf{B} = \mathbf{C}$. Thus the existence of $\mathbf{A}^{-1}$ plays the role for $\mathbf{AB} = \mathbf{AC}$ that $a \neq 0$ plays for $ab = ac$. The corresponding cancellation law holds for $\mathbf{BA} = \mathbf{CA}$. The next theorem is still another application.

Theorem 0.8.2. *Suppose* $\mathbf{A}$ *is invertible. Then*

$$\mathbf{Ax} = \mathbf{b}$$

has one and only one solution,

$$\mathbf{x} = \mathbf{A}^{-1}\mathbf{b}. \tag{0.8.9}$$

Proof: We see that $\mathbf{A}^{-1}\mathbf{b}$ is a solution by verification:

$$\mathbf{A}\left(\mathbf{A}^{-1}\mathbf{b}\right) = \mathbf{Ib} = \mathbf{b}$$

That it is the only solution can be shown equally easily. Let $\mathbf{x}_0$ be a solution of $\mathbf{Ax} = \mathbf{b}$. Then, since $\mathbf{Ax}_0 = \mathbf{b}$ by hypothesis and $\mathbf{I} = \mathbf{A}^{-1}\mathbf{A}$,

$$\mathbf{x}_0 = \mathbf{Ix}_0 = \mathbf{A}^{-1}(\mathbf{Ax}_0) = \mathbf{A}^{-1}\mathbf{b}$$

which shows that $\mathbf{x}_0 = \mathbf{A}^{-1}\mathbf{b}$. □

This important theorem implies that $\mathbf{Ax} = \mathbf{0}$ has as its only solution the vector $\mathbf{x} = \mathbf{0}$ whenever $\mathbf{A}$ is not singular. If $\mathbf{A}$ is singular, then $\mathbf{Ax} = \mathbf{0}$ will have infinitely many solutions.

Corollary 0.8.1. *The system* $\mathbf{Ax} = \mathbf{0}$ *has solutions* $\mathbf{x} \neq \mathbf{0}$ *if and only if* $\mathbf{A}$ *is singular. The system has only the solution* $\mathbf{x} = \mathbf{0}$ *if and only if* $\mathbf{A}$ *is invertible.*

• EXERCISES

1. If $\mathbf{A}^{-1}$ exists, $\mathbf{A}$ has size $n \times n$ and $\mathbf{AC} = \mathbf{B}$, then $\mathbf{A}^{-1}\mathbf{B}$ exists. Explain.
2. Show by example that $(\mathbf{A} + \mathbf{B})^{-1} \neq \mathbf{A}^{-1} + \mathbf{B}^{-1}$.
3. Verify $\mathbf{AA}^{-1} = \mathbf{I}$ for the matrix (0.8.6).
4. Verify that the matrix $\mathbf{A}$ just before Example 0.8.1 is singular.
5. Show that $(\mathbf{A}^{-1}\mathbf{BA})^2 = \mathbf{A}^{-1}\mathbf{B}^2\mathbf{A}$. Generalize to $(\mathbf{A}^{-1}\mathbf{BA})^n$.
6. Show that $(\mathbf{A}^2)^{-1} = (\mathbf{A}^{-1})^2$. Generalize to $(\mathbf{A}^n)^{-1} = (\mathbf{A}^{-1})^n$.

7. Set $\mathbf{u} = \begin{bmatrix} u_1 & u_2 & \cdots & u_n \end{bmatrix}^T$. Write out $\mathbf{uu}^T$ to show that $\mathbf{uu}^T$ is singular, assuming $n \geq 2$.
8. It is possible to prove that $\mathbf{A}$ is invertible if and only if there exists $\mathbf{B}$ such that $\mathbf{AB} = \mathbf{I}$ or there exists $\mathbf{C}$ such that $\mathbf{CA} = \mathbf{I}$. Accepting the theorem, establish this theorem: If $\mathbf{AB}$ is invertible, then so are $\mathbf{A}$ and $\mathbf{B}$, assuming that $\mathbf{A}$ and $\mathbf{B}$ are square. *Hint*: Use Corollary 0.8.1.
9. Under the same hypotheses in Problem 8, establish this theorem: If $\mathbf{A}$ is singular, so is $\mathbf{AB}$ and $\mathbf{BA}$ for every $\mathbf{B}$, assuming that $\mathbf{A}$ and $\mathbf{B}$ are square.

0.9 The Computation of $\mathbf{A}^{-1}$

Given $\mathbf{A}$, we wish to find $\mathbf{X}$, if possible, so that $\mathbf{AX} = \mathbf{I}$. Denote the ith column of $\mathbf{X}$ by $\mathbf{x}_i$, $i = 1, 2, \ldots, n$, so that

$$\mathbf{X} = \begin{bmatrix} \mathbf{x}_1 & \mathbf{x}_2 & \cdots & \mathbf{x}_n \end{bmatrix} \tag{0.9.1}$$

By the definition of matrix multiplication, it follows that

$$\mathbf{AX} = \begin{bmatrix} \mathbf{Ax}_1 & \mathbf{Ax}_2 & \cdots & \mathbf{Ax}_n \end{bmatrix} = \mathbf{I} \tag{0.9.2}$$

and hence, the columns of $\mathbf{Ax}$ must be the columns of $\mathbf{I}$. So

$$\mathbf{Ax}_1 = \mathbf{e}_1, \quad \mathbf{Ax}_2 = \mathbf{e}_2, \quad \ldots, \quad \mathbf{Ax}_n = \mathbf{e}_n \tag{0.9.3}$$

So finding $\mathbf{x}$ amounts to solving n systems, each with n unknowns, and each with the same coefficient matrix $\mathbf{A}$. We can solve all n systems at once by forming the "large" augmented matrix $[\mathbf{A}|\,\mathbf{I}]$ and using Gaussian elimination. An example will illustrate this idea.

Example 0.9.1. *Find the inverse of* $\mathbf{A} = \begin{bmatrix} 1 & 2 \\ -1 & 1 \end{bmatrix}$.

Solution: Here we have

$$\begin{bmatrix} 1 & 2 \\ -1 & 1 \end{bmatrix} \mathbf{X} = \begin{bmatrix} 1 & 2 \\ -1 & 1 \end{bmatrix} \begin{bmatrix} \mathbf{x}_1 & \mathbf{x}_2 \end{bmatrix} = \mathbf{I}$$

For this problem, the augmented matrix $[\mathbf{A}|\,\mathbf{I}]$ is given by

$$\mathbf{B} = [\mathbf{A}|\,\mathbf{I}] = \left[\begin{array}{cc|cc} 1 & 2 & 1 & 0 \\ -1 & 1 & 0 & 1 \end{array}\right]$$

Now we apply elementary row operations on $\mathbf{B}$ until the identity matrix appears in place of $\mathbf{A}$:

$$\mathbf{B} = \left[\begin{array}{cc|cc} 1 & 2 & 1 & 0 \\ -1 & 1 & 0 & 1 \end{array}\right] \to \left[\begin{array}{cc|cc} 1 & 2 & 1 & 0 \\ 0 & 3 & 1 & 1 \end{array}\right] \to \left[\begin{array}{cc} 1 & 0 \\ 0 & 1 \end{array}\right| \left|\begin{array}{cc} 1/3 & -2/3 \\ 1/3 & 1/3 \end{array}\right]$$

Thus, we can deduce that the columns of $\mathbf{X}$ must be

$$\mathbf{x}_1 = \begin{bmatrix} 1/3 \\ 1/3 \end{bmatrix} \quad \text{and} \quad \mathbf{x}_2 = \begin{bmatrix} -2/3 \\ 1/3 \end{bmatrix}$$

and hence,

$$\mathbf{X} = \begin{bmatrix} 1/3 & -2/3 \\ 1/3 & 1/3 \end{bmatrix}$$

We leave it to the reader to verify that $\mathbf{XA} = \mathbf{AX} = \mathbf{I}$.

This technique yields $\mathbf{X}$ such that $\mathbf{AX} = \mathbf{I}$. It is a theorem of matrix theory that $\mathbf{AX} = \mathbf{I}$ implies $\mathbf{XA} = \mathbf{I}$ for square matrices $\mathbf{A}$. For this reason, we need not check both $\mathbf{AX} = \mathbf{I}$ and $\mathbf{XA} = \mathbf{I}$ to verify that we do indeed have an inverse.

It is also important to note that this method works if the entries of $\mathbf{A}$ are parameters or functions. Here is an interesting example to illustrate this point. We will find many other examples when studying systems of differential equations.

Example 0.9.2. *Find the inverse of* $\mathbf{A} = \begin{bmatrix} a & b \\ c & d \end{bmatrix}$, $ad - bc = 1$, *and compare with* (0.8.6).

Solution: Consider the following arrow diagram in which we assume that $a \neq 0$:

$$\left[\begin{array}{cc|cc} a & b & 1 & 0 \\ c & d & 0 & 1 \end{array}\right] \to \left[\begin{array}{cc|cc} a & b & 1 & 0 \\ ac & ad & 0 & a \end{array}\right] \to \left[\begin{array}{cc|cc} a & b & 1 & 0 \\ 0 & 1 & -c & a \end{array}\right]$$

Then, since $ad - bc = 1$ we have $ad = 1 + bc$ which leads in turn to

$$\left[\begin{array}{cc|cc} a & b & 1 & 0 \\ c & d & 0 & 1 \end{array}\right] \to \left[\begin{array}{cc|cc} 1 & 0 & d & -b \\ 0 & 1 & -c & a \end{array}\right]$$

Hence

$$\begin{bmatrix} a & b \\ c & d \end{bmatrix}^{-1} = \begin{bmatrix} d & -b \\ -c & a \end{bmatrix}$$

If $a = 0$ we need to make a preliminary adjustment in order to find $\mathbf{A}^{-1}$; we interchange rows 1 and 2 so that c is in the (1,1) position. If c is also zero then $\mathbf{A}$ has a column of zeros and $\mathbf{A}$ is singular.

Example 0.9.3. *Invert the matrix* $\mathbf{A} = \begin{bmatrix} 1 & 0 & 1 \\ 2 & 1 & 1 \\ 1 & 1 & 2 \end{bmatrix}$.

Solution: An arrow diagram that yields $\mathbf{A}^{-1}$ follows:

$$[\mathbf{A}|\,\mathbf{I}] = \left[\begin{array}{ccc|ccc} 1 & 0 & 1 & 1 & 0 & 0 \\ 2 & 1 & 1 & 0 & 1 & 0 \\ 1 & 1 & 2 & 0 & 0 & 1 \end{array}\right] \to \cdots \to \left[\begin{array}{ccc|ccc} 1 & 0 & 1 & 1 & 0 & 0 \\ 0 & 1 & -1 & -2 & 1 & 0 \\ 0 & 0 & 2 & 1 & -1 & 1 \end{array}\right]$$

$$\to \left[\begin{array}{ccc|ccc} 1 & 0 & 0 & 1/2 & 1/2 & -1/2 \\ 0 & 1 & 0 & -3/2 & 1/2 & 1/2 \\ 0 & 0 & 1 & 1/2 & -1/2 & 1/2 \end{array}\right]$$

Thus,

$$\mathbf{A}^{-1} = \begin{bmatrix} 1/2 & 1/2 & -1/2 \\ -3/2 & 1/2 & 1/2 \\ 1/2 & -1/2 & 1/2 \end{bmatrix} = \frac{1}{2}\begin{bmatrix} 1 & 1 & -1 \\ -3 & 1 & 1 \\ 1 & -1 & 1 \end{bmatrix}$$

Example 0.9.4. *Show that* $\mathbf{A} = \begin{bmatrix} 2 & 2 & 1 \\ 3 & 3 & -2 \\ 1 & 1 & -3 \end{bmatrix}$ *is singular.*

Solution: We attempt to find $\mathbf{X}$ such that $\mathbf{AX} = \mathbf{I}$. The following arrow diagram shows why $\mathbf{A}^{-1}$ cannot exist. (Begin by adding row 1 to row 3.)

$$[\mathbf{A}|\,\mathbf{I}] = \left[\begin{array}{ccc|ccc} 2 & 2 & 1 & 1 & 0 & 0 \\ 3 & 3 & -2 & 0 & 1 & 0 \\ 1 & 1 & -3 & 0 & 0 & 1 \end{array}\right] \to \cdots \to \left[\begin{array}{ccc|ccc} 2 & 2 & 1 & 1 & 0 & 0 \\ 1 & 1 & -3 & -1 & 1 & 0 \\ 0 & 0 & 0 & 1 & -1 & 1 \end{array}\right]$$

Thus, $\mathbf{AX} = \mathbf{I}$ is equivalent to

$$\begin{bmatrix} 2 & 2 & 1 \\ 1 & 1 & -3 \\ 0 & 0 & 0 \end{bmatrix}\mathbf{X} = \begin{bmatrix} 1 & 0 & 0 \\ -1 & 1 & 0 \\ 1 & -1 & 1 \end{bmatrix}$$

However, no $\mathbf{X}$ can solve this system because the product of the two matrices on the left is a matrix whose third row is a row of zeros. In fact this is a general principle: Whenever we encounter a row of zeros in the row reduction of $\mathbf{A}$, it is impossible to invert $\mathbf{A}$; no matrix $\mathbf{X}$ can solve $\mathbf{AX} = \mathbf{I}$, and therefore, no inverse can exist for $\mathbf{A}$.

•EXERCISES

Use row operations on the augmented matrix $[\mathbf{A}|\mathbf{I}]$ to decide whether the matrices given in Problems 1—15 are singular. Whenever the inverse exists, find it.

1. $\begin{bmatrix} 1 & -1 \\ 1 & 1 \end{bmatrix}$.
2. $\begin{bmatrix} 2 & 6 \\ 1 & 3 \end{bmatrix}$.
3. $\begin{bmatrix} 2 & 0 \\ 0 & 1 \end{bmatrix}$.
4. $\begin{bmatrix} 1 & 2 \\ 0 & 0 \end{bmatrix}$.
5. $\begin{bmatrix} 1 & 2 \\ -2 & 1 \end{bmatrix}$.
6. $\begin{bmatrix} \cos\theta & \sin\theta \\ -\sin\theta & \cos\theta \end{bmatrix}$.
7. $\begin{bmatrix} -1 & 0 & 1 \\ 1 & 0 & 0 \\ 0 & 0 & 1 \end{bmatrix}$.
8. $\begin{bmatrix} 2 & 0 & 1 \\ 0 & 3 & 4 \\ 0 & 0 & 3 \end{bmatrix}$.
9. $\begin{bmatrix} 2 & 0 & 0 \\ 4 & -1 & 0 \\ 0 & 1 & -1 \end{bmatrix}$.
10. $\begin{bmatrix} 0 & 0 & 1 \\ 0 & 1 & 0 \\ 1 & 0 & 0 \end{bmatrix}$.
11. $\begin{bmatrix} 1 & 0 & 2 \\ 0 & 1 & 0 \\ 0 & 5 & 0 \end{bmatrix}$.
12. $\begin{bmatrix} 1 & 0 & 2 \\ 2 & 1 & 1 \\ 1 & 1 & 1 \end{bmatrix}$.
13. $\begin{bmatrix} 1 & 2 & 2 \\ 1 & 1 & 2 \\ 1 & -2 & 2 \end{bmatrix}$.
14. $\begin{bmatrix} 1 & 1 & 2 \\ -1 & 2 & 1 \\ 0 & 1 & 1 \end{bmatrix}$.
15. $\begin{bmatrix} 0 & 1 & 1 & 0 \\ 0 & 0 & 1 & 1 \\ 1 & 0 & 1 & 1 \\ 1 & 1 & 1 & 1 \end{bmatrix}$.

16. Find the inverse, if it exists, of each of the following symmetric matrices. Note that each inverse is also symmetric.

(a) $\begin{bmatrix} 2 & 1 \\ 1 & 1 \end{bmatrix}$.

(b) $\begin{bmatrix} 2 & 1 & 2 \\ 1 & 0 & 1 \\ 2 & 1 & -2 \end{bmatrix}$.

17. Suppose $\mathbf{AX} = \mathbf{B}$. Explain why $\mathbf{B}$ has a row of zeros if $\mathbf{A}$ does.

MATLAB

18. Repeat Problems 2–6 by using “inv(A)”, and “rref(A,eye(2))”.
19. Repeat Problems 8–15 by using “inv(A)” and “rref(A,eye(3))”.
20. Repeat Problems 16 by using “inv(A)” and “rref(A,eye(4))”.
21. Compute “inv(magic(3))” and “inv(hilb(4))”.
22. Compute “R = rand(4)” and “inv(R)”. What does “inv(inv(R))” return?

0.10 Determinants

0.10.1 Definitions and Fundamental Theorems

Determinants play a major role in our studies of matrices, for, among other uses, the determinant of a matrix vanishes if and only if the matrix is singular. Time and space constraints prohibit us from developing the theory of determinants from first principles. Instead, we shall present a brief survey of the relevant facts.

First of all, the *determinant* of $\mathbf{A}$, written $\det \mathbf{A}$, is a scalar defined only for square matrices. It is a sum of $n!$ terms, each term being plus or minus the product of entries of $\mathbf{A}$, one entry from each row and column. Each term has the form

$$(-1)^k a_{1*} a_{2*} \cdots a_{n*} \tag{0.10.1}$$

where the second subscript, indicated by the $*$, is one of the numbers $\{1, 2, \ldots, n\}$, no one of which is used twice. The exponent k is the total number of inversions[3] of the second subscript. Thus,

$$\det \mathbf{A} = \sum (-1)^k a_{1*} a_{2*} \cdots a_{n*} \tag{0.10.2}$$

Since there is one term for each permutation of the numbers $\{1, 2, \ldots, n\}$, the sum in (0.10.2) contains $n!$ terms, a number that grows very rapidly with n. For this reason, we seldom use (0.10.2) for computational purposes. If $n = 2$, however, the definition is easy to use and results in the familiar formula

$$\det \begin{bmatrix} a_{11} & a_{12} \\ a_{21} & a_{22} \end{bmatrix} = a_{11}a_{22} - a_{12}a_{21} \tag{0.10.3}$$

There are two common methods of evaluating $\det \mathbf{A}$. In this subsection we explore the method most suitable to efficient calculation. Our method depends on two fundamental theorems, the second of which we offer without proof.

Theorem 0.10.1. *If* $\mathbf{A}$ *is upper or lower triangular with diagonal entries* $a_{11}, a_{22}, \ldots, a_{nn}$, *then*

$$\det \mathbf{A} = a_{11}a_{22} \cdots a_{nn} \tag{0.10.4}$$

[3] By an inversion, we mean the number of pairs of elements in which a larger number precedes a smaller one; for example, the numbers (1,5,2,4,3) form the inversions (5,2), (5,4), (5,3), and (4,3).

Proof: Each term in the definition of det **A** will contain a zero factor, except the term containing only the diagonal entries and there are no inversions. □

Theorem 0.10.2. *Let* **A** *be a square matrix.*

(1) *If two rows of* **A** *are interchanged to form* **B**, *then*

$$\det \mathbf{A} = -\det \mathbf{B}$$

(2) *If a row of* **A** *is multiplied by* k *to form* **B**, *then*

$$k \det \mathbf{A} = \det \mathbf{B}$$

(3) *If a multiple of one row of* **A** *is added to a different row of* **A** *to form* **B**, *then*

$$\det \mathbf{A} = \det \mathbf{B}$$

These two theorems provide an efficient and effective means for calculating the determinant. Note in particular that Theorem 0.10.2 describes the effects of an elementary row operation on det **A**. Here are two examples illustrating this point.

Example 0.10.1. *Calculate the determinant of* $\mathbf{A} = \begin{bmatrix} 0 & 1 & 1 \\ 1 & 0 & 0 \\ 0 & 0 & 1 \end{bmatrix}$.

Solution: The following arrow diagram represents the result of interchanging rows 1 and 2:

$$\mathbf{A} = \begin{bmatrix} 0 & 1 & 1 \\ 1 & 0 & 0 \\ 0 & 0 & 1 \end{bmatrix} \to \begin{bmatrix} 1 & 0 & 0 \\ 0 & 1 & 1 \\ 0 & 0 & 1 \end{bmatrix} = \mathbf{B}$$

By Theorem 0.10.2 (1), $\det \mathbf{A} = -\det \mathbf{B}$. By Theorem 0.10.1 applied to **B**

$$\det \mathbf{B} = \det \begin{bmatrix} 1 & 0 & 0 \\ 0 & 1 & 1 \\ 0 & 0 & 1 \end{bmatrix} = 1$$

so $\det \mathbf{A} = -1$.

We can compute the result of elementary row operations rapidly and accurately. So for matrices with known constant entries, row reduction to triangular form is the preferred method for computing the determinant. For matrices with parametric entries, other methods are often used.

Example 0.10.2. *Calculate* det $\mathbf{A}$, *where* $\mathbf{A} = \begin{bmatrix} 1 & 2 & 1 & 3 \\ -1 & 1 & 3 & 2 \\ 1 & 0 & 2 & 3 \\ -1 & 1 & 1 & 4 \end{bmatrix}$.

Solution: We apply elementary row operations to $\mathbf{A}$ to obtain an upper triangular matrix $\mathbf{B}$ as follows:

$$\mathbf{A} = \begin{bmatrix} 1 & 2 & 1 & 3 \\ -1 & 1 & 3 & 2 \\ 1 & 0 & 2 & 3 \\ -1 & 1 & 1 & 4 \end{bmatrix} \rightarrow \begin{bmatrix} 1 & 2 & 1 & 3 \\ 0 & 3 & 4 & 5 \\ 0 & -2 & 1 & 0 \\ 0 & 3 & 2 & 7 \end{bmatrix} \rightarrow \begin{bmatrix} 1 & 2 & 1 & 3 \\ 0 & 3 & 4 & 5 \\ 0 & 0 & 11/3 & 10/3 \\ 0 & 0 & -2 & 2 \end{bmatrix}$$

$$\begin{bmatrix} 1 & 2 & 1 & 3 \\ 0 & 3 & 4 & 5 \\ 0 & 0 & 11/3 & 10/3 \\ 0 & 0 & 0 & 42/11 \end{bmatrix} = \mathbf{B}$$

Therefore, by Theorems 0.10.1 and 0.10.2, $\det \mathbf{A} = \det \mathbf{B} = 1 \cdot 3 \cdot \frac{11}{3} \cdot \frac{42}{11} = 42$.

Example 0.10.3. *Show that* $\det(k\mathbf{A}) = k^n \det \mathbf{A}$, *where* $\mathbf{A}$ *is* $n \times n$.

Solution: Since each row of $\mathbf{A}$ is multiplied by the same scalar k, and since there are n rows in $\mathbf{A}$, the proof follows by applying Theorem 0.10.2, Part (2), n times.

In view of the complicated definitions of matrix multiplication and determinants, it is surprising that a simple relationship exists between the determinant of a product and the product of determinants:

Theorem 0.10.3. *If* $\mathbf{A}$ *and* $\mathbf{B}$ *are square matrices, then*

$$\det(\mathbf{AB}) = \det \mathbf{A} \det \mathbf{B} \tag{0.10.5}$$

Example 0.10.4. *Show that* $\det(\mathbf{AB}) = \det \mathbf{A} \det \mathbf{B}$, *where*

$$\mathbf{A} = \begin{bmatrix} 1 & 3 \\ 0 & 2 \end{bmatrix}, \qquad \mathbf{B} = \begin{bmatrix} 1 & 1 \\ -1 & 1 \end{bmatrix}$$

Solution: We have

$$\mathbf{AB} = \mathbf{C} = \begin{bmatrix} -2 & 4 \\ -2 & 2 \end{bmatrix}$$

It is easy to verify that $\det \mathbf{A} = \det \mathbf{B} = 2$ and $\det \mathbf{C} = 4$.

Example 0.10.5. *Show that* $\det \mathbf{A} \det \mathbf{A}^{-1} = 1$.

Solution: Since $\mathbf{AA}^{-1} = \mathbf{I}$, it follows from Theorem 0.10.3 that

$$\det \mathbf{A} \det \mathbf{A}^{-1} = \det\left(\mathbf{AA}^{-1}\right) = \det \mathbf{I}$$

From Theorem 0.10.2, $\det \mathbf{I} = 1$. Hence, $\det \mathbf{A} \det \mathbf{A}^{-1} = 1$.

Example 0.10.5 shows that nonsingular matrices have nonzero determinants. Hence, if $\det \mathbf{A} = 0$, then $\mathbf{A}$ must be singular. The converse is also true, although we do not prove this here: If $\mathbf{A}$ is singular then $\det \mathbf{A} = 0$. This relationship between the vanishing of the determinant of a matrix and the singularity of the matrix is an important tool in the study of systems.

Theorem 0.10.4. *A necessary and sufficient condition that* $\mathbf{A}$ *be singular is that* $\det \mathbf{A} = 0$.

Theorem 0.10.5. *For every square matrix* $\mathbf{A}$

$$\det \mathbf{A} = \det\left(\mathbf{A}^T\right) \tag{0.10.6}$$

Since the columns of $\mathbf{A}$ are the rows of $\mathbf{A}^T$, Theorem 0.10.5 enables us to replace "rows" with "columns" in each part of Theorem 0.10.2.

Example 0.10.6. *Show that the following matrices are singular:*

$$\mathbf{A} = \begin{bmatrix} 2 & 3 & 0 \\ -1 & 0 & 0 \\ 1 & 2 & 0 \end{bmatrix} \qquad \textit{and} \qquad \mathbf{B} \begin{bmatrix} 3 & 1 & 6 \\ 2 & 0 & 4 \\ -1 & 2 & -2 \end{bmatrix}$$

Solution: The first matrix has a column of zeros. Its transpose, therefore, has a row of zeros. Hence, its transpose is singular, and because of this, $\det \mathbf{A}^T = 0$. Theorem 0.10.5 shows that $\det \mathbf{A} = 0$. In the same manner, we see that the second matrix $\mathbf{B}$ has a zero determinant because its transpose has its first and third rows proportional. So $\det \mathbf{B} = \det \mathbf{B}^T = 0$.

• EXERCISES

1. Show that (a) $\det(\mathbf{S}^{-1}\mathbf{A}\mathbf{S}) = \det \mathbf{A}$ and (b) $\det \mathbf{A}^m = (\det \mathbf{A})^m$.
2. Find any $\mathbf{A}$ and $\mathbf{B}$ that illustrates $\det(\mathbf{A} + \mathbf{B}) \neq \det \mathbf{A} + \det \mathbf{B}$.

Evaluate the determinants of the matrices in Problems 3–10.

3. $\begin{bmatrix} 3 & 1 & 3 \\ 2 & 0 & 4 \\ -1 & 2 & -2 \end{bmatrix}$.

4. $\begin{bmatrix} 2 & 3 & 4 \\ -1 & 0 & 3 \\ 1 & 2 & 3 \end{bmatrix}$.

5. $\begin{bmatrix} 1 & 1 & 1 \\ -1 & -2 & 2 \\ 1 & 2 & 3 \end{bmatrix}$.

6. $\begin{bmatrix} 3 & 1 & 0 \\ 1 & 3 & -1 \\ 3 & -3 & 2 \end{bmatrix}$.

7. $\begin{bmatrix} 3 & 1 & -1 & 0 \\ 2 & 2 & 2 & 1 \\ -1 & 3 & 0 & 4 \\ 8 & 6 & -2 & 2 \end{bmatrix}$.

8. $\begin{bmatrix} 1 & 1 & 1 & 1 \\ -2 & -3 & 1 & 0 \\ 4 & 3 & 8 & 1 \\ 7 & 5 & -2 & 0 \end{bmatrix}$.

9. $\begin{bmatrix} 2 & -1 & 6 & 3 \\ -2 & 4 & 5 & -1 \\ 3 & 4 & 3 & 2 \\ 1 & -1 & 2 & 3 \end{bmatrix}$.

10. $\begin{bmatrix} 0 & 0 & 0 & a \\ 0 & 0 & b & 0 \\ 0 & c & 0 & 0 \\ d & 0 & 0 & 0 \end{bmatrix}$.

11. Verify Theorem 0.10.3 using the product

$$\begin{bmatrix} 2 & 3 & 1 \\ -1 & 0 & 2 \\ 3 & 4 & 1 \end{bmatrix} \begin{bmatrix} 2 & -1 & 3 \\ 1 & 0 & -1 \\ 3 & 4 & 2 \end{bmatrix}.$$

12. Prove that $\det \mathbf{A}^n = (\det \mathbf{A})^n$.
13. Use Theorems 0.10.4 and 0.10.5 to prove that $\mathbf{A}^T$ is singular if and only if $\mathbf{A}$ is singular.

0.10.2 Minors and Cofactors

The *minor* of the entry a_{ij} of $\mathbf{A}$ is the determinant of the matrix formed by striking out the ith row and j^{th} column of $\mathbf{A}$. For example, if

$$\mathbf{A} = \begin{bmatrix} 1 & -1 & 2 \\ 0 & -2 & 3 \\ 4 & -4 & 6 \end{bmatrix} \tag{0.10.7}$$

then the three minors of the entries in the first row are, respectively,

$$\det \begin{bmatrix} -2 & 3 \\ -4 & 6 \end{bmatrix} = 0, \quad \det \begin{bmatrix} 0 & 3 \\ 4 & 6 \end{bmatrix} = -12, \quad \det \begin{bmatrix} 0 & -2 \\ 4 & -4 \end{bmatrix} = 8$$

The *cofactor* of a_{ij} is written A_{ij} and is $(-1)^{i+j}$ times the minor of a_{ij}. Hence, the cofactors of the entries in the first row are $A_{11} = 0, A_{12} = 12, A_{13} = 8$. The importance of the cofactors is due to the following remarkable theorem:

Theorem 0.10.6. *For each i and j,*

$$\det \mathbf{A} = a_{i1}A_{i1} + a_{i2}A_{i2} + \cdots + a_{in}A_{in} \tag{0.10.8}$$

$$\det \mathbf{A} = a_{1j}A_{1j} + a_{2j}A_{2j} + \cdots + a_{nj}A_{nj} \tag{0.10.9}$$

Remark: This theorem explains the name "expansion by cofactors", a terminology used to describe equations (0.10.8) and (0.10.9). Note that these sums are comprised of products of entries along a row (or column) with the entries cofactor.

Example 0.10.7. *Find* det $\mathbf{A}$ *by using* Theorem 0.10.6 *and expanding by cofactors of the entries in the first row of* $\mathbf{A}$. *Use*

$$\mathbf{A} = \begin{bmatrix} 3 & 2 & 1 \\ -1 & 0 & 1 \\ 1 & 2 & 2 \end{bmatrix}$$

Solution: Expanding by the entries in the first row, we obtain

$$\begin{aligned} \det \mathbf{A} &= 3 \cdot \det \begin{bmatrix} 0 & 1 \\ 2 & 2 \end{bmatrix} - 2 \cdot \det \begin{bmatrix} -1 & 1 \\ 1 & 2 \end{bmatrix} + 1 \cdot \det \begin{bmatrix} -1 & 0 \\ 1 & 2 \end{bmatrix} \\ &= 3(-2) - 2(-3) + (-2) = -2 \end{aligned}$$

Note that the cofactor expansion about the second row or column would involve one less determinant to evaluate.

• EXERCISES

1. Use Theorem 0.10.6 to evaluate

$$\det\begin{bmatrix} 3 & 2 & -1 \\ 3 & 0 & 3 \\ -1 & 2 & 1 \end{bmatrix}$$

(a) Expand by entries in the first row.
(b) Expand by the entries in the second row.
(c) Expand by the entries in the first column.
(d) Expand by the entries in the second column.

2. Find

$$\det\begin{bmatrix} 2 & 0 & 8 & 6 \\ -1 & 4 & 2 & 0 \\ 0 & -1 & 3 & 0 \\ 3 & 5 & 7 & 3 \end{bmatrix}$$

by using Theorem 0.10.6 and expanding by its entries in the first row.

3. Use row operations on the matrix in Problem 2 to obtain a row or column with three zeros, and then use Theorem 0.10.6 with cofactors of this row (or column).

4. Use Theorem 0.10.6 to find the determinants of the following matrices.

(a) $\begin{bmatrix} 2 & 0 \\ 0 & 1 \end{bmatrix}$.

(b) $\begin{bmatrix} 1 & 2 \\ 0 & 0 \end{bmatrix}$.

(c) $\begin{bmatrix} 0 & 2 & 3 \\ 2 & 0 & 2 \\ 3 & 2 & 0 \end{bmatrix}$.

(d) $\begin{bmatrix} 1 & 0 & 2 \\ 2 & 1 & 1 \\ 1 & 1 & 1 \end{bmatrix}$.

(e) $\begin{bmatrix} 0 & 0 & a \\ 0 & b & 0 \\ c & 0 & 0 \end{bmatrix}$.

(f) $\begin{bmatrix} 1 & 2 & 3 \\ 2 & 3 & 1 \\ 3 & 1 & 2 \end{bmatrix}$.

(g) $\begin{bmatrix} 2 & 1 & 1 & 1 \\ 1 & 2 & 0 & 0 \\ 1 & 0 & 0 & 1 \\ 1 & 0 & 1 & 2 \end{bmatrix}$.

(h) $\begin{bmatrix} 0 & 1 & 1 & 0 \\ 0 & 0 & 1 & 1 \\ 1 & 0 & 1 & 1 \\ 1 & 1 & 1 & 1 \end{bmatrix}$.

MATLAB

5. Write an M-file that computes A_{ij} given $\mathbf{A}$.

6. Write an M-file that computes $\det \mathbf{A}$ using (0.10.8).

7. Write an M-file that computes $\det \mathbf{A}$ using (0.10.9).
8. The "classical" *adjoint* of $\mathbf{A}$ is defined $\operatorname{adj}(\mathbf{A}) = [A_{ij}]^T$. That is, the (i, j) entry of $\operatorname{adj}(\mathbf{A})$ is the cofactor A_{ij}. It is a theorem of matrix theory — a consequence of Theorem 0.10.6)— that $\mathbf{A}^{-1} \det \mathbf{A} = \operatorname{adj}(\mathbf{A})$. Write an M-file which computes $\operatorname{adj}(\mathbf{A})$. Use this file to verify between the adjoint and the inverse of $\mathbf{A}$.

0.11 Linear Independence

It is possible that the system $\mathbf{Ax} = \mathbf{0}$ has infinitely many solutions. In order to characterize the set of all solutions of such systems, we must first understand the notion of linear independence. This is the topic to which this section is dedicated.

Suppose we are given lists of k vectors $\mathbf{a}_1, \mathbf{a}_2, \ldots, \mathbf{a}_k$ and k scalars $c_1, c_2, \ldots, c_k$. Consider the expression

$$c_1 \mathbf{a}_1 + c_2 \mathbf{a}_2 + \cdots + c_k \mathbf{a}_k = \mathbf{0} \tag{0.11.1}$$

If (0.11.1) holds for some set of scalars, not all zero, then the list of vectors $\mathbf{a}_1, \mathbf{a}_2, \ldots, \mathbf{a}_k$ is said to be *linearly dependent*, (0.11.1) is called a *dependency relationship*, and the scalars $c_1, c_2, \ldots, c_k$ are called the *weights* in the dependency relationship. The reason for this terminology is seen by considering the case in which $c_1 \neq 0$. Then (0.11.1) implies that

$$\mathbf{a}_1 = (-c_2/c_1)\, \mathbf{a}_2 + (-c_3/c_1)\, \mathbf{a}_3 + \cdots + (-c_k/c_1)\, \mathbf{a}_k \tag{0.11.2}$$

Equation (0.11.2) shows that $\mathbf{a}_1$ is a "weighted sum" of the other vectors in the list and in this sense "depends" on them. For example,

$$1 \cdot \begin{bmatrix} 2 \\ -1 \\ 0 \end{bmatrix} + (-2) \cdot \begin{bmatrix} 1 \\ 0 \\ 0 \end{bmatrix} + 1 \cdot \begin{bmatrix} 0 \\ 1 \\ -3 \end{bmatrix} + 3 \cdot \begin{bmatrix} 0 \\ 0 \\ 1 \end{bmatrix} = \begin{bmatrix} 0 \\ 0 \\ 0 \end{bmatrix}$$

is a dependency relationship with weights $c_1 = c_3 = 1, c_2 = -2$, and $c_4 = 3$, and therefore, the vectors

$$\begin{bmatrix} 2 \\ -1 \\ 0 \end{bmatrix}, \begin{bmatrix} 0 \\ 1 \\ -3 \end{bmatrix}, \begin{bmatrix} 1 \\ 0 \\ 0 \end{bmatrix}, \begin{bmatrix} 0 \\ 0 \\ 1 \end{bmatrix}$$

are linearly dependent.

Example 0.11.1. *Demonstrate the linear dependence of the following lists of vectors by finding a dependency relationship.*

(a) $\begin{bmatrix} 1 \\ 1 \\ 0 \end{bmatrix}, \begin{bmatrix} -1 \\ 1 \\ 0 \end{bmatrix}, \begin{bmatrix} 0 \\ 1 \\ 0 \end{bmatrix}$

(b) $\mathbf{0}, \mathbf{a}_1, \mathbf{a}_2, \mathbf{a}_3$

(c) $\mathbf{a}_1, \mathbf{a}_2 - \mathbf{a}_1, \mathbf{a}_2 + 2\mathbf{a}_1$

Solution: (a) We attempt to find weights c_1, c_2, and c_3, not all zero, such that

$$c_1 \begin{bmatrix} 1 \\ 1 \\ 0 \end{bmatrix} + c_2 \begin{bmatrix} -1 \\ 1 \\ 0 \end{bmatrix} + c_3 \begin{bmatrix} 0 \\ 1 \\ 0 \end{bmatrix} = \mathbf{0}$$

By inspection, $c_1 = 1, c_2 = 1$, and $c_3 = -2$, and a dependency relationship is given by

$$\begin{bmatrix} 1 \\ 1 \\ 0 \end{bmatrix} + \begin{bmatrix} -1 \\ 1 \\ 0 \end{bmatrix} - 2 \begin{bmatrix} 0 \\ 1 \\ 0 \end{bmatrix} = \mathbf{0}$$

(b) $1\mathbf{0} + 0\mathbf{a}_1 + 0\mathbf{a}_2 + 0\mathbf{a}_3 = \mathbf{0}$, with weights 1, 0, 0, 0, respectively.

(c) $3\mathbf{a}_1 + (\mathbf{a}_2 - \mathbf{a}_1) - (\mathbf{a}_2 + 2\mathbf{a}_1) = \mathbf{0}$.

and the weights are, respectively, 3, 1, -1.

If a given list of vectors is not linearly dependent, then it is called *linearly independent.* Since there can be no dependency relationship for a linearly independent set, (0.11.1) implies that all the scalar coefficients must be zero if the list is linearly independent.

Example 0.11.2. *Show that the following list of vectors is linearly independent:*

$$\begin{bmatrix} 1 \\ 0 \\ 0 \end{bmatrix}, \begin{bmatrix} 1 \\ 1 \\ 0 \end{bmatrix}, \begin{bmatrix} 1 \\ 1 \\ 1 \end{bmatrix}$$

Solution: We attempt to find c_1, c_2 and c_3 such that

$$c_1 \begin{bmatrix} 1 \\ 0 \\ 0 \end{bmatrix} + c_2 \begin{bmatrix} 1 \\ 1 \\ 0 \end{bmatrix} + c_3 \begin{bmatrix} 1 \\ 1 \\ 1 \end{bmatrix} = \begin{bmatrix} 0 \\ 0 \\ 0 \end{bmatrix}$$

From the third row, it follows that $c_3 = 0$. Hence, we are left with solving

$$c_1 \begin{bmatrix} 1 \\ 0 \\ 0 \end{bmatrix} + c_2 \begin{bmatrix} 1 \\ 1 \\ 0 \end{bmatrix} = \begin{bmatrix} 0 \\ 0 \\ 0 \end{bmatrix}$$

from which it follows that $c_2 = 0$. These facts lead to $c_1 = 0$. Hence, no dependency relationship is possible.

Generally, we cannot expect it to be so easy to decide whether a list of vectors is linearly independent nor to find the specific weights in a dependency relationship if the list is linearly dependent. A helpful tool for distinguishing between these cases can be deduced by rewriting (0.11.1) in matrix-vector form. To do so, define a matrix $\mathbf{A}$ whose columns are the vectors $\mathbf{a}_1, \mathbf{a}_2, \ldots, \mathbf{a}_k$ and a vector whose entries are the weights $c_1, c_2, \ldots, c_k$. Thus,

$$\mathbf{A} = \begin{bmatrix} \mathbf{a}_1 & \mathbf{a}_2 & \ldots & \mathbf{a}_k \end{bmatrix}, \quad \mathbf{c} = \begin{bmatrix} c_1 \\ c_2 \\ \vdots \\ c_k \end{bmatrix} \tag{0.11.3}$$

The definition of matrix multiplication implies that

$$\mathbf{A}\mathbf{c} = c_1\,\mathbf{a}_1 + c_2\,\mathbf{a}_2 + \cdots + c_k\,\mathbf{a}_k \tag{0.11.4}$$

Hence, the columns of $\mathbf{A}$ are linearly independent if and only if there exists a solution of $\mathbf{Ac} = \mathbf{0}$ other than the trivial $\mathbf{c} = \mathbf{0}$. If such is the case, then the entries of $\mathbf{c}$ are the weights, and $\mathbf{Ac} = \mathbf{0}$ is a dependency relationship. A convenient way to find a dependency relationship (if one exists) is to exploit the arrow diagram technique on $\mathbf{A}$. Example 0.11.3 will illustrate this idea. First we treat an example in which the dependency can be obtained without much effort.

Example 0.11.3. *Show that the following list of vectors is linearly dependent:*

$$\begin{bmatrix} 1 \\ 1 \end{bmatrix}, \begin{bmatrix} 1 \\ 0 \end{bmatrix}, \begin{bmatrix} 1 \\ -1 \end{bmatrix}$$

Solution: This particular example can be solved by inspection, as in Example 0.11.1:

$$\begin{bmatrix} 1 & 1 & 1 \\ 1 & 0 & -1 \end{bmatrix} \begin{bmatrix} c_1 \\ c_2 \\ c_3 \end{bmatrix} = c_1 \begin{bmatrix} 1 \\ 1 \end{bmatrix} + c_2 \begin{bmatrix} 1 \\ 0 \end{bmatrix} + c_3 \begin{bmatrix} 1 \\ -1 \end{bmatrix} = \begin{bmatrix} 0 \\ 0 \end{bmatrix}$$

is a dependency relationship with weights $c_1 = 1, c_2 = -2, c_3 = 1$. Note that any nonzero multiple of these weights would do as well.

Example 0.11.4. *Show that the following list is linearly dependent.*

$$\begin{bmatrix} 1 \\ 1 \\ 0 \\ 1 \end{bmatrix}, \begin{bmatrix} 1 \\ 0 \\ 0 \\ 1 \end{bmatrix}, \begin{bmatrix} 1 \\ -1 \\ 1 \\ 0 \end{bmatrix}, \begin{bmatrix} 0 \\ 1 \\ -1 \\ 1 \end{bmatrix}$$

Solution: In this example the weights are not easy to find by inspection. So instead, we attempt to find a nontrivial solution of $\mathbf{Ac} = \mathbf{0}$ by row reducing $\mathbf{A}$ (this is equivalent to Gaussian elimination, except we do not augment $\mathbf{A}$ with the right-hand side column of zeros):

$$\mathbf{A} = \begin{bmatrix} 1 & 1 & 1 & 0 \\ 1 & 0 & -1 & 1 \\ 0 & 0 & 1 & -1 \\ 1 & 1 & 0 & 1 \end{bmatrix} \rightarrow \begin{bmatrix} 1 & 1 & 1 & 0 \\ 0 & -1 & -2 & 1 \\ 0 & 0 & 1 & -1 \\ 0 & 0 & -1 & 1 \end{bmatrix}$$

$$\rightarrow \begin{bmatrix} 1 & 0 & -1 & 1 \\ 0 & -1 & -2 & 1 \\ 0 & 0 & 1 & -1 \\ 0 & 0 & -1 & 1 \end{bmatrix} \rightarrow \begin{bmatrix} 1 & 0 & 0 & 0 \\ 0 & 1 & 0 & 1 \\ 0 & 0 & 1 & -1 \\ 0 & 0 & 0 & 0 \end{bmatrix}$$

The third row shows that $c_3 = c_4$. We pick $c_4 = 1$ so that $c_3 = 1$. The second row shows that $c_2 = -c_4$, and the first row implies that $c_1 = 0$. These are the weights in the dependency relationship

$$-\begin{bmatrix} 1 \\ 0 \\ 0 \\ 1 \end{bmatrix} + \begin{bmatrix} 1 \\ -1 \\ 1 \\ 0 \end{bmatrix} + \begin{bmatrix} 0 \\ 1 \\ -1 \\ 1 \end{bmatrix} = \begin{bmatrix} 0 \\ 0 \\ 0 \\ 0 \end{bmatrix}$$

If, as is often the case, the matrix $\mathbf{A}$ is square, then a criterion based on the determinant for linear dependency is possible and convenient. Indeed, we make use of this idea in the theory of systems of differential equations. The basic idea is this: If $\mathbf{A}$ is invertible, then, by Corollary 0.8.1, the system $\mathbf{Ac} = \mathbf{0}$ has only the trivial solution because

$$\mathbf{c} = \mathbf{Ic} = \mathbf{A}^{-1}\mathbf{Ac} = \mathbf{A}^{-1}\mathbf{0} = \mathbf{0}$$

shows that $\mathbf{c}$ must be $\mathbf{0}$. The following theorem connects $\det \mathbf{A}$ with the linear independence of the rows and columns of $\mathbf{A}$.

Theorem 0.11.1. *The rows (columns) of the $n \times n$ matrix $\mathbf{A}$ are linearly independent if and only if*

$$\det \mathbf{A} \neq 0 \tag{0.11.5}$$

Proof: By Theorem 0.10.4, $\det \mathbf{A} \neq 0$ if and only if $\mathbf{A}$ is invertible. This is precisely the condition that $\mathbf{Ac} = \mathbf{0}$ has only the trivial solution. Thus the columns of $\mathbf{A}$ are linearly independent if and only if $\det \mathbf{A} \neq 0$. Since $\det \mathbf{A} = \det \mathbf{A}^T$, the same observation holds for the rows of $\mathbf{A}$. □

Example 0.11.5. *Show that the following list is linearly independent:*

$$\begin{bmatrix} 1 \\ 2 \\ 1 \end{bmatrix}, \begin{bmatrix} 0 \\ 1 \\ 1 \end{bmatrix}, \begin{bmatrix} 2 \\ 1 \\ 1 \end{bmatrix}$$

Solution: Since

$$c_1 \begin{bmatrix} 1 \\ 2 \\ 1 \end{bmatrix} + c_2 \begin{bmatrix} 0 \\ 1 \\ 1 \end{bmatrix} + c_3 \begin{bmatrix} 2 \\ 1 \\ 1 \end{bmatrix} = \begin{bmatrix} 1 & 0 & 2 \\ 2 & 1 & 1 \\ 1 & 1 & 1 \end{bmatrix} \mathbf{c} = \mathbf{0}$$

we need only show that the matrix is nonsingular. We use Theorem 0.11.1:

$$\det \begin{bmatrix} 1 & 0 & 2 \\ 2 & 1 & 1 \\ 1 & 1 & 1 \end{bmatrix} = 2$$

and hence, the given list of vectors is linearly independent.

Example 0.11.6. *Show that the following list is linearly dependent:*

$$\begin{bmatrix} 1 \\ 1 \\ 1 \end{bmatrix}, \begin{bmatrix} 1 \\ -2 \\ 4 \end{bmatrix}, \begin{bmatrix} 1 \\ 4 \\ -2 \end{bmatrix}$$

Solution: For this example,

$$\det \mathbf{A} = \det \begin{bmatrix} 1 & 1 & 1 \\ 1 & -2 & 4 \\ 1 & 4 & -2 \end{bmatrix} = 0$$

and therefore, the list is linearly dependent. However, it is often just as important to find the dependency relationship as to learn that a list is linearly dependent. In this regard, the determinant technique is inferior to the method discussed earlier in this chapter.

• EXERCISES

1. Which of the following lists are linearly independent? For those that are linearly dependent, determine a dependency relationship.

(a) $\begin{bmatrix} 1 \\ 0 \\ 1 \end{bmatrix}, \begin{bmatrix} 1 \\ 1 \\ -1 \end{bmatrix}, \begin{bmatrix} -1 \\ 1 \\ -3 \end{bmatrix}.$ (b) $\begin{bmatrix} 1 \\ 0 \\ 0 \end{bmatrix}, \begin{bmatrix} 1 \\ 1 \\ 0 \end{bmatrix}, \begin{bmatrix} 1 \\ 1 \\ 1 \end{bmatrix}.$

(c) $\begin{bmatrix} 1 \\ 0 \\ 1 \end{bmatrix}, \begin{bmatrix} 1 \\ 1 \\ -1 \end{bmatrix}, \begin{bmatrix} 1 \\ 1 \\ -3 \end{bmatrix}.$ (d) $\begin{bmatrix} 2 \\ 0 \\ 1 \end{bmatrix}, \begin{bmatrix} 4 \\ -2 \\ 3 \end{bmatrix}, \begin{bmatrix} 1 \\ -1 \\ 1 \end{bmatrix}.$

(e) $\begin{bmatrix} 2 \\ 0 \\ 1 \end{bmatrix}, \begin{bmatrix} 1 \\ 1 \\ 0 \end{bmatrix}, \begin{bmatrix} 0 \\ 0 \\ 1 \end{bmatrix}, \begin{bmatrix} 1 \\ -1 \\ 1 \end{bmatrix}.$ (f) $\begin{bmatrix} 1 \\ 1 \\ 3 \end{bmatrix}, \begin{bmatrix} -1 \\ 0 \\ -2 \end{bmatrix}, \begin{bmatrix} 1 \\ 2 \\ 4 \end{bmatrix}.$

(g) $\begin{bmatrix} 1 \\ 1 \\ 0 \\ 1 \end{bmatrix}, \begin{bmatrix} -1 \\ 2 \\ 0 \\ 0 \end{bmatrix}, \begin{bmatrix} 0 \\ 0 \\ 1 \\ 0 \end{bmatrix}, \begin{bmatrix} 1 \\ -1 \\ -1 \\ 1 \end{bmatrix}.$ (h) $\begin{bmatrix} -1 \\ 0 \\ 0 \\ 1 \end{bmatrix}, \begin{bmatrix} 2 \\ -1 \\ 1 \\ 1 \end{bmatrix}, \begin{bmatrix} 0 \\ -1 \\ 1 \\ 3 \end{bmatrix}.$

2. Find an integer k so that the following vectors are linearly dependent:

$$\begin{bmatrix} k \\ -1 \\ 1 \end{bmatrix}, \begin{bmatrix} 1 \\ k \\ -1 \end{bmatrix}, \begin{bmatrix} -1 \\ 1 \\ k \end{bmatrix}$$

(*Hint*: Find the determinant whose columns are the given list of vectors.)

3. Find a dependency relationship among the columns of these matrices:

(a) $\mathbf{A} = \begin{bmatrix} 1 & 1 & -1 & 1 \\ 1 & 0 & 1 & 2 \\ 2 & 1 & 0 & 3 \\ -1 & 1 & 0 & 0 \end{bmatrix}$. (b) $\mathbf{B} = \begin{bmatrix} 1 & 1 & 2 & -1 \\ 1 & 0 & 1 & 1 \\ -1 & 1 & 0 & 0 \\ 1 & 2 & 3 & 0 \end{bmatrix}$.

4. Which of the lists of vectors are linearly independent?

(a) $\begin{bmatrix} 1 \\ 0 \\ 0 \end{bmatrix}, \begin{bmatrix} * \\ 1 \\ 0 \end{bmatrix}, \begin{bmatrix} * \\ * \\ 1 \end{bmatrix}$. (b) $\begin{bmatrix} 0 \\ 1 \\ 1 \end{bmatrix}, \begin{bmatrix} 1 \\ 1 \\ 0 \end{bmatrix}, \begin{bmatrix} 1 \\ 0 \\ 1 \end{bmatrix}$.

(c) $\begin{bmatrix} 1 \\ 0 \\ 0 \\ 0 \\ * \\ * \end{bmatrix}, \begin{bmatrix} 0 \\ 1 \\ 0 \\ 0 \\ * \\ * \end{bmatrix}, \begin{bmatrix} 1 \\ 2 \\ 1 \\ 0 \\ * \\ * \end{bmatrix}, \begin{bmatrix} 1 \\ 2 \\ 1 \\ 1 \\ * \\ * \end{bmatrix}$. (d) $\begin{bmatrix} * \\ 0 \\ 0 \\ 1 \\ * \\ 0 \end{bmatrix}, \begin{bmatrix} * \\ 1 \\ 0 \\ 0 \\ * \\ 0 \end{bmatrix}, \begin{bmatrix} * \\ 2 \\ 1 \\ 0 \\ * \\ 2 \end{bmatrix}, \begin{bmatrix} * \\ 1 \\ 1 \\ 0 \\ * \\ 0 \end{bmatrix}$.

(e) $\begin{bmatrix} 1 \\ 0 \\ * \\ * \\ 0 \end{bmatrix}, \begin{bmatrix} 0 \\ 1 \\ * \\ * \\ 0 \end{bmatrix}, \begin{bmatrix} 0 \\ 0 \\ * \\ * \\ 1 \end{bmatrix}$.

5. If $\mathbf{x}_1, \mathbf{x}_2, \mathbf{x}_3, \mathbf{x}_4$ is linearly independent, so is $\mathbf{x}_1, \mathbf{x}_2, \mathbf{x}_3$. Why?

6. If $\mathbf{x}_1, \mathbf{x}_2, \mathbf{x}_3$ is linearly dependent, so is the list $\mathbf{x}_1, \mathbf{x}_2, \mathbf{x}_3, \mathbf{x}_4$ for any $\mathbf{x}_4$. Why?

7. Prove that $\mathbf{x}_1$ is linearly independent if and only if $\mathbf{x}_1 \neq \mathbf{0}$.

8. If $\mathbf{x}_1, \mathbf{x}_2, \ldots, \mathbf{x}_k$ is linearly dependent and $k > 1$, show that at least one of the vectors in this list is a weighted sum of the others.

9. If $\mathbf{x}_0$ is a weighted sum of $\mathbf{x}_1, \mathbf{x}_2, \ldots, \mathbf{x}_k$, show that the list $\mathbf{x}_0, \mathbf{x}_1, \mathbf{x}_2, \ldots, \mathbf{x}_k$ is linearly dependent.

10. Suppose no member of the linearly dependent list $\mathbf{x}_1, \mathbf{x}_2, \ldots, \mathbf{x}_k$ is $\mathbf{0}$ and $k > 1$. If $c_1\mathbf{x}_1 + c_2\mathbf{x}_2, \cdots + c_k\mathbf{x}_k = \mathbf{0}$, show that at least two weights are nonzero.

11. Suppose that $\mathbf{A}$ is a square matrix whose columns form a linearly independent list. Show that the columns of $\mathbf{A}^n$ form a linearly independent list.

12. Suppose that $\mathbf{A}$ is a square matrix. Suppose $\mathbf{B}$ is another square matrix whose columns form a linearly dependent list. Show that the columns of $\mathbf{AB}$ and $\mathbf{BA}$ form linearly dependent lists.

Chapter 1

First-Order Differential Equations

1.1 Preliminaries

Differential equations play a vital role in the solution of many problems encountered when modeling physical phenomena. All the disciplines in the physical sciences, each with its own unique physical situations, require that the student be able to derive the necessary differential equations and then solve them. We shall consider a variety of physical situations that lead to differential equations and present the theoretical and practical means for obtaining their solutions.

Here is an illustrative example. A city's water reservoir contains a solvent that if present in sufficient concentration, can cause illness. As water is withdrawn from the reservoir, water containing this potentially hazardous solvent is added. How does the concentration of solvent in the reservoir vary with time? Let us be more specific. Suppose our fictitious reservoir has volume V (in cubic meters, m^3) of water with an initial concentration C_0 (in grams per cubic meter, $\mathrm{g/m}^3$) of salt, the solvent. Such a salt and water mixture is called a brine. A brine containing concentration C_1 is flowing into the tank at a constant rate q (in cubic meters per second, m^3/s), while an equal flow of the mixture is issuing from the tank. See Figure 1.1.

The salt concentration is kept uniform throughout by continual stirring. Let $C(t)$ be the concentration of salt in the tank as a function of time and C_1 be the concentration of the entering brine.

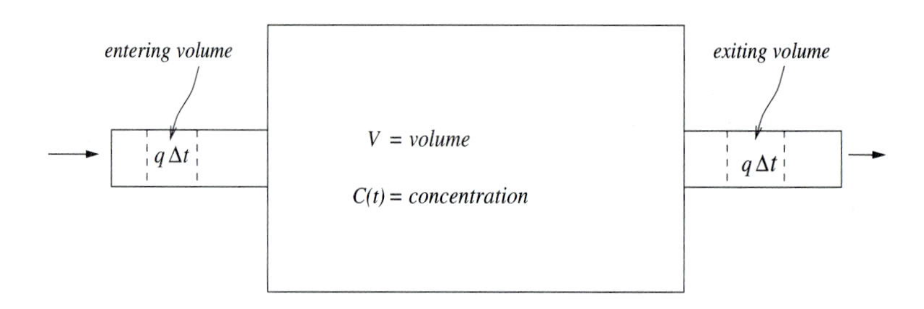

Figure 1.1: The Reservoir

The change in salt content in the tank is equal to the amount of salt in, less the amount out. This "conservation law" leads to a "balance" equation. For a small time increment $\Delta t, C_1 q \Delta t$ is the amount of salt entering during the time span Δt. If Δt is sufficiently small, we may assume that $C(t)$ is essentially constant, and hence, $C(t)q\Delta t$ is the amount which leaves. The balance equation is, therefore,

$$C_1\, q\, \Delta t - C\, q\, \Delta t = \left[C\left(t + \Delta t\right) - C(t)\right] V \tag{1.1.1}$$

provided we assume that the concentration of the solution leaving is equal to the concentration $C(t)$ in the tank. The volume V of solution is constant, since the outgoing flow rate is assumed to be equal to the incoming flow rate. Writing $\Delta C = C\left(t + \Delta t\right) - C(t)$, we can arrange (1.1.1) to give

$$q\left(C_1 - C(t)\right) = V \frac{\Delta C(t)}{\Delta t} \tag{1.1.2}$$

Now let $\Delta t \to 0$. By the definition of the derivative,

$$\lim_{\Delta t \to 0} = \frac{\Delta C}{\Delta t} = \frac{dC}{dt} = C'(t) \tag{1.1.3}$$

and (1.1.2) becomes the "rate equation"

$$C'(t) + \frac{q}{V} C(t) = \frac{q}{V} C_1 \tag{1.1.4}$$

We have written this in standard form: The coefficient of the derivative is unity, and any term that does not contain the unknown function is on the right-hand side.

To be specific, let us suppose that the initial concentration in a $10\,\mathrm{m}^3$ tank is $C_0 = 0.02\,\mathrm{g/m}^3$ and that a brine with a concentration of $0.01\,\mathrm{g/m}^3$ flows into the tank at a rate of $2\,\mathrm{m}^3/\,\mathrm{s}$. Assuming the outflow equals the inflow, what is the concentration of salt in the tank when $t = 0.1$ s? Later in this chapter we will solve the differential equation (1.1.4) and from the solution determine an exact value for $C(0.1)$. We can, however, obtain an approximation to $C(t)$ without actually solving (1.1.4) by using (1.1.1) in this form:

$$C(t + \Delta t) = \left(1 - \frac{q}{V}\Delta t\right) C(t) + \frac{q}{V}\Delta t C_1 \tag{1.1.5}$$

Now replace the parameters q, V, C_1 by their given values to obtain

$$C(t + \Delta t) = C(t) - 0.2C(t)\Delta t + 0.002\Delta t \tag{1.1.6}$$

Note that a knowledge of $C(t)$ at $t = t_1$ determines an approximation for the value of $C(t)$ at $t = t_1 + \Delta t$. We pick $\Delta t = 0.1$ seconds, and use a computer or calculator to find the values in the following table:

t	$C(t)$
0.00	0.02000
0.01	0.01998
0.02	0.01996
0.03	0.01994
0.04	0.01992
0.05	0.01990
0.06	0.01988
0.07	0.01986
0.08	0.01984
0.09	0.01982
0.10	0.01980

Thus, an approximation to $C(0.1)$ is 0.0198.

In general, an equation relating an unknown function to its various derivatives is a *differential equation*. Two examples are[1]

$$u' = u \tag{1.1.7}$$

$$v'' + tv' = t \tag{1.1.8}$$

Equations (1.1.4), (1.1.7), and (1.1.8) are examples of differential equations in one unknown function $C(t)$, $u(t)$, and $v(t)$, respectively. The system

$$\begin{aligned} x' &= x + 3y \\ y' &= x - y \end{aligned} \tag{1.1.9}$$

[1]Throughout this book, we write u for $u(t)$, u' for $u'(t)$, and u'' for $u''(t)$.

presents the problem of determining two functions $x(t)$ and $y(t)$ satisfying both equations simultaneously.

A *solution* of a differential equation is a differentiable function that satisfies the equation on some interval (α, β). For example, $u(t) = e^t$ satisfies $u' = u$ for all t because $u'(t) = e^t$ and $u(t) = e^t$. For this reason, e^t is called a solution of $u' = 0$. Likewise, $v(t) = t$ is a solution of (1.1.8) for all t because $(t)'' + t(t)' = 0 + t \cdot 1 = t$. Finally, we easily verify that the functions $x(t) = 3e^{2t} + e^{-2t}, y(t) = e^{2t} - e^{-2t}$, solve the system given in (1.1.9).

Often it is not possible to express a solution in terms of combinations of "elementary" functions. Such is the case[2] with the following minor modification of (1.1.8):

$$v'' + tv = t \tag{1.1.10}$$

In these circumstances, we turn to alternative methods for describing the solutions. These include, to mention just a few possibilities, presenting the solution in tabular form, or as a graph, or as a power series. In this book we confine ourselves primarily to equations for which the solutions can be expressed in terms of the functions of elementary calculus. The chapters on power series and numerical methods, however, present important exceptions to this statement.

It is helpful in understanding the rationale for the classification of differential equations to consider for a moment a sampling of equations that have played significant roles in applications.

The study of the diffusion of heat in a circular cylinder with radius r leads to the *Bessel equation*

$$T'' + \frac{1}{r}T' + \mu^2 T = 0 \tag{1.1.11}$$

The study of vacuum tubes leads to the *Van der Pol equation*

$$u'' + \left(1 - u^2\right) u' - u = 0 \tag{1.1.12}$$

The motion of the moon under the influence of the earth's and the sun's gravitational attraction leads to *Hill's equation*

$$u'' + \cos t\, u = 0 \tag{1.1.13}$$

Under certain reasonable biological assumptions a population of predators and prey satisfies the system

$$\begin{aligned} u' &= u\,(r - a)\,v \\ v' &= v\,(bu - s) \end{aligned} \tag{1.1.14}$$

[2]This fact is not obvious.

where a, b, r, and s are parameters relating to various biological factors such as birth and death rates of the predators and prey.

1.2 Definitions

An *ordinary differential equation* (written ODE) is a differential equation in which the *dependent variable* (the unknown function and its derivatives) are functions of one *independent variable* (the quantity upon which the dependent variable depends). The examples given in Section 1.1 are all ODE's. A *partial differential equation* (PDE) is a differential equation in which the dependent variable is a function of more than one independent variable. *Laplace's equation,*

$$\frac{\partial^2 u}{\partial x^2} + \frac{\partial^2 u}{\partial y^2} = 0 \tag{1.2.1}$$

in which the dependent variable is $u(x, y)$, is one important example. In this book we confine ourselves to ordinary differential equations in Chapters 1 through 7 and partial differential equations in Chapter 8.

The dependent variable usually models either the unknown quantity sought after in some physical problem or a quantity closely related to it. For example, if the lift on an airfoil is desired, the laws of fluid mechanics lead to a partial differential equation with unknown function (the velocity) $v(x, y, t)$. If v (the dependent variable) can be found, then we can calculate the pressure on the airfoil and consequently the lift.

The *order* of a differential equation is the order of the highest derivative of the unknown function that appears in the equation. The orders of (1.1.4) and (1.1.7) is one, and that of (1.1.8) through (1.1.10) is two. The order of the equation

$$u''' + t^4\, u^5\, u' = \sin t \tag{1.2.2}$$

is three. The appearance of t^4 and u^5 is irrelevant in determining the order of the equation.

We consider an important special class of equations that, if they can be written in one of the following forms, are called *linear*:

$$\text{order 1}: \quad u' + p(t)u = f(t) \tag{1.2.3}$$

$$\text{order 2}: \quad u'' + p(t)u' + q(t)u = f(t) \tag{1.2.4}$$

The general n^{th} order linear equation is given by

$$u^{(n)} + p_1(t)u^{(n-1)} + \cdots + p_{n-1}(t)u^{(1)} + p_n(t)u = f(t) \tag{1.2.5}$$

Equations that are not linear are *nonlinear.* Equation (1.2.2) is nonlinear, as are $u'' + \sin u = 0$ and $u' + u^{1/2} = 0$. If $f(t) = 0$ in (1.2.5), the differential equation is *homogeneous*; otherwise it is *nonhomogeneous.* Also, note that (1.2.5) is written in *standard form*: The coefficient of the highest order derivative is unity, and all terms that do not contain the dependent variable are on the right-hand side.

With the exceptions of this section and Section 1.1, the principle goal of this chapter is to study first-order ODE's. Sections 1.1 to 1.4 are devoted to the linear equation. An introduction to first-order equations which are not linear is reserved for Section 1.5. A more extensive analysis of nonlinear first-order equations is reserved for Chapter 7, in which we consider numerical and qualitative methods. Special cases of linear systems are considered in Chapters 2, 4, and 5.

Some linear equations are particularly easy to solve. For example, the first-order nonhomogeneous equation

$$u' = \cos t \tag{1.2.6}$$

has infinitely many solutions

$$u(t) = \int \cos t\, dt = \sin t + K \tag{1.2.7}$$

one for each choice of the constant K. The third-order equation

$$u''' = 0 \tag{1.2.8}$$

has the solutions

$$u(t) = K_0 + K_1 t + K_2 t^2 \tag{1.2.9}$$

obtained by integrating (1.2.8) three times. The three arbitrary constants $K_0, K_1,$ and K_2 are the constants of integration. The ODE's considered in this and subsequent chapters possess solutions that will be obtained with more difficulty than these; however, there are times when simple equations such as $u^{(n)} = f(t)$ model phenomena of interest.

As we shall see later on, every linear ODE possesses a family of solutions. A specific member of this family is determined by supplementing the ODE by conditions on the solutions. For example, the second-order linear ODE

$$u'' + 4u' + 4u = 0 \tag{1.2.10}$$

has the family of solutions[3]

$$u(t) = K_1 e^{-2t} + K_2 t e^{-2t} \tag{1.2.11}$$

Exactly one member of this set is determined by requiring $u(t)$ to meet the initial conditions, say, $u(0) = 1, u'(0) = -1$. Indeed, since $u(0) = K_1$, we have $K_1 = 1$. And since

$$u'(t) = -2K_1 e^{-2t} + K_2 e^{-2t} - 2K_2 t e^{-2t} \tag{1.2.12}$$

and $K_1 = 1$, we have

$$u'(0) = -2 + K_2 = -1 \tag{1.2.13}$$

Thus, $K_2 = 1$. These values select the unique solution

$$u(t) = e^{-2t} + te^{-2t} \tag{1.2.14}$$

from the family given in (1.2.11). Different initial conditions select different solutions from this family. See Problem 5.

Suppose that $b_1, b_2, \ldots, b_n$ are n fixed real numbers. Then the *initial-value problem* (for the nth-order linear ODE) is given by

$$u^{(n)} + p_1(t)u^{(n-1)} + \cdots + p_{n-1}(t)u^{(1)} + p_n(t)u = f(t) \tag{1.2.15}$$

$$u(t_0) = b_1, \quad u'(t_0) = b_2, \quad \ldots, \quad u^{(n-1)}(t_0) = b_n \tag{1.2.16}$$

Note these important requirements: (1) t_0 is the value of the independent variable at which all initial conditions are imposed. (2) There are as many initial conditions as the order of the differential equation.

If conditions are given at more than one value of t, a *boundary-value problem* results. For a second-order linear ODE, a boundary-value problem is

$$u'' + p(t)u' + q(t)u = f(t) \tag{1.2.17}$$

$$u(t_0) = b_1, \quad u(t_1) = b_2$$

In general, boundary-value problems are much more difficult to solve than initial-value problems.

One last remark: If the solutions of the equation are confined to an interval[4] (α, β), then we shall always assume that the points t_0 and t_1 are in (α, β).

---

[3]These solutions were obtained by using a technique introduced in Section 3.6.

[4]We shall often write (α, β) for the open interval $\alpha < t < \beta$ and $[\alpha, \beta]$ for the closed interval $\alpha \leq t \leq \beta$.

We shall most often assume that $t_0 = 0$. This is not a real restriction because it is always possible to transform the independent variable from t to τ, so that $t = t_0$ corresponds to $\tau = 0$. This is done by defining $\tau = t - t_0$. Then $dt = d\tau$, and

$$u(t) = u(\tau + t_0) = v(\tau) \tag{1.2.18}$$

defines a new unknown function $v(\tau)$. Hence, $u(t_0) = v(0) = b_1$, and by the chain rule,

$$u'(t) = \frac{du}{dt} = \frac{dv}{d\tau}\frac{d\tau}{dt} = v'(\tau) \tag{1.2.19}$$

We are now able to write the initial-value problem given by (1.2.15) as

$$v' + p(\tau + t_0)v = f(\tau + t_0), \quad v(0) = b_1 \tag{1.2.20}$$

Note that the solution of (1.2.20) is $v(\tau)$. The function $u(t)$ is recaptured by the transformation $v(\tau) = v(t - t_0) = u(t)$.

Example 1.2.1. *Write the initial-value problem*

$$u' + (\cos t)u = \sin t, \quad u(\pi/2) = 0$$

so that the initial condition is at 0.

Solution: Let $\tau = t - \pi/2$. Then $u(t) = u(\tau + \pi/2) = v(\tau), d\tau/dt = 1$, and $u(\pi/2) = v(0)$. By the chain rule,

$$u'(t) = \frac{dv}{d\tau}\frac{d\tau}{dt} = v'(\tau)$$

Therefore, the given initial-value problem in t is now an initial-value problem in τ,

$$v' - (\sin\tau)v = \cos\tau, \quad v(0) = 0$$

since $\sin(\tau + \pi/2) = \cos\tau$ and $\cos(\tau + \pi/2) = -\sin\tau$.

•EXERCISES

1. Find the order and state which of the following equations are linear and which are homogeneous?

(a) $u'/u = 1 + t$. (b) $u'u = 1 + t$.
(c) $\sin u' = u$. (d) $u'' = u$.
(e) $u'' = t^2$. (f) $(u^2)' = -u + 1$.

2. Repeat Problem 1 for these equations.
(a) $u'' = u^2$. (b) $u'' - 2u' + u = \cos t$.

3. Find a family of solutions of each differential equation.
(a) $u' = t^2 + 2$. (b) $u' = \sin t + e^t$.
(c) $u' = t + \cos^2 t$. (d) $u'' = 2t$.
(e) $u^{(3)} = t^2$. (f) $u^{(4)} = t - 2$.

4. Change the dependent variable so that each equation is an initial-value problem with its initial condition given at 0.
(a) $u' + 2tu = 3t, \quad u(2) = 0$.
(b) $tu' + u = t\sin\, t, \quad u(1) = 2$.
(c) $tu' + t^2 u = e^t, \quad u(1) = 0$.

5. Find the solution from among those given in (1.2.11) that meets the initial conditions $u(0) = 0, u'(0) = 1$.

1.3 The First-Order Linear Equation

Consider the standard form of the first-order equation,

$$u' + p(t)u = f(t) \tag{1.3.1}$$

where $p(t)$ and $f(t)$ are arbitrary functions defined in (α, β). We call $f(t)$ the *forcing function.* If the forcing function is zero, the associated *homogeneous* equation is:

$$u' + p(t)u = 0 \tag{1.3.2}$$

This is also called the equation *complementary* to (1.3.1). One of the remarkable facts about linear ODE's is that the solutions of the nonhomogeneous equation can always be constructed from the solutions of the complementary equation.[5] For this reason, we begin our study of (1.3.1) by devoting a subsection to the study of its complementary equation.

[5]This is shown in Subsection 1.3.2.

1.3.1 Homogeneous Equations

Consider the homogeneous equation (1.3.2). Write the equation in the form

$$\frac{du}{u} = -p(t)\,dt \tag{1.3.3}$$

This is integrated to provide

$$\ln u = -\int p(t)\,dt + C \tag{1.3.4}$$

or, equivalently,

$$u(t) = e^{-\int p(t)\,dt + C} = e^{C}\,e^{-\int p(t)\,dt} \tag{1.3.5}$$

If we let

$$P(t) = \int p(t)\,dt \tag{1.3.6}$$

then (1.3.5) becomes

$$u(t) = Ke^{-P(t)} \tag{1.3.7}$$

where $K = e^C$ and $P(t)$ does not contain a constant of integration. (Indeed, because $e^{-P(t)+c} = ke^{-P(t)}$, constants of integration can be absorbed into K.) Equation (1.3.7) provides us with the solution to all first order, linear, homogeneous differential equations. Here are some examples.

Example 1.3.1. *Find solutions of* $u' + pu = 0$, *where* p *is constant.*

Solution: For this problem

$$P(t) = \int p\,dt = pt$$

Therefore, using (1.3.7), we obtain

$$u(t) = Ke^{-pt}$$

which is a solution for each K.

Example 1.3.2. *Use* (1.3.7) *to find solutions of* $t^2u' = u$, $t > 0$.

Solution: In order to use (1.3.7), we must first divide the given equation by t^2, so that the coefficient of u' is 1 and then bring u to the left-hand side. When this is done the given equation is

$$u' - t^{-2}u = 0$$

so that the coefficient of u is $p(t) = -t^{-2}$. Then

$$P(t) = -\int t^{-2}\,dt = t^{-1}$$

Therefore, from (1.3.7),

$$u(t) = Ke^{-1/t}$$

is a solution for $t > 0$. It is also a solution for $t < 0$ but not for any interval containing $t = 0$.

Example 1.3.3. *Find solutions of*

$$\left(1 - t^2\right)u' + u = 0, \quad -1 < t < 1$$

Solution: In this example, as in the previous one, we obtain $p(t)$ by first dividing the given equation by $(1 - t^2)$ obtaining

$$u' + \left(1 - t^2\right)^{-1} u = 0$$

Now,

$$P(t) = \int \left(1 - t^2\right)^{-1} dt = \ln\left(\frac{1+t}{1-t}\right)^{1/2}$$

Hence for each K and each t in $-1 < t < 1$, we have the solution

$$u(t) = K\left(\frac{1-t}{1+t}\right)^{1/2}$$

where we have used $e^{\ln x} = x$, which is valid as long as $x > 0$.

Example 1.3.4. *Find solutions of* $u' + (\tan t\,)u = 0$ *valid in* $(-\pi/2, \pi/2)$.

Solution: We integrate to find

$$P(t) = \int \tan t \, dt = -\ln(\cos t)$$

which is valid in the interval $(-\pi/2, \pi/2)$, since $\cos t$ is positive in that interval. Consequently, the solution is

$$u_1(t) = Ke^{\ln(\cos t)} = K \cos t$$

With these examples as guides, we now consider the general homogeneous initial-value problem

$$u' + p(t)u = 0, \quad u(t_0) = b$$

where $p(t)$ is continuous in (α, β) and $\alpha < t_0 < \beta$. The following theorem is fundamental. It asserts that there is one and only one solution of the general initial-value problem (1.3.8). It is the fact that there is only one solution which is the difficult part of the proof. Here's the theorem.

Theorem 1.3.1. *For every real constant b, the initial-value problem*

$$u' + p(t)u = 0, \quad u(t_0) = b \tag{1.3.8}$$

has the unique solution

$$u(t) = be^{-P(t)} \tag{1.3.9}$$

where

$$P(t) = \int_{t_0}^{t} p(s) \, ds \tag{1.3.10}$$

Proof: First of all, $P(t_0) = 0$ so that $u(t_0) = b$. Secondly, (1.3.8) is the derivative of (1.3.9) and these facts demonstrate that the function given by (1.3.9) is a solution of this initial-value problem. So all that remains to prove is that there are no other solutions, and this is the challenging part of the argument. Assume the contrary. Suppose there exists another solution, say, $v(t)$. Observe that for all t in (α, β)

$$v'(t) + p(t)v(t) = 0 \tag{1.3.11}$$

$$u'(t) + p(t)u(t) = 0 \tag{1.3.12}$$

is a consequence of the fact that we are assuming $u(t)$ and $v(t)$ are solutions of (1.3.8). Now multiply (1.3.11) by $u(t)$ and (1.3.12) by $v(t)$, and subtract the resulting identities to obtain

$$v(t)u'(t) - v'(t)u(t) = 0 \tag{1.3.13}$$

We use the quotient rule for derivatives and (1.3.13) to deduce

$$\frac{d}{dt}\frac{u(t)}{v(t)} = \frac{v(t)\,u'(t) - v'(t)u(t)}{v^2(t)} = 0 \tag{1.3.14}$$

But the only continuous differentiable function with a zero derivative is the constant function. Using this observation in (1.3.14) leads to $u(t)/v(t) = C$. Thus, $u(t) = Cv(t)$, and because $u(t) = be^{-P(t)}$

$$v(t) = be^{-P(t)}/C \tag{1.3.15}$$

By hypothesis, $v(t_0) = b$. However, when t is set to t_0 in (1.3.15), we have $b = b/C$ since $P(t_0) = 0$. So replace b/C by b in (1.3.15) to deduce

$$v(t) = be^{-P(t)} \tag{1.3.16}$$

But then $v(t) = u(t)$. $\square$

Example 1.3.5. *Find the solution of the initial-value problem*

$$u' + tu = 0, \quad u(0) = 1$$

Solution: In view of (1.3.10), we find $P(t) = t^2/2$. By Theorem 1.3.1,

$$u(t) = be^{-t^2/2}$$

Since $u(0) = 1$, we have $u(0) = 1 = be^0 = b$. Hence,

$$u(t) = e^{-t^2/2}$$

is the unique solution of the given initial-value problem.

Corollary 1.3.1. *Under the hypothesis of* Theorem 1.3.1, *the initial-value problem*

$$u' + p(t)u = 0, \quad u(t_0) = 0 \tag{1.3.17}$$

has the unique solution $u(t) = 0$.

Proof: Since every initial-value problem has exactly one solution (Theorem 1.3.1), and since $u(t) = 0$ is a solution, it is the only one. □

Corollary 1.3.2. *Under the hypothesis of* Theorem 1.3.1, *the set* $Ke^{-P(t)}$ *contains all the solutions of*

$$u' + p(t)u = 0, \quad \alpha < t < \beta$$

Proof: Suppose $v(t)$ is any solution of $u' + p(t)u = 0$. Let t_0 be any point in $\alpha < t < \beta$ and set $b = u(t_0)$. Then $u(t) = be^{-P(t)}$ is the unique solution of $u' + p(t)u = 0$, $t_0 = b$. Hence, $u(t) = be^{-P(t)}$ and the corollary is proved by taking $K = b$. □

It is because of Corollary 1.3.2 that we call $Ke^{-P(t)}$ the general solution of $u' + p(t)u = 0$. This conclusion is in sharp contrast to the state of affairs for nonlinear equations. For example, it is easy to check that the family of functions $u(t) = Kt + K^2$ consists of solutions of

$$(u')^2 + tu' - u = 0$$

for all K. But the solution $u(t) = -t^2/4$ is not a member of this family! Moreover, $1 - t$ and $-t^2/4$ are two solutions of the initial-value problem

$$(u')^2 + tu' - u = 0, \quad u(2) = -1$$

The existence of two solutions for a linear initial-value problem is impossible.

The notion of a general solution of a nonlinear equation is less useful than for a linear equation for two reasons: First, it is often impossible to write an explicit formula for even one solution. Second, as the previous example illustrates, even when we can find a family of solutions, it is often the case that this family misses some solutions.

•EXERCISES

Determine $P(t)$, and by using (1.3.9), find the unique solution for each of the following initial-value problems in Problems 1–11.

1. $u' = 10u, \quad u(0) = 1.$
2. $u' = u, \quad u(0) = -1.$
3. $u' - 2u = 0, \quad u(1) = 1.$
4. $u' = \sin tu, \quad u(\pi) = 1.$
5. $tu' = u, \quad u(1) = -1.$
6. $u' = 2tu, \quad u(0) = c.$
7. $udt - tdu = 0, \quad u(1) = -3.$
8. $-dt + tdu, \quad u(1) = 1/2, \quad t > 0.$
9. $-dt + tdu, \quad u(-1) = 1/2, \quad t < 0.$
10. $(2 + t^2)\, u' + 2tu = 0, \quad u(0) = 1.$
11. $u' + 2\cos 2tu = 0, \quad u(\pi/4) = 1, \quad \pi/8 < t < 3\pi/8.$

12. Find a family of solutions of the equation in Example 1.3.1 by using $u(0) = 1$. Are the resulting sets of solutions different?

13. Find a family of solutions of the equation in Example 1.3.2 except that $t < 0$. Why is $t_0 = 0$ an inappropriate choice for t_0?

14. Consider the differential equation given in Example 1.3.3 and the family of functions

$$u(t) = K\left|\frac{1-t}{1+t}\right|^{1/2}$$

Verify that $u(t)$ is a solution in every interval that does not contain $t = 1$ or $t = -1$.

15. Show that $1-t$ and $-t^2/4$ are two solutions of the initial-value problem $(u')^2 + tu' - u = 0$, $u(2) = -1$.

1.3.2 Nonhomogeneous Equations

We now consider the general nonhomogeneous linear ODE

$$u' + p(t)u = f(t) \tag{1.3.18}$$

Suppose $u_h(t)$ is any solution of the complementary equation

$$u' + p(t)u = 0 \tag{1.3.19}$$

This means that

$$u_h'(t) + p(t)u_h(t) = 0 \tag{1.3.20}$$

Next, suppose $u_p(t)$ is any solution of (1.3.18). This means that

$$u_p'(t) + p(t)u_p(t) = f(t) \tag{1.3.21}$$

Now consider

$$u(t) = u_p(t) + Ku_h(t) \tag{1.3.22}$$

We shall now show that $u(t)$ is a solution of (1.3.18) for any choice of K. By definition of $u(t)$ as given in (1.3.22), $u'(t) = u_p'(t) + Ku_h'(t)$. Now substitute $u(t) = u_p(t) + Ku_h(t)$ and $u'(t) = u_p'(t) + Ku_h'(t)$ into (1.3.18):

$$\begin{aligned} u'(t) + p(t)u(t) &= u_p'(t) + Ku_h'(t) + p(t)\left(u_p(t) + Ku_h(t)\right) \\ &= \Big(u_p'(t) + p(t)u_p(t)\Big) + K\left(u_h'(t) + p(t)u_h(t)\right) \end{aligned}$$

But

$$K\left(u_h'(t) + p(t)u_h(t)\right) = 0$$

and

$$u_p'(t) + p(t)u_p(t) = f(t)$$

by (1.3.20) and (1.3.21), respectively. Thus $u'(t) + p(t)u(t) = f(t)$. This proves our assertion.

The point of this result is that we can obtain a family of solutions of (1.3.18) by adding any specific solution of that equation to a general solution of its complementary equation. We shall see later that $u(t)$, so constructed, is a general solution of (1.3.18) in that all solutions may be found in this set of solutions. Because of this and the fact that we have already found a general solution of the complementary equation by Theorem 1.3.1, we devote this section to obtaining particular solutions of (1.3.18). There are many methods for doing so. We choose one that generalizes to higher order linear ODE's and to linear systems. This method is called "variation of parameters", although the name is seldom used for a single first-order ODE. Here is an outline of the method:

Step 1: Find the general solution $u_h(t)$ of the complementary equation and then replace the arbitrary parameter K by the function $v(t)$. Now define the function $u_p(t)$ by[6]

$$u_p = v(t)e^{-P(t)} \tag{1.3.23}$$

---

[6]The name "variation of parameters" arises because the trial function $u(t) = v(t)\exp\left(-P(t)\right)$ is constructed from the general solution of the homogeneous equation by replacing the parameter K by the function $v(t)$.

Step 2: Find $v(t)$ so that $u_p(t)$, as given by (1.3.23), is a solution of (1.3.18).

This is done by substituting $u_p(t)$ and $u_p'(t)$ into (1.3.18). Since $P'(t) = p(t)$, we have

$$\begin{aligned} u_p' &= v'e^{-P(t)} - vP'(t)e^{-P(t)} \\ &= v'e^{-P(t)} - vp(t)e^{-P(t)} = v'e^{-P(t)} - p(t)u_p(t) \end{aligned} \tag{1.3.24}$$

where the last equality follows from (1.3.23). Now bring $-p(t)u_p(t)$ to the left-hand side of (1.3.24) to obtain

$$u_p' + p(t)u_p = v'e^{-P(t)} \tag{1.3.25}$$

So $u_p(t)$ is a solution of (1.3.18) if and only if

$$v'e^{-P(t)} = f(t) \tag{1.3.26}$$

Now integrate (1.3.26) to find

$$v(t) = \int e^{P(t)} f(t)\, dt \tag{1.3.27}$$

Step 3: Form $u_p(t)$ from (1.3.23) using $v(t)$ found in (1.3.27).

This is the method of *variation of parameters* for the first-order linear ODE. Note that the constant of integration is not needed in (1.3.27), since that would lead to the term $Ce^{-P(t)}$ (the homogeneous solution), which is already included in (1.3.22).

Example 1.3.6. *Use the method of variation of parameters to find a solution of* $u' + 2u = 4$.

Solution: A general solution of the complementary equation is easily found; it is $u_h(t) = Ke^{-2t}$. Since $f(t) = 4$, it follows from (1.3.27) that

$$v(t) = \int 4e^{2t}\, dt = 2e^{2t}$$

We obtain a solution $u_p(t)$ from (1.3.23):

$$u_p(t) = v(t)e^{-P(t)} = 2e^{2t}e^{-2t} = 2$$

So from (1.3.22),

$$u(t) = u_h(t) + u_p(t) = Ke^{-2t} + 2$$

Example 1.3.7. *For constants $p, A,$ and w, find a solution of*

$$u' + pu = Ae^{wt}$$

Solution: A general solution of the complementary equation is $u_h(t) = Ke^{-pt}$ and we search for a solution of the form $u_p(t) = v(t)e^{-pt}$. By (1.3.27),

$$v(t) = A\int e^{wt}e^{pt}\,dt = A\int e^{(w+p)t}\,dt$$

There are two alternatives: Either $p+w \neq 0$, or $p+w=0$. Each case leads to a different family of solutions. First suppose that $p+w \neq 0$. Then

$$v(t) = A\frac{e^{(p+w)t}}{p+w}$$

and therefore,

$$u_p(t) = A\frac{e^{(p+w)t}}{p+w}e^{-pt} = \frac{A}{p+w}e^{wt}$$

So,

$$u(t) = u_h(t) + u_p(t) = \frac{A}{p+w}e^{wt} + Ke^{-pt}$$

Now suppose that $p+w=0$, so $p=-w$ and the given ODE is now

$$u' + pu = Ae^{-pt}$$

Then $v(t) = At$, and in this case $u_p(t) = Ate^{-pt}$, and

$$u(t) = u_h(t) + u_p(t) = Ke^{-pt} + Ate^{-pt}$$

is a family of solutions, one for each choice of K.

The general case follows the same pattern as the illustrative examples. We start with (1.3.27) and write a solution for $v(t)$ in the form

$$v(t) = \int f(t)e^{P(t)}\,dt \tag{1.3.28}$$

Since $u_p(t) = v(t)e^{-P(t)}$ we have

$$u_p(t) = e^{-P(t)}\int f(t)e^{P(t)}\,dt \tag{1.3.29}$$

and

$$u(t) = Ke^{-P(t)} + e^{-P(t)} \int f(t)e^{P(t)}\,dt \tag{1.3.30}$$

We now establish that $u(t)$, as given in (1.3.30), is a general solution of (1.3.18) in the sense that every solution of (1.3.18) can be obtained by choosing K appropriately. First, for reference, we state our second theorem.

Theorem 1.3.2. *The linear, first-order equation*

$$u' + p(t)u = f(t)$$

has the solution

$$u(t) = Ke^{-P(t)} + e^{-P(t)} \int f(t)e^{P(t)}\,dt$$

for each choice of K.

Theorem 1.3.3. *The initial-value problem*

$$u' + p(t)u = f(t), \quad u(t_0) = b \tag{1.3.31}$$

has the unique solution

$$u(t) = Ke^{-P(t)} + e^{-P(t)} \int_{t_0}^{t} f(s)e^{P(s)}\,ds \tag{1.3.32}$$

where $K = be^{P(t_0)}$.

Proof: By Theorem 1.3.2, $u(t)$ is a solution of (1.3.31) obtainable from (1.3.30) by letting $t = t_0$ and choosing $K = be^{P(t_0)}$. Next, when t is equal to t_0 in (1.3.32), the integral term is zero, and we obtain $u(t_0) = b$. Thus, $u(t)$ is indeed a solution of the initial-value problem (1.3.31). That it is the only such solution follows by assuming the contrary. Accordingly, that $u(t)$ and $v(t)$ are both solutions of (1.3.31). Define $w(t) = u(t) - v(t)$. Then

$$\begin{aligned} w'(t) = u'(t) - v'(t) &= (f(t) - p(t)u(t)) - (f(t) - p(t)v(t)) \\ &= -p(t)(u(t) - v(t)) = -p(t)w(t) \end{aligned}$$

which shows that $w(t)$ is a solution of $w' + p(t)w = 0$. Also, at $t = t_0$ we find $u(t_0) = v(t_0) = b$. Hence, $w(t_0) = u(t_0) - v(t_0) = 0$. Thus, $w(t)$ is a solution of the homogeneous initial-value problem

$$w' - p(t)w = 0, \quad w(t_0) = 0$$

By Corollary 1.3.1 of Theorem 1.3.1, the only solution of this problem is the identically zero function. Thus, $w(t) = 0$ and hence $v(t) = u(t)$. □

We call

$$u(t) = e^{-P(t)} \int f(t)\, e^{P(t)}\, dt + Ke^{-P(t)} \tag{1.3.33}$$

a *general solution* of $u' + p(t)u = f(t)$ because every solution of this ODE can be constructed from $u(t)$ by choosing K appropriately. The proof is almost line for line the proof for the analogous results for homogeneous ODE's. (See Corollary 1.3.2 of Theorem 1.3.1.)

Example 1.3.8. *Find the solution of the initial-value problem*

$$u' + (\tan t)u = \cos t, \quad u(0) = 1, \quad -\pi/2 < t < \pi/2$$

Solution: The complementary equation has the general solution $K \cos t$ as we have seen in Example 1.3.4. So by variation of parameters,

$$v(t) = \int \cos t\, e^{-\ln(\cos t)}\, dt = t$$

because

$$e^{-\ln(\cos t)} = e^{\ln(\cos t)^{-1}} = (\cos t)^{-1} = 1/\cos t$$

and the integrand in the integral for $v(t)$ is 1. From this, it follows that

$$u(t) = (t + K) \cos t$$

We determine K by setting $t = 0$ and $u(0) = 1$ in the foregoing family of solutions. This leads to $K = 1$. Hence, the desired solution is $u(t) = (t + 1) \cos t$.

Example 1.3.9. *Find the solution of the initial-value problem*

$$u' + 2u = 4, \quad u(0) = 1$$

Solution: We have found the solutions of $u' + 2u = 4$ in Example 1.3.6, namely, $u(t) = 2 + Ke^{-2t}$. So K is found from the initial condition $u(0) = 1$ without any need to invoke any further steps. We have $u(0) = 2 + K = 1$, which implies $K = -1$. Hence, $u(t) = 2 - e^{-2t}$ is the desired solution.

Example 1.3.10. *Find the solution of the initial-value problem*

$$u'' + (\tan t)\, u' = \cos\, t, \quad u(0) = 1 = u'(0), \quad -\pi/2 < t < \pi/2$$

Solution: Although this is a second-order equation in u, it is first order in u'. So set $w = u'$. Then the given ODE is

$$w' + (\tan t)\, w = \cos t$$

whose general solution was derived in Example 1.3.8, to wit,

$$w(t) = t\, \cos t + C_1\, \cos t$$

Hence, $u(t)$ is found by integrating $w(t)$. We have

$$u(t) = t \sin\, t + \cos t + C_1 \sin t + C_2$$

Thus, $u(0) = C_2 + 1 = 1$, and $u'(0) = C_1 = 1$. The desired solution is then

$$u(t) = t\, \sin\, t + \cos\, t + \sin\, t$$

•EXERCISES

1. Find a general solution for the ODE's that follow. In some problems it may be necessary to leave the answer in terms of integrals.
   (a) $u' + u = 2$. (b) $u' + 2u = 2t$.
   (c) $u' + tu = 10$. (d) $u' - 2u = e^t$.
   (e) $u' + u = te^{-t}$. (f) $u' - u = \cos t$.
   (g) $tu' - 2u = te^t$. (h) $t^2u' - u = 2\sin\left(t^{-1}\right), \quad t > 0$.
2. Find the solutions of these initial-values problems.
   (a) $u' = 2u - 1, \quad u(0) = 2$.
   (b) $u' \tan t = u + 1, \quad u(\pi/2) = 0, \quad 0 < t < \pi$.
   (c) $tu' + u = 2t, \quad u(1) = 10 \quad 0 < t$.
   (d) $(1 + t^2)u' + 2tu = 2t, \quad u(0) = 1$.

3. Find the solutions of these initial-values problems.
(a) $u' + 2u = 2e^{-2t}, \quad u(0) = 2.$
(b) $u' + tu = \exp(-t^2/2), \quad u(1) = 0.$
(c) $u' - u = t, \quad u(0) = 1.$
(d) $u' - 2u = 4, \quad u(0) = 0.$

4. Show by substitution that
$$u(t) = (A(p+w)^{-1}e^{(p+w)t} + K)e^{-pt}$$
is a solution of the equation
$$u' + pu = Ae^{wt}, \quad p \neq -w$$

5. Verify by substitution that $4e^{-t}\sinh t$ and $2 - 2e^{-2t}$ are both solutions of the initial-value problem $u' + 2u = 4, u(0) = 0$. Does this contradict Theorem 1.3.2? Explain.

6. What theorems of calculus permit the deduction $\frac{d}{dt}e^{-P(t)} = -p(t)e^{-P(t)}$?

7. Use the product rule and Problem 6 to deduce
$$u_p'(t) = \left(\frac{d}{dt}e^{-P(t)} \int f(t)e^{P(t)}\,dt\right) = f(t) - p(t)u_p(t)$$

8. Let $u_p(t)$ be any particular solution of $u' + p(t)u = f(t)$ and $u_h(t)$ be any solution of its complementary equation. Show that $u_p(t) + u_h(t)$ is a solution of $u' + p(t)u = f(t)$.

9. Find an expression for a general solution of $u' + 2u = f(t)$.

10. Use the fact that $u_p(t) = a$ is a solution of $u' + p(t)u = ap(t)$ and the result of Problem 7 to find the general solution of $u' + p(t)u = ap(t)$

11. Use the idea presented in Example 1.3.10 and Theorem 1.3.2 to write the general solution of $u'' + p(t)u' = f(t)$.

1.4 Applications of First-Order Linear Equations

There are remarkably many different and important physical problems that lend themselves to modeling by first-order linear ODE's. In this section we explore a variety of these applications.

1.4.1 Simple Electrical Networks

Consider the circuit in Figure 1.2 containing a resistance R, an inductance L, and a capacitance C in series. A known voltage $v(t)$ is impressed across the terminals. The ODE relating the current $i(t)$ to the voltage may be found by applying Kirchoff's first law, which states that the voltage impressed on a closed loop is equal to the sum of the voltage drops in the rest of the loop.

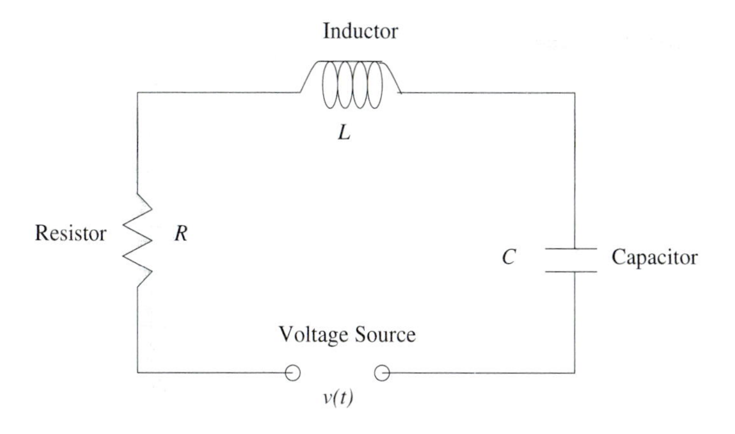

Figure 1.2: A Simple Electric Circuit

Experimental observations have shown that the voltage drops across each of the elements of the circuit in the figure are given by

$$\begin{aligned} \text{voltage drop across a resistor } &= Ri(t) \\ \text{voltage drop across a capacitor } &= C^{-1}q(t) \\ \text{voltage drop across an inductor } &= Li'(t) \end{aligned} \tag{1.4.1}$$

Kirchhoff's law and (1.4.1) result in the differential equation

$$v = Li' + Ri + C^{-1}q \tag{1.4.2}$$

where the values $q(t), v(t), i(t), L, R$, and C are in physically consistent units: coulombs, volts, amps, henrys, ohms, and farads, respectively. The function $q(t)$ is the charge on the capacitor and is related to the current $i(t)$ flowing through it by $i(t) = q'(t)$. The three-component circuit in Figure 1.2 leads to a second-order ODE. By omitting one of these elements, (1.4.2) reduces to a first-order ODE.

Example 1.4.1. *Find the current $i(t)$ flowing through the electric circuit of Figure 1.2 if the voltage is given by $v(t) = V \sin\, \Omega t$ and there is no capacitor. Use $i(0) = i_0$.*

Solution: Since there is no capacitor in this circuit (1.4.2) becomes

$$Li' + Ri = V \sin\, \Omega$$

This is a nonhomogeneous linear ODE. In the notation of Section 1.3,

$$p(t) = \frac{R}{L}, \qquad f(t) = \frac{V}{L} \sin\, \Omega t$$

Hence, $P(t) = \frac{R}{L}t$. From (1.3.33),

$$i(t) = e^{-(R/L)t} \int e^{(R/L)t} \frac{V}{L} \sin\, \Omega t\, dt + Ke^{-(R/L)t}$$

By referring to a table of integrals or by performing two integrations by parts, the reader can verify that

$$i(t) = V \left(\frac{R\sin\, \Omega t - \Omega L\cos\, \Omega t}{R^2 + \Omega^2 L^2} \right) + \left(i_0 + \frac{V \Omega L}{R^2 + \Omega^2 L^2} \right) e^{-(R/L)t}$$

where we have evaluated K by letting $i(0) = i_0$. Note that for sufficiently large t, the exponential function is near enough to zero to be neglected, and only the trigonometric term remains.

Example 1.4.2. *Find the current flowing through the circuit in Figure 1.2 and the voltage drop across the resistor if no inductor is present and the voltage is given by* $v(t) = 0$. *Use* $i(0) = i_0$.

Solution: Since there is no inductor and $v(t) = 0$, (1.4.2) becomes $Ri + c^{-1}q = 0$. Differentiate this equation and use $i(t) = q'(t)$ to obtain

$$Ri' + \frac{1}{C} i = 0$$

In terms of the parameters of this example, and because of the given initial condition, we have the following initial-value problem to solve:

$$Ri' + \frac{1}{C} i = 0, \quad i(0) = i_0$$

This problem has the solution given by (1.3.7).

$$i(t) = Ke^{-t/RC}$$

The initial condition requires $i(0) = K = i_0$. Hence, $i(t) = i_0 e^{-t/RC}$. This solution shows that the current generated by the discharge of the capacitor is an exponentially decreasing function. The voltage drop across the resistor is given by $Ri(t)$, so $V(t) = Ri(t) = Ri_0 e^{-t/RC}$. The voltage drop also decreases exponentially.

•EXERCISES

1. A voltage $v(t)$ is impressed on a series circuit composed of a 10 Ω resistor and a 10^{-4} H inductor. Determine the current as a function of time if the current is zero at $t = 0$ for each of the impressed voltages.
   (a) 1.12. (b) $10\sin 60t$.
   (c) $20\cos 30t$. (d) $0.2e^{2t}$.
   (e) $10e^{-2t}$. (f) $12t$.
2. A voltage $v(t)$ is impressed on a series circuit containing a 20 Ω resistor and a 10^{-3} H inductor. Calculate the current after 20 μs using $i = 0$ at $t = 0$ for each of the following voltages.
   (a) 120 V. (b) $-2\sin 10t$.
   (c) $120e^{-2t}$. (d) $120\left(e^{t} - 1\right)$.
3. A series circuit composed of a 50 Ω resistor and a 10^{-7} F capacitor is excited by the voltage $v(t)$. What is the general expression for the charge on the capacitor for the following voltages?
   (a) $12\sin 2t$. (b) $20\cos 60t$.
   (c) $100e^{-2t}$. (d) $10e^{0.1t}$.
4. A $v(t)$ is impressed on a series circuit containing a 200 Ω resistance and a 10^{-6} F capacitor. Determine the current after 10 μs using $i = 0$ at $t = 0$ for each of the following voltages.
   (a) $12\sin 2t$. (b) $20\cos 60t$.
   (c) $100e^{-2t}$. (d) $10e^{0.1t}$.

1.4.2 Linear Rate Equations

Recall the "rate equation" which we derived in Section 1.1:

$$C'(t) + \frac{q}{V}C(t) = \frac{q}{V}C_1 \qquad (1.4.3)$$

We are now in a position to find its general solution and solve a variety of problems relating to the phenomena it models. Here are two examples.

Example 1.4.3. *Suppose the initial concentration in a* $10\,\text{m}^3$ *tank is* $C_0 = 0.02\,\text{g/m}^3$. *If pure water flows into the tank at a rate of* $2\,\text{m}^3/s$, *and if the outflow equals the inflow, determine the time necessary for the solution to reach a concentration of* $0.011\,\text{g/m}^3$.

Solution: We can determine the time at which the concentration of brine in the tank is 0.011 g/m^3 by using (1.4.3) with $C_1 = 0$ (since C_1 is the concentration of salt in the incoming fluid, in this case pure water). Using the parameters of this example, (1.4.3) is now

$$C' + 0.2C = 0, \qquad C(0) = 0.02$$

The solution of this initial-value problem is given by (1.3.31). Since $P(t) = 0.2t$ we have $C(t) = 0.02e^{-0.2t}$. Assume that $t = \tau$ marks the time at which the concentration is 0.011, that is , $C(\tau) = 0.011$. Then, $C(\tau) = 0.011 = 0.02e^{-0.2\tau}$. Using logarithms we obtain $\ln(0.55) = -0.2\tau$. Thus, $\tau = 2.99$ s.

Example 1.4.4. *Redo Example 1.4.3, except assume that a brine flows into the tank. The brine has a concentration of* $0.01\,\mathrm{g/m^3}$.

Solution: Equation (1.4.3) now takes the form

$$C'(t) + 0.2C(t) = 0.002, \qquad C(0) = 0.02$$

We may solve this by applying the techniques of Section 1.3 or by noting that this is a special case of the equation solved in Example 1.3.7. In that example, the general solution of $u' + pu = Ae^{\Omega t}$ was found to be

$$u(t) = \frac{A}{p + \Omega} e^{\Omega t} + Ke^{-pt}$$

We set $p = 0.2$, $\Omega = 0$, and $A = 0.002$ to obtain

$$C(t) = 0.002\frac{1}{0.2} + Ke^{-0.2t} = 0.01 + Ke^{-0.2t}$$

Now set $t = 0$ to obtain K. We find that $C(0) = 0.02 = 0.01 + K$. Hence, $K = 0.01$, and therefore,

$$C(t) = 0.01 + 0.01e^{-0.2t}$$

Now we find $t = \tau$ such that $C(\tau) = 0.011$:

$$C(\tau) = 0.011 = 0.01\left(1 + e^{-0.2\tau}\right)$$

Hence, $0.1 = e^{-0.2\tau}$, and we find that $\tau = 11.51$ s.

Since a brine solution dilutes the contents of the tank at a slower rate than pure water, the τ for this example is greater than τ for the previous example. Note that the value of $C(t)$ at $t = 0.1$ (see the solution in Section 1.1) can be obtained from $C(t) = 0.01 + 0.01e^{-0.2t}$ and is $C(.1) = 0.01 + 0.01e^{-0.02} = 0.01980$, the same as that obtained is Section 1.1.

•EXERCISES

In Problems 1–4, assume the initial concentration of salt in 10 m^3 of solution is 0.2 g/m^3. Brine flows into the tank at a rate of 0.1 m^3/s until the volume is 20 m^3, at which time t_f the solution flows out of the tank at the same rate as it flows into the tank. Find an expression for the concentration $C(t)$ given the following concentrations of inflowing brine. Find one function for (a) $t < t_f$ and another for (b) $t > t_f$.

1. $0\ g/m^3$
2. $0.02\,g/m^3$
3. $0.04e^{-0.002t}\,g/m^3$
4. $0.01e^{0.01t}\,g/m^3$
5. An average person takes 18 breaths per minute and each breath exhales $0.0016\,m^3$ of air containing 4% more CO_2 than was inhaled. At the start of a seminar with 300 participants, the room air contains 0.4% CO_2. The ventilation system delivers 10 m^3 of fresh air per minute to the 1500 m^3 room. Find an expression for the concentration level of CO_2 in the room as a function of time. Air is assumed to leave the room at the same rate that it enters.

1.4.3 Fluid Flow

In the absence of viscous effects, it has been observed that a liquid (water, for example) will flow from a hole with velocity depending on the height of the free surface of the liquid above the hole. Let $v(t)$ be the velocity and $h(t)$ the height. This observation is made precise by the equation

$$v(t) = \sqrt{2gh(t)} \qquad \text{m/s} \tag{1.4.4}$$

where $g = 9.81\,\text{m/s}^2$ is the local acceleration due to gravity.

We now develop an ODE whose solution will allow us to determine how long it will take to empty a particular reservoir. Assume that a hole with diameter d is located at the bottom of a cylindrical tank of diameter D, and that the initial water height is h_0 meters above the hole - Figure 1.3.

The incremental volume ΔV is given by

$$\Delta V = Av\Delta t = \frac{1}{4}\pi d^2\sqrt{2gh}\,\Delta t \tag{1.4.5}$$

This change in volume ΔV must equal the volume lost in the tank due to the decrease in liquid level Δh. That is,

$$\Delta V = -\frac{1}{4}\pi D^2 \Delta h \tag{1.4.6}$$

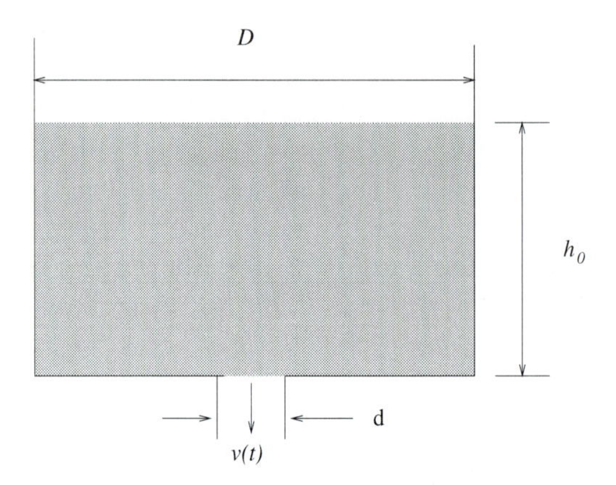

Figure 1.3: The Draining of a Reservoir

Equating the last two displayed equations and taking the limit as $\Delta t \to 0$ leads to

$$h'(t) = -\frac{d^2}{D^2}\sqrt{2gh(t)} \tag{1.4.7}$$

In differential form, (1.4.7) may be written as

$$\frac{dh}{h^{1/2}} = -\frac{d^2}{D^2}\sqrt{2g}\,dt \tag{1.4.8}$$

It is not difficult to integrate (1.4.8) to obtain the solution:

$$h^{1/2}(t) = -\frac{d^2}{D^2}\sqrt{\frac{g}{2}}\,t + C \tag{1.4.9}$$

So,

$$h(t) = \left(-\frac{d^2}{D^2}\sqrt{\frac{g}{2}}\,t + C\right)^2 \tag{1.4.10}$$

Setting $t = 0$ and using $h(0) = h_0$ now leads to $C^2 = h_0$, so that

$$h(t) = \left(-\frac{d^2}{D^2}\sqrt{\frac{g}{2}}\,t + \sqrt{h_0}\right)^2 \tag{1.4.11}$$

Let $t = t_0$ be the time needed to drain the tank completely. The value of t_0 is determined from the equation $h(t_0) = 0$. That is,

$$\frac{d^2}{D^2}\sqrt{\frac{g}{2}}\,t_0 = \sqrt{h_0} \tag{1.4.12}$$

which results in

$$t_0 = \frac{D^2}{d^2}\left(\frac{2}{g}h_0\right)^{1/2} \tag{1.4.13}$$

We illustrate this result with an example.

Example 1.4.5. *Determine an expression for the height of water in the funnel shown in* Figure 1.4. *How much time is necessary to drain the funnel?*

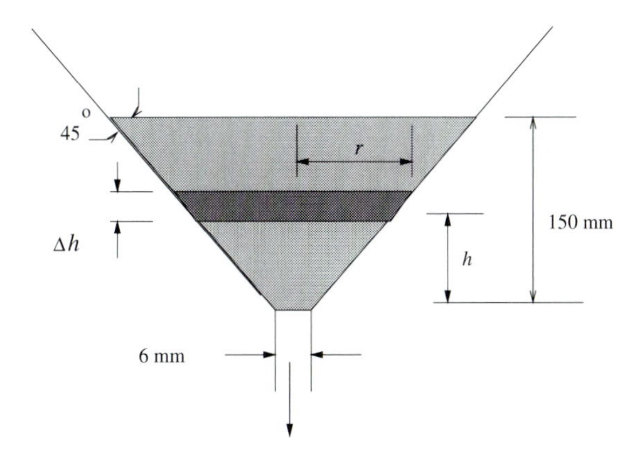

Figure 1.4: The Draining of a Funnel

Solution: The volume of fluid leaving the funnel during the time increment Δt is given by (1.4.6):

$$\Delta V = vA\Delta t = \frac{\pi}{4}\sqrt{2gh}\, d^2 \Delta t$$

The change in volume due to a liquid-level decrease of Δh is

$$\Delta V = -\pi r^2 \Delta h = -\pi h^2 \Delta h$$

because r is approximately h. Equate these two expressions for ΔV, divide by Δt, and let $\Delta t \to 0$. Then

$$-h^{3/2}h' = \frac{1}{4}d^2\sqrt{2g}$$

or

$$-h^{3/2}dh = \frac{1}{4}d^2\sqrt{2g}\, dt$$

An integration results in $h^{5/2} = -\frac{5}{8}d^2\sqrt{2g}\,t + C$. So

$$h(t) = \left(-\frac{5}{8}d^2\sqrt{2g}\,t + C\right)^{2/5}$$

When $t = 0, h = 0.15\,\text{m}$. Hence $0.15 = C^{2/5}$ and therefore $C = 0.008714$. Setting $t = \tau$ and letting $h = 0$ results in

$$0.008714 = \frac{5}{8}\sqrt{2 \times 9.81}\,(0.006)^2\,\tau$$

Thus the time needed to drain the funnel is $\tau = 87.44\,\text{s}$.

•EXERCISES

1. A square tank 2 m on a side contains water to a depth of 1.2 m. How long will it take to drain the tank if the tank contains a 20-mm-diameter hole in the bottom?
2. A funnel has an included angle of 60° at the bottom, where a 10-mm hole exists. If the funnel is filled with water to a depth of 20 cm above the drain hole, calculate the time needed to drain the funnel.
3. A square tank 3 m on a side is filled with water to a depth of 2 m. A vertical slot 6 mm wide from the top to the bottom allows the water to drain out. Determine the height of the water level as a function of time and how much time is necessary for one half of the water to drain out.

1.4.4 Radioactive Decay

A radioactive substance emits particles and in doing so decreases its mass. This process is known as *radioactive decay.* For example, the radioactive isotope carbon 14 emits particles and loses half its mass in 5,730 years. The instantaneous rate of decay is proportional to the mass of the radioactive substance at that instant. At time t suppose $M(t)$ is the mass of a radioactive substance. Then assuming an initial mass M_0,

$$M' + kM = 0, \quad M(0) = M_0 \tag{1.4.14}$$

is the initial-value problem describing this simple model of radioactive decay; in this equation, k is a *rate constant* that depends on the substance. Since in this model the mass diminishes, we have that $k > 0$. The solution of (1.4.14) is given by

$$M(t) = M_0 e^{-kt} \tag{1.4.15}$$

Let τ be the time at which $M(\tau) = \frac{1}{2}M_0$. Then τ is called the *half-life* of the radioactive substance. From (1.4.15) we find that

$$M(\tau) = \frac{1}{2}M_0 = M_0 e^{-k\tau} \tag{1.4.16}$$

Hence,

$$-k\tau = \ln 0.5 = -0.69315$$

implying a half life of $\tau = 0.69315/k$.

Example 1.4.6. *The half-life of* carbon 14 *is* $\tau = 5730$ *years. Calculate the percentage of* carbon 14 *that remains after* 50,000 *years.*

Solution: Equation (1.4.16) provides us with

$$k = 0.69315/5730 = 1.2097 \times 10^{-4}\ \text{yr}^{-1}$$

Equation (1.4.15) gives the mass after $t_1 = 50,000$ years:

$$M(t_1) = M_0 e^{-1.2097\times 10^{-4}\times 5\times 10^4} = 2.3616 \times 10^{-3} M_0$$

So the percentage of the original mass to the mass remaining is

$$\frac{M(t_1)}{M_0} \times 100 = 0.23616\,\%$$

•EXERCISES

1. The half-life of a radioactive element is 2,000 years. What percentage of its original mass is left after 10,000 years?
2. The evaporation rate of moisture from a sheet hung on a clothesline is proportional to the sheet's moisture content. If one-half of the moisture evaporates in the first 30 minutes, how long will it take for 95% of the moisture to evaporate?
3. The half-life of a radioactive element is 30 years. How long will it take for a 10-kg sample to be reduced to 1 kg?

1.4.5 Population Growth

Experiments show that a population with zero death rate (say, a colony of bacteria with sufficient food and no predators) grows at a rate proportional to the size of its population. If the size of the population is $P(t)$, then we have $P'(t) = kP(t)$ for some constant $k > 0$, which is an experimentally derived constant of proportionality. Hence, $P(t)$ satisfies

$$P' - kP = 0, \quad P(0) = P_0 \tag{1.4.17}$$

where P_0 is the size of the population at time $t = 0$. Thus, the solution

$$P(t) = P_0 e^{kt} \tag{1.4.18}$$

is a function reminiscent of the function obtained for radioactive decay except that k replaces $-k$. Clearly, "doubling time" is a parameter that is useful and its determination is completely analogous with the computations for half-life given by (1.4.16).

Example 1.4.7. *A certain bacteria culture is observed to increase by* 10% *during the first* 10 *hours. How long will it take for the population to increase by a factor of* 1,000?

Solution: We are able to calculate k from (1.4.18) and the given data. We have $P(t_1) = 1.1P_0$, where $t_1 = 10$ hours. Therefore,

$$P(t_1) = 1.1P_0 = P_0 e^{10k}$$

and consequently, $\ln 1.1 = 10k$. Thus, $k = 0.009531 \text{ hr}^{-1}$. The time τ for $P(t)$ to reach $1,000P_0$ is then found from the equation $P(\tau) = 1,000P_0 = P_0 e^{0.009531\tau}$ from which we determine that $\ln 1,000 = 0.009531\tau$. Thus, $\tau = 724.8$ hr.

•EXERCISES

1. A population of 200 million people is observed to increase 2% each year. How many years will it take for the population to double?
2. A bacteria culture doubles in 2 hours. How long will it take for the population to increase by a factor of 100?
3. A population grows initially with a growth constant $k = 0.001$ /hr and subsequently with a growth constant of 0.002 hr. How long will it take the population to double if equal time is spent in each growth period?

1.4.6 Compound Interest

It is common knowledge that A_0 dollars given today is worth more than A_0 promised t years from now, if for no other reason than A_0 invested for t years earns interest. Under assumptions commonly used in the financial community, the growth of money under interest is exponential. Let $u(t)$ represent the value at time t of an original investment of one dollar. The function $u(t)$

is the *interest accumulation function.* It is assumed in the elementary theory of interest that $u(t)$ increases monotonically and is differentiable. Also, since an investment of one dollar is worth $u(t)$ in t years, it is reasonable to suppose that A_0 is worth $A(t) = A_0u(t)$ in t years. The *force of interest* $p(t)$ is defined by

$$p(t) = \frac{A'(t)}{A(t)}$$

In the most common circumstances $p(t) = p$ is constant. Hence, $A(t)$ is the solution of the following initial-value problem:

$$A' - pA = 0, \quad A(0) = A_0 \tag{1.4.19}$$

Thus,

$$A(t) = A_0e^{pt} \tag{1.4.20}$$

We define the annual interest rate i by the equation $1 + i = e^p$. Then $\ln(1 + i) = p$, and hence,

$$e^{pt} = e^{t\ \ln(1+i)} = (1 + i)^t \tag{1.4.21}$$

Substituting (1.4.21) into (1.4.20) yields the more familiar formula for the growth of money under the assumption of compound interest:

$$A(t) = A_0(1 + i)^t \tag{1.4.22}$$

It is standard practice to assume that A is measured in dollars and t in years, and to quote i as a percentage per year. For example, \$12 invested now in a savings account that earns 7% per year for two years is worth $A(2) = 12(1 + 0.07)^2 = \$13.74$ two years from now.

Example 1.4.8. *Determine the time τ required for money to double under annual interest rates of $i = 2, 5, 7,$ and 10 %. Then compute the product τi.*

Solution: From (1.4.22), it follows that

$$A(\tau) = 2A_0 = A_0(1 + i)^\tau$$

Hence, $2 = (1 + i)^\tau$, so that $\tau = \ln 2/\ln(1 + i)$. Using the interest rates $i = 0.02, 0.05, 0.07,$ and 0.1, we find that $\tau = 35.0, 14.2, 10.2$ and 7.27 years, respectively. The product τi, given in percentages, is $\tau i = 70$, 71, 71.4, and 72.7, respectively. This explains the so-called "rule of 72": $\tau i = 72$ used by investors to approximate the time for money to double in value.

•EXERCISES

1. A person wishes to have \$200,000 when retiring at age 65 by investing at age 25 a lump sum in a bank paying 8% compounded yearly. How much money must be deposited initially?
2. A person invests \$1,000 every January 1 for six years at 6% interest, compounded annually, for six years. How much money does the person have December 31 of the sixth year?
3. At what interest rate should money be invested if it is to double after (a) 12 years, (b) 8 years, and (c) 6 years.? Find the exact answer and its approximation by the rule of 72.

1.4.7 Newton's Law of Cooling

Suppose that $T(t)$ is the temperature of a body immersed in a cooler surrounding liquid. Newton postulated that the warmer body would lose heat at a rate proportional to the difference between its present temperature and the temperature of the liquid. So if we assume that the temperature of the liquid is constant, say, T_m, and initially the cooling body's temperature is T_0, that is, $T(0) = T_0$, then *Newton's law of cooling* is

$$T' = -k\left(T - T_m\right), \quad T(0) = T_0 \tag{1.4.23}$$

In standard form,

$$T' + kT = kT_m, \quad T(0) = T_0 \tag{1.4.24}$$

The solution of (1.4.24) is

$$T(t) = \left(T_0 - T_m\right)e^{-kt} + T_m \tag{1.4.25}$$

Suppose we assume that $t = \tau$ is the time at which $T(t) = T_1$, that is, $T(\tau) = T_1$. Then we can compute the physical parameter k as follows: Use $t = \tau$ in (1.4.25) so that

$$T_1 = \left(T_0 - T_m\right)e^{-k\tau} + T_m \tag{1.4.26}$$

which implies that

$$\ln\left(\frac{T_1 - T_m}{T_0 - T_m}\right) = -k\tau \tag{1.4.27}$$

from which k is easily determined. Now set $R = (T_1 - T_m) / (T_0 - T_m)$. Then $k > 0$ if and only if $R < 1$, and this occurs if and only if temperature T_1 is closer to T_m than temperature T_0. (See Problem 1).

Returning to the solution (1.4.25), note that as $t \to \infty, T(t) \to T_m$, a result that is intuitively appealing: The temperature of a body immersed in surrounding liquid held at a constant temperature will eventually reach the temperature of the liquid.

•EXERCISES

1. Let $R = (T_1 - T_m) / (T_0 - T_m)$. Then (1.4.27) may be written, $\ln R = -k\tau$. Use these relationships to show that the temperature of the body is approaching the temperature of the medium if and only if $k > 0$.
2. An object at a temperature of 80° C is placed in a refrigerator maintained at 5° C. Estimate the time for the temperature of this object to reach 8°C if $k = 0.2/\text{min}$.

1.5 Nonlinear Equations of First Order

Nonlinear ODE's are significantly more difficult to handle than linear ODE's for a variety of reasons of which the most important is the possibility of the failure of uniqueness for the solution of initial-value problems. As we have seen, the initial-value problem

$$u'^2 + tu' - u = 0, \quad u(2) = -1 \tag{1.5.1}$$

has the solutions $u_1(t) = 1 - t$ and $u_2(t) = -\frac{1}{4}t^2$. The existence of two solutions to this nonlinear initial-value problem is in stark contrast to the uniqueness theorem for linear initial-value problems. Besides this difficulty, there are initial conditions for which no real solution of (1.5.1) exists. We can see why by setting $t = 0$ and $u(0) = -1$ in (1.5.1). Then (1.5.1) reduces to $u'(0)^2 = -1$, which is impossible. Again, contrast this with the linear problem, which allows a solution for any choice of the initial condition. Thus the motivating factor for introducing the idea of general solution, namely, that all initial value problems are solvable by a suitable choice of the constants of integration in the general solution, does not obtain for the nonlinear equation.

In the next two subsections we study two kinds of nonlinear equations characterized by the fact that they may be solved by elementary methods.

1.5.1 Separable Equations

Some first-order equations can be reduced to

$$h(u)u' = g(t) \tag{1.5.2}$$

which, in differential form, is

$$h(u)du = g(t)dt \tag{1.5.3}$$

This equation is called *separable*: The independent variable and its differential are on one side, and the dependent variable and its differential are on the other. Separable equations are soluble by an integration:

$$\int h(u)\, du = \int g(t)\, dt + K \tag{1.5.4}$$

Unless the function $h(u)$ is a particularly simple function of u, (1.5.4) is not an improvement over (1.5.3). For example,

$$\int \sin \sqrt{u}\, du = \int g(t)\, dt \tag{1.5.5}$$

defines u with no more insight than its differential counterpart,

$$\sin \sqrt{u}\, du = g(t)\, dt \tag{1.5.6}$$

The following example is more satisfactory.

Example 1.5.1. *Find a family of solutions of*

$$tu' + u^2 = 4$$

and a particular solution meeting the initial condition $u(1) = 1$.

Solution: This nonlinear ODE is separable because it can be written in the form $\left(4 - u^2\right)^{-1} du = t^{-1}dt$. Hence,

$$\int \frac{du}{4 - u^2} = \int \frac{dt}{t}$$

providing us with

$$\frac{1}{4}\ln\frac{2+u}{2-u} = \ln t + C$$

Therefore,

$$\ln\left(\frac{2+u}{2-u}\right)^{1/4} = \ln t + C = \ln(Kt)$$

where we have set $C = \ln K$. By taking exponentials and solving for $u(t)$ we obtain

$$u(t) = 2\,\frac{(K\,t)^4 - 1}{(K\,t)^4 + 1}$$

It is an exercise in differentiation to verify that this function is a solution for all K and all t. Now, $u(1) = 2\,(K^4 - 1)\,/\,(K^4 + 1) = 1$ implies $K^4 = 3$. So $u(t) = 2(3t^4 - 1)/(3t^4 + 1)$ is the desired solution as we may verify by substitution.

Example 1.5.2. *Show that the following ODE is separable*

$$u' + p(t)u = 0$$

Solution: This ODE may be written as $u^{-1}du = -p(t)dt$, which shows that it is separable. Hence, $\ln u(t) = -\int p(t)\,dt + C$. Taking exponentials leads to the family of solutions obtained in Section 1.3.

Certain ODE's that are not separable can be transformed into separable equations by a change of variables. One such class is the equations of the form

$$u' = f\,(u/t) \tag{1.5.7}$$

Define v by the relationship $v = u/t$. Then $f(u/t) = f(v)$ and by differentiating $u = tv$, we obtain

$$u' = tv' + v \tag{1.5.8}$$

Substituting this expression into (1.5.7) leads to

$$tv' + v = f(v) \tag{1.5.9}$$

This ODE is separable. In fact,

$$\frac{dv}{f(v) - v} = \frac{dt}{t} \tag{1.5.10}$$

and integrating both sides leads to an expression relating $v(t)$ and t, namely,

$$\int \frac{dv}{f(v) - v} = \ln\,(Kt) \tag{1.5.11}$$

where the constant of integration is again selected as $\ln K$.

Example 1.5.3. *Find solutions of*

$$u' = \frac{u}{t} + \frac{t}{u}$$

Solution: Set $u = tv$ and use (1.5.8). Then the given ODE is transformed into a separable equation in the dependent variable v:

$$tv' + v = v + \frac{1}{v}$$

Hence,

$$vdv = \frac{1}{t}dt$$

and therefore,

$$v^2(t) = 2\ \ln(Ct) = \ln{(Ct)}^2$$

Finally, we obtain two solution families,

$$u(t) = tv(t) = \pm t\left(\ln\ Kt^2\right)^{1/2}$$

where $K = C^2$.

•EXERCISES

Find all solutions to the equations in Problems 1–7.

1. $u' = 10u^2$.
2 $u' = \cot u\ \sin t$.
3. $t^2u' + u^2 = 1$.
4. $t\,(t+2)\,u' = u^2$.
5. $t^2u' = tu + u^2$.
6. $t^3 + u^3 - tu^2u' = 0$.
7. $tu' - u + u\ln t = u\ln u$.
8. Verify by substitution that the two solution families given in Example 1.5.3 are indeed solutions.
9. Solve $u' + pu = 0$, $p =$ constant, by separating the variables.
10. Find solutions of $tu' = (t-u)^2 + u$ by setting $v = t - u$.
11. Find solutions of $(t+2u+1)^{-1}\,u' = t + 2u + 4$ by setting $v = t + 2u + 1$.

Solve the initial-value problems given as Problems 12 and 13.

12. $tu' = (t-u)^2 + u$, $u(1) = 2$. (See Problem 10).
13. $tuu' - u^2 = t^2$, $u(1) = 0$. (See Example 1.5.3.)

14. In Example 1.5.1 we have found a family of solutions
$$u(t) = 2\,\tfrac{(K\,t)^4-1}{(K\,t)^4+1}$$
of the nonlinear ODE $tu' + u^2 = 4$.
   (a) Show that $u(t) = 2$ and $u(t) = -2$ are two solutions of this equation.
   (b) Show that $u(t) = 2$ and $u(t) = -2$ are not obtainable by any selection of K in the family of solutions.
   (c) Find b so that there are no solutions from this family which meet the initial condition $u(1) = b$.

1.5.2 Exact Equations

Suppose $\phi = \phi(t, u)$ is a function whose second-order partial derivatives, $\phi_{tt}, \phi_{tu}, \phi_{ut}, \phi_{uu}$ exist and are continuous.[7] Under the assumptions we have made on ϕ and its derivatives, we can conclude that

$$\phi_{tu} = \phi_{ut} \tag{1.5.12}$$

and the *total differential* of ϕ, written $d\phi$, exists. Recall that $d\phi$ is defined as

$$d\phi = \phi_t\,dt + \phi_u\,du \tag{1.5.13}$$

For example, if

$$\phi(t, u) = e^t \sin u$$

then

$$d\phi = e^t \sin u\,dt + e^t \cos u\,du$$

Also, $\phi(t, u) = K$ if and only if $d\phi = 0$.

Now consider the ODE

$$\frac{du}{dt} = \frac{M(t, u)}{N(t, u)}$$

which we write in the form

$$M dt + N du = 0 \tag{1.5.14}$$

[7] The subscript notation is short for the more familiar notation

$$\phi_t = \frac{\partial\phi}{\partial t}, \phi_u = \frac{\partial\phi}{\partial u}, \phi_{tt} = \frac{\partial^2\phi}{\partial t^2}, \phi_{tu} = \frac{\partial^2\phi}{\partial t\partial u}, \phi_{ut} = \frac{\partial^2\phi}{\partial u\partial t}, \phi_{uu} = \frac{\partial^2\phi}{\partial u^2}.$$

If there is a function ϕ whose total differential is the left-hand side of (1.5.14), that is, if

$$d\phi = Mdt + Ndu = 0 \tag{1.5.15}$$

then $\phi(t,u) = K$, from which we hope to determine $u(t)$. It is for this reason that we study the class of equations for which ϕ satisfies the condition, (1.5.15). We say that (1.5.14) is *exact* if (1.5.15) holds for some ϕ. The following theorem asserts that the only exact equations are those for which $M_u = N_t$. That is to say, if this condition fails, then no ϕ can exist for which (1.5.15) holds. The converse of this theorem is also true, but it is much harder to prove.

Theorem 1.5.1. *If $M(t,u)$ and $N(t,u)$ have continuous first-order partial derivatives, and if the equation*

$$M(t,u)dt + N(t,u)du = 0 \tag{1.5.16}$$

is exact, then

$$M_u = N_t \tag{1.5.17}$$

Proof: Suppose (1.5.16) is exact. Then, from the definition of ϕ and (1.5.13) and (1.5.15),

$$\phi_t = M(t,u), \qquad \phi_u = N(t,u) \tag{1.5.18}$$

and hence,

$$\phi_{tu} = M_u, \qquad \phi_{ut} = N_t$$

But $\phi_{tu} = \phi_{ut}$ by (1.5.12). Hence, $M_u = N_t$. □

Once we have excluded the possibility that the equation is not exact, we can use (1.5.18) to obtain ϕ and then, hopefully, u. The following three examples illustrate how this may be done.

Example 1.5.4. *Find ϕ, and thus verify that*

$$-\frac{u}{t^2}dt + \frac{1}{t}du = 0$$

is exact. Find solutions by determining $u(t)$ from $\phi(t,u) = K$.

Solution: By definition, $M = -u/t^2$ and $N = 1/t$. We verify (1.5.17):

$$M_u = -t^{-2} = N_t$$

Assuming the existence of ϕ, we use (1.5.18) as follows:

$$\phi_t = \frac{\partial \phi}{\partial t} = M = -\frac{u}{t^2}$$

Now we multiply by dt and integrate to obtain

$$\phi(t,u) = -\int \frac{u}{t^2}\,dt = \frac{u}{t} + C(u)$$

where $C(u)$ is an arbitrary differentiable function of u. Using this expression for ϕ, it follows that

$$\phi_u(t,u) = \frac{\partial}{\partial u}\left[\frac{u}{t} + C(u)\right] = \frac{1}{t} + C'(u) = \frac{1}{t}$$

But this implies that $C'(u) = 0$ and hence that $C(t) = C$. (We may as well choose $C = 0$, since we only need one ϕ.) Hence, (1.5.15) implies that $\phi(t,u) = u(t)/t = K$; that is, $u(t) = Kt$. Let's verify that this function is indeed a solution of the given ODE. This is easy: $du = K dt$, and hence,

$$-\frac{u}{t^2}dt + \frac{1}{t}du = -\frac{Kt}{t^2}dt + \frac{K}{t}dt = 0$$

Example 1.5.5. *Find a solution of*

$$tu^2 dt + \left(2 + t^2 u\right) du = 0, \quad u(1) = 2$$

Solution: Here, $M = tu^2$ and $N = 2 + t^2 u$. Hence,

$$M_u = 2tu = N_t$$

and the given ODE is exact. We compute ϕ by noting that

$$\phi_t(t,u) = M(t,u) = tu^2$$

So, with u held constant

$$\phi(t,u) = u^2 \int t\,dt = \frac{1}{2}t^2u^2 + C(u)$$

Differentiating this last result partially with respect to u leads to

$$\phi_u(t,u) = t^2u + C'(u) = N(t,u) = 2 + t^2u$$

Hence, $C'(u) = 2$, and therefore, $C(u) = 2u$. It now follows that the only possible ϕ are

$$\phi(t,u) = \frac{1}{2}t^2u^2 + C(u) = \frac{1}{2}t^2u^2 + 2u$$

and those obtained from this ϕ by adding arbitrary constants. Now, setting $\phi(t,u) = K$, we have

$$\frac{1}{2}t^2u^2 + 2u = K$$

The initial condition, $u(1) = 2$, leads to $K = 6$. So solving for $u(t)$ by invoking the quadratic formula, we get

$$u(t) = -2t^{-2}\left(1 - \sqrt{1+3t^2}\right)$$

We use the negative root in order that $u(1) = 2$. Implicit differentiation of $\frac{1}{2}t^2u^2 + 2u = K$ is probably the easiest way to verify that $u(t)$ is indeed a solution of the given initial-value problem.

Example 1.5.6. *Show that the following ODE is not exact:*

$$tu' = u$$

Solution: First, put the equation in the form of (1.5.16). It is $-udt + tdu = 0$. Here, $M = -u$, $N = t$, and $M_u \neq N_t$. By Theorem 1.5.1, the given equation is not exact. However, it is separable!

The point of the preceding example is that being exact is a "fragile" property; the simple act of multiplying the exact equation of Example 1.5.4 by t^2 results in the ODE of Example 1.5.6, which is not exact. The other side of this coin is that it may be possible to convert an ODE which is not exact into one which is by a clever choice of a multiplying function.

The function $Q(t,u)$ is called an *integrating factor* of the equation

$$Mdt + Ndu = 0 \tag{1.5.19}$$

if the equation

$$QM dt + QN du = 0 \tag{1.5.20}$$

is exact. Using the test for exactness on (1.5.20) leads to

$$\frac{\partial (QM)}{\partial u} = \frac{\partial (QN)}{\partial t} \tag{1.5.21}$$

from which it is sometimes possible to find $Q(t, u)$. The nonhomogeneous, first-order, linear ODE is a case in point. (See Problem 3.)

•EXERCISES

1. Verify that each equation is exact and find a family of solutions.
   (a) $(2 + t^2) u' + 2tu = 0.$ (b) $3u^2 u' + t^2 = 0.$
   (c) $\sin 2tu' + 2u \cos 2t = 0.$ (d) $e^t (u' + u) = 0.$
   (e) $e^t (\sin t\, u' + u \sin t + u \cos t) = 0.$
2. Solve each of the following initial-value problems by showing that the equations are exact and finding ϕ.
   (a) $(1 + t^2) u' + 2tu = 0, \quad u(0) = 1.$
   (b) $(u + t) u' + u = t, \quad u(1) = 0.$
   (c) $e^t (u' + u) = 0, \quad u(0) = 1.$
3. Write the general first-order, nonhomogeneous equation in the form

$$du + (p(t)u - f(t))dt = 0$$

Show that there exists an integrating factor $Q(t)$ for this equation that is a function of t alone. Find $Q(t)$, the resulting $\phi(t, u)$, and then the solution of this linear equation.

PROJECTS

1. Transforming Second-Order Equations into First-Order Equations.

Consider the general second-order ODE

$$u'' = f(t, u, u') \tag{1}$$

(a) If f in (1) is missing u, show that the substitution $v = u'$ reduces (1) to a first-order ODE in v.

(b) If f in (1) is missing t, let $v = u'$ and consider u as the independent variable. Show that $u'' = vdv/du$ and hence, (1) is $vdv/du = f(u, v)$.

2. Equations Linear in t and u.

Consider $u = tP(u') + Q(u')$.

(a) Show that

$$u' = P(u') + tP'(u')\,u'' + Q'(u')\,u'' \tag{1}$$

(b) Let $v = u'$ and $w = t$. Show that (1) may be written in the form

$$(v - p(v))\frac{dw}{dv} - wP'(v) = Q'(v) \tag{2}$$

(c) Write the general solution of (2).

There are very few changes of variables which are effective in simplifying nonlinear ODE's. The changes that are available are specific to the equations under study. The next two projects illustrate important special cases.

3. Riccati's Equation.

Consider the *Riccati equation*

$$u' + p(t)u + q(t)u^2 = f(t) \tag{1}$$

The study of this equation is attributed to the Italian mathematician Jacopo Riccati (1667 - 1748) . The Riccati equation can be transformed into a linear ODE by the change of dependent variable $u\,v\,q(t) = v'$.

(a) Define v by $u\,v\,q(t) = v'$. Differentiate $vq(t)u = v'$ to obtain

$$v'' = (vq(t)u)' = vq(t)u' + v'q(t)u + vq'(t)u$$

(b) Now multiply (1) by $q(t)v$ and use the above relationship for the term $vq(t)u'$ and $vq(t)u = v'$ for the term $vq^2(t)u^2$ to obtain

$$vq(t)u' + vq(t)p(t)u + vq^2(t)u^2 = v'' + \left(p(t) - \frac{q'(t)}{q(t)}\right)v'$$

(c) Use these results to show that

$$v'' + \left(p(t) - \frac{q'(t)}{q(t)}\right)v' - vq(t)f(t) = 0$$

(d) Solve (1) in the special case that $f(t) = 0$.

4. Bernoulli's Equation.

The brothers James (1654-1705) and John (1667-1748) Bernoulli contributed to the solution of

$$u' + p(t)u = q(t)u^n, \qquad n \neq 1 \tag{1}$$

the so called *Bernoulli equation* . The method shown here is due to John. Like the Riccati equation , (1) can be transformed into a linear ODE by a clever choice of change of variables.

(a) Show that the change of variables $u^{1-n} = v$ transforms (1) into a linear ODE. *Hint*: Multiply (1) by u^{-n} to obtain

$$u^{-n}u' + p(t)u^{1-n} = q(t) \tag{2}$$

Now differentiate $v = u^{1-n}$ to obtain

$$v' = (1-n)\,u^{-n}u'$$

Use this expression for $u^{-n}u'$ in (2) to obtain

$$v' + (1-n)p(t)v = (1-n)q(t)$$

a linear first-order nonhomogeneous ODE.

(b) For example, if $p(t) = a$ and $q(t) = b$, (1) is

$$u' + au + bu^n = 0$$

Find and solve the resulting equation in v. Indeed, show that

$$(u(t))^{1-n} = \frac{b}{a} + Ke^{(n-1)at}$$

In the special cases $n = 1$ or $q(t) = 0$, (1) is linear and if $p(t) = 0$, it is separable. In neither of these cases is a change of variables necessary.

5. The Logistic Equation.

A special case of the Bernoulli equation in which

$$p(t) = -a, \quad q(t) = -s \text{ and } n = 2$$

is given by

$$u' - au + su^2 = 0, \quad u(0) = P_0 \tag{1}$$

This equation is called the *logistic equation* and has application in population dynamics.

(a) Show that

$$u(t) = aP_0 \frac{e^{at}}{sP_0 e^{at} + a - sP_0} \tag{2}$$

is a solution of (1) for all choices of P_0.

(b) Set $P_0 = 1$, and $r = a/s$ in (2) to obtain

$$u(t) = \frac{re^{at}}{e^{at} + r - 1}$$

Plot this solution for $a = \frac{1}{2}$ and $r = 2$ and 15.

Chapter 2

Linear Systems with Constant Coefficients

2.1 Introduction

Many important physical and biological systems involve several interrelated variables, the magnitude of currents and voltages in various branches of an electric circuit, the position of masses in a spring network, and the concentration of solutes in the cells of an organism are just a few. In some cases, linear ODE's may approximate very closely the laws that govern the behavior of these systems. Linear models assume that the rate of change $x_i'(t)$ of each variable is a linear combination of the other variables:

$$x_i'(t) = a_{i1}x_1(t) + a_{i2}x_2(t) + \cdots + a_{in}x_n(t) \tag{2.1.1}$$

The parameters a_{ij}, $j = 1, 2, \ldots, n$ are constants determined by physical considerations, and n is the number of dependent variables in the system modeled by the equations. Consider, as an example, the system

$$\begin{aligned} x'(t) &= x(t) + y(t) \\ y'(t) &= 4x(t) + y(t) \end{aligned} \tag{2.1.2}$$

where $x_1(t) = x(t)$ and $x_2(t) = y(t)$. This system can be written in the matrix vector form

$$\begin{bmatrix} x'(t) \\ y'(t) \end{bmatrix} = \begin{bmatrix} 1 & 1 \\ 4 & 1 \end{bmatrix} \begin{bmatrix} x(t) \\ y(t) \end{bmatrix} \tag{2.1.3}$$

Define $\mathbf{A}, \mathbf{x(t)}$ and $\mathbf{x'(t)}$ by

$$\mathbf{A} = \begin{bmatrix} 1 & 1 \\ 4 & 1 \end{bmatrix}, \quad \mathbf{x}(t) = \begin{bmatrix} x(t) \\ y(t) \end{bmatrix}, \quad \mathbf{x}'(t) = \begin{bmatrix} x'(t) \\ y'(t) \end{bmatrix} \tag{2.1.4}$$

Then (2.1.3) may be written

$$\mathbf{x}'(t) = \mathbf{A}\mathbf{x}(t) \tag{2.1.5}$$

This symbolic, compact form may be used to represent (2.1.1) as well. In that case $\mathbf{A}$ is $n \times n$, the components of $\mathbf{x}(t)$ are the n unknown functions $x_i(t)$, and the components of $\mathbf{x}'(t)$ are their derivatives.

The resemblance between (2.1.5) and the scalar system

$$x'(t) = ax(t) \tag{2.1.6}$$

is so striking, that without the convention that vectors are printed in boldface type, the scalar and vector equations are indistinguishable. The similarities between the system (2.1.5) and the scalar equation (2.1.6) go deeper than this notational resemblance. Indeed, just as e^{at} is a solution of (2.1.6), $e^{\mathbf{A}t}$ can be interpreted as a solution of (2.1.5). (See Project No. 4 at the end of this chapter.) Before we can hope to understand how this can be done — after all, it is not obvious what is meant by $e^{\mathbf{A}t}$ — it is necessary to define the eigenvalues and eigenvectors of a matrix. This theory is developed in Section 2.2. The remaining sections in this chapter use these results to solve various problems that arise in the study of systems of linear ODE's represented by (2.1.5).

2.2 Eigenvalues and Eigenvectors

Consider once again the system (2.1.2):

$$\begin{aligned} x'(t) &= x(t) + y(t) \\ y'(t) &= 4x(t) + y(t) \end{aligned} \tag{2.2.1}$$

What functions are likely to be solutions of this system? After some experimentation, and from the insight gained when solving the linear equation with constant coefficients, one might think of trying

$$\begin{aligned} x(t) &= u_1 e^{\lambda t} \\ y(t) &= u_2 e^{\lambda t} \end{aligned} \tag{2.2.2}$$

where u_1, u_2 and λ are constants chosen, if possible, so that $x(t)$ and $y(t)$ are solutions of (2.2.1). We compute $x'(t)$ and $y'(t)$, then substitute $x(t), y(t), x'(t)$, and $y'(t)$ into (2.2.1) and obtain

$$\begin{aligned} \lambda u_1 e^{\lambda t} &= u_1 e^{\lambda t} + u_2 e^{\lambda t} \\ \lambda u_2 e^{\lambda t} &= 4u_1 e^{\lambda t} + u_2 e^{\lambda t} \end{aligned} \tag{2.2.3}$$

Since $e^{\lambda t} \neq 0$ for all λ and t, (2.2.3) simplifies into

$$\begin{aligned} \lambda u_1 &= u_1 + u_2 \\ \lambda u_2 &= 4u_1 + u_2 \end{aligned} \tag{2.2.4}$$

If we can satisfy these two equations with proper choices of λ, u_1, and u_2, then (2.2.2) represents a solution of (2.2.1). In matrix form (2.2.4) becomes

$$\begin{bmatrix} 1 & 1 \\ 4 & 1 \end{bmatrix} \begin{bmatrix} u_1 \\ u_2 \end{bmatrix} = \lambda \begin{bmatrix} u_1 \\ u_2 \end{bmatrix} \tag{2.2.5}$$

This system is an example of an algebraic *eigenvalue-eigenvector* problem which can be stated in general as we do now.

For the square matrix $\mathbf{A}$, find a scalar λ and associated nonzero vectors $\mathbf{u}$ such that

$$\mathbf{Au} = \lambda \mathbf{u} \tag{2.2.6}$$

The vector $\mathbf{u}$ is assumed to be nonzero, for if $\mathbf{u} = \mathbf{0}$ were permitted, every scalar λ would satisfy (2.2.6) for $\mathbf{u} = \mathbf{0}$, and every scalar would be an eigenvalue of $\mathbf{A}$.

Although we derived (2.2.6) from the particular system (2.2.1), it also arises from the assumption that $\mathbf{x}(t) = \mathbf{u}e^{\lambda t}$ is a solution of the $n \times n$ system $\mathbf{x}' = \mathbf{Ax}$, because under this assumption, we have

$$\mathbf{x}'(t) = \lambda \mathbf{u} e^{\lambda t}, \quad \mathbf{Ax}(t) = \mathbf{Au}e^{\lambda t}$$

and hence,

$$\lambda \mathbf{u} e^{\lambda t} = \mathbf{Au} e^{\lambda t} \tag{2.2.7}$$

Thus (2.2.6) is a result of cancelling $e^{\lambda t}$ in (2.2.7).

The scalar λ is called an *eigenvalue* of $\mathbf{A}$, and any nonzero vector that solves (2.2.6) for this λ is called an *eigenvector* of $\mathbf{A}$. The pair $(\lambda, \mathbf{u})$ is called an *eigenpair* of $\mathbf{A}$. It follows from $\mathbf{Au} = \lambda\mathbf{u}$ that $\mathbf{Au} - \lambda\mathbf{u} = \mathbf{0}$, and since $\mathbf{Iu} = \mathbf{u}$, this in turn leads to the equation[1]

$$(\mathbf{A} - \lambda\mathbf{I})\,\mathbf{u} = \mathbf{0} \tag{2.2.8}$$

The solution to this equation (see the corollary to Theorem 0.8.2) is

$$\mathbf{u} = (\mathbf{A} - \lambda\mathbf{I})^{-1}\,\mathbf{0} = \mathbf{0}$$

if $\mathbf{A} - \lambda\mathbf{I}$ is invertible. In order for $\mathbf{u}$ to be nonzero, $\mathbf{A} - \lambda\mathbf{I}$ must be singular. Hence, by Theorem 0.10.4,

$$\det(\mathbf{A} - \lambda\mathbf{I}) = 0 \tag{2.2.9}$$

Since the only unknown in (2.2.9) is λ, we can use (2.2.9) to find λ, and knowing λ, solve (2.2.8) for $\mathbf{u}$. These arguments lead to an important theorem and corollary:

Theorem 2.2.1. *A necessary and sufficient condition that the scalar* λ *is an eigenvalue of* $\mathbf{A}$ *is* $\det(\mathbf{A} - \lambda\mathbf{I}) = 0$.

Corollary 2.2.1. *A necessary and sufficient condition that the scalar* λ *is an eigenvalue of* $\mathbf{A}$ *is that* $\mathbf{A} - \lambda\mathbf{I}$ *be singular.*

Proof: By Theorem 2.2.1, $\mathbf{A} - \lambda\mathbf{I}$ is singular if and only if $\det(\mathbf{A} - \lambda\mathbf{I}) = 0$. □

Example 2.2.1. *Find the eigenvalues of* $\mathbf{A} = \begin{bmatrix} 1 & 1 \\ 4 & 1 \end{bmatrix}$.

Solution: This is the coefficient matrix for the system (2.2.1). We compute

$$\det(\mathbf{A} - \lambda\mathbf{I}) = \det\begin{bmatrix} 1-\lambda & 1 \\ 4 & 1-\lambda \end{bmatrix} = (\lambda - 3)(\lambda + 1)$$

So $\det(\mathbf{A} - \lambda\mathbf{I}) = 0$ if and only if $\lambda = 3$ or $\lambda = -1$. These are the eigenvalues of $\mathbf{A}$.

---

[1]It is necessary to replace $\mathbf{u}$ by $\mathbf{Iu}$ since we cannot factor $\mathbf{u}$ and write $(\mathbf{A} - \lambda)\,\mathbf{u} = \mathbf{0}$. The quantity $\mathbf{A} - \lambda$ in the parentheses is not defined; $\mathbf{A} - \lambda\mathbf{I}$ is defined. The identity matrix $\mathbf{I}$ is quite handy!

Example 2.2.2. *Find the eigenvalues of* $\mathbf{A} = \begin{bmatrix} 1 & 1 \\ 0 & 1 \end{bmatrix}$.

Solution: For this matrix,

$$\det(\mathbf{A} - \lambda\mathbf{I}) = \det \begin{bmatrix} 1-\lambda & 1 \\ 0 & 1-\lambda \end{bmatrix} = (1-\lambda)^2 = 0$$

if and only if $\lambda = 1$. So $\lambda = 1$ is the only eigenvalue of $\mathbf{A}$.

Example 2.2.3. *Find the eigenvalues of* $\mathbf{A} = \begin{bmatrix} -1 & 2 & 2 \\ 2 & 2 & 2 \\ -3 & -6 & -6 \end{bmatrix}$.

Solution: For this $\mathbf{A}$, (2.2.9) is

$$\det(\mathbf{A} - \lambda\mathbf{I}) = \det \begin{bmatrix} -1-\lambda & 2 & 2 \\ 2 & 2-\lambda & 2 \\ -3 & -6 & -6-\lambda \end{bmatrix} = -\lambda(\lambda+2)(\lambda+3) = 0$$

(We leave it to the reader to verify this computation.) So the eigenvalues of $\mathbf{A}$ are $\lambda_1 = 0, \lambda_2 = -2$, and $\lambda_3 = -3$. Note that the matrix $\mathbf{A} - \lambda\mathbf{I}$ is singular for these values of λ because its determinant is zero.

These examples suggest that $\det(\mathbf{A} - \lambda\mathbf{I})$ is a polynomial of degree n in λ if $\mathbf{A}$ is $n \times n$. Indeed, if we define $C(\lambda)$ by

$$C(\lambda) = \det(\mathbf{A} - \lambda\mathbf{I}) \tag{2.2.10}$$

then

$$\begin{aligned} C(\lambda) &= \det \begin{bmatrix} a_{11}-\lambda & a_{12} & \cdots & a_{1n} \\ a_{21} & a_{22}-\lambda & \cdots & a_{2n} \\ \vdots & \vdots & \vdots & \vdots \\ a_{n1} & a_{n2} & \cdots & a_{nn}-\lambda \end{bmatrix} \\ &= (-\lambda)^n + c_{n-1}(-\lambda)^{n-1} + \cdots + c_1(-\lambda) + c_0 \end{aligned} \tag{2.2.11}$$

We call $C(\lambda)$ the *characteristic polynomial* of $\mathbf{A}$. The equation $C(\lambda) = 0$ is the *characteristic equation* of $\mathbf{A}$. The characteristic polynomial is always of degree n and has n roots, some or all of which may be repeated, and some or all of which may not be real. In all the applications we will consider, the coefficients of $C(\lambda)$ will be real, and hence, the eigenvalues of $\mathbf{A}$ (the roots of $C(\lambda)$) will be real or come in complex conjugate pairs.

Once an eigenvalue has been determined, say, $\lambda = \lambda_1$, then $(\mathbf{A} - \lambda_1 \mathbf{I})$ is a specific singular matrix, and the homogeneous system

$$(\mathbf{A} - \lambda_1 \mathbf{I})\,\mathbf{u} = \mathbf{0} \qquad (2.2.12)$$

may be solved for corresponding eigenvectors $\mathbf{u} \neq \mathbf{0}$. Since (2.2.12) is a homogeneous system, multiples of solutions are also solutions. We will often choose this multiple so that the resulting eigenvector has a convenient form.

Example 2.2.4. *Find an eigenpair of* $\mathbf{A} = \begin{bmatrix} 1 & 1 \\ 0 & 1 \end{bmatrix}$ *for each eigenvalue.*

Solution: From Example 2.2.2, the only eigenvalue is $\lambda = 1$. Therefore, the eigenvectors satisfy

$$(\mathbf{A} - \mathbf{I})\,\mathbf{u} = \begin{bmatrix} 0 & 1 \\ 0 & 0 \end{bmatrix} \begin{bmatrix} u_1 \\ u_2 \end{bmatrix} = \mathbf{0}$$

This system represents the simultaneous equations

$$\begin{aligned} 0u_1 + 1u_2 &= 0 \\ 0u_1 + 0u_2 &= 0 \end{aligned}$$

with infinitely many solutions $u_1 = \alpha, \alpha$ arbitrary, and $u_2 = 0$. Thus, $\mathbf{u}_1 = [\alpha, 0]^T$ is an eigenvector for each $\alpha \neq 0$, and the eigenpairs are

$$(\lambda, \mathbf{u}_1) = \left(1, \begin{bmatrix} \alpha \\ 0 \end{bmatrix}\right) = \left(1, \alpha \begin{bmatrix} 1 \\ 0 \end{bmatrix}\right), \quad \alpha \neq 0$$

A convenient representative of the set of eigenpairs in this example is obtained by setting $\alpha = 1$,

$$(\lambda, \mathbf{u}_1) = \left(1, \begin{bmatrix} 1 \\ 0 \end{bmatrix}\right)$$

All other eigenvectors corresponding to $\lambda = 1$ are multiples of $\mathbf{u}_1$.

Example 2.2.5. *Find eigenpairs of the identity matrix* $\mathbf{I}_2$.

Solution: Even though the matrix in this example is different from the matrix in the previous example, the characteristic polynomials are identical, and thus, the only eigenvalue of $\mathbf{I}$ is $\lambda = 1$. By contrast, in this example there are two linearly independent eigenvectors associated with $\lambda = 1$. This is easily seen: The system

$$(\mathbf{I} - 1\mathbf{I})\,\mathbf{u} = \begin{bmatrix} 0 & 0 \\ 0 & 0 \end{bmatrix} \begin{bmatrix} u_1 \\ u_2 \end{bmatrix} = \mathbf{0}$$

represents the simultaneous equations

$$\begin{aligned} 0u_1 + 0u_2 &= 0 \\ 0u_1 + 0u_2 &= 0 \end{aligned}$$

for which $u_1 = \alpha$ and $u_2 = \beta$, α and β arbitrary, is clearly a solution. Thus,

$$\mathbf{u} = \begin{bmatrix} \alpha \\ \beta \end{bmatrix} = \alpha \begin{bmatrix} 1 \\ 0 \end{bmatrix} + \beta \begin{bmatrix} 0 \\ 1 \end{bmatrix}$$

is an eigenvector for every choice of α, β except when $\alpha = \beta = 0$. One eigenpair is obtained by setting $\alpha = 1, \beta = 0$ so that

$$(\lambda_1, \mathbf{u}_1) = \left(1, \begin{bmatrix} 1 \\ 0 \end{bmatrix}\right) = (1, \mathbf{e}_1)$$

A second eigenpair arises by choosing $\alpha = 0, \beta = 1$, leading to

$$(\lambda_1, \mathbf{u}_2) = \left(1, \begin{bmatrix} 0 \\ 1 \end{bmatrix}\right) = (1, \mathbf{e}_2)$$

Example 2.2.6. *Find the eigenpairs of* $\mathbf{A} = \begin{bmatrix} -1 & 2 & 2 \\ 2 & 2 & 2 \\ -3 & -6 & -6 \end{bmatrix}$.

Solution: In Example 2.2.3 we found that $0, -2, -3$ are the eigenvalues of $\mathbf{A}$. So we have three cases, one for each eigenvalue. Rather than solve the simultaneous equations represented by the system $(\mathbf{A} - \lambda\mathbf{I})\,\mathbf{u} = \mathbf{0}$, we row reduce the coefficient matrix $(\mathbf{A} - \lambda\mathbf{I})$. This is equivalent to using Gaussian elimination, except we do not write the column of zeros in an augmented matrix.

(1) $\lambda_1 = 0$:

$$\mathbf{A} - 0\mathbf{I} = \mathbf{A} = \begin{bmatrix} -1 & 2 & 2 \\ 2 & 2 & 2 \\ -3 & -6 & -6 \end{bmatrix} \cdots \to \begin{bmatrix} 1 & 0 & 0 \\ 0 & 1 & 1 \\ 0 & 0 & 0 \end{bmatrix}$$

Hence, $u_1 = 0, u_2 = -u_3$, and a representative eigenpair is obtained by setting $u_3 = 1$ (or we could have set $u_2 = 1$ if we desired):

$$(\lambda_1, \mathbf{u}_1) = \left(0, \begin{bmatrix} 0 \\ -1 \\ 1 \end{bmatrix}\right)$$

(2) $\lambda_2 = -2$:

$$\mathbf{A} + 2\mathbf{I} = \begin{bmatrix} 1 & 2 & 2 \\ 2 & 4 & 2 \\ -3 & -6 & -4 \end{bmatrix} \cdots \rightarrow \begin{bmatrix} 1 & 2 & 0 \\ 0 & 0 & 1 \\ 0 & 0 & 0 \end{bmatrix}$$

Hence, $u_3 = 0, u_1 = -2u_2$, and a representative eigenpair is obtained by selecting $u_2 = 1$:

$$(\lambda_2, \mathbf{u}_2) = \left(-2, \begin{bmatrix} -2 \\ 1 \\ 0 \end{bmatrix}\right)$$

(3) $\lambda_2 = -3$:

$$\mathbf{A} + 3\mathbf{I} = \begin{bmatrix} 2 & 2 & 2 \\ 2 & 5 & 2 \\ -3 & -6 & -3 \end{bmatrix} \cdots \rightarrow \begin{bmatrix} 1 & 0 & 1 \\ 0 & 1 & 0 \\ 0 & 0 & 0 \end{bmatrix}$$

Now $u_2 = 0, u_1 = -u_3$, and by selecting $u_3 = 1$, we obtain a representative eigenpair

$$(\lambda_3, \mathbf{u}_3) = \left(-3, \begin{bmatrix} -1 \\ 0 \\ 1 \end{bmatrix}\right)$$

Example 2.2.7. *Find the eigenpairs of* $\mathbf{A} = \begin{bmatrix} 1 & 1 \\ 4 & 1 \end{bmatrix}$.

Solution: We have already determined the eigenvalues of this matrix in Example 2.2.1. They are $\lambda_1 = -1$ and $\lambda_2 = 3$. We must find two eigenpairs, one for each eigenvalue, using (2.2.8).

(1) $\lambda_1 = -1$: We will present the details for this first eigenvalue. For $\lambda = -1$, (2.2.8) becomes

$$(\mathbf{A} + \mathbf{I}) = \mathbf{0}$$

In expanded form. this is written as

$$\begin{bmatrix} 2 & 1 \\ 4 & 2 \end{bmatrix} \begin{bmatrix} u_1 \\ u_2 \end{bmatrix} = \begin{bmatrix} 0 \\ 0 \end{bmatrix}$$

or in component form,

$$\begin{aligned} 2u_1 + u_2 &= 0 \\ 4u_1 + 2u_2 &= 0 \end{aligned}$$

Multiply the first equation by (-2) and add:

$$\begin{aligned} 2u_1 + u_2 &= 0 \\ 0u_1 + 0u_2 &= 0 \end{aligned}$$

These equations require only that $u_2 = -2u_1$. We may select any value for u_1, say, $u_1 = 1$. Then $u_2 = -2$, and the eigenpair is

$$(\lambda_1, \mathbf{u}_1) = \left(-1, \begin{bmatrix} 1 \\ -2 \end{bmatrix}\right)$$

Alternatively, we use the technique of Gaussian elimination. Since there are zeroes on the right hand side of (2.2.8), we simply row-reduce the matrix $(\mathbf{A} - \lambda\mathbf{I})$ as follows:

$$\mathbf{A} + \mathbf{I} = \begin{bmatrix} 2 & 1 \\ 4 & 2 \end{bmatrix} \rightarrow \cdots \rightarrow \begin{bmatrix} 2 & 1 \\ 0 & 0 \end{bmatrix}$$

So $u_2 = -2u_1$, and picking $u_1 = 1$, leads to the same eigenpair.
(2) $\lambda_2 = 3$: Using row reduction, there results

$$\mathbf{A} - 3\mathbf{I} = \begin{bmatrix} -2 & 1 \\ 4 & -2 \end{bmatrix} \rightarrow \begin{bmatrix} -2 & 1 \\ 0 & 0 \end{bmatrix}$$

So $u_2 = 2u_1$, and choosing $u_1 = 1$, we get the eigenpair

$$(\lambda_2, \mathbf{u}_2) = \left(3, \begin{bmatrix} 1 \\ 2 \end{bmatrix}\right)$$

It should be pointed out that we could select any value for u_1 (or u_2 and then calculate u_1). For example, if we desired the eigenvector to be a unit vector, we would demand that $u_1^2 + u_2^2 = 1$; then, for $\mathbf{u}_1$, we would find $u_1 = 1/\sqrt{5}$ and $u_2 = -2/\sqrt{5}$. Selecting $u_2 = 1$ so that $u_1 = -1/2$ would also be acceptable. We like to avoid fractions and radicals, so we made the selections as previously shown.

•EXERCISES

Find the characteristic polynomial and the eigenpairs for each matrix in Problems 1–22.

1. $\begin{bmatrix} 1 & 4 \\ 2 & 3 \end{bmatrix}$.

2. $\begin{bmatrix} 0 & 4 \\ 1 & 0 \end{bmatrix}$.

3. $\begin{bmatrix} 2 & 0 \\ 0 & -1 \end{bmatrix}$.

4. $\begin{bmatrix} 0 & 3 \\ 3 & 8 \end{bmatrix}$

5. $\begin{bmatrix} 2 & 2 \\ -1 & -1 \end{bmatrix}$.

6. $\begin{bmatrix} 5 & 4 \\ 4 & -1 \end{bmatrix}$.

7. $\begin{bmatrix} 2 & -2 \\ 2 & 2 \end{bmatrix}$.

8. $\begin{bmatrix} 3 & 1 \\ 5 & -1 \end{bmatrix}$.

9. $\begin{bmatrix} 0 & -1 \\ ab & a+b \end{bmatrix}$.

10. $\begin{bmatrix} 0 & ab \\ -1 & a+b \end{bmatrix}$.

11. $\begin{bmatrix} 2 & 2 & 0 \\ 1 & 2 & 1 \\ 1 & 2 & 1 \end{bmatrix}$.

12. $\begin{bmatrix} 1 & 2 & 4 \\ 0 & 1 & 0 \\ 0 & 2 & 1 \end{bmatrix}$.

13. $\begin{bmatrix} 1 & 1 & 0 \\ 0 & 1 & 0 \\ 0 & 0 & 1 \end{bmatrix}$.

14. $\begin{bmatrix} 3 & 0 & 1 \\ 0 & 2 & 0 \\ 5 & 0 & -1 \end{bmatrix}$.

15. $\begin{bmatrix} 10 & 8 & 0 \\ 8 & -2 & 0 \\ 0 & 0 & 4 \end{bmatrix}$.

16. $\begin{bmatrix} 1 & 1 & 0 \\ 0 & 1 & 1 \\ 0 & 0 & 1 \end{bmatrix}$.

17. $\begin{bmatrix} -1 & 3 & 0 \\ 3 & 7 & 0 \\ 0 & 0 & 6 \end{bmatrix}$.

18. $\begin{bmatrix} 1 & 1 & 1 \\ 1 & 1 & 1 \\ 1 & 1 & 1 \end{bmatrix}$.

19. $\begin{bmatrix} 0 & 0 & 0 \\ 0 & 0 & 0 \\ 0 & 0 & 0 \end{bmatrix}$.

20. $\begin{bmatrix} 0 & 1 & 0 \\ 0 & 0 & 1 \\ 2 & 1 & -2 \end{bmatrix}$.

21. $\begin{bmatrix} \cos\theta & \sin\theta \\ -\sin\theta & \cos\theta \end{bmatrix}$.

22. $\begin{bmatrix} \cos\theta & \sin\theta \\ \sin\theta & \cos\theta \end{bmatrix}$.

23. Show that $\lambda = 0$ is an eigenvalue of $\mathbf{A}$ if and only if $\mathbf{A}$ is singular.

24. Consider the matrices

(a) $\begin{bmatrix} 2 & 1 & 0 \\ 0 & 2 & 1 \\ 0 & 0 & 2 \end{bmatrix}$ (b) $\begin{bmatrix} 2 & 0 & 0 \\ 0 & 2 & 1 \\ 0 & 0 & 2 \end{bmatrix}$ (c) $\begin{bmatrix} 2 & 0 & 0 \\ 0 & 2 & 0 \\ 0 & 0 & 2 \end{bmatrix}$

Show that all three have characteristic polynomials

$$C(\lambda) = (2-\lambda)^2$$

and find all the eigenvectors of each matrix.

25. How are the eigenvalues of $\mathbf{A} - \mu\mathbf{I}$ related to those of $\mathbf{A}$?
26. How are the eigenvalues of $\mu\mathbf{A}$ related to those of $\mathbf{A}$?
27. How are the eigenvalues of $\mathbf{A}^n$ related to those of $\mathbf{A}$?
28. Show that the characteristic polynomials of $\mathbf{A}$ and $\mathbf{A}^T$ are the same.
29. Show that the eigenpairs of $\mathbf{A}$ and $\mathbf{A}^T$ need not be the same.
30. How are the eigenvalues of $\mathbf{A}^{-1}$ related to those of $\mathbf{A}$?
31. Show that the characteristic polynomials of $\mathbf{A}$ and $\mathbf{S}^{-1}\mathbf{AS}$ are identical.

MATLAB

32. Repeat Problems 1–8 using "[V,D] = eig(A)".
33. Repeat Problems 11–20 using "[V,D] = eig(A)".
34. Repeat Problem 24 by experimenting with various matrices.
35. Repeat Problem 25 by experimenting with various matrices.
36. Repeat Problem 26 by experimenting with various matrices.
37. Repeat Problem 27 by experimenting with various matrices.
38. Repeat Problem 28 by experimenting with various matrices.

2.3 Systems of Linear, First-Order, Homogeneous Differential Equations

We will illustrate later the fact that an nth-order ODE can be represented by a system of n first-order ODE's. Because of this, the study of first-order systems takes on far more significance than it might seem at first glance. Indeed, most computer programs for solving ODE's assume that the equation is written in its equivalent system form. So it is particularly important that we treat first-order systems thoroughly and rigorously.

We begin with some notational conventions. Let

$$\mathbf{x} = \begin{bmatrix} x_1(t) \\ x_2(t) \\ \vdots \\ x_n(t) \end{bmatrix}, \quad \mathbf{x}' = \begin{bmatrix} x_1'(t) \\ x_2'(t) \\ \vdots \\ x_n'(t) \end{bmatrix} \tag{2.3.1}$$

and

$$\mathbf{A} = \begin{bmatrix} a_{11} & a_{12} & \dots & a_{1n} \\ a_{21} & a_{22} & \dots & a_{2n} \\ \vdots & \vdots & & \dots \\ a_{n1} & a_{n2} & \dots & a_{nn} \end{bmatrix} \tag{2.3.2}$$

Then

$$\mathbf{x}' = \mathbf{A}\mathbf{x} \tag{2.3.3}$$

represents a system of n first-order ODE's in the n unknown functions x_i, $i = 1, 2, \dots, n$. The matrix $\mathbf{A}$ is the *coefficient matrix* of this system. Note that the system of linear homogeneous first-order ODE's

$$\begin{aligned} x_1'(t) &= x_1(t) + x_2(t) \\ x_2'(t) &= 4x_1(t) + x_2(t) \end{aligned} \tag{2.3.4}$$

is a special case of (2.3.3), where $n = 2$ and $\mathbf{A} = \begin{bmatrix} 1 & 1 \\ 4 & 1 \end{bmatrix}$.

The general system (2.3.3) is called *homogeneous.* If (2.3.3) is adjoined with the initial condition $\mathbf{x}(0) = \mathbf{b}$, the resulting problem

$$\mathbf{x}' = \mathbf{A}\mathbf{x}, \qquad \mathbf{x}(0) = \mathbf{b} \tag{2.3.5}$$

is called an *Initial-value problem.* The vector $\mathbf{b}$ is called the *initial vector.*

In this section we set ourselves the task of finding solutions $\mathbf{x}(t)$ of the initial-value problem (2.3.5). Although every such initial-value problem has a unique solution, proving this fact and obtaining the solution are significant undertakings. We content ourselves with a modest, but very useful and important compromise: We shall solve (2.3.5) for those $\mathbf{A}$ for which $\mathbf{b}$ can be written as a linear combination of the eigenvectors of $\mathbf{A}$. [2]

In the previous section we confirmed by example that the function $\mathbf{x}(t) = \mathbf{u}e^{\lambda t}$ is a solution of (2.3.3) if $(\lambda, \mathbf{u})$ is an eigenpair of $\mathbf{A}$. We now set out to prove this result for the system (2.3.5).

Theorem 2.3.1. *The function* $\mathbf{x}(t) = \mathbf{u}e^{\lambda t}$ *is a solution of the system* $\mathbf{x}' = \mathbf{A}\mathbf{x}$ *if and only if* $(\lambda, \mathbf{u})$ *is an eigenpair of the coefficient matrix* $\mathbf{A}$.

[2]This case includes the common situation in which $\mathbf{A}$ has n linearly independent eigenvectors, for then every $\mathbf{b}$ is some linear combination of the eigenvectors.

Proof: We have already shown that if $\mathbf{x}(t) = \mathbf{u}e^{\lambda t}$ is a solution of the system $\mathbf{x}' = \mathbf{A}\mathbf{x}$, then $(\lambda, \mathbf{u})$ is an eigenpair of $\mathbf{A}$. The converse is also true. For suppose that $(\lambda, \mathbf{u})$ is an eigenpair of $\mathbf{A}$. Then, writing $\mathbf{x}(t) = \mathbf{u}e^{\lambda t}$, we have $\mathbf{x}'(t) = \lambda \mathbf{u}e^{\lambda t}$ and, since $\mathbf{A}\mathbf{u} = \lambda\mathbf{u}$,

$$\mathbf{A}\mathbf{x}(t) = \mathbf{A}\mathbf{u}e^{\lambda t} = \lambda \mathbf{u}e^{\lambda t} = \mathbf{x}'(t)$$

□

Suppose $(\lambda_1, \mathbf{u}_1), (\lambda_2, \mathbf{u}_2), \ldots, (\lambda_k, \mathbf{u}_k)$ are eigenpairs of $\mathbf{A}$, where $1 \leq k \leq n$. (We do not assume that the eigenvalues $\lambda_1, \lambda_2, \ldots, \lambda_k$ are real or distinct.) Define[3] $\mathbf{x}_1(t) = \mathbf{u}_1 \exp(\lambda_1 t)$, and set

$$\mathbf{x}(t) = c_1\mathbf{x}_1(t) + c_2\mathbf{x}_2(t) + \cdots + c_k\mathbf{x}_k(t) \tag{2.3.6}$$

Theorem 2.3.2. *For each choice of scalars $c_1, c_2, \ldots, c_k$, the function $\mathbf{x}(t)$ as defined by (2.3.6) is a solution of $\mathbf{x}' = \mathbf{A}\mathbf{x}$.*

Proof: We have, using $\mathbf{x}_i = \mathbf{u}_i \exp(\lambda_i t)$,

$$\mathbf{x}_i'(t) = \lambda_1 \mathbf{u}_i \exp(\lambda_i t) = \lambda_i \mathbf{x}_i(t) \tag{2.3.7}$$

and, since $\mathbf{A}\mathbf{u}_i = \lambda_i \mathbf{u}_i$,

$$\begin{aligned} \mathbf{A}\mathbf{x}_i(t) &= \mathbf{A}\mathbf{u}_i \exp(\lambda_i t) \\ &= \lambda_i \mathbf{u}_i \exp(\lambda_i t) = \lambda_i \mathbf{x}_i(t) \end{aligned} \tag{2.3.8}$$

If we differentiate (2.3.6) and use (2.3.7), we obtain

$$\mathbf{x}'(t) = c_1\lambda_1\mathbf{x}_1(t) + c_2\lambda_2\mathbf{x}_2(t) + \cdots + c_k\lambda_k\mathbf{x}_k(t)$$

However, from (2.3.6) and (2.3.8),

$$\begin{aligned} \mathbf{A}\mathbf{x}(t) &= c_1\mathbf{A}\mathbf{x}_1(t) + c_2\mathbf{A}\mathbf{x}_2(t) + \cdots + c_k\mathbf{A}\mathbf{x}_k(t) \\ &= c_1\lambda_1\mathbf{x}_1(t) + c_2\lambda_2\mathbf{x}_2(t) + \cdots + c_k\lambda_k\mathbf{x}_k(t) \\ &= \mathbf{x}'(t) \end{aligned}$$

□

[3] It is often convenient to write $\exp(z)$ for e^z especially when z itself is a compound expression.

Corollary 2.3.1. *Under the assumptions and notation of Theorem 2.3.2, if, for some choice of constants* $c_1, c_2, \ldots, c_k$,

$$\mathbf{b} = c_1\mathbf{u}_1 + c_2\mathbf{u}_2 + \cdots + c_k\mathbf{u}_k \tag{2.3.9}$$

then

$$\mathbf{x}(t) = c_1\mathbf{x}_1(t) + c_2\mathbf{x}_2(t) + \cdots + c_k\mathbf{x}_k(t) \tag{2.3.10}$$

is a solution of the initial-value problem (2.3.5).

Proof: By Theorem 2.3.2, $\mathbf{x}(t)$ is a solution of $\mathbf{x}' = \mathbf{A}\mathbf{x}$. In view of (2.3.9) and (2.3.10),

$$\begin{aligned}\mathbf{x}(0) &= c_1\mathbf{u}_1 \exp(0) + c_2\mathbf{u}_2 \exp(0) + \cdots + c_k\mathbf{x}_k \exp(0) \\ &= c_1\mathbf{u}_1 + c_2\mathbf{u}_2 + \cdots + c_k\mathbf{u}_k = \mathbf{b}\end{aligned}$$

□

Note that Corollary 2.3.1 does not require the existence of n linearly independent eigenvectors of $\mathbf{A}$. All that is required is that the initial vector $\mathbf{b}$ be a linear combination of some of the eigenvectors of $\mathbf{A}$.

Corollary 2.3.2. *If* $\mathbf{A}$ *has* n *distinct eigenvalues, then*

$$\mathbf{x}' = \mathbf{A}\mathbf{x}, \qquad \mathbf{x}(0) = \mathbf{b}$$

has a solution of the form

$$\mathbf{x}(t) = c_1\mathbf{x}_1(t) + c_2\mathbf{x}_2(t) + \cdots + c_n\mathbf{x}_n(t)$$

for some choice of constants, c_1, c_2, $\ldots$, c_n.

Proof: We are tacitly assuming that $\mathbf{A}$ is $n \times n$. Since there are n distinct eigenvalues of $\mathbf{A}$, there are n linearly independent eigenvectors of $\mathbf{A}$, one for each eigenvalue. But any vector $\mathbf{b}$ is a linear combination of n linearly independent vectors so that the hypotheses of Corollary 2.3.1 hold. □

Example 2.3.1. *Find the solution of* $\mathbf{x}' = \begin{bmatrix} 1 & 1 \\ 4 & 1 \end{bmatrix}\mathbf{x}, \quad \mathbf{x}(0) = \begin{bmatrix} 1 \\ 1 \end{bmatrix}.$

Solution: We have already found eigenpairs for the coefficient matrix of the system in Example 2.2.7, namely,

$$\left(-1, \begin{bmatrix} 1 \\ -2 \end{bmatrix}\right) \quad \text{and} \quad \left(3, \begin{bmatrix} 1 \\ 2 \end{bmatrix}\right)$$

To exploit Corollary 2.3.1, we find c_1 and c_2 such that (2.3.9) is satisfied. That is,

$$c_1 \begin{bmatrix} 1 \\ -2 \end{bmatrix} + c_2 \begin{bmatrix} 1 \\ 2 \end{bmatrix} = \begin{bmatrix} 1 \\ 1 \end{bmatrix}$$

This equation may be solved by inspection or by row reductions:

$$\left[\begin{array}{cc|c} 1 & 1 & 1 \\ -2 & 2 & 1 \end{array}\right] \to \left[\begin{array}{cc|c} 1 & 1 & 1 \\ 0 & 4 & 3 \end{array}\right]$$

Thus, $c_1 = \frac{1}{4}, c_2 = \frac{3}{4}$, and, from Corollary 2.3.1, it follows that the solution is

$$\mathbf{x}(t) = \frac{1}{4}\begin{bmatrix} 1 \\ -2 \end{bmatrix} e^{-t} + \frac{3}{4}\begin{bmatrix} 1 \\ 2 \end{bmatrix} e^{3t}$$

Even though the eigenvectors in Example 2.3.1 are known only up to a multiplicative constant, the solution is unique. For example, had we chosen

$$\begin{bmatrix} -1 \\ 2 \end{bmatrix} \quad \text{and} \quad \begin{bmatrix} 2 \\ 4 \end{bmatrix}$$

as the two eigenvectors, the constants would be $c_1 = \frac{1}{4}, c_2 = \frac{3}{8}$ and the resulting solution would remain unchanged, as it must for the initial-value problem. Finally, if the initial condition is

$$\mathbf{b} = \begin{bmatrix} 2 \\ 4 \end{bmatrix}$$

then the solution involves only the eigenpair

$$\left(3, \begin{bmatrix} 1 \\ 2 \end{bmatrix}\right)$$

and is given by

$$\mathbf{x}(t) = 2\begin{bmatrix} 1 \\ 2 \end{bmatrix} e^{3t}$$

This is the case because the initial-condition vector is a multiple of the second eigenvector; it has no component in the direction of the first eigenvector.

Example 2.3.2. *Find the solution of* $\mathbf{x}' = \begin{bmatrix} 0 & 1 & 0 \\ 0 & 0 & 1 \\ 2 & 1 & -2 \end{bmatrix}, \quad \mathbf{x}(0) = \begin{bmatrix} 1 \\ 0 \\ 1 \end{bmatrix}.$

Solution: After some labor (see Problem 20 of the previous exercise set), we find the eigenpairs

$$\left(1, \begin{bmatrix} 1 \\ 1 \\ 1 \end{bmatrix}\right), \quad \left(-2, \begin{bmatrix} 1 \\ -2 \\ 4 \end{bmatrix}\right), \quad \left(-1, \begin{bmatrix} 1 \\ -1 \\ 1 \end{bmatrix}\right)$$

Our object is to solve (2.3.9), written as

$$\begin{bmatrix} 1 & 1 & 1 \\ 1 & -2 & -1 \\ 1 & 4 & 1 \end{bmatrix} \mathbf{c} = \begin{bmatrix} 1 \\ 0 \\ 1 \end{bmatrix}$$

Using Gaussian elimination, we deduce from the arrow diagram of the augmented matrix

$$\left[\begin{array}{ccc|c} 1 & 1 & 1 & 1 \\ 1 & -2 & -1 & 0 \\ 1 & 4 & 1 & 1 \end{array}\right] \to \cdots \to \left[\begin{array}{ccc|c} 1 & 0 & 0 & 1/2 \\ 0 & 1 & 0 & 0 \\ 0 & 0 & 1 & 1/2 \end{array}\right]$$

that

$$c_1 = \frac{1}{2}, \quad c_2 = 0, \quad \text{and} \quad c_3 = \frac{1}{2}$$

and therefore, by (2.3.10),

$$\mathbf{x}(t) = \frac{1}{2}\begin{bmatrix} 1 \\ 1 \\ 1 \end{bmatrix} e^t + \frac{1}{2}\begin{bmatrix} 1 \\ -1 \\ 1 \end{bmatrix} e^{-t} = \begin{bmatrix} \cosh t \\ \sinh t \\ \cosh t \end{bmatrix}$$

Example 2.3.3. *Find the solution of*

$$\mathbf{x}' = \begin{bmatrix} 2 & 1 & 0 \\ 0 & 2 & 0 \\ 0 & 0 & -1 \end{bmatrix} \mathbf{x}, \quad \mathbf{x}(0) = \begin{bmatrix} 2 \\ 0 \\ -3 \end{bmatrix}$$

Solution: The characteristic polynomial is $(2-\lambda)^2(1+\lambda)$ with roots $\lambda_1 = \lambda_2 = 2, \lambda_3 = -1$. For the root $\lambda = -1$, an eigenpair and a corresponding solution are

$$\left(-1, \begin{bmatrix} 0 \\ 0 \\ 1 \end{bmatrix}\right), \quad \mathbf{x}_1(t) = \begin{bmatrix} 0 \\ 0 \\ 1 \end{bmatrix} e^{-t}$$

The repeated root $\lambda = 2$ has only one linearly independent eigenvector associated with it. An eigenpair and its corresponding solution are

$$\left(2, \begin{bmatrix} 1 \\ 0 \\ 0 \end{bmatrix}\right), \quad \mathbf{x}_2(t) = \begin{bmatrix} 1 \\ 0 \\ 0 \end{bmatrix} e^{2t}$$

We combine $\mathbf{x}_1(t)$ with $\mathbf{x}_2(t)$ to form

$$\mathbf{x}(t) = c_1 \begin{bmatrix} 0 \\ 0 \\ 1 \end{bmatrix} e^{-t} + c_2 \begin{bmatrix} 1 \\ 0 \\ 0 \end{bmatrix} e^{2t}$$

The constants c_1 and c_2 are found from

$$\mathbf{x}(0) = c_1 \begin{bmatrix} 0 \\ 0 \\ 1 \end{bmatrix} + c_2 \begin{bmatrix} 1 \\ 0 \\ 0 \end{bmatrix} = \begin{bmatrix} 2 \\ 0 \\ -3 \end{bmatrix}$$

Thus, $c_1 = -3$, and $c_2 = 2$ and the solution is then

$$\mathbf{x}(t) = -3 \begin{bmatrix} 0 \\ 0 \\ 1 \end{bmatrix} e^{-t} + 2 \begin{bmatrix} 1 \\ 0 \\ 0 \end{bmatrix} e^{2t}$$

Example 2.3.4. *Show that there are no solutions of the form* (2.3.10) *for*

$$\mathbf{x}' = \begin{bmatrix} 2 & 1 & 0 \\ 0 & 2 & 0 \\ 0 & 0 & -1 \end{bmatrix} \mathbf{x}, \quad \mathbf{x}(0) = \begin{bmatrix} 1 \\ 1 \\ 1 \end{bmatrix}$$

Solution: From the work in Example 2.3.3, we have

$$\mathbf{x}(t) = c_1 \begin{bmatrix} 0 \\ 0 \\ 1 \end{bmatrix} e^{-t} + c_2 \begin{bmatrix} 1 \\ 0 \\ 0 \end{bmatrix} e^{2t}$$

so that

$$\mathbf{x}(0) = c_1 \begin{bmatrix} 0 \\ 0 \\ 1 \end{bmatrix} + c_2 \begin{bmatrix} 1 \\ 0 \\ 0 \end{bmatrix} = \begin{bmatrix} 1 \\ 1 \\ 1 \end{bmatrix}$$

which has a contradiction in the second component.

Example 2.3.5. *Find α and β such that the following initial-value problem has a solution of the form (2.3.10), and find all such solutions*

$$\mathbf{x}' = \begin{bmatrix} 1 & 1 \\ -1 & -1 \end{bmatrix} \mathbf{x}, \quad \mathbf{x}(0) = \begin{bmatrix} \alpha \\ \beta \end{bmatrix}$$

Solution: The eigenvalues of the coefficient matrix are $\lambda_1 = 0, \lambda_2 = 0$ and an eigenpair is

$$\left(0, c \begin{bmatrix} 1 \\ -1 \end{bmatrix}\right)$$

So we are left with the problem of finding c such that

$$c \begin{bmatrix} 1 \\ -1 \end{bmatrix} = \begin{bmatrix} \alpha \\ \beta \end{bmatrix}$$

Clearly, this is possible if and only if $\alpha = -\beta = c$. So, if $\alpha \neq -\beta$, then the given system has no solution among the family of functions $\mathbf{u}e^{\lambda t}$. On the other hand, if $\alpha = -\beta = c$, the solutions are

$$\mathbf{x}(t) = \alpha \begin{bmatrix} 1 \\ -1 \end{bmatrix} e^{0t} = \alpha \begin{bmatrix} 1 \\ -1 \end{bmatrix}$$

It is easy to misconstrue the point of Examples 2.3.4 and 2.3.5. The conclusion is not that there are no solutions of these initial-value problems. Indeed,

$$\mathbf{x}(t) = \begin{bmatrix} \alpha + (\alpha + \beta)\, t \\ \beta - (\alpha + \beta)\, t \end{bmatrix} \tag{2.3.11}$$

is a solution of Example 2.3.5 for all α and β, as we can verify directly. Rather, the point is that the coefficient matrices of these systems have too few eigenpairs for us to construct a solution of the form (2.3.10) and there are too few eigenpairs only when the roots of the characteristic polynomial are repeated. [4] We can overcome this difficulty by using a different approach. In Chapter 5 we introduce one such method, the Laplace transform, which does not depend directly upon the nature of the eigenvectors of $\mathbf{A}$. The method of power series, explored in Chapter 6, is yet another approach to solving systems without explicit regard to the eigenvectors of $\mathbf{A}$.

It is comforting to know that most physical problems result in matrices which have distinct eigenvalues, so that this difficulty does not often arise.

2.3.1 Complex Eigenvalues

Since the eigenvalues of $\mathbf{A}$ are the roots of an nth degree polynomial, it may happen that some eigenvalues are complex. If this is the case, the solution $\mathbf{x}(t) = \mathbf{u}e^{\lambda t}$ is complex-valued. We now show that each complex-valued solution leads to two real-valued solutions, $\operatorname{Re}\mathbf{x}(t)$ and $\operatorname{Im}\mathbf{x}(t)$. We continue to assume that the elements of $\mathbf{A}$ are real.

It is not difficult to prove that $\operatorname{Re}\mathbf{x}' = (\operatorname{Re}\mathbf{x})'$ and, likewise, that $\operatorname{Im}\mathbf{x}' = (\operatorname{Im}\mathbf{x})'$. From these facts, it follows that

$$\operatorname{Re}\mathbf{x}' = (\operatorname{Re}\mathbf{x})' = \operatorname{Re}(\mathbf{A}\mathbf{x}) = \mathbf{A}\operatorname{Re}\mathbf{x} \tag{2.3.12}$$

and

$$\operatorname{Im}\mathbf{x}' = (\operatorname{Im}\mathbf{x})' = \operatorname{Im}(\mathbf{A}\mathbf{x}) = \mathbf{A}\operatorname{Im}\mathbf{x} \tag{2.3.13}$$

These equations show that $\operatorname{Re}\mathbf{x}(t)$ and $\operatorname{Im}\mathbf{x}(t)$ are real solutions of $\mathbf{x}' = \mathbf{A}\mathbf{x}$ whenever $\mathbf{x}(t)$ is. This is the crucial observation in the proof of the next theorem. (If $\mathbf{x}(t)$ is real, then $\operatorname{Re}\mathbf{x}(t) = \mathbf{x}(t)$ and $\operatorname{Im}\mathbf{x}(t) = \mathbf{0}$, and this idea generates no new solutions.)

Theorem 2.3.3. *If* $(\lambda, \mathbf{u})$ *is an eigenpair of* $\mathbf{A}$ *and* $\mathbf{x}(t) = \mathbf{u}e^{\lambda t}$, *then* $\operatorname{Re}\mathbf{x}(t)$ *and* $\operatorname{Im}\mathbf{x}(t)$ *are solutions of* $\mathbf{x}' = \mathbf{A}\mathbf{x}$.

Proof: The function $\mathbf{x}(t) = \mathbf{u}e^{\lambda t}$ is a solution of $\mathbf{x}' = \mathbf{A}\mathbf{x}$. The theorem is proved by reference to (2.3.12) and (2.3.13). □

The functions $\operatorname{Re}\mathbf{x}(t)$ and $\operatorname{Im}\mathbf{x}(t)$ are real-valued. In fact, we may use

[4] In order to solve all systems by the methods we have presented, it is necessary to generalize the idea of an eigenvector. We choose not to explore the idea here because the algebraic details would take us too far into a theory best left to more advanced texts.

Euler's formula,

$$e^{i\theta} = \cos\theta + i\sin\theta$$

to find the explicit forms of $\operatorname{Re}\mathbf{x}(t)$ and $\operatorname{Im}\mathbf{x}(t)$. To do this, suppose that $\lambda = a + ib$ is a complex eigenvalue of $\mathbf{A}$, where a and $b \neq 0$ are real numbers and a corresponding eigenvector is $\mathbf{u} = \mathbf{v} + i\mathbf{w}$. Then

$$\begin{aligned} \mathbf{x}(t) &= \mathbf{u}e^{\lambda t} = (\mathbf{v} + i\mathbf{w})\, e^{(a+ib)t} = (\mathbf{v} + i\mathbf{w})\, e^{at} e^{ibt} \qquad (2.3.14) \\ &= e^{at}\,(\mathbf{v} + i\mathbf{w})\,(\cos bt + i\sin bt) \\ &= e^{at}\,(\mathbf{v}\cos bt - \mathbf{w}\sin bt) + ie^{at}\,(\mathbf{w}\cos bt + \mathbf{v}\sin bt) \end{aligned}$$

Therefore, collecting real and imaginary parts, we obtain

$$\operatorname{Re}\mathbf{x}(t) = e^{at}\,(\mathbf{v}\cos bt - \mathbf{w}\sin bt) \qquad (2.3.15)$$

$$\operatorname{Im}\mathbf{x}(t) = e^{at}\,(\mathbf{w}\cos bt + \mathbf{v}\sin bt) \qquad (2.3.16)$$

In any specific problem, it is generally easier to compute $\operatorname{Re}\mathbf{x}(t)$ and $\operatorname{Im}\mathbf{x}(t)$ directly from $\mathbf{x}(t)$ rather than using (2.3.15) and (2.3.16).

We illustrate Theorem 2.3.3 in the following examples. Note that each complex eigenpair $(\lambda, \mathbf{u})$ generates two solutions and that these solutions always involve trigonometric functions.

Example 2.3.6. *Find two real solutions of the system* $\mathbf{x}' = \begin{bmatrix} 1 & 1 \\ -1 & 1 \end{bmatrix}\mathbf{x}$.

Solution: The characteristic polynomial is $\lambda^2 - 2\lambda + 2 = 0$ with roots $\lambda_1 = 1 + i, \lambda_2 = 1 - i$. An eigenvector corresponding to λ_1 is obtained by solving

$$(\mathbf{A} - (1+i)\mathbf{I})\mathbf{u} = \begin{bmatrix} -i & 1 \\ -1 & -i \end{bmatrix}\mathbf{u} = \mathbf{0}$$

The arrow diagram is as effective for complex matrices as it is for real matrices, only the arithmetic may be a bit more tedious:

$$\begin{bmatrix} -i & 1 \\ -1 & -i \end{bmatrix} \to \begin{bmatrix} -i & 1 \\ 0 & 0 \end{bmatrix}$$

This implies that $iu_1 = u_2$ and, after setting $u_1 = 1$, leads to the eigenpair

$$\left(1 + i, \begin{bmatrix} 1 \\ i \end{bmatrix}\right)$$

It also implies that the given system has the complex-valued solution

$$\begin{aligned}\mathbf{x}(t) &= \begin{bmatrix} 1 \\ i \end{bmatrix} e^{(1+i)t} = e^t \left(\begin{bmatrix} 1 \\ 0 \end{bmatrix} + i \begin{bmatrix} 0 \\ 1 \end{bmatrix} \right) (\cos t + i \sin t) \\ &= e^t \left(\begin{bmatrix} 1 \\ 0 \end{bmatrix} \cos t - \begin{bmatrix} 0 \\ 1 \end{bmatrix} \sin t \right) + i e^t \left(\begin{bmatrix} 1 \\ 0 \end{bmatrix} \sin t + \begin{bmatrix} 0 \\ 1 \end{bmatrix} \cos t \right) \\ &= e^t \begin{bmatrix} \cos t \\ -\sin t \end{bmatrix} + i e^t \begin{bmatrix} \sin t \\ \cos t \end{bmatrix}\end{aligned}$$

So,

$$\operatorname{Re} \mathbf{x}(t) = e^t \begin{bmatrix} \cos t \\ -\sin t \end{bmatrix}, \quad \operatorname{Im} \mathbf{x}(t) = e^t \begin{bmatrix} \sin t \\ \cos t \end{bmatrix}$$

are both real solutions. We form a general solution by combining these solutions:

$$\mathbf{x}(t) = c_1 \operatorname{Re} \mathbf{x}(t) + c_2 \operatorname{Im} \mathbf{x}(t)$$

(Verify that using the eigenvalue $\lambda_2 = 1 - i$ leads to the same general solution.)

Example 2.3.7. *Find the solution of* $\mathbf{x}' = \begin{bmatrix} 0 & 1 \\ -1 & 0 \end{bmatrix} \mathbf{x}, \quad \mathbf{x}(0) = \begin{bmatrix} 1 \\ 1 \end{bmatrix}.$

Solution: Here the characteristic equation is $\lambda^2 + 1 = 0$, and hence, the eigenvalues are $\lambda_1 = i, \lambda_2 = -i$. Corresponding to the eigenvalue $\lambda_1 = i$ is an eigenpair $\left(i, \begin{bmatrix} 1 \\ i \end{bmatrix} \right)$ and a solution

$$\begin{aligned}\mathbf{x}(t) &= \begin{bmatrix} 1 \\ i \end{bmatrix} e^{it} = \begin{bmatrix} 1 \\ i \end{bmatrix} (\cos t + i \sin t) \\ &= \left(\begin{bmatrix} 1 \\ 0 \end{bmatrix} + i \begin{bmatrix} 0 \\ 1 \end{bmatrix} \right) (\cos t + i \sin t) \\ &= \left(\begin{bmatrix} 1 \\ 0 \end{bmatrix} \cos t - \begin{bmatrix} 0 \\ 1 \end{bmatrix} \sin t \right) + i \left(\begin{bmatrix} 1 \\ 0 \end{bmatrix} \sin t + \begin{bmatrix} 0 \\ 1 \end{bmatrix} \cos t \right) \\ &= \begin{bmatrix} \cos t \\ -\sin t \end{bmatrix} + i \begin{bmatrix} \sin t \\ \cos t \end{bmatrix}\end{aligned}$$

Two real solutions follow from this expression:

$$\operatorname{Re} \mathbf{x}(t) = \begin{bmatrix} \cos t \\ -\sin t \end{bmatrix}, \quad \operatorname{Im} \mathbf{x}(t) = \begin{bmatrix} \sin t \\ \cos t \end{bmatrix}$$

The general solution is then written as $\mathbf{x}(t) = c_1 \mathrm{Re}\,\mathbf{x}(t) + c_2 \mathrm{Im}\,\mathbf{x}(t)$. The initial condition demands that

$$c_1 \mathrm{Re}\,\mathbf{x}(0) + c_2 \mathrm{Im}\,\mathbf{x}(0) = c_1 \begin{bmatrix} 1 \\ 0 \end{bmatrix} + c_2 \begin{bmatrix} 0 \\ 1 \end{bmatrix} = \begin{bmatrix} 1 \\ 1 \end{bmatrix}$$

where upon we find that $c_1 = c_2 = 1$. Hence, the desired solution is

$$\mathbf{x}(t) = \begin{bmatrix} \cos t \\ -\sin t \end{bmatrix} + \begin{bmatrix} \sin t \\ \cos t \end{bmatrix} = \begin{bmatrix} \cos t + \sin t \\ \cos t - \sin t \end{bmatrix}$$

(Verify that the eigenpair $\left(-i, \begin{bmatrix} 1 \\ -i \end{bmatrix}\right)$ leads to the same solution.)

Example 2.3.8. *Find real solutions of* $\mathbf{x}' = \begin{bmatrix} 0 & 1 & 0 \\ 0 & 0 & 1 \\ 0 & -1 & 0 \end{bmatrix} \mathbf{x}$.

Solution: It is not difficult to see that the characteristic polynomial is $-\lambda^3 - \lambda$. The roots of this polynomial are $0, i$ and, $-i$. We leave it to the reader to show that the eigenpairs are

$$\left(0, \begin{bmatrix} 1 \\ 0 \\ 0 \end{bmatrix}\right), \quad \left(i, \begin{bmatrix} 1 \\ i \\ -1 \end{bmatrix}\right), \quad \left(-i, \begin{bmatrix} 1 \\ -i \\ -1 \end{bmatrix}\right)$$

The real eigenpair defines the real solution

$$\mathbf{x}_1(t) = \begin{bmatrix} 1 \\ 0 \\ 0 \end{bmatrix} e^{0t} = \begin{bmatrix} 1 \\ 0 \\ 0 \end{bmatrix}$$

The second eigenpair in this list defines the complex-valued solution

$$\begin{aligned} \mathbf{x}(t) &= \begin{bmatrix} 1 \\ i \\ -1 \end{bmatrix} e^{it} = \begin{bmatrix} 1 \\ i \\ -1 \end{bmatrix} (\cos t + i \sin t) \\ &= \begin{bmatrix} \cos t + i \sin t \\ -\sin t + i \cos t \\ -\cos t - i \sin t \end{bmatrix} = \begin{bmatrix} \cos t \\ -\sin t \\ -\cos t \end{bmatrix} + i \begin{bmatrix} \sin t \\ \cos t \\ -\sin t \end{bmatrix} \end{aligned}$$

from which we obtain the two real solutions

$$\mathbf{x}_2(t) = \mathrm{Re}\,\mathbf{x}(t) = \begin{bmatrix} \cos t \\ -\sin t \\ -\cos t \end{bmatrix}, \quad \mathbf{x}_3(t) = \mathrm{Im}\,\mathbf{x}(t) = \begin{bmatrix} \sin t \\ \cos t \\ -\sin t \end{bmatrix}$$

2.3.2 Repeated Eigenvalues

If the roots of the characteristic polynomial are repeated, we may not be able to obtain sufficiently many eigenpairs to construct a solution which meets the given initial condition. We encountered this situation in Examples 2.3.4 and 2.3.5. This case can only occur when at least one eigenvalue is repeated, that is, the eigenvalues are not distinct. However, even when the eigenvalues are not distinct, we may be able to solve the initial-value problem. In this section, we explore the special case in which there are n linearly independent eigenvectors even though some of the eigenvalues are repeated roots of the characteristic polynomial. Here are some examples illustrating this point.

Example 2.3.9. *Find a solution of* $\mathbf{x}' = \begin{bmatrix} 2 & 0 & 1 \\ 0 & 2 & 1 \\ 0 & 0 & -1 \end{bmatrix} \mathbf{x}, \quad \mathbf{x}(0) = \begin{bmatrix} 1 \\ 1 \\ 1 \end{bmatrix}.$

Solution: Here the characteristic polynomial is $(2-\lambda)^2(1+\lambda)$ with roots $\lambda_1 = -1, \lambda_2 = 2, \lambda_3 = 2$. First we find an eigenvector corresponding to the eigenvalue $\lambda_1 = -1$. To do this, we solve

$$\begin{bmatrix} 3 & 0 & 1 \\ 0 & 3 & 1 \\ 0 & 0 & 0 \end{bmatrix} \mathbf{u}_1 = \mathbf{0}$$

One eigenpair and a corresponding solution is

$$\left(-1, \begin{bmatrix} 1 \\ 1 \\ -3 \end{bmatrix}\right), \quad \mathbf{x}_1(t) = \begin{bmatrix} 1 \\ 1 \\ -3 \end{bmatrix} e^{-t}$$

Next we consider the repeated eigenvalue $\lambda_2 = \lambda_3 = 2$. The equation for the corresponding eigenvectors is

$$\begin{bmatrix} 0 & 0 & 1 \\ 0 & 0 & 1 \\ 0 & 0 & -3 \end{bmatrix} \mathbf{u} = \mathbf{0}$$

Each of the three scalar equations demands that $u_3 = 0$ leaving u_1 and u_2 arbitrary. A convenient way to choose u_1 and u_2 so that we can obtain two independent eigenvectors is $u_1 = 1$ and $u_2 = 0$ for the first eigenvector. Then an eigenpair and solution are

$$\left(2, \begin{bmatrix} 1 \\ 0 \\ 0 \end{bmatrix}\right), \quad \mathbf{x}_2(t) = \begin{bmatrix} 1 \\ 0 \\ 0 \end{bmatrix} e^{2t}$$

A second solution may be obtained by setting $u_1 = 0$ and $u_2 = 1$. Then an eigenpair and a corresponding solution are

$$\left(2, \begin{bmatrix} 0 \\ 1 \\ 0 \end{bmatrix}\right), \quad \mathbf{x}_3(t) = \begin{bmatrix} 0 \\ 1 \\ 0 \end{bmatrix} e^{2t}$$

We combine $\mathbf{x}_1(t)$ with $\mathbf{x}_2(t)$ and $\mathbf{x}_3(t)$ to form a family of solutions with three arbitrary constants,

$$\mathbf{x}(t) = c_1 \begin{bmatrix} 1 \\ 1 \\ -3 \end{bmatrix} e^{-t} + c_2 \begin{bmatrix} 1 \\ 0 \\ 0 \end{bmatrix} e^{2t} + c_3 \begin{bmatrix} 0 \\ 1 \\ 0 \end{bmatrix} e^{2t}$$

Next, we solve a nonhomogeneous system to determine c_1, c_2, c_3:

$$\mathbf{x}(0) = c_1 \begin{bmatrix} 1 \\ 1 \\ -3 \end{bmatrix} + c_2 \begin{bmatrix} 1 \\ 0 \\ 0 \end{bmatrix} + c_3 \begin{bmatrix} 0 \\ 1 \\ 0 \end{bmatrix} = \begin{bmatrix} 1 & 1 & 0 \\ 1 & 0 & 1 \\ -3 & 0 & 0 \end{bmatrix} \mathbf{c} = \begin{bmatrix} 1 \\ 1 \\ 1 \end{bmatrix}$$

The augmented matrix of this system reduces to

$$\left[\begin{array}{ccc|c} 1 & 1 & 0 & 1 \\ 1 & 0 & 1 & 1 \\ -3 & 0 & 0 & 1 \end{array}\right] \to \cdots \to \left[\begin{array}{ccc|c} 1 & 0 & 0 & -1/3 \\ 0 & 1 & 0 & 4/3 \\ 0 & 0 & 1 & 4/3 \end{array}\right]$$

from which we determine that $c_1 = -1/3, c_2 = c_3 = 4/3$. Therefore, the solution is given by

$$\begin{aligned} \mathbf{x}(t) &= -\frac{1}{3} \begin{bmatrix} 1 \\ 1 \\ -3 \end{bmatrix} e^{-t} + \frac{4}{3} \begin{bmatrix} 1 \\ 0 \\ 0 \end{bmatrix} e^{2t} + \frac{4}{3} \begin{bmatrix} 0 \\ 1 \\ 0 \end{bmatrix} e^{2t} \\ &= -\frac{1}{3} \begin{bmatrix} 1 \\ 1 \\ -3 \end{bmatrix} e^{-t} + \frac{4}{3} \begin{bmatrix} 1 \\ 1 \\ 0 \end{bmatrix} e^{2t} \end{aligned}$$

Contrast the solution of Example 2.3.9 with the closely related problem illustrated in the next example. Note particularly that the characteristic equations are identical in these two examples.

Example 2.3.10. *Find α, β, γ such that the following initial-value problem has a solution of the form (2.3.10):*

$$\mathbf{x}' = \begin{bmatrix} 2 & 1 & 0 \\ 0 & 2 & 0 \\ 0 & 0 & -1 \end{bmatrix} \mathbf{x}, \quad \mathbf{x}(0) = \begin{bmatrix} \alpha \\ \beta \\ \gamma \end{bmatrix}$$

Solution: The coefficient matrix is the matrix in Example 2.3.3. From the work in that example, we know that

$$\mathbf{x}(t) = c_1 \begin{bmatrix} 0 \\ 0 \\ 1 \end{bmatrix} e^{-t} + c_2 \begin{bmatrix} 1 \\ 0 \\ 0 \end{bmatrix} e^{2t}$$

The constants c_1 and c_2 are found, if possible, so that

$$\mathbf{x}(0) = c_1 \begin{bmatrix} 0 \\ 0 \\ 1 \end{bmatrix} + c_2 \begin{bmatrix} 1 \\ 0 \\ 0 \end{bmatrix} = \begin{bmatrix} a \\ \beta \\ \gamma \end{bmatrix}$$

This system is consistent if and only if $\beta = 0$ in which case $c_1 = \gamma$ and $c_2 = \alpha$. The relevant solution is then

$$\mathbf{x}(t) = \gamma \begin{bmatrix} 0 \\ 0 \\ 1 \end{bmatrix} e^{-t} + \alpha \begin{bmatrix} 1 \\ 0 \\ 0 \end{bmatrix} e^{2t}$$

In this last example we have seen that $\beta \neq 0$ leads to an initial-value problem whose solution is not a combination of functions of the form $\mathbf{u}e^{\lambda t}$. This failure is due to the fact that the coefficient matrix has *only* two linearly independent eigenvectors. In general, if $\mathbf{A}$ has $k < n$ linearly independent eigenvectors, the system $\mathbf{Ac} = \mathbf{x}(0) = \mathbf{b}$ for the determination of $\mathbf{c}$ has n equations in k unknowns, and such systems may be inconsistent (as in the previous example when $\beta \neq 0$).

Theorem 2.3.4 is an example of a *superposition theorem*; a theorem in which solutions are combined in a way that generates new solutions.

Theorem 2.3.4. *Suppose* $\mathbf{x}_1(t)$ *and* $\mathbf{x}_2(t)$ *are solutions of the initial-value problems*

$$\mathbf{x}' = \mathbf{Ax}, \quad \mathbf{x}(0) = \mathbf{b}_1$$

and

$$\mathbf{x}' = \mathbf{Ax}, \quad \mathbf{x}(0) = \mathbf{b}_2$$

respectively. Then

$$\alpha\mathbf{x}_1(t) + \beta\mathbf{x}_2(t)$$

is the solution of the initial-value problem

$$\mathbf{x}' = \mathbf{Ax}, \quad \mathbf{x}(0) = \alpha\mathbf{b}_1 + \beta\mathbf{b}_2$$

Proof: Set $\mathbf{z}(t) = \alpha \mathbf{x}_1(t) + \beta \mathbf{x}_2(t)$. We shall first prove that $\mathbf{z}(t)$ is a solution of $\mathbf{x}' = \mathbf{A}\mathbf{x}$. We have

$$\mathbf{z}' = \alpha \mathbf{x}_1'(t) + \beta \mathbf{x}_2'(t)$$

and, in view of the fact that $\mathbf{A}\mathbf{x}_1(t) = \mathbf{x}_1'(t)$ and $\mathbf{A}\mathbf{x}_2(t) = \mathbf{x}_2'(t)$,

$$\mathbf{A}\mathbf{z}(t) = \alpha \mathbf{A}\mathbf{x}_1(t) + \beta \mathbf{A}\mathbf{x}_2(t) = \alpha \mathbf{x}_1'(t) + \beta \mathbf{x}_2'(t)$$

We also have

$$\mathbf{z}(0) = \alpha \mathbf{x}_1(0) + \beta \mathbf{x}_2(0) = \alpha \mathbf{b}_1 + \beta \mathbf{b}_2$$

a consequence of the hypothesis that $\mathbf{x}_1(0) = \mathbf{b}_1$ and $\mathbf{x}_2(0) = \mathbf{b}_2$. Note that the coefficient matrix $\mathbf{A}$ is the same in all three initial-value problems but that $\mathbf{b}_1$, $\mathbf{b}_2$, α and β are arbitrary □

•EXERCISES

In Problems 1–8, use Corollary 2.3.1 (and, where appropriate, Theorem 2.3.4) to find the solution of the initial-value problem $\mathbf{x}' = \mathbf{A}\mathbf{x}$, $\mathbf{x}(0) = \mathbf{b}$ for each of the following $\mathbf{b}$:

(a) $\begin{bmatrix} 1 \\ 1 \end{bmatrix}$ (b) $\begin{bmatrix} -1 \\ 1 \end{bmatrix}$ (c) $\begin{bmatrix} 0 \\ 1 \end{bmatrix}$

(d) $\begin{bmatrix} 1 \\ 0 \end{bmatrix}$ (e) $\begin{bmatrix} 2 \\ -1 \end{bmatrix}$ (f) $\begin{bmatrix} -1 \\ -1 \end{bmatrix}$

1. $\mathbf{A} = \begin{bmatrix} 2 & -2 \\ -1 & 1 \end{bmatrix}$.

2. $\mathbf{A} = \begin{bmatrix} 2 & 2 \\ 1 & 1 \end{bmatrix}$.

3. $\mathbf{A} = \begin{bmatrix} 1 & 1 \\ 1 & 1 \end{bmatrix}$.

4. $\mathbf{A} = \begin{bmatrix} 1 & 3 \\ 1 & -1 \end{bmatrix}$.

5. $\mathbf{A} = \begin{bmatrix} 2 & 1 \\ 2 & 3 \end{bmatrix}$.

6. $\mathbf{A} = \begin{bmatrix} 0 & 1 \\ 1 & 1 \end{bmatrix}$.

7. $\mathbf{A} = \begin{bmatrix} 1 & -1 \\ 0 & 1 \end{bmatrix}$.

8. $\mathbf{A} = \begin{bmatrix} 0 & 1 \\ 1 & 0 \end{bmatrix}$.

In Problems 9–12, use these initial conditions for each matrix:

(a) $\begin{bmatrix} 1 \\ 0 \\ 0 \end{bmatrix}$ (b) $\begin{bmatrix} 0 \\ 1 \\ 0 \end{bmatrix}$ (c) $\begin{bmatrix} 0 \\ 0 \\ 1 \end{bmatrix}$

(d) $\begin{bmatrix} 1 \\ 1 \\ 0 \end{bmatrix}$ (e) $\begin{bmatrix} 1 \\ 1 \\ 1 \end{bmatrix}$ (f) $\begin{bmatrix} 1 \\ -1 \\ 1 \end{bmatrix}$

9. $\mathbf{A} = \begin{bmatrix} 7 & 2 & -3 \\ 4 & 6 & -4 \\ 5 & 2 & -1 \end{bmatrix}$.

10. $\mathbf{A} = \begin{bmatrix} 1 & 0 & 0 \\ 0 & 0 & 2 \\ 0 & -2 & 0 \end{bmatrix}$.

11. $\mathbf{A} = \begin{bmatrix} 1 & 0 & 0 \\ 0 & 0 & 2 \\ 0 & 2 & 0 \end{bmatrix}$.

12. $\mathbf{A} = \begin{bmatrix} 4 & -3 & -2 \\ 2 & -1 & -2 \\ 3 & -3 & -1 \end{bmatrix}$.

13. Formulate an analog of Corollary 2.3.1 if the initial condition is given as $\mathbf{x}(t_0) = \mathbf{b}$. (*Hint*: Let $\tau = t - t_0$.)

14. Show by substitution that $\operatorname{Re}\mathbf{x}(t)$ and $\operatorname{Im}\mathbf{x}(t)$, as given by (2.3.15) and (2.3.16), are solutions of $\mathbf{x}' = \mathbf{A}\mathbf{x}$.

15. Verify that

$$\mathbf{x}(t) = \begin{bmatrix} \alpha + (\alpha + \beta)\, t \\ \beta - (\alpha + \beta)\, t \end{bmatrix}$$

is a solution of Example 2.3.5 for every choice of scalars α and β.

16. Verify that

$$\mathbf{x}(t) = \begin{bmatrix} \alpha \cos t + \beta \sin t \\ -\alpha \sin t + \beta \cos t \end{bmatrix}$$

is a solution of Example 2.3.6 for every choice of scalars α and β.

17. Verify that

$$\mathbf{x}(t) = \begin{bmatrix} \cos t + \sin t \\ \cos t - \sin t \end{bmatrix}$$

is a solution of Example 2.3.7.

18. Verify that

$$\mathbf{x}(t) = -\tfrac{1}{3}\begin{bmatrix} 1 \\ 1 \\ -3 \end{bmatrix} e^{-t} + \tfrac{4}{3}\begin{bmatrix} 1 \\ 1 \\ 0 \end{bmatrix} e^{2t}$$

is a solution of Example 2.3.9.

19. Redo Example 2.3.6 using the eigenvalue $\lambda_2 = 1 - i$.

20. The proof of Theorem 2.3.3 requires that $\mathbf{A}$ be real. Why?

21. Find vectors $\mathbf{x_1}$ and $\mathbf{x_2}$ so that

$$\mathbf{x}_3(t) = (\mathbf{x_1} + \mathbf{x_2} t)\, e^{2t}$$

is a solution of

$$\mathbf{x}' = \begin{bmatrix} 2 & 1 & 0 \\ 0 & 2 & 0 \\ 0 & 0 & -1 \end{bmatrix} \mathbf{x}$$

22. Use Problem 21 to solve the initial-value problem Example 2.3.10.

23. Why does Theorem 2.3.3 yields only one solution if λ is real?

MATLAB

24. Use (2.3.15) and (2.3.16) to solve Problems 1–12.

2.4 Solution and Fundamental Solution Matrices

If $\Phi(t)$ is an $n \times n$ matrix whose columns are solutions of $\mathbf{x}' = \mathbf{A}\mathbf{x}$, then we call $\Phi(t)$ a *solution matrix* of this system. Solution matrices play a major role in the theory of systems of ODE's. In Section 2.6, for example, we show how to use solution matrices to find solutions of nonhomogeneous systems.

Example 2.4.1. *Find a solution matrix for the systems in*
(a) *Example 2.3.1.* (b) *Example 2.3.2.*
(c) *Example 2.3.6.* (d) *Example 2.3.9.*

Solution: (a) In Example 2.3.1 we found two eigenpairs,

$$\left(-1, \begin{bmatrix} 1 \\ -2 \end{bmatrix}\right) \quad \text{and} \quad \left(3, \begin{bmatrix} 1 \\ 2 \end{bmatrix}\right)$$

By Theorem 2.3.1, these lead to the linearly independent solutions

$$\mathbf{x}_1(t) = e^{-t}\begin{bmatrix} 1 \\ -2 \end{bmatrix}, \quad \mathbf{x}_2(t) = e^{3t}\begin{bmatrix} 1 \\ 2 \end{bmatrix}$$

Therefore, a solution matrix is

$$\Phi(t) = [\, \mathbf{x}_1(t) \;\; \mathbf{x}_2(t) \,] = \begin{bmatrix} e^{-t} & e^{3t} \\ -2e^{-t} & 2e^{3t} \end{bmatrix}$$

(b) The eigenpairs listed in Example 2.3.2 are

$$\left(1, \begin{bmatrix} 1 \\ 1 \\ 1 \end{bmatrix}\right), \quad \left(-2, \begin{bmatrix} 1 \\ -2 \\ 4 \end{bmatrix}\right), \quad \left(-1, \begin{bmatrix} 1 \\ -1 \\ 1 \end{bmatrix}\right)$$

The corresponding linearly independent solutions are

$$\mathbf{x}_1(t) = e^{t}\begin{bmatrix} 1 \\ 1 \\ 1 \end{bmatrix}, \quad \mathbf{x}_2(t) = e^{-2t}\begin{bmatrix} 1 \\ -2 \\ 4 \end{bmatrix}, \quad \mathbf{x}_3(t) = e^{-t}\begin{bmatrix} 1 \\ -1 \\ 1 \end{bmatrix}$$

Thus, a solution matrix is

$$\Phi(t) = \begin{bmatrix} e^{t} & e^{-2t} & e^{-t} \\ e^{t} & -2e^{-2t} & -e^{-t} \\ e^{t} & 4e^{-2t} & e^{-t} \end{bmatrix}$$

(c) In Example 2.3.6, we computed the two solutions

$$\mathbf{x}_1(t) = e^t \begin{bmatrix} \cos t \\ -\sin t \end{bmatrix}, \quad e^t \begin{bmatrix} \sin t \\ \cos t \end{bmatrix}$$

So a solution matrix is

$$\Phi(t) = \begin{bmatrix} e^t \cos t & e^t \sin t \\ -e^t \sin t & e^t \cos t \end{bmatrix}$$

(d) For the system in Example 2.3.9, a solution matrix is

$$\Phi(t) = \begin{bmatrix} e^{-t} & e^{2t} & 0 \\ e^{-t} & 0 & e^{2t} \\ -3e^{-t} & 0 & 0 \end{bmatrix}$$

Because the columns of solution matrices are solutions of $\mathbf{x}' = \mathbf{A}\mathbf{x}$, it is not surprising that solution matrices are solutions of a related matrix differential system. In fact, (2.4.1) in the following theorem could have been used to define solution matrices.

Theorem 2.4.1. *$\Phi(t)$ is a solution matrix of the system $\mathbf{x}' = \mathbf{A}\mathbf{x}$ if and only if*

$$\Phi'(t) = \mathbf{A}\Phi(t) \tag{2.4.1}$$

Proof: Let $\Phi(t) = \begin{bmatrix} \mathbf{x}_1(t) & \mathbf{x}_2(t) & \dots & \mathbf{x}_n(t) \end{bmatrix}$. Assume $\Phi(t)$ is a solution matrix. The following equations hold because the columns of $\Phi(t)$ are solutions of $\mathbf{x}' = \mathbf{A}\mathbf{x}$.

$$\begin{aligned} \Phi'(t) &= \begin{bmatrix} \mathbf{x}_1'(t) & \mathbf{x}_2'(t) & \dots & \mathbf{x}_n'(t) \end{bmatrix} \\ &= \begin{bmatrix} \mathbf{A}\mathbf{x}_1(t) & \mathbf{A}\mathbf{x}_2(t) & \dots & \mathbf{A}\mathbf{x}_n(t) \end{bmatrix} \\ &= \mathbf{A}\begin{bmatrix} \mathbf{x}_1(t) & \mathbf{x}_2(t) & \dots & \mathbf{x}_n(t) \end{bmatrix} = \mathbf{A}\Phi(t) \end{aligned}$$

The converse is proved by observing that the argument goes both ways. □

Corollary 2.4.1. *If $\Phi(t)$ is a solution matrix of $\mathbf{x}' = \mathbf{A}\mathbf{x}$, then for every square, constant matrix $\mathbf{C}$, $\Phi(t)\mathbf{C}$ is a solution matrix for the same system.*

Proof: Since $\Phi(t)$ is a solution matrix, $\Phi'(t) = \mathbf{A}\Phi(t)$, and it follows that $(\Phi(t)\mathbf{C})' = \Phi'(t)\mathbf{C} = (\mathbf{A}\Phi(t))\,\mathbf{C} = \mathbf{A}\,(\Phi(t)\mathbf{C})$. □

Corollary 2.4.2. *If $\Phi(t)$ is a solution matrix of $\mathbf{x}' = \mathbf{A}\mathbf{x}$, then for every constant vector $\mathbf{c}$, $\Phi(t)\mathbf{c}$ is a solution of the same system.*

Proof: The argument is the same as the one given for Corollary 2.4.1. We check that $\Phi(t)\mathbf{c}$ is a solution of $\mathbf{x}' = \mathbf{A}\mathbf{x}$. Set $\mathbf{x} = \Phi\mathbf{c}$. Then, we have

$$\begin{aligned}\mathbf{x}' &= (\Phi(t)\mathbf{c})' \\ &= \Phi'(t)\mathbf{c} = \mathbf{A}\,(\Phi(t)\mathbf{c})) \\ &= \mathbf{A}\mathbf{x}\end{aligned}$$

□

Corollary 2.4.2 asserts that for each vector of constants $\mathbf{c}$, the vector $\mathbf{x}(t) = \Phi(t)\mathbf{c}$ is a solution of $\mathbf{x}' = \mathbf{A}\mathbf{x}$. Under what circumstances is it possible to find $\mathbf{c}$ so that $\mathbf{x}(t) = \Phi(t)\mathbf{c}$ satisfies $\mathbf{x}(t_0) = \mathbf{b}$, that is, so that the initial condition can be satisfied? Since $\mathbf{x}(t) = \Phi(t)\mathbf{c}$ is a solution, the initial condition implies that

$$\mathbf{x}(t_0) = \Phi(t_0)\mathbf{c} = \mathbf{b} \tag{2.4.2}$$

We will succeed if we can solve (2.4.2). But we have no way of knowing if such a system has a solution, because there is no apriori reason for believing that (2.4.2) is consistent. In fact, the definition of a solution matrix does not even exclude the extreme (and useless) case $\Phi(t) = \mathbf{O}$. To remedy this, we add an extra requirement on solution matrices that will guarantee (as we shall see) that $\Phi(t)$ will be invertible for all t.

Let us call a solution matrix $\Phi(t)$ a *fundamental solution matrix* if $\Phi(t_0) = \mathbf{I}$ for some t_0. Of the four solution matrices constructed in Example 2.4.1 only (c) is a fundamental solution matrix. The other solution matrices are not fundamental. Unless there is a need to the contrary we shall assume that $t_0 = 0$ and denote fundamental solution matrices by $\Psi(t)$. In any case, it is important to keep clearly in mind that fundamental solution matrices are, first and foremost, solution matrices. The requirement that $\Psi(0) = \mathbf{I}$ is simply a way of normalizing solution matrices.

Every invertible solution matrix $\Phi(t)$ can be converted into a fundamental solution matrix by the following device: We simply form $\Psi(t) = \Phi(t)\Phi^{-1}(0)$. We state and prove this in the next corollary.

Corollary 2.4.3. *If $\Phi(t)$ is a solution matrix of $\mathbf{x}' = \mathbf{A}\mathbf{x}$ and $\Phi(0)$ is invertible, then*

$$\Psi(t) = \Phi(t)\Phi^{-1}(0) \tag{2.4.3}$$

is a fundamental solution matrix of this system and

$$\mathbf{x}(t) = \Psi(t)\mathbf{b} \tag{2.4.4}$$

solves the initial-value problem

$$\mathbf{x}' = \mathbf{A}\mathbf{x}, \quad \mathbf{x}(0) = \mathbf{b} \tag{2.4.5}$$

Proof: Consider the general initial-value problem (2.4.5), where $\mathbf{b}$ is an arbitrary initial vector. By Corollary 2.4.1 $\Phi(t)\Phi^{-1}(0)$ is a solution matrix. Clearly, $\Psi(0) = \Phi(0)\Phi^{-1}(0) = \mathbf{I}$. Finally, by Corollary 2.4.2, $\mathbf{x}(t) = \Psi(t)\mathbf{b}$ is a solution of (2.4.4), and $\mathbf{x}(0) = \Phi(0)\Phi^{-1}(0)\mathbf{b} = \mathbf{b}$. □

The next theorem shows how to construct fundamental solution matrices for those systems whose coefficient matrices do have n linearly independent eigenvectors.

Theorem 2.4.2. *Suppose, $(\lambda_1, \mathbf{u}_1), (\lambda_2, \mathbf{u}_2), \ldots, (\lambda_n, \mathbf{u}_n)$ are eigenpairs of $\mathbf{A}$ and that $\{\mathbf{u}_1, \mathbf{u}_2, \ldots, \mathbf{u}_n\}$ is linearly independent. Set*

$$\mathbf{x}_i = \mathbf{x}_i(t) = \mathbf{u}_i \exp(\lambda_i t) \tag{2.4.6}$$

Then $\Phi(t) = [\,\mathbf{x}_1\ \mathbf{x}_2\ \ldots\ \mathbf{x}_n\,]$ is an invertible solution matrix of $\mathbf{x}' = \mathbf{A}\mathbf{x}$ at $t = 0$, and

$$\Psi(t) = \Phi(t)\Phi^{-1}(0)$$

is a fundamental solution matrix.

Proof: It should be clear that $\Phi(t)$ is a solution matrix and that

$$\Phi(0) = [\,\mathbf{u}_1\ \mathbf{u}_2\ \ldots\ \mathbf{u}_n\,]$$

is a matrix whose columns are n linearly independent eigenvectors of $\mathbf{A}$. But then $\Phi(0)$ is invertible. Now, $\Psi(t) = \Phi(t)\Phi^{-1}(0)$ is a fundamental solution matrix by Corollary 2.4.3 of Theorem 2.4.1. □

Note that this theorem does not assert that these matrices have real entries. In fact, $\Phi(t)$ will not have real entries when even one of the eigenvalues is not real. However, it is possible to show that $\Psi(t)$ has only real entries. (See Corollary 2.5.6.)

There are a variety of matrices for which the hypothesis of Theorem 2.4.2 holds. We state without proof that symmetric matrices and $n \times n$ matrices with n distinct eigenvalues satisfy these conditions. It is important to note that matrices with repeated eigenvalues may or may not have n linearly independent eigenvectors. Each case must be treated separately. Review Examples 2.3.9 and 2.3.10.

The existence of a fundamental solution matrix amounts to being able to solve every initial-value problem whose initial conditions are given at $t = 0$. An important question is this: How do we find fundamental solution matrices when $\mathbf{A}$ does not possess n linearly independent eigenvectors? We show in Chapter 5 that the Laplace transform can be used to find fundamental solution matrices even when $\mathbf{A}$ has fewer than n linearly independent eigenvectors.

Example 2.4.2. *Find a solution to the following system by using* (2.4.3):

$$\mathbf{x}' = \begin{bmatrix} 1 & 1 \\ 4 & 1 \end{bmatrix} \mathbf{x}, \quad \mathbf{x}(0) = \begin{bmatrix} 1 \\ 0 \end{bmatrix}$$

Solution: We found a solution matrix for this system in Example 2.4.1(a):

$$\Phi(t) = \begin{bmatrix} e^{-t} & e^{3t} \\ -2e^{-t} & 2e^{3t} \end{bmatrix}$$

We can readily compute

$$\Phi(0) = \begin{bmatrix} 1 & 1 \\ -2 & 2 \end{bmatrix}, \qquad \Phi^{-1}(0) = \frac{1}{4}\begin{bmatrix} 2 & -1 \\ 2 & 1 \end{bmatrix}$$

The fundamental solution matrix $\Psi(t) = \Phi(t)\Phi^{-1}(0)$ is given by

$$\Psi(t) = \frac{1}{4}\begin{bmatrix} e^{-t} & e^{3t} \\ -2e^{-t} & 2e^{3t} \end{bmatrix}\begin{bmatrix} 2 & -1 \\ 2 & 1 \end{bmatrix}$$

So

$$\mathbf{x}(t) = \frac{1}{4}\begin{bmatrix} e^{-t} & e^{3t} \\ -2e^{-t} & 2e^{3t} \end{bmatrix}\begin{bmatrix} 2 & -1 \\ 2 & 1 \end{bmatrix}\begin{bmatrix} 1 \\ 0 \end{bmatrix} = \frac{1}{4}\begin{bmatrix} e^{-t} & e^{3t} \\ -2e^{-t} & 2e^{3t} \end{bmatrix}\begin{bmatrix} 2 \\ 2 \end{bmatrix}$$
$$= \frac{1}{2}\begin{bmatrix} e^{-t} + e^{3t} \\ -2e^{-t} + 2e^{3t} \end{bmatrix}$$

By using $e^{-t}+e^{3t} = e^t\left(e^{-2t} + e^{2t}\right) = 2e^t \cosh 2t$ and $-e^{-t}+e^{3t} = 2e^t \sinh 2t$, we can put the solution in the more compact form

$$\Psi(t) = e^t\begin{bmatrix} \cosh 2t \\ 2\sinh 2t \end{bmatrix}$$

Example 2.4.3. *Use* (2.4.3) *to find the real solution of*

$$\mathbf{x}' = \begin{bmatrix} 0 & -1 \\ 1 & 0 \end{bmatrix}\mathbf{x}, \quad \mathbf{x}(0) = \begin{bmatrix} 1 \\ 1 \end{bmatrix}$$

Solution: The eigenpairs are

$$\left(i, \begin{bmatrix} i \\ 1 \end{bmatrix}\right) \quad \text{and} \quad \left(-i, \begin{bmatrix} -i \\ 1 \end{bmatrix}\right)$$

In terms of Theorem 2.4.2, the solution matrix is given by

$$\Phi(t) = \begin{bmatrix} ie^{it} & -ie^{-it} \\ e^{it} & e^{-it} \end{bmatrix}$$

Taking real (or imaginary) parts of $\Phi(t)$ does not provide a suitable real solution matrix because the resulting matrix is singular for all t. (See Problem 28.) However, (2.4.3) provides a real fundamental solution matrix. We have

$$\Phi^{-1}(0) = \begin{bmatrix} i & -i \\ 1 & 1 \end{bmatrix}^{-1} = \frac{1}{2}\begin{bmatrix} -i & 1 \\ i & 1 \end{bmatrix}$$

so that

$$\Psi(t) = \Phi(t)\Phi^{-1}(0) = \frac{1}{2}\begin{bmatrix} ie^{it} & -ie^{-it} \\ e^{it} & e^{-it} \end{bmatrix}\begin{bmatrix} -i & 1 \\ i & 1 \end{bmatrix}$$
$$= \frac{1}{2}\begin{bmatrix} e^{it} + e^{-it} & i\left(e^{it} - e^{-it}\right) \\ i\left(-e^{it} + e^{-it}\right) & e^{it} + e^{-it} \end{bmatrix}$$
$$= \begin{bmatrix} \cos t & -\sin t \\ \sin t & \cos t \end{bmatrix}$$

which is nonsingular and a fundamental solution matrix! (It is nonsingular because $\det \Psi \neq 0$.) Finally,

$$\mathbf{x}(t) = \begin{bmatrix} \cos t & -\sin t \\ \sin t & \cos t \end{bmatrix} \begin{bmatrix} 1 \\ 1 \end{bmatrix} = \begin{bmatrix} \cos t - \sin t \\ \sin t + \cos t \end{bmatrix}$$

Example 2.4.4. *Show that* $\mathbf{x}' = \begin{bmatrix} 1 & 1 \\ -1 & -1 \end{bmatrix} \mathbf{x}$ *has the fundamental solution matrix*

$$\Psi(t) = \begin{bmatrix} t+1 & t \\ -t & 1-t \end{bmatrix}$$

Solution: (See Example 2.3.5 and the discussion that follows it.) For $\Psi(t)$ to be a fundamental solution matrix, then according to Theorem 2.4.1, it must satisfy

$$\Psi'(t) = \begin{bmatrix} 1 & 1 \\ -1 & -1 \end{bmatrix} \Psi(t)$$

But differentiating the given solution matrix gives

$$\Psi'(t) = \begin{bmatrix} 1 & 1 \\ -1 & -1 \end{bmatrix}$$

and

$$\mathbf{A}\Psi(t) = \begin{bmatrix} 1 & 1 \\ -1 & -1 \end{bmatrix} \begin{bmatrix} t+1 & t \\ -t & 1-t \end{bmatrix} = \begin{bmatrix} 1 & 1 \\ -1 & -1 \end{bmatrix}$$

Hence, $\Psi' = \mathbf{A}\Psi$. The observation $\Psi(0) = \mathbf{I}$ completes the argument.

Example 2.4.5. *Use* (2.4.3) *to find the solution of*

$$\mathbf{x}' = \begin{bmatrix} 2 & 0 & 1 \\ 0 & 2 & 1 \\ 0 & 0 & -1 \end{bmatrix} \mathbf{x}, \quad \mathbf{x}(0) = \begin{bmatrix} 1 \\ -1 \\ 0 \end{bmatrix}$$

Solution: In Example 2.4.1(d) we found the solution matrix

$$\Phi(t) = \begin{bmatrix} e^{-t} & e^{2t} & 0 \\ e^{-t} & 0 & e^{2t} \\ -3e^{-t} & 0 & 0 \end{bmatrix}$$

Since

$$\Phi^{-1}(0) = \frac{1}{3}\begin{bmatrix} 0 & 0 & -1 \\ 3 & 0 & 1 \\ 0 & 3 & 1 \end{bmatrix}$$

we have

$$\begin{aligned}\Psi(t) &= \frac{1}{3}\begin{bmatrix} e^{-t} & e^{2t} & 0 \\ e^{-t} & 0 & e^{2t} \\ -3e^{-t} & 0 & 0 \end{bmatrix}\begin{bmatrix} 0 & 0 & -1 \\ 3 & 0 & 1 \\ 0 & 3 & 1 \end{bmatrix} \\ &= \frac{1}{3}\begin{bmatrix} 3e^{2t} & 0 & e^{2t}-e^{-t} \\ 0 & 3e^{2t} & e^{2t}-e^{-t} \\ 0 & 0 & 3e^{-t} \end{bmatrix}\end{aligned}$$

and hence (2.4.4) implies that

$$\mathbf{x}(t) = \frac{1}{3}\begin{bmatrix} 3e^{2t} & 0 & e^{2t}-e^{-t} \\ 0 & 3e^{2t} & e^{2t}-e^{-t} \\ 0 & 0 & 3e^{-t} \end{bmatrix}\begin{bmatrix} 1 \\ -1 \\ 0 \end{bmatrix} = \begin{bmatrix} 1 \\ -1 \\ 0 \end{bmatrix} e^{2t}$$

•EXERCISES

Find a fundamental solution matrix of $\mathbf{x}' = \mathbf{A}\mathbf{x}$ given the following coefficient matrices.

1. $\mathbf{A} = \begin{bmatrix} 1 & 3 \\ 1 & -1 \end{bmatrix}$.
2. $\mathbf{A} = \begin{bmatrix} 1 & 1 \\ 3 & -1 \end{bmatrix}$.
3. $\mathbf{A} = \begin{bmatrix} 2 & 1 \\ 2 & 3 \end{bmatrix}$.
4. $\mathbf{A} = \begin{bmatrix} 1 & 1 \\ -1 & 1 \end{bmatrix}$.
5. $\mathbf{A} = \begin{bmatrix} 0 & 2 \\ 1 & -1 \end{bmatrix}$.
6. $\mathbf{A} = \begin{bmatrix} 2 & 0 \\ 1 & -1 \end{bmatrix}$.
7. $\mathbf{A} = \begin{bmatrix} 1 & 1 \\ -1 & 1 \end{bmatrix}$.
8. $\mathbf{A} = \begin{bmatrix} 2 & 1 \\ 3 & 0 \end{bmatrix}$.
9. $\mathbf{A} = \begin{bmatrix} 2 & -1 \\ 0 & 1 \end{bmatrix}$.
10. $\mathbf{A} = \begin{bmatrix} 1 & 2 \\ 1 & 0 \end{bmatrix}$.
11. $\mathbf{A} = \begin{bmatrix} 2 & 1 \\ 3 & 4 \end{bmatrix}$.
12. $\mathbf{A} = \begin{bmatrix} 1 & \alpha^2 - 1 \\ 1 & -1 \end{bmatrix}$.
13. $\mathbf{A} = \begin{bmatrix} 4 & -3 & -2 \\ 2 & -1 & -2 \\ 3 & -3 & -1 \end{bmatrix}$.
14. $\mathbf{A} = \begin{bmatrix} 1 & 1 & 1 \\ 0 & 2 & 1 \\ 0 & 0 & 0 \end{bmatrix}$.

15. $\mathbf{A} = \begin{bmatrix} 1 & 1 & 1 \\ 1 & 1 & 1 \\ 0 & 0 & 0 \end{bmatrix}$.

16. $\mathbf{A} = \begin{bmatrix} 1 & -1 & -1 \\ 0 & 0 & 1 \\ 0 & -2 & -3 \end{bmatrix}$.

Solve these initial-value problems by using fundamental solution matrices.

17. Problem 1, $\mathbf{b} = \begin{bmatrix} 1 \\ 0 \end{bmatrix}$.

18. Problem 1, $\mathbf{b} = \begin{bmatrix} 0 \\ 1 \end{bmatrix}$.

19. Problem 1, $\mathbf{b} = \begin{bmatrix} 1 \\ 1 \end{bmatrix}$.

20. Problem 1, $\mathbf{b} = \begin{bmatrix} 0 \\ 0 \end{bmatrix}$.

21. Problem 16, $\mathbf{b} = \begin{bmatrix} 0 \\ 1 \\ 0 \end{bmatrix}$.

22. Problem 16, $\mathbf{b} = \begin{bmatrix} 1 \\ 0 \\ 0 \end{bmatrix}$.

23. Problem 16, $\mathbf{b} = \begin{bmatrix} 0 \\ 0 \\ 1 \end{bmatrix}$.

24. Problem 16, $\mathbf{b} = \begin{bmatrix} -1 \\ 2 \\ -2 \end{bmatrix}$.

25. Find a connection between the answer to Problem 24 and those to Problems 21, 22, and 23.

26. Show that $\Psi = e^t\mathbf{I}$ is a fundamental solution matrix of $\mathbf{x}' = \mathbf{x}$.

27. Find the inverse of the solution matrix of Example 2.4.3:

$$\Phi(t) = \begin{bmatrix} ie^{it} & -ie^{-it} \\ e^{it} & e^{-it} \end{bmatrix}$$

28. Show that the real and imaginary parts of the solution matrix of Example 2.4.3 are solution matrices but that neither are invertible for any t.

2.5 Some Fundamental Theorems

The point of this section is to exploit the fact that fundamental solution matrices exist, are unique, and are invertible for all t, a result whose (partial) proof we reserve for Section 2.8.

Theorem 2.5.1. *Fundamental solution matrices are nonsingular for all* t.

Proof: The proof of this key theorem is reserved for Section 2.8. We accept the theorem and the fact that all systems have fundamental solution matrices in order not to interrupt the flow of arguments in this section. □

Corollary 2.5.1. *Solution matrices invertible at $t = 0$ are invertible for all t.*

Proof: Suppose $\Phi(t)$ is an invertible solution matrix at $t = 0$. Then by Corollary 2.4.1,

$$\Psi(t) = \Phi(t)\Phi^{-1}(0)$$

is a fundamental solution matrix and is invertible (for all t) by Theorem 2.5.1. But $\Psi(t) = \Phi(t)\Phi^{-1}(0)$ implies that

$$\Phi(t) = \Psi(t)\Phi(0) \tag{2.5.1}$$

so $\Phi(t)$ is the product of invertible matrices and is therefore invertible for each t. □

Theorem 2.5.2. *If $\mathbf{z}(t)$ is any solution of $\mathbf{x}' = \mathbf{A}\mathbf{x}$ and $\Psi(t)$ is a fundamental solution matrix of this system, then*

$$\mathbf{z}(t) = \Psi(t)\mathbf{z}(0) \tag{2.5.2}$$

Proof: By Theorem 2.5.1, fundamental solution matrices are nonsingular for all t. This enables us to define $\mathbf{v}(t)$ by

$$\mathbf{z}(t) = \Psi(t)\mathbf{v}(t) \tag{2.5.3}$$

Now differentiate this expression to obtain

$$\mathbf{z}'(t) = \Psi'(t)\mathbf{v}(t) + \Psi(t)\mathbf{v}'(t) \tag{2.5.4}$$

Since $\Psi(t)$ is a solution matrix and $\mathbf{z}(t)$ is a solution of $\mathbf{x}' = \mathbf{A}\mathbf{x}$, we have $\Psi'(t) = \mathbf{A}\Psi(t)$ and $\mathbf{z}'(t) = \mathbf{A}\mathbf{z}$. We use these facts and (2.5.3) to simplify (2.5.4):

$$\mathbf{A}\mathbf{z}(t) = \mathbf{A}\Psi(t)\mathbf{v}(t) + \Psi(t)\mathbf{v}'(t) = \mathbf{A}\mathbf{z}(t) + \Psi(t)\mathbf{v}'(t)$$

Hence,

$$\mathbf{0} = \Psi(t)\mathbf{v}'(t) \tag{2.5.5}$$

Since $\Psi(t)$ is invertible, (2.5.5) implies that $\mathbf{v}'(t) = \mathbf{0}$, from which we deduce that $\mathbf{v}(t) = \mathbf{c}$, where $\mathbf{c}$ is a constant vector. So (2.5.3) becomes

$$\mathbf{z}(t) = \Psi(t)\mathbf{c} \tag{2.5.6}$$

To evaluate $\mathbf{c}$, we use $t = 0$ in (2.5.6) and the fact that $\Psi(0) = \mathbf{I}$:

$$\mathbf{z}(0) = \Psi(0)\mathbf{c} = \mathbf{I}\mathbf{c} = \mathbf{c} \tag{2.5.7}$$

Replacing $\mathbf{c}$ by $\mathbf{z}(0)$ in (2.5.6) leads to (2.5.2). □

Corollary 2.5.2. *If $\mathbf{z}(t)$ is any solution of $\mathbf{x}' = \mathbf{A}\mathbf{x}$ and $\Phi(t)$ is an invertible solution matrix of this system, then there exists a $\mathbf{c}$ such that*

$$\mathbf{z}(t) = \Phi(t)\mathbf{c} \tag{2.5.8}$$

Proof: We know that $\Psi(t) = \Phi(t)\Phi^{-1}(0)$ is a fundamental solution matrix, and by Theorem 2.5.2,

$$\mathbf{z}(t) = \Psi(t)\mathbf{z}(0) \tag{2.5.9}$$

Now replace $\Psi(t)$ by $\Phi(t)\Phi^{-1}(0)$ in (2.5.9) to get

$$\mathbf{z}(t) = \Phi(t)\left(\Phi^{-1}(0)\mathbf{z}(0)\right) = \Phi(t)\mathbf{c}$$

So $\mathbf{c} = \Phi^{-1}(0)\mathbf{z}(0)$. □

The point of this corollary is that every solution of $\mathbf{x}' = \mathbf{A}\mathbf{x}$ can be found from the family of solutions $\Phi(t)\mathbf{c}$ by choosing $\mathbf{c}$ appropriately. It is for this reason that we call $\Phi(t)\mathbf{c}$ a *general solution* of $\mathbf{x}' = \mathbf{A}\mathbf{x}$.

Corollary 2.5.3. *The function $\mathbf{x}(t) = \mathbf{0}$ is the unique solution of the initial-value problem*

$$\mathbf{x}' = \mathbf{A}\mathbf{x}, \quad \mathbf{x}(0) = \mathbf{0} \tag{2.5.10}$$

Proof: From Theorem 2.5.2, every solution of (2.5.11) is given by $\mathbf{x}(t) = \Psi(t)\mathbf{x}(0) = \Psi(t)\mathbf{0} = \mathbf{0}$. □

Theorem 2.5.3. *If* $\mathbf{y}(t)$ *and* $\mathbf{z}(t)$ *are solutions of*

$$\mathbf{x}' = \mathbf{A}\mathbf{x}, \quad \mathbf{x}(0) = \mathbf{b} \tag{2.5.11}$$

then $\mathbf{y}(t) = \mathbf{z}(t)$.

Proof: Define $\mathbf{w}(t) = \mathbf{y}(t) - \mathbf{z}(t)$. Then $\mathbf{w}(t)$ is a solution of (2.5.11) and $\mathbf{w}(0) = \mathbf{0}$. Hence, $\mathbf{w}(t) = \mathbf{0}$ follows from Corollary 2.5.3, and $\mathbf{y}(t) = \mathbf{z}(t)$. □

Corollary 2.5.4. *The fundamental solution matrix of* $\mathbf{x}' = \mathbf{A}\mathbf{x}$ *at* $t = 0$ *is unique.*

Proof: Let $\mathbf{e}_i$ be the ith column of $\mathbf{I}$. Then the ith column of a fundamental solution matrix satisfies $\mathbf{x}(t) = \mathbf{A}\mathbf{x}(t), \mathbf{x}(0) = \mathbf{e}_i$. By Theorem 2.5.3, there is only one solution of this initial-value problem. Hence, the columns of the fundamental solution matrix are uniquely determined. □

Do not misconstrue the content of this corollary. It does not assert that there is only one fundamental solution matrix of $\mathbf{x}' = \mathbf{A}\mathbf{x}$. The fundamental solution matrix is a function of t_0 as well as $\mathbf{A}$. (See Problem 7.)

Up to this point we have been considering systems of ODE's in which our objective is to find a vector of solutions. For the next theorem, we look for solutions of the matrix ODE $\mathbf{X}' = \mathbf{A}\mathbf{X}$.

Corollary 2.5.5. *Let* $\Phi(t)$ *be an invertible solution matrix for the system* $\mathbf{x}' = \mathbf{A}\mathbf{x}$. *Then*

$$\mathbf{X}(t) = \Phi(t)\Phi^{-1}(0)\mathbf{B} \tag{2.5.12}$$

is the unique solution of the initial-value problem

$$\mathbf{X}' = \mathbf{A}\mathbf{X}, \quad \mathbf{X}(0) = \mathbf{B} \tag{2.5.13}$$

Proof: The matrix $\mathbf{X}(t)$ is surely a solution of (2.5.13), and it should be equally clear that $\mathbf{X}(0) = \Phi(0)\Phi^{-1}(0)\mathbf{B} = \mathbf{B}$. The uniqueness follows by the same argument used in Corollary 2.5.4 □

Corollary 2.5.6. *Fundamental solution matrices are real for all t.*

Proof: First we recall that $\Psi'(t) = \mathbf{A}\Psi(t), \Psi(0) = \mathbf{I}$. From this, it follows that

$$(\operatorname{Im}\Psi)' = \mathbf{A}\operatorname{Im}\Psi, \quad \operatorname{Im}\Psi(0) = \mathbf{0}$$

because $\mathbf{A}$ is real and $\operatorname{Im}\mathbf{I} = \mathbf{0}$. But $\mathbf{X}(t) = \mathbf{0}$ also solves

$$\mathbf{X}' = \mathbf{AX}, \quad \mathbf{X}(0) = \mathbf{0}$$

By Corollary 2.5.5, $\operatorname{Im}\Psi(t) = \mathbf{0}$ for all t. Hence,

$$\Psi(t) = \operatorname{Re}\Psi(t) + i\operatorname{Im}\Psi(t) = \operatorname{Re}\Psi(t)$$

□

Corollary 2.5.7. *If $\Phi(t)$ is a solution matrix that is real and invertible at $t = 0$, then $\Phi(t)$ is real for all t.*

Proof: This corollary follows from Corollary 2.5.6 by noting that $\Phi(t)\Phi^{-1}(0)$ is a fundamental solution matrix. □

•EXERCISES

Show that each of the following solution matrices are invertible for all t.

1. $\Phi(t) = \begin{bmatrix} e^{-t} & e^{3t} \\ -2e^{-t} & 2e^{3t} \end{bmatrix}$, arising in Example 2.4.1 (a).

2. $\Phi(t) = e^{t}\begin{bmatrix} \cos t & \sin t \\ -\sin t & \cos t \end{bmatrix}$, arising in Example 2.4.1 (c).

3. $\Phi(t) = \begin{bmatrix} e^{t} & e^{-2t} & e^{-t} \\ e^{t} & -2e^{-2t} & -e^{-t} \\ e^{t} & 4e^{-2t} & e^{-t} \end{bmatrix}$, arising in Example 2.4.1 (b).

4. $\Phi(t) = e^{2t}\begin{bmatrix} e^{-3t} & 1 & 0 \\ e^{-3t} & 0 & 1 \\ -3e^{-3t} & 0 & 0 \end{bmatrix}$, arising in Example 2.4.1 (d).
5. $\Phi(t) = \begin{bmatrix} t+1 & t \\ -t & 1-t \end{bmatrix}$, arising in Example 2.4.4.
6. Suppose Φ_1 and Φ_2 are two solution matrices of $\mathbf{x}' = \mathbf{A}\mathbf{x}$. Show that $c_1\Phi_1 + c_2\Phi_2$ is also a solution matrix of this system.
7. Give examples to show that if $\Psi(t)$ is a fundamental solution matrix of $\mathbf{x}' = \mathbf{A}\mathbf{x}$ at $t = 0$, and $\mathbf{Z}(t)$ is a fundamental solution matrix of this same system at $t_0 \neq 0$, it does not follow that $\Psi(t) = \mathbf{Z}(t)$.

2.6 Solutions of Nonhomogeneous Systems

In this section we study techniques for solving the nonhomogeneous system

$$\mathbf{x}' = \mathbf{A}\mathbf{x} + \mathbf{f}(t) \tag{2.6.1}$$

where

$$\mathbf{f}(t) = \begin{bmatrix} f_1(t) \\ f_2(t) \\ \vdots \\ f_n(t) \end{bmatrix} \neq \mathbf{0}$$

is a vector, each of whose entries $f_i(t)$ is sectionally continuous for all t in $\alpha < t < \beta$. We call $\mathbf{f}(t)$ a *forcing function* and the system

$$\mathbf{x}' = \mathbf{A}\mathbf{x} \tag{2.6.2}$$

the *complementary system* of $\mathbf{x}' = \mathbf{A}\mathbf{x} + \mathbf{f}(t)$. The initial-value problem is defined by the equations

$$\mathbf{x}' = \mathbf{A}\mathbf{x} + \mathbf{f}(t), \quad \mathbf{x}(0) = \mathbf{b} \tag{2.6.3}$$

Theorem 2.6.1. *If $\Psi(t)$ is a fundamental solution matrix of $\mathbf{x}' = \mathbf{A}\mathbf{x}$ and $\mathbf{x}_p(t)$ is a solution of $\mathbf{x}' = \mathbf{A}\mathbf{x} + \mathbf{f}(t)$, then*

$$\mathbf{x}(t) = \Psi(t)\left(\mathbf{b} - \mathbf{x}_p(t)\right) + \mathbf{x}_p(t) \tag{2.6.4}$$

is the unique solution of the initial-value problem

$$\mathbf{x}' = \mathbf{A}\mathbf{x} + \mathbf{f}(t), \quad \mathbf{x}(0) = \mathbf{b} \tag{2.6.5}$$

Proof: First, we show that

$$\mathbf{x}(t) = \Psi(t)\mathbf{c} + \mathbf{x}_p(t)$$

is a solution of (2.6.5) for every choice of $\mathbf{c}$. We do this by substitution. Recall that $\Psi(t)\mathbf{c}$ solves $\mathbf{x}' = \mathbf{A}\mathbf{x}$, so that

$$\Psi'(t)\mathbf{c} = \mathbf{A}\Psi(t)\mathbf{c} \tag{2.6.6}$$

Also, by hypothesis, $\mathbf{x}_p(t)$ satisfies $\mathbf{x}' = \mathbf{A}\mathbf{x} + \mathbf{f}(t)$; that is

$$\mathbf{x}_p'(t) = \mathbf{A}\mathbf{x}_p(t) + \mathbf{f}(t) \tag{2.6.7}$$

Then differentiating $\mathbf{x}(t) = \Psi(t)\mathbf{c} + \mathbf{x}_p(t)$ and using (2.6.6) and (2.6.7), we have

$$\begin{aligned}\mathbf{x}'(t) &= \Psi'(t)\mathbf{c} + \mathbf{x}_p'(t)\\ &= \mathbf{A}\Psi(t)\mathbf{c} + \mathbf{A}\mathbf{x}_p(t) + \mathbf{f}(t)\\ &= \mathbf{A}\mathbf{x}(t) + \mathbf{f}(t)\end{aligned}$$

Second, because $\Psi(0) = \mathbf{I}$, we have

$$\mathbf{x}(0) = \Psi(0)\left(\mathbf{b} - \mathbf{x}_p(0)\right) + \mathbf{x}_p(0) = \mathbf{I}\left(\mathbf{b} - \mathbf{x}_p(0)\right) + \mathbf{x}_p(0) = \mathbf{b}$$

Now we establish that (2.6.4) provides the only solution. Suppose the contrary; that is, suppose $\mathbf{z}(t)$ is another solution of (2.6.5). Set

$$\mathbf{w}(t) = \mathbf{z}(t) - \mathbf{x}(t)$$

where $\mathbf{x}(t)$ is given by (2.6.4). It is easy to verify that $\mathbf{w}(t)$ satisfies

$$\mathbf{w}' = \mathbf{A}\mathbf{w}, \quad \mathbf{w}(0) = \mathbf{0}$$

By Corollary 2.5.3, it follows that $\mathbf{w}(t) = \mathbf{0}$. Thus, $\mathbf{x}(t) = \mathbf{z}(t)$. □

Corollary 2.6.1. *If $\mathbf{z}(t)$ is a solution of $\mathbf{x}' = \mathbf{A}\mathbf{x} + \mathbf{f}(t)$ then under the hypothesis of Theorem 2.6.1,*

$$\mathbf{z}(t) = \Psi(t)\left(\mathbf{z}(0) - \mathbf{x}_p(0)\right) + \mathbf{x}_p(t)$$

Proof: By hypothesis, $\mathbf{z}(t)$ satisfies the initial-value problem

$$\mathbf{x}' = \mathbf{A}\mathbf{x} + \mathbf{f}(t), \quad \mathbf{x}(0) = \mathbf{z}(0)$$

But by Theorem 2.6.1,

$$\mathbf{x}(t) = \Psi(t)\left(\mathbf{z}(0) - \mathbf{x}_p(0)\right) + \mathbf{x}_p(t)$$

is the unique solution of this problem. So $\mathbf{z}(t) = \mathbf{x}(t)$. □

Corollary 2.6.1 shows that the family of functions

$$\mathbf{x}(t) = \Phi(t)\mathbf{c} + \mathbf{x}_p(t) \tag{2.6.8}$$

is extensive enough to include every solution of $\mathbf{x}' = \mathbf{Ax} + \mathbf{f}(t)$, provided that Φ is an invertible solution matrix of $\mathbf{x}' = \mathbf{Ax}$. For this reason, we call $\mathbf{x}(t)$ in (2.6.8) a *general solution* of $\mathbf{x}' = \mathbf{Ax} + \mathbf{f}(t)$. However, since (2.6.1) has infinitely many solutions, any one of which may be used in (2.6.8), there are infinitely many general solutions. For example, let $\mathbf{c} = \mathbf{k} + \mathbf{c}_1$, where $\mathbf{k}$ is arbitrary, but $\mathbf{c}_1$ is any specific vector of constants. Then $\Phi(t)\mathbf{c}_1 + \mathbf{x}_p(t)$ is an alternative particular solution, and $\mathbf{x}(t) = \Phi(t)\mathbf{k} + \Phi(t)\mathbf{c}_1 + \mathbf{x}_p(t)$ is another general solution.

Theorem 2.6.2. *Suppose that* $\mathbf{x}_1(t)$ *and* $\mathbf{x}_2(t)$ *are solutions of*

$$\mathbf{x}' = \mathbf{Ax} + \mathbf{f}_1(t) \quad \textit{and} \quad \mathbf{x}' = \mathbf{Ax} + \mathbf{f}_2(t)$$

respectively. Then $\mathbf{x}(t) = \mathbf{x}_2(t) + \mathbf{x}_1(t)$ *is a solution of*

$$\mathbf{x}' = \mathbf{Ax} + \mathbf{f}_1(t) + \mathbf{f}_2(t) \tag{2.6.9}$$

Proof: This theorem is another example of a superposition theorem. The proof, once again, is a matter of substitution. First of all,

$$\mathbf{x}'(t) = \mathbf{x}_2'(t) + \mathbf{x}_1'(t) = (\mathbf{Ax}_2 + \mathbf{f}_2(t)) + (\mathbf{Ax}_1 + \mathbf{f}_1(t))$$

Then we compute

$$\begin{aligned}\mathbf{x}'(t) &= \mathbf{Ax}(t) + \mathbf{f}_1(t) + \mathbf{f}_2(t)\\ &= \mathbf{A}\left(\mathbf{x}_1(t) + \mathbf{x}_2(t)\right) + \mathbf{f}_1(t) + \mathbf{f}_2(t)\\ &= (\mathbf{Ax}_1 + \mathbf{f}_1(t)) + (\mathbf{Ax}_2 + \mathbf{f}_2(t))\end{aligned}$$

□

A general solution of the nonhomogeneous problem requires some particular solution and a general solution of its complementary equation. We know how to find a general solution of the complementary equation, so we now turn our attention to finding particular solutions of nonhomogeneous systems. There are two methods. The first is essentially educated guesswork and is called the method of undetermined coefficients. The second is a method that works for all "suitable" forcing functions and has profound theoretical implications; it is known as variation of parameters.

2.6.1 Undetermined Coefficients

It sometimes happens that we may discover a solution by inspection. A systematic approach utilizing this idea is called the method of *undetermined coefficients.* Its success is intimately connected to the structure of the forcing function $\mathbf{f}(t)$. Indeed, there are only a handful of forcing functions for which this method works. The ones we treat are of the form $\mathbf{k}e^{rt}$, which includes the constant forcing function $\mathbf{f}(t) = \mathbf{k}$ and the forcing functions $\mathbf{k}e^{at}\cos bt$ and $\mathbf{k}e^{at}\sin bt$. Forcing functions that are linear combinations of these functions are handled by superposition. (See Theorem 2.6.2.)

Theorem 2.6.3. *If r is not an eigenvalue of $\mathbf{A}$, then the system*

$$\mathbf{x}' = \mathbf{A}\mathbf{x} + \mathbf{k}e^{rt} \tag{2.6.10}$$

has a particular solution

$$\mathbf{x}_p(t) = \mathbf{c}e^{rt} \tag{2.6.11}$$

where $\mathbf{c} = -(\mathbf{A} - r\mathbf{I})^{-1}\mathbf{k}$.

Proof: The argument consists of substituting $\mathbf{x}_p(t) = \mathbf{c}e^{rt}$ into (2.6.10) and showing that it is always possible to find $\mathbf{c}$. Indeed,

$$\mathbf{x}_p'(t) = r\mathbf{c}e^{rt} = \mathbf{A}\mathbf{c}e^{rt} + \mathbf{k}e^{rt}$$

By collecting terms and cancelling e^{rt}, we find that $\mathbf{c}$ is uniquely determined (if r is not an eigenvalue) by

$$(\mathbf{A} - r\mathbf{I})\,\mathbf{c} = -\mathbf{k} \tag{2.6.12}$$

$$\mathbf{c} = -\,(\mathbf{A} - r\mathbf{I})^{-1}\,\mathbf{k} \tag{2.6.13}$$

because $(\mathbf{A} - r\mathbf{I})$ is invertible. □

Example 2.6.1. *Find a particular solution of* $\mathbf{x}' = \begin{bmatrix} 0 & -1 \\ -1 & 0 \end{bmatrix} \mathbf{x} + \begin{bmatrix} 1 \\ 1 \end{bmatrix}$.

Solution: The trial function is $\mathbf{x}_p(t) = \mathbf{c}e^{0t} = \mathbf{c}$. Then, using $r = 0$ in (2.6.12), we have

$$(\mathbf{A} - 0\mathbf{I})\,\mathbf{c} = \begin{bmatrix} 0 & -1 \\ -1 & 0 \end{bmatrix} \mathbf{c} = -\begin{bmatrix} 1 \\ 1 \end{bmatrix} = \begin{bmatrix} -1 \\ -1 \end{bmatrix}$$

This algebraic system has the unique solution $\mathbf{c} = [1,\ 1]^T$. Thus, $\mathbf{x}_p(t) = \mathbf{c}$.

Example 2.6.2. *Find a particular solution of the system*

$$\mathbf{x}' = \begin{bmatrix} 0 & -1 \\ -1 & 0 \end{bmatrix} \mathbf{x} + 3e^{2t} \begin{bmatrix} 1 \\ 1 \end{bmatrix}$$

Solution: The trial function is $\mathbf{x}_p(t) = \mathbf{c}e^{2t}$ so $r = 2$. Then, (2.6.12) provides

$$(\mathbf{A} - 2\mathbf{I})\,\mathbf{c} = \begin{bmatrix} -2 & -1 \\ -1 & -2 \end{bmatrix} \mathbf{c} = -3 \begin{bmatrix} 1 \\ 1 \end{bmatrix}$$

This algebraic system has the unique solution $\mathbf{c} = [1,\ 1]^T$, and therefore, a particular solution is

$$\mathbf{x}_p(t) = \begin{bmatrix} 1 \\ 1 \end{bmatrix} e^{2t}$$

Sometimes we are able to obtain $\mathbf{c}$ even when r is a root of $C(\lambda)$. Here is an example.

Example 2.6.3. *Find a particular solution of*

$$\mathbf{x}' = \begin{bmatrix} 1 & 1 \\ 4 & 1 \end{bmatrix} \mathbf{x} + ae^{-t} \begin{bmatrix} 1 \\ 2 \end{bmatrix}$$

Solution: Using (2.6.12), we have

$$(\mathbf{A} + 1\mathbf{I})\,\mathbf{c} = \begin{bmatrix} 2 & 1 \\ 4 & 2 \end{bmatrix} \mathbf{c} = -a \begin{bmatrix} 1 \\ 2 \end{bmatrix}$$

However, $\mathbf{A}+\mathbf{I}$ is singular because $r=-1$ is an eigenvalue of $\mathbf{A}$. We resort to the arrow diagram

$$\left[\begin{array}{cc|c} 2 & 1 & -a \\ 4 & 2 & -2a \end{array}\right] \rightarrow \left[\begin{array}{cc|c} 2 & 1 & -a \\ 0 & 0 & 0 \end{array}\right]$$

to obtain $2c_1+c_2=-a$ and $0c_1+0c_2=0$. A convenient choice is $c_1=0$ and $c_2=-a$. Hence,

$$\mathbf{c}=\left[\begin{array}{c} 0 \\ -a \end{array}\right]$$

which leads to a particular solution

$$\mathbf{x}_p(t)=-a\left[\begin{array}{c} 0 \\ 1 \end{array}\right]e^{-t}$$

Example 2.6.4. *If $b\neq 2$, show that there is no solution of the form $\mathbf{c}e^{-t}$ for the system*

$$\mathbf{x}'=\left[\begin{array}{cc} 1 & 1 \\ 4 & 1 \end{array}\right]\mathbf{x}+ae^{-t}\left[\begin{array}{c} 1 \\ b \end{array}\right]$$

Solution: We repeat the steps of Example 2.6.3 and find that

$$\left[\begin{array}{cc|c} 2 & 1 & -a \\ 4 & 2 & -2a \end{array}\right] \rightarrow \left[\begin{array}{cc|c} 2 & 1 & -a \\ 0 & 0 & -ab+2a \end{array}\right]$$

which is contradictory for $b\neq 2$ because no choice of c_1, c_2 can satisfy this system. (We are tacitly assuming $a\neq 0$.) Hence, our implimentation of the method of undetermined coefficients does not work for this problem.

Example 2.6.5. *Find a particular solution of* $\mathbf{x}'=\left[\begin{array}{cc} 1 & 0 \\ -1 & 3 \end{array}\right]\mathbf{x}+\left[\begin{array}{c} e^{2t} \\ 3 \end{array}\right]$.

Solution: We write the forcing function as a sum of two functions,

$$\left[\begin{array}{c} e^{2t} \\ 3 \end{array}\right]=e^{2t}\left[\begin{array}{c} 1 \\ 0 \end{array}\right]+\left[\begin{array}{c} 0 \\ 3 \end{array}\right]$$

and treat the system as two separate problems,

$$\mathbf{x}'=\mathbf{A}\mathbf{x}+e^{2t}\left[\begin{array}{c} 1 \\ 0 \end{array}\right] \qquad (1)$$

$$\mathbf{x}' = \mathbf{A}\mathbf{x} + \begin{bmatrix} 0 \\ 3 \end{bmatrix} \tag{2}$$

These nonhomogeneous ODE's have particular solutions, $\mathbf{x}_1(t) = \mathbf{c}_1 e^{2t}$ and $\mathbf{x}_2(t) = \mathbf{c}_2$, respectively. For (1), $r = 2$, and we find $\mathbf{c}_1$ from

$$\begin{bmatrix} -1 & 0 \\ -1 & 1 \end{bmatrix} \mathbf{c}_1 = -\begin{bmatrix} 1 \\ 0 \end{bmatrix}$$

Thus,

$$\mathbf{c}_1 = \begin{bmatrix} 1 \\ 1 \end{bmatrix}$$

and hence,

$$\mathbf{x}_1(t) = e^{2t} \begin{bmatrix} 1 \\ 1 \end{bmatrix}$$

For (2), $r = 0$, and it is easy to find

$$\mathbf{x}_2(t) = \begin{bmatrix} 0 \\ -1 \end{bmatrix}$$

By the principle of superposition, Theorem 2.6.2, a solution of the original system is the sum of the solutions just obtained:

$$\mathbf{x}_p(t) = e^{2t} \begin{bmatrix} 1 \\ 1 \end{bmatrix} + \begin{bmatrix} 0 \\ -1 \end{bmatrix}$$

For forcing functions that are products of exponentials and trigonometric functions, such as $\mathbf{k}e^{at}\cos bt$ or $\mathbf{k}e^{at}\sin bt$, we use $\mathbf{k}e^{(a+ib)t}$ and the fact that

$$\operatorname{Re}\left(\mathbf{k}e^{(a+ib)t}\right) = \mathbf{k}e^{at}\cos bt \tag{2.6.14}$$

$$\operatorname{Im}\left(\mathbf{k}e^{(a+ib)t}\right) = \mathbf{k}e^{at}\sin bt \tag{2.6.15}$$

assuming $\mathbf{k}$ is real.

The motivation for (2.6.14) and (2.6.15) is that, with $\mathbf{A}$ real,

$$\begin{aligned} \operatorname{Re}\mathbf{x}'(t) &= \frac{d}{dt}\operatorname{Re}\mathbf{x}(t) = \mathbf{A}\operatorname{Re}\mathbf{x}(t) + \operatorname{Re}\mathbf{k}e^{(a+ib)t} \\ &= \mathbf{A}\operatorname{Re}\mathbf{x}(t) + \mathbf{k}e^{at}\cos bt \end{aligned} \tag{2.6.16}$$

And similarly,

$$\begin{aligned}\operatorname{Im}\mathbf{x}'(t) &= \frac{d}{dt}\operatorname{Im}\mathbf{x}(t) = \mathbf{A}\operatorname{Im}\mathbf{x}(t) + \operatorname{Im}\mathbf{k}e^{(a+ib)t} \\ &= \mathbf{A}\operatorname{Im}\mathbf{x}(t) + \mathbf{k}e^{at}\sin bt \end{aligned} \tag{2.6.17}$$

From this, it follows that the real and imaginary parts of the solution of

$$\mathbf{x}' = \mathbf{A}\mathbf{x} + \mathbf{k}e^{(a+ib)t} \tag{2.6.18}$$

are, respectively, the real solutions of $\mathbf{x}' = \mathbf{A}\mathbf{x} + \mathbf{k}e^{at}\cos bt$ and $\mathbf{x}' = \mathbf{A}\mathbf{x} + \mathbf{k}e^{at}\sin bt$. We illustrate this in the next example.

Example 2.6.6. *Find a particular solution of*

$$\mathbf{x}' = \begin{bmatrix} 1 & 1 \\ -1 & 1 \end{bmatrix}\mathbf{x} + \cos t\begin{bmatrix} -1 \\ 2 \end{bmatrix}$$

Solution: Comparing the given forcing function with the model given in (2.6.14), we see that $a = 0$ and $b = 1$. So we replace the given equation by

$$\mathbf{z}' = \begin{bmatrix} 1 & 1 \\ -1 & 1 \end{bmatrix}\mathbf{z} + e^{it}\begin{bmatrix} -1 \\ 2 \end{bmatrix} \tag{1}$$

solve (1), and take the real part of this solution, as in (2.6.16). Since $r = i$, (2.6.12) becomes

$$\begin{bmatrix} 1-i & 1 \\ -1 & 1-i \end{bmatrix}\mathbf{c} = -\begin{bmatrix} -1 \\ 2 \end{bmatrix}$$

In expanded form, this system represents the simultaneous equations

$$\begin{aligned}(1-i)c_1 + c_2 &= 1 \\ -c_1 + (1-i)c_2 &= -2\end{aligned}$$

whose solution is found to be $c_1 = 1+i, c_2 = -1$. So,

$$\mathbf{c} = \begin{bmatrix} 1+i \\ -1 \end{bmatrix}$$

Then $\mathbf{z}_p(t) = \mathbf{c}e^{it}$ is a particular solution of (1). In expanded form

$$\begin{aligned}\mathbf{z}_p(t) &= \begin{bmatrix} 1+i \\ -1 \end{bmatrix}(\cos t + i\sin t) \\ &= \begin{bmatrix} \cos t - \sin t \\ -\cos t \end{bmatrix} + i\begin{bmatrix} \cos t + \sin t \\ -\sin t \end{bmatrix}\end{aligned}$$

Hence, the desired particular solution is

$$\mathbf{x}_p(t) = \operatorname{Re}\mathbf{z}_p(t) = \begin{bmatrix} \cos t - \sin t \\ -\cos t \end{bmatrix}$$

•EXERCISES

Write each of the forcing functions in Problems 1–5 as a sum of functions of the form $\mathbf{k}e^{rt}$, where r may be zero or complex.

1. $\mathbf{f}(t) = (1 + \cos t)\begin{bmatrix} 1 \\ 1 \end{bmatrix}$.

2. $\mathbf{f}(t) = \begin{bmatrix} 1 - e^t \\ 1 \end{bmatrix}$.

3. $\mathbf{f}(t) = \begin{bmatrix} e^{-t} \\ e^t \end{bmatrix}$.

4. $\mathbf{f}(t) = e^{2t}\sin t\begin{bmatrix} e^t \\ 1 \end{bmatrix}$.

5. $\mathbf{f}(t) = \begin{bmatrix} \cos 3t \\ \sin 2t \end{bmatrix}$.

Use undetermined coefficients to find a particular solution to each of the systems in Problems 6–13, $\mathbf{x}' = \mathbf{A}\mathbf{x} + \mathbf{f}(t)$.

6. $\mathbf{A} = \begin{bmatrix} 1 & 1 \\ -1 & 1 \end{bmatrix}$, $\mathbf{f}(t) = e^{2t}\begin{bmatrix} 1 \\ 0 \end{bmatrix}$.

7. $\mathbf{A} = \begin{bmatrix} 0 & 2 \\ 1 & -1 \end{bmatrix}$, $\mathbf{f}(t) = \sin t\begin{bmatrix} 4 \\ 2 \end{bmatrix}$.

8. $\mathbf{A} = \begin{bmatrix} 0 & 2 \\ 1 & -1 \end{bmatrix}$, $\mathbf{f}(t) = (1 + \cos t)\begin{bmatrix} 4 \\ 2 \end{bmatrix}$.

9. $\mathbf{A} = \begin{bmatrix} 1 & 0 \\ -1 & 3 \end{bmatrix}$, $\mathbf{f}(t) = \begin{bmatrix} 2e^{-t} \\ 1 \end{bmatrix}$.

10. $\mathbf{A} = \begin{bmatrix} 2 & 1 \\ 3 & 0 \end{bmatrix}$, $\mathbf{f}(t) = -e^{-t}\cos 2t\begin{bmatrix} 6 \\ 2 \end{bmatrix}$.

11. $\mathbf{A} = \begin{bmatrix} 2 & -1 \\ 0 & 1 \end{bmatrix}$, $\mathbf{f}(t) = e^{-2t}\sin t\begin{bmatrix} 10 \\ 10 \end{bmatrix}$.

12. $\mathbf{A} = \begin{bmatrix} 1 & 2 \\ 1 & 0 \end{bmatrix}$, $\mathbf{f}(t) = e^{-t}\sin t\begin{bmatrix} 4 \\ 1 \end{bmatrix}$.

13. $\mathbf{A} = \begin{bmatrix} 1 & 2 \\ 1 & 0 \end{bmatrix}$, $\quad \mathbf{f}(t) = e^{-t}\sin t \begin{bmatrix} -1 \\ 1 \end{bmatrix}$.

14. Show that
$$\mathbf{k} = -\tfrac{a}{2}\begin{bmatrix} 1 \\ 0 \end{bmatrix}$$
is a solution of
$$\begin{bmatrix} 2 & 1 \\ 4 & 2 \end{bmatrix}\mathbf{k} = -a\begin{bmatrix} 1 \\ 2 \end{bmatrix}$$
Use this result to find a particular solution of
$$\mathbf{x}' = \begin{bmatrix} 1 & 1 \\ 4 & 1 \end{bmatrix}\mathbf{x} + e^{-t}\begin{bmatrix} a \\ 2a \end{bmatrix}$$
distinct from the one derived in Example 2.6.3.

15. Show that
$$\mathbf{x}_p(t) = \tfrac{1}{5}\begin{bmatrix} (\beta - 3\alpha)\cos t - (2\beta - \alpha)\sin t \\ -(\alpha + 3\beta)\cos t + (2\alpha + \beta)\sin t \end{bmatrix}$$
is a particular solution of
$$\mathbf{x}' = \begin{bmatrix} 1 & 1 \\ -1 & 1 \end{bmatrix}\mathbf{x} + \cos t\begin{bmatrix} \alpha \\ \beta \end{bmatrix}$$

16. Show that
$$\mathbf{x}_p(t) = \tfrac{1}{5}\begin{bmatrix} (\beta - 3\alpha)\sin t + (2\beta - \alpha)\cos t \\ -(\alpha + 3\beta)\sin t - (2\alpha + \beta)\cos t \end{bmatrix}$$
is a particular solution of
$$\mathbf{x}' = \begin{bmatrix} 1 & 1 \\ -1 & 1 \end{bmatrix}\mathbf{x} + \sin t\begin{bmatrix} \alpha \\ \beta \end{bmatrix}$$

17. Find a particular solution of the system
$$\mathbf{x}' = \begin{bmatrix} 1 & 1 \\ 1 & -1 \end{bmatrix}\mathbf{x} + \begin{bmatrix} \cos t \\ -\sin t \end{bmatrix}$$

MATLAB

18. Use "-(A-r eye(n)) (k)" to solve Problems 6—9.
19. Use "-(A-r eye(n)) (k)" to solve Problems 10—13.

2.6.2 Variation of Parameters

It is a remarkable aspect of linear ODE's that a solution of a nonhomogeneous system can always be determined using the general solution of the complementary system. The method for doing so is called the method of *variation of parameters.* We discuss this important idea in this section.

We begin by assuming that we have an $n \times n$ solution matrix $\Phi(t)$ for the system $\mathbf{x}' = \mathbf{A}\mathbf{x}$, complementary to $\mathbf{x}' = \mathbf{A}\mathbf{x} + \mathbf{f}(t)$. Suppose also that

$\Phi^{-1}(0)$ exists. Define a function $\mathbf{v}(t)$ by

$$\mathbf{x}(t) = \Phi(t)\mathbf{v}(t) \tag{2.6.19}$$

and assume $\mathbf{x}(t)$ is a solution of $\mathbf{x}' = \mathbf{A}\mathbf{x} + \mathbf{f}(t)$. Indeed, since $\Phi(t)$ is invertible, $\mathbf{v}(t) = \Phi^{-1}(t)\mathbf{x}(t)$. (The parallel between (2.6.19) and the change of variables (1.3.23) introduced to solve the first-order scalar ODE $u' + p(t)u = f(t)$ is so striking and suggestive that the reader is urged to review Section 1.3 before continuing on.)

Differentiating (2.6.19) leads to

$$\mathbf{x}'(t) = \Phi'(t)\mathbf{v}(t) + \Phi(t)\mathbf{v}'(t) \tag{2.6.20}$$

However, because $\Phi'(t) = \mathbf{A}\Phi(t)$, (See Theorem 2.4.1), (2.6.20) may be written as

$$\mathbf{x}'(t) = \mathbf{A}\Phi(t)\mathbf{v}(t) + \Phi(t)\mathbf{v}'(t) \tag{2.6.21}$$

Substituting the expressions for $\mathbf{x}(t)$ and $\mathbf{x}'(t)$ given by (2.6.19) and (2.6.20), into (2.6.21) leads to

$$\mathbf{A}\Phi(t)\mathbf{v}(t) + \Phi(t)\mathbf{v}'(t) = \mathbf{A}\Phi(t)\mathbf{v}(t) + \mathbf{f}(t) \tag{2.6.22}$$

And this expression simplifies to

$$\Phi(t)\mathbf{v}'(t) = \mathbf{f}(t) \tag{2.6.23}$$

Therefore,

$$\mathbf{v}'(t) = \Phi^{-1}(t)\mathbf{f}(t) \tag{2.6.24}$$

the crucial equation in the method. We can integrate (2.6.24) to obtain

$$\mathbf{v}(t) = \int \Phi^{-1}(t)\mathbf{f}(t)\,dt + \mathbf{c} \tag{2.6.25}$$

where $\mathbf{c}$ is the constant (vector) of integration. We now use (2.6.25) in (2.6.19) to obtain

$$\mathbf{x}(t) = \Phi(t)\int \Phi^{-1}(t)\mathbf{f}(t)\,dt + \Phi(t)\mathbf{c} \tag{2.6.26}$$

It is interesting to note that the first term on the right of (2.6.26) is a particular solution of $\mathbf{x}' = \mathbf{A}\mathbf{x} + \mathbf{f}(t)$, while the second term is a general solution of its complementary equation $\mathbf{x}' = \mathbf{A}\mathbf{x}$. Consequently, we write the particular solution as

$$\mathbf{x}_p(t) = \Phi(t)\int \Phi^{-1}(t)\mathbf{f}(t)\,dt \tag{2.6.27}$$

Here are a few examples that illustrate the procedure.

Example 2.6.7. *Find a particular solution of* $\mathbf{x}' = \begin{bmatrix} 1 & 1 \\ 4 & 1 \end{bmatrix}\mathbf{x} + e^t\begin{bmatrix} 1 \\ 1 \end{bmatrix}$.

Solution: A solution matrix for this system was found in Example 2.4.1:

$$\Phi(t) = \begin{bmatrix} e^{-t} & e^{3t} \\ -2e^{-t} & 2e^{3t} \end{bmatrix}$$

We can solve $\Phi(t)\mathbf{v}'(t) = \mathbf{f}(t)$ for $\mathbf{v}'(t)$ by Gaussian elimination (See Section 0.4):

$$\left[\begin{array}{cc|c} e^{-t} & e^{3t} & e^t \\ -2e^{-t} & 2e^{3t} & e^t \end{array}\right] \to \left[\begin{array}{cc|c} 1 & e^{4t} & e^{2t} \\ -2 & 2e^{4t} & e^{2t} \end{array}\right] \to \left[\begin{array}{cc|c} 1 & e^{4t} & e^{2t} \\ 0 & 4e^{4t} & 3e^{2t} \end{array}\right]$$

$$\to \left[\begin{array}{cc|c} 1 & 0 & \frac{1}{4}e^{2t} \\ 0 & 4e^{4t} & 3e^{2t} \end{array}\right] \to \left[\begin{array}{cc|c} 1 & 0 & \frac{1}{4}e^{2t} \\ 0 & 1 & \frac{3}{4}e^{-2t} \end{array}\right]$$

So,

$$\mathbf{v}'(t) = \frac{1}{4}\begin{bmatrix} e^{2t} \\ 3e^{-2t} \end{bmatrix}$$

The vector $\mathbf{v}(t)$ is found by integrating the preceding equation;

$$\mathbf{v}(t) = \frac{1}{8}\begin{bmatrix} e^{2t} \\ -3e^{-2t} \end{bmatrix}$$

We can now form a particular solution. From (2.6.27),

$$\mathbf{x}_p(t) = \Phi(t)\mathbf{v}(t) = -\frac{1}{8}\begin{bmatrix} e^{-t} & e^{3t} \\ -2e^{-t} & 2e^{3t} \end{bmatrix}\begin{bmatrix} -e^{2t} \\ 3e^{-2t} \end{bmatrix} = -\frac{1}{4}\begin{bmatrix} 1 \\ 4 \end{bmatrix}e^t$$

The skeptical reader may wish to verify that $\mathbf{x}_p(t)$ does indeed solve the given system.

Example 2.6.8. *Find a particular solution of*

$$\mathbf{x}' = \begin{bmatrix} 1 & 1 \\ -1 & 1 \end{bmatrix}\mathbf{x} + \begin{bmatrix} \cos t \\ -\sin t \end{bmatrix}$$

Solution: Most of the tedious computations have already been done in Example 2.3.6, in which we found two solutions for this system, namely,

$$\mathbf{x}_1(t) = e^t\begin{bmatrix} \cos t \\ -\sin t \end{bmatrix}, \quad \mathbf{x}_2(t) = e^t\begin{bmatrix} \sin t \\ \cos t \end{bmatrix}$$

From these solutions, we have a solution matrix

$$\Phi(t) = e^t \begin{bmatrix} \cos t & \sin t \\ -\sin t & \cos t \end{bmatrix}$$

Now,

$$\begin{bmatrix} \cos t & \sin t \\ -\sin t & \cos t \end{bmatrix}^{-1} = \begin{bmatrix} \cos t & -\sin t \\ \sin t & \cos t \end{bmatrix}$$

and therefore,

$$\mathbf{v}'(t) = \Phi^{-1}(t)\mathbf{f}(t) = e^{-t} \begin{bmatrix} \cos t & -\sin t \\ \sin t & \cos t \end{bmatrix} \begin{bmatrix} \cos t \\ -\sin t \end{bmatrix} = e^{-t} \begin{bmatrix} 1 \\ 0 \end{bmatrix}$$

Integrate $\mathbf{v}'(t)$ to find

$$\mathbf{v}(t) = -e^{-t} \begin{bmatrix} 1 \\ 0 \end{bmatrix}$$

and it follows that a particular solution is

$$\mathbf{x}_p(t) = \Phi(t)\mathbf{v}(t) = - \begin{bmatrix} \cos t & \sin t \\ -\sin t & \cos t \end{bmatrix} \begin{bmatrix} 1 \\ 0 \end{bmatrix} = \begin{bmatrix} -\cos t \\ \sin t \end{bmatrix}$$

Example 2.6.9. *Find a particular solution of*

$$\mathbf{x}'(t) = \begin{bmatrix} 1 & 1 \\ 4 & 1 \end{bmatrix} \mathbf{x}(t) + e^{-t} \begin{bmatrix} 0 \\ 4 \end{bmatrix}$$

Solution: This system is a special case of the system in Example 2.6.4. For this particular forcing function, we found that there does not exist a solution of the form $\mathbf{k}e^{-t}$ because $r = -1$ is an eigenvalue of the coefficient matrix. The method of variation of parameters is not limited by this fact. In Example 2.6.7 we found the solution matrix for the complementary system; it and its inverse are

$$\Phi(t) = \begin{bmatrix} e^{-t} & e^{3t} \\ -2e^{-t} & 2e^{3t} \end{bmatrix}, \quad \Phi^{-1}(t) = \frac{1}{4} \begin{bmatrix} 2e^{t} & -e^{t} \\ 2e^{-3t} & e^{-3t} \end{bmatrix}$$

Thus,

$$\mathbf{v}'(t) = \frac{1}{4} \begin{bmatrix} 2e^{t} & -e^{t} \\ 2e^{-3t} & e^{-3t} \end{bmatrix} \begin{bmatrix} 0 \\ 4e^{-t} \end{bmatrix} = \begin{bmatrix} -1 \\ e^{-4t} \end{bmatrix}$$

From this, we obtain

$$\mathbf{v}(t) = -\frac{1}{4} \begin{bmatrix} 4t \\ e^{-4t} \end{bmatrix}$$

A particular solution is therefore

$$\mathbf{x}_p(t) = \Phi(t)\mathbf{v}(t) = -\frac{1}{4}\begin{bmatrix} e^{-t} & e^{3t} \\ -2e^{-t} & 2e^{3t} \end{bmatrix}\begin{bmatrix} 4t \\ e^{-4t} \end{bmatrix}$$

$$= -\frac{e^{-t}}{4}\begin{bmatrix} 1+4t \\ 2-8t \end{bmatrix}$$

•EXERCISES

Use the method of variation of parameters to find a particular solution of the following nonhomogeneous systems. Compare your answers with the solutions obtained by undetermined coefficients in Section 2.6.1, where appropriate.

1. $\mathbf{x}' = \begin{bmatrix} 1 & 1 \\ -1 & 1 \end{bmatrix}\mathbf{x} + e^{2t}\begin{bmatrix} 1 \\ 0 \end{bmatrix}$, where $\Phi(t) = e^t\begin{bmatrix} \cos t & \sin t \\ -\sin t & \cos t \end{bmatrix}$.

2. $\mathbf{x}' = \begin{bmatrix} 0 & 2 \\ 1 & -1 \end{bmatrix}\mathbf{x} + \sin t\begin{bmatrix} 4 \\ 2 \end{bmatrix}$, where $\Phi(t) = e^t\begin{bmatrix} 2e^t & -e^{-2t} \\ e^t & e^{-2t} \end{bmatrix}$.

3. $\mathbf{x}' = \begin{bmatrix} 0 & 2 \\ 1 & -1 \end{bmatrix}\mathbf{x} + (1+\cos t)\begin{bmatrix} 4 \\ 2 \end{bmatrix}$ (see Problem 2).

4. $\mathbf{x}' = \begin{bmatrix} 1 & 0 \\ -1 & 3 \end{bmatrix}\mathbf{x} + \begin{bmatrix} 2e^{-t} \\ 1 \end{bmatrix}$, where $\Phi(t) = \begin{bmatrix} 2e^t & 0 \\ e^t & e^{3t} \end{bmatrix}$.

5. $\mathbf{x}' = \begin{bmatrix} 2 & 1 & 0 \\ 0 & 2 & 0 \\ 0 & 0 & -1 \end{bmatrix}\mathbf{x} + \begin{bmatrix} e^t \\ 1 \\ 0 \end{bmatrix}$, where $\Phi(t) = e^{2t}\begin{bmatrix} 1 & t & 0 \\ 0 & 1 & 0 \\ 0 & 0 & e^{-3t} \end{bmatrix}$.

6. $\mathbf{x}' = \begin{bmatrix} 2 & 1 \\ 3 & 0 \end{bmatrix}\mathbf{x} - e^{-t}\cos 2t\begin{bmatrix} 6 \\ 2 \end{bmatrix}$, where $\Phi(t) = \begin{bmatrix} e^{3t} & -e^{-t} \\ e^{3t} & 3e^{-t} \end{bmatrix}$.

7. $\mathbf{x}' = \begin{bmatrix} 2 & -1 \\ 0 & 1 \end{bmatrix}\mathbf{x} + e^{-2t}\sin t\begin{bmatrix} 10 \\ 10 \end{bmatrix}$, where $\Phi(t) = \begin{bmatrix} e^t & e^{2t} \\ e^t & 0 \end{bmatrix}$.

8. $\mathbf{x}' = \begin{bmatrix} 1 & 2 \\ 1 & 0 \end{bmatrix}\mathbf{x} + e^{-t}\sin t\begin{bmatrix} -1 \\ 1 \end{bmatrix}$, where $\Phi(t) = \begin{bmatrix} 2e^{2t} & -e^{-t} \\ e^{2t} & e^{-t} \end{bmatrix}$.

9. $\mathbf{x}' = \begin{bmatrix} 1 & 2 \\ 1 & 0 \end{bmatrix}\mathbf{x} + e^{-t}\sin t\begin{bmatrix} 4 \\ 1 \end{bmatrix}$ (see Problem 8).

Find a solution to Problems 10—14 using variation of parameters.

10. $\mathbf{x}' = \begin{bmatrix} 2 & 1 \\ 3 & 0 \end{bmatrix}\mathbf{x} + t\begin{bmatrix} 10 \\ 10 \end{bmatrix}$ (See Problem 6).

11. $\mathbf{x}' = \begin{bmatrix} 1 & 1 \\ -1 & 1 \end{bmatrix}\mathbf{x} + (t+1)\begin{bmatrix} 2 \\ 0 \end{bmatrix}$ (see Problem 1).

12. $\mathbf{x}' = \begin{bmatrix} 0 & 1 \\ a^2 & 0 \end{bmatrix} \mathbf{x} + t^2 \begin{bmatrix} 1 \\ 0 \end{bmatrix}, \qquad a \neq 0.$

13. $\mathbf{x}' = \mathbf{0x} + \mathbf{f}(t)$. (*Hint*: Show that $\Psi(t) = \mathbf{I}$.)

14. $\mathbf{x}' = (a\mathbf{I})\,\mathbf{x} + \mathbf{f}(t)$. (*Hint*: Show that $\Psi(t) = e^{at}\mathbf{I}$.)

15. Verify that a solution of

$$\mathbf{x}' = \begin{bmatrix} 1 & 1 \\ 4 & 1 \end{bmatrix} \mathbf{x} + e^{-t} \begin{bmatrix} 0 \\ 4 \end{bmatrix}$$

is

$$\mathbf{x}_p(t) = -\frac{e^{-t}}{4} \begin{bmatrix} 1 + 4t \\ 2 - 8t \end{bmatrix}$$

16. Verify by substitution that the particular solution given in Example 2.6.7 is indeed a solution of the given system.

2.7 Nonhomogeneous Initial-Value Problems

Recall that a general solution of $\mathbf{x}' = \mathbf{Ax} + \mathbf{f}(t)$ is the sum of a general solution of its complementary equation and any particular solution. Indeed, (2.6.8) presents a general solution:

$$\mathbf{x}(t) = \Phi(t)\mathbf{c} + \mathbf{x}_p(t) \tag{2.7.1}$$

where $\Phi(t)$ is any invertible solution matrix for the complementary system. In the last section (2.6.27) provides

$$\mathbf{x}_p(t) = \Phi(t) \int \Phi^{-1}(t)\mathbf{f}(t)\, dt \tag{2.7.2}$$

If we use (2.7.2) in (2.7.1), we obtain an explicit formula for the general solution, a result we record as a theorem.

Theorem 2.7.1. *Assuming that $\Phi(t)$ is an invertible solution matrix of $\mathbf{x}' = \mathbf{Ax}$, then*

$$\mathbf{x}(t) = \Phi(t) \left(\mathbf{c} + \int \Phi^{-1}(t)\mathbf{f}(t)\, dt \right) \tag{2.7.3}$$

is a general solution of

$$\mathbf{x}' = \mathbf{Ax} + \mathbf{f}(t)$$

Proof: This is just a restatement of Theorem 2.6.1 with the particular solution obtained from variation of parameters. □

Theorem 2.7.2. *Assuming that* $\Phi(t)$ *is an invertible solution matrix of* $\mathbf{x}' = \mathbf{A}\mathbf{x}$, *then*

$$\mathbf{x}(t) = \Phi(t)\left(\Phi^{-1}(0)\mathbf{b} + \int_0^t \Phi^{-1}(s)\mathbf{f}(s)\,ds\right) \tag{2.7.4}$$

is the unique solution of the initial-value problem

$$\mathbf{x}' = \mathbf{A}\mathbf{x} + \mathbf{f}(t), \quad \mathbf{x}(0) = \mathbf{b} \tag{2.7.5}$$

Proof: Equation 2.7.4 is just (2.7.3), where we use a definite integral to represent a particular solution and $\mathbf{c} = \Phi^{-1}(0)\mathbf{b}$. All we need to do is show that $\mathbf{x}(t)$ as given by (2.7.4) satisfies the initial condition $\mathbf{x}(0) = \mathbf{b}$. But this follows immediately because

$$\mathbf{x}(0) = \Phi(0)\left(\Phi^{-1}(0)\mathbf{b} + \int_0^0 \Phi^{-1}(s)\mathbf{f}(s)\,ds\right) = \Phi(0)\Phi^{-1}(0)\mathbf{b} = \mathbf{b}$$

□

An alternative way of expressing the content of Theorem 2.7.2 is this: We find a general solution of the complementary system and choose the arbitrary constant $\mathbf{c}$ so that the initial condition is met. Then we add to this solution a particular solution that vanishes at zero; that is,

$$\mathbf{x}(t) = \Phi(t)\Phi^{-1}(0)\mathbf{b} + \Phi(t)\int_0^t \Phi^{-1}(s)\mathbf{f}(s)\,ds \tag{2.7.6}$$

In (2.7.6), $\Phi(t)\Phi^{-1}(0)\mathbf{b}$ solves $\mathbf{x}' = \mathbf{A}\mathbf{x}$, $\mathbf{x}(0) = \mathbf{b}$, and $\Phi(t)\int_0^t \Phi^{-1}(s)\mathbf{f}(s)\,ds$ is a particular solution of $\mathbf{x}' = \mathbf{A}\mathbf{x} + \mathbf{f}(t)$ that vanishes at $t = 0$.

Corollary 2.7.1. *Under the hypothesis of the theorem, the function*

$$\mathbf{x}(t) = \Phi(t)\int_0^t \Phi^{-1}(s)\mathbf{f}(s)\,ds \tag{2.7.7}$$

is the unique solution of the initial-value problem

$$\mathbf{x}' = \mathbf{A}\mathbf{x} + \mathbf{f}(t), \quad \mathbf{x}(0) = \mathbf{0} \tag{2.7.8}$$

•EXERCISES

In Problems 1–3, find the solution of $\mathbf{x}' = \mathbf{A}\mathbf{x} + \mathbf{f}(t)$, $\mathbf{x}(0) = \mathbf{b}$, for each $\mathbf{f}(t)$ and $\mathbf{b}$, given the solution matrix $\Psi(t)$ and coefficient matrix $\mathbf{A}$:

$$\Psi(t) = e^t \begin{bmatrix} 1-t & t \\ -t & 1+t \end{bmatrix}, \quad \mathbf{A} = \begin{bmatrix} 0 & 1 \\ -1 & 2 \end{bmatrix}$$

1. $\mathbf{f}(t) = \begin{bmatrix} 0 \\ 1 \end{bmatrix}, \quad \mathbf{b} = \begin{bmatrix} 0 \\ 1 \end{bmatrix}$

2. $\mathbf{f}(t) = \begin{bmatrix} 1 \\ 1 \end{bmatrix}, \quad \mathbf{b} = \begin{bmatrix} 0 \\ 1 \end{bmatrix}$

3. $\mathbf{f}(t) = \begin{bmatrix} 1 \\ 1 \end{bmatrix} e^{-t}, \quad \mathbf{b} = \begin{bmatrix} 0 \\ 0 \end{bmatrix}$

4. Find the general solution of $\mathbf{x}' = \mathbf{A}\mathbf{x} + \mathbf{f}(t)$, $\mathbf{x}(t_0) = \mathbf{b}$. (*Hint*: Let $\tau = t - t_0$, and use Theorem 2.7.2 or its corollary.)

2.8 Fundamental Solution Matrices

In this section[5] we establish a number of interesting results in the theory of fundamental solution matrices. We assume that every system $\mathbf{x}' = \mathbf{A}\mathbf{x}$ has a fundamental solution matrix at some fixed $t = t_0$.

Theorem 2.8.1. *If $\Psi(t)$ and $\Theta(t)$ are fundamental solution matrices of $\mathbf{x}' = \mathbf{A}\mathbf{x}$ and $\mathbf{x}' = -\mathbf{A}^T\mathbf{x}$, respectively, then*

$$\Psi^{-1}(t) = \Theta^T(t) \tag{2.8.1}$$

Proof: Since $\Theta(t)$ is differentiable, so is its transpose, and therefore, so is the product $\Theta^T(t)\Psi(t)$. In fact,

$$\begin{aligned} \left(\Theta^T(t)\Psi(t)\right)' &= \left(\Theta^T(t)\right)'\Psi(t) + \Theta^T(t)\Psi'(t) \qquad (2.8.2) \\ &= (\Theta')^T\Psi + \Theta^T\Psi' \\ &= \left(-\mathbf{A}^T\Theta\right)^T\Psi + \Theta^T\mathbf{A}\Psi \\ &= -\Theta^T\mathbf{A}\Psi + \Theta^T\mathbf{A}\Psi = \mathbf{0} \end{aligned}$$

[5]The section may be skipped without any loss in continuity.

Hence, by integration of (2.8.2),

$$\Theta^T(t)\Psi(t) = \mathbf{K}$$

where $\mathbf{K}$ is a constant matrix. To evaluate $\mathbf{K}$, choose $t = t_0$. Then

$$\Theta^T(t_0)\,\Psi(t_0) = \mathbf{I} = \mathbf{K}$$

from which it follows that $\Theta^T(t)\Psi(t) = \mathbf{I}$. Thus, for each t, $\Psi(t)$ is invertible and in fact, $\Psi^{-1}(t) = \Theta^T(t)$. □

Not only does Theorem 2.8.1 prove that the inverse of every fundamental solution matrix for $\mathbf{x}' = \mathbf{A}\mathbf{x}$ exists, but also it provides the form for the inverse, namely, the transpose of the fundamental solution matrix of $\mathbf{x}' = -\mathbf{A}^T\mathbf{x}$. Indeed, if $t_0 = 0$, we shall show in Corollary 2.8.1, to follow, that $\Psi^{-1}(t) = \Psi(-t)$.

Theorem 2.8.2. *If $\Psi(t)$ is a fundamental solution matrix at $t = 0$, then for each t and t_0,*

$$\Psi(t + t_0) = \Psi(t)\Psi(t_0) \tag{2.8.3}$$

Proof: Let t_0 be some fixed value of t. By Corollary 2.4.1 it follows that $\Phi(t) = \Psi(t)\Psi(t_0)$ is a solution matrix for $\mathbf{x}' = \mathbf{A}\mathbf{x}$. Since $\Phi(0) = \Psi(0)\Psi(t_0) = \Psi(t_0)$, the solution matrix $\Phi(t) = \Psi(t)\Psi(t_0)$ is the unique solution of

$$\mathbf{X}' = \mathbf{A}\mathbf{X}, \qquad \mathbf{X}(0) = \Psi(t_0) \tag{2.8.4}$$

Now consider the matrix $\Theta(t) = \Psi(t + t_0)$. By the chain rule,

$$\Theta'(t) = \frac{d}{dt}\Psi(t + t_0) = \Psi'(t + t_0)\,\frac{d}{dt}(t + t_0) = \Psi'(t + t_0)$$

which shows that $\Psi(t + t_0)$ is also a solution of (2.8.4). By Corollary 2.5.5,

$$\Phi(t) = \Psi(t)\Psi(t_0) = \Psi(t + t_0)$$

□

Corollary 2.8.1. *If $\Psi(t)$ is a fundamental solution matrix at $t = 0$ then*

$$(1) \quad \Psi^{-1}(t) = \Psi(-t) \tag{2.8.5}$$
$$(2) \quad \Psi(a+b) = \Psi(a)\Psi(b) \tag{2.8.6}$$

Proof: (1). Choose $t = -t_0$ in (2.8.3). Then, for all t_0,

$$\mathbf{I} = \Psi(-t_0 + t_0) = \Psi(-t_0)\Psi(t_0)$$

(2). Set $a = t_0$ and $b = t$ in (2.8.3). Then $\Psi(a+b) = \Psi(a)\Psi(b)$ and $\Psi(a+b) = \Psi(b+a) = \Psi(b)\Psi(a)$. □

Corollary 2.8.2. *If $\Phi(t)$ is a solution matrix invertible at $t = 0$, then*

$$\Phi^{-1}(t) = \Phi^{-1}(0)\Phi(-t)\Phi^{-1}(0) \tag{2.8.7}$$

Proof: Set $\Psi(t) = \Phi(t)\Phi^{-1}(0)$. Then, $\Psi(t)$ is a fundamental solution matrix and

$$\Phi(t) = \Psi(t)\Phi(0) \tag{2.8.8}$$

From (2.8.8), it follows that

$$\Phi^{-1}(t) = \Phi^{-1}(0)\Psi^{-1}(t) \tag{2.8.9}$$

Now use (2.8.5) to write (2.8.9) as

$$\Phi^{-1}(t) = \Phi^{-1}(0)\Psi(-t) \tag{2.8.10}$$

However, from (2.8.8),

$$\Phi(-t) = \Psi(-t)\Phi(0) \tag{2.8.11}$$

and therefore, solving for $\Psi(-t)$ in (2.8.11), we obtain

$$\Psi(-t) = \Phi(-t)\Phi^{-1}(0) \tag{2.8.12}$$

Now substitute (2.8.12) into (2.8.10) to obtain (2.8.7). □

Properties (1) and (2) of Corollary 2.8.1 are reminiscent of two properties of the exponential function, namely,

$$\exp(t)^{-1} = \exp(-t)$$

and

$$\exp(a + b) = \exp(a)\exp(b)$$

•EXERCISES

1. Prove Part(2) of Corollary 2.8.1.

2. Suppose that $\Psi(t)$ is a fundamental solution matrix of the system $\mathbf{x}' = \mathbf{A}\mathbf{x}$. Show that $\mathbf{x}_p(t) = \int_0^t \Phi(t-s)\mathbf{f}(s)\,ds$ is a solution of $\mathbf{x}' = \mathbf{A}\mathbf{x} + \mathbf{f}(t)$. (*Hint*: The derivative contains two terms!) Use these properties of fundamental solution matrices: $\Psi(0) = \mathbf{I}$, $\Psi^{-1}(t) = \Psi(-t)$, and $\Psi(t + t_0) = \Psi(t)\Psi(t_0)$.

3. Suppose $\Psi(t)$ is a fundamental solution matrix for $\mathbf{x}' = \mathbf{A}\mathbf{x}$. Show that the solution of the initial-value problem $\mathbf{x}' = \mathbf{A}\mathbf{x} + \mathbf{f}(t)$, $\mathbf{x}(0) = \mathbf{b}$ is given by $\mathbf{x}(t) = \Psi(t)\mathbf{b} + \int_0^t \Phi(t-s)\mathbf{f}(s)\,ds$. (See Problem 2.)

PROJECTS

1. Uncoupling Systems

Consider $\mathbf{x}' = \mathbf{A}\mathbf{x}$ in which $(\lambda_1, \mathbf{x}_1), (\lambda_2, \mathbf{x}_2), \ldots, (\lambda_n, \mathbf{x}_n)$ are n eigenpairs associated with $\mathbf{A}$. Assume that $\lambda_i \neq \lambda_j$ for $i \neq j$. Let

$$\mathbf{S} = \begin{bmatrix} \mathbf{x}_1 & \mathbf{x}_2 & \ldots & \mathbf{x}_n \end{bmatrix}$$

(a) Show that $\Lambda = \mathbf{S}^{-1}\mathbf{A}\mathbf{S} = \begin{bmatrix} \lambda_1 & 0 & \ldots & 0 \\ 0 & \lambda_2 & \ldots & 0 \\ \vdots & & & \\ 0 & 0 & \ldots & \lambda_n \end{bmatrix}$

(b) Let $\mathbf{y} = \mathbf{S}\mathbf{x}$. Show that $\mathbf{x}' = \mathbf{A}\mathbf{x}$ transforms into $\mathbf{y}' = \Lambda\mathbf{y}$. Explain why this system in $\mathbf{y}$ is called "uncoupled" and find n linearly independent solutions. Use these solutions to find n solutions of $\mathbf{x}' = \mathbf{A}\mathbf{x}$.

2. Repeated Eigenvalues

In this project, we study three examples that show some of the complexities involved in solving systems with repeated roots. Note that all three systems have the characteristic equation $C(\lambda) = (1-\lambda)^4 = 0$, and therefore, all three systems have the eigvalue $\lambda = 1$ repeated three times.

(a) Consider the system $\mathbf{x}' = \begin{bmatrix} 1 & 1 & 0 & 0 \\ 0 & 1 & 0 & 0 \\ 0 & 0 & 1 & 0 \\ 0 & 0 & 0 & 1 \end{bmatrix} \mathbf{x}.$

Find three solutions of the form $\mathbf{x}(t) = \mathbf{u}e^t$ and one of the form $\mathbf{x}(t) = (\mathbf{v} + \mathbf{w}t)\, e^t, \mathbf{w} \neq \mathbf{0}$.

(b) Consider the system $\mathbf{x}' = \begin{bmatrix} 1 & 1 & 0 & 0 \\ 0 & 1 & 0 & 0 \\ 0 & 0 & 1 & 1 \\ 0 & 0 & 0 & 1 \end{bmatrix} \mathbf{x}.$

Find two solutions of the form $\mathbf{x}(t) = \mathbf{u}e^t$ and two of the form $\mathbf{x}(t) = (\mathbf{v} + \mathbf{w}t)\, e^t, \mathbf{w} \neq \mathbf{0}$.

(c) Consider the system $\mathbf{x}' = \begin{bmatrix} 1 & 1 & 0 & 0 \\ 0 & 1 & 1 & 0 \\ 0 & 0 & 1 & 0 \\ 0 & 0 & 0 & 1 \end{bmatrix} \mathbf{x}.$

Find a solution of the form $\mathbf{x}(t) = (\mathbf{v} + \mathbf{w}t + \mathbf{z}t^2)\, e^t$, where $\mathbf{w} \neq \mathbf{0}$ and $\mathbf{z} \neq \mathbf{0}$.

3. **Some Theorems on Fundamental Solution Matrices**

Let $\Psi(t)$ be a fundamental solution matrix of $\mathbf{x}' = \mathbf{A}\mathbf{x}$ at $t = 0$.

(a) Show that $\mathbf{A}\Psi(t) = \Psi(t)\mathbf{A}$. (*Hint*: Show that both matrices solve the same initial-value problem.)

(b) Show that $\Psi(-t)$ is a fundamental solution matrix for $\mathbf{x}' = -\mathbf{A}\mathbf{x}$ at $t = 0$. (*Hint*: $\Psi(-t)$ satisfies $\mathbf{X}' = -\mathbf{A}\mathbf{X}, \quad \mathbf{X}(0) = \mathbf{I}$.)

(c) Show that $\Psi^T(t)$ is a fundamental solution matrix of $\mathbf{x}' = \mathbf{A}^T\mathbf{x}$.

(d) Suppose $\Theta(t)$ is a fundamental solution matrix at $t = t_0$. Show that $\Theta^{-1}(t) = \Theta(t_0 - t)$.

(e) Show that $\Psi^{(k)}(t) = \mathbf{A}^k\Psi(t)$. Hence establish that $\Psi^{(k)}(t)$ is an invertible solution matrix of $\mathbf{X}' = \mathbf{A}^k\mathbf{X}$.

(f) Show that $\Psi^k(t) = \Psi(kt)$ is a fundamental solution matrix of $\mathbf{x}' = k\mathbf{A}\mathbf{x}$.

4. **The Fundamental Solution Matrix as an Exponential**

Suppose, as usual, that $\Psi(t)$ denotes the fundamental solution matrix of $\mathbf{x}' = \mathbf{A}\mathbf{x}$ at $t = 0$. Then $\Psi(t)$ satisfies

$$\Psi'(t) = \mathbf{A}\Psi(t), \quad \Psi(0) = \mathbf{I}$$

This result leads to the following expression for the higher derivatives of $\Psi(t)$:

$$\Psi^{(2)}(t) = \mathbf{A}\Psi'(t) = \mathbf{A}^2\Psi(t)$$

$$\Psi^{(3)}(t) = \mathbf{A}\Psi^{(2)}(t) = \mathbf{A}^3\Psi(t)$$

and, in general,

$$\Psi^{(k)}(t) = \mathbf{A}^k\Psi(t)$$

Thus, $\Psi^{(k)}(0) = \mathbf{A}^k$. Taylor's theorem can be applied to each entry of $\Psi^{(k)}(t)$ to yield the following expansion:

$$\begin{aligned}\Psi(t) &= \mathbf{I} + \mathbf{A}t + \frac{1}{2!}\mathbf{A}^2t^2 + \cdots + \frac{1}{k!}\mathbf{A}^kt^k + \ldots \\ &= \mathbf{I} + \sum_{k=1}^{\infty} \frac{1}{k!}\mathbf{A}^kt^k \end{aligned} \tag{1}$$

This series is meaningless without some conditions that indicate for which t the series converges. For the purpose of this problem assume that each of the n^2 series in (1) converges on $-a < t < a$. In view of the familiar Taylor series expansion of e^{at}, namely

$$e^{at} = 1 + \sum_{k=1}^{\infty} \frac{1}{k!} a^k t^k$$

it seems quite reasonable to define $\Psi(t) = e^{\mathbf{A}t}$ by the series (1). If this is done, and if we may assume that the derivative of a series with matrix coefficients is the term-by-term derivative of the series, then, as $\mathbf{A}$ is a constant matrix, differentiating (1) leads to

$$\begin{aligned}\frac{d}{dt}\Psi(t) &= \mathbf{A} + \mathbf{A}^2 t + \frac{1}{2!}\mathbf{A}^3 t^2 + \cdots + \frac{1}{(k-1)!}\mathbf{A}^k t^k + \ldots \\ &= \mathbf{A}\left(\mathbf{I} + \mathbf{A}t + \frac{1}{2!}\mathbf{A}^2 t^2 + \cdots + \frac{1}{k!}\mathbf{A}^k t^k + \ldots\right) \\ &= \mathbf{A}\Psi(t) \qquad (2)\end{aligned}$$

So at least formally, the series appears to represent a fundamental solution matrix. The series (2) can sometimes be used to construct $\Psi(t)$. What is required is that $\mathbf{A}^k$ be easy to determine.

(a) Suppose $\mathbf{A}^2 = a\mathbf{A}$. Find $\Psi(t)$ for the system $\mathbf{x} = \mathbf{A}\mathbf{x}$.

(b) Suppose $\mathbf{A}^2 = a^2\mathbf{I}$. Find $\Psi(t)$ for the system $\mathbf{x} = \mathbf{A}\mathbf{x}$.

5. **Some Properties of Eigenvalues and Eigenvectors**

Prove the following theorems.

(a) If $(\lambda_1, \mathbf{u})$ and $(\lambda_1, \mathbf{v})$ are eigenpairs of $\mathbf{A}$, then for each choice of scalars α and β, $(\lambda_1, \alpha\mathbf{u} + \beta\mathbf{v})$ is also an eigenpair, provided that $\alpha\mathbf{u} + \beta\mathbf{v} \neq \mathbf{0}$.

(b) If $(\lambda, \mathbf{u})$ and $(\mu, \mathbf{v})$ are eigenpairs of $\mathbf{A}$ with $\lambda \neq \mu$, then $\{\mathbf{u}, \mathbf{v}\}$ is linearly independent.

(c) If $(\lambda, \mathbf{u})$ and $(\mu, \mathbf{v})$ are eigenpairs of $\mathbf{A}$ with $\lambda \neq \mu$, then there cannot exist scalars α and β such that $\alpha\mathbf{u} + \beta\mathbf{v}$ is an eigenvector of $\mathbf{A}$. Contrast this conclusion with (a).

(d) If $\mathbf{A}$ is real and λ is a nonreal eigenvalue of $\mathbf{A}$, then $\mathbf{A}\mathbf{u} = \lambda\mathbf{u}$, $\mathbf{u} \neq \mathbf{0}$, implies that $\mathbf{u}$ cannot be real.

Chapter 3

Second-Order Linear Equations

3.1 Introduction

The second-order equation is the most common and most important equation arising in the applications of ODE's to the physical sciences. In Sections 3.1—3.5, we present aspects of the general theory of the second-order linear equation. In the remaining sections, we study the important special case in which the equations have constant coefficients. The theory for the nth-order linear equation is presented in Chapter 4.

3.2 Sectionally Continuous Functions

The general second-order linear ODE is given by

$$u'' + p(t)u' + q(t)u = f(t) \qquad (3.2.1)$$

The function $f(t)$ is called a *forcing function*, and $p(t), q(t)$ are *coefficient functions*. Although we shall assume that the coefficient functions are continuous, the forcing function often has jump discontinuities. Figures 3.1-3.3 illustrate three common forcing functions found in engineering practice, the "Unit Step", "Saw Tooth", and "Square Wave". Each has jump discontinuities.

The graphs in Figures 3.1—3.3 suggest what is meant by a jump discontinuity and how we might define the value of these functions at the jump:

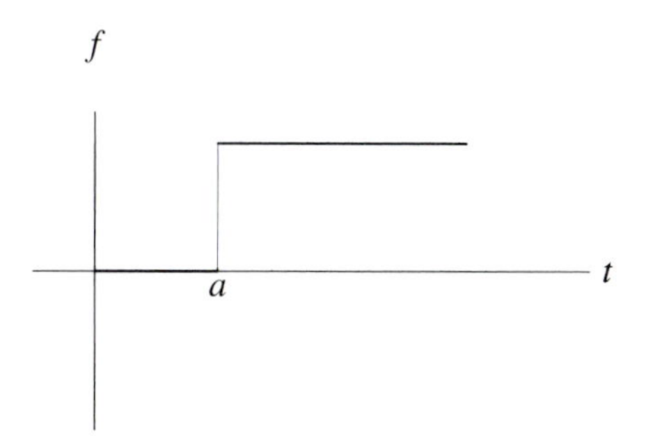

Figure 3.1: The Graph of a Unit Step Function

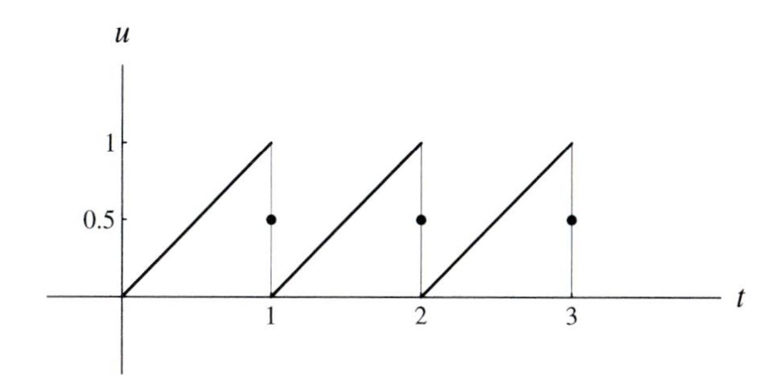

Figure 3.2: The Graph of a Saw Tooth Function

First, we need a notation for one-sided limits. We write

$$f(t_0^-) = \lim_{t \to t_0} f(t), \quad t < t_0 \tag{3.2.2}$$

$$f(t_0^+) = \lim_{t \to t_0} f(t), \quad t > t_0 \tag{3.2.3}$$

We say that f has a *jump discontinuity* at $t = t_0$ if the limits $f(t_0^-)$ and $f(t_0^+)$ both exist. The amount of the jump is $|f\left(t_0^+\right) - f\left(t_0^-\right)|$. Although we do not require $f(t)$ to have a value at t_0, we will define $f(t_0)$ as the average of limits from the right and the left:

$$f(t_0) = \tfrac{1}{2}\left[f(t_0^+) + f(t_0^-)\right] \tag{3.2.4}$$

Figures 3.1—3.3 illustrates these points. The value of the functions at their jumps is portrayed by a dot in these graphs.

Two ways in which a function can have a discontinuity that is not a jump are illustrated in Figures 3.4 and 3.5. Figure 3.4 is the graph of $u(t) = 1/t$; Figure 3.5 is the graph of $\sin \pi/t$. Neither function has a jump discontinuity at $t = 0$.

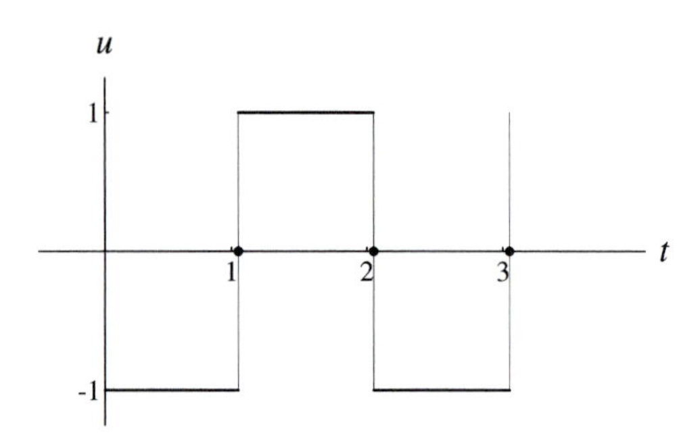

Figure 3.3: The Graph of a Square Wave

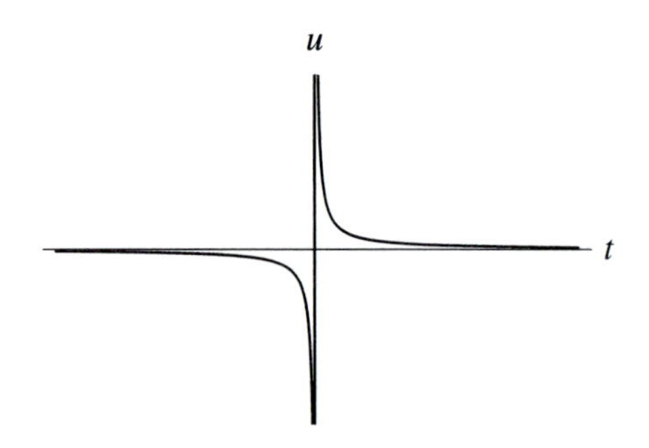

Figure 3.4: An Infinite Jump at $t = 0$

The function $f(t)$ is said to be *sectionally continuous* in (α, β) if it has only a finite number of jump discontinuities in each finite subinterval of (α, β). (We allow $\alpha = -\infty$ and/or $\beta = \infty$.) Thus, the "Unit Step," the "Sawtooth," and the "Square Wave" functions portrayed in Figures 3.1 — 3.3 are each sectionally continuous in $(-\infty, \infty)$. (Of course, if $f(t)$ has no discontinuities, it is continuous.) If α is finite, we require that $f(\alpha^+)$ exist. If β is finite, we require that $f(\beta^-)$ exist. The graphs of the functions sketched in Figures 3.4 and 3.5 are not sectionally continuous in any interval containing $t = 0$ or in which $t = 0$ is an end point.

We assume throughout[1] that the coefficient functions $p(t)$ and $q(t)$ are continuous in (α, β) and that the forcing function $f(t)$ is sectionally continuous in (α, β). Under these hypotheses, (3.2.1) and the initial-value problem

$$\begin{aligned} &u'' + p(t)u' + q(t)u = f(t) \\ &u(t_0) = b_1, \quad u'(t_0) = b_2 \end{aligned} \tag{3.2.5}$$

is called standard. Here is the fundamental theorem for standard second-order, linear initial-value problems.

[1]We remind the reader that (α, β) refers to the open interval $\alpha < t < \beta$.

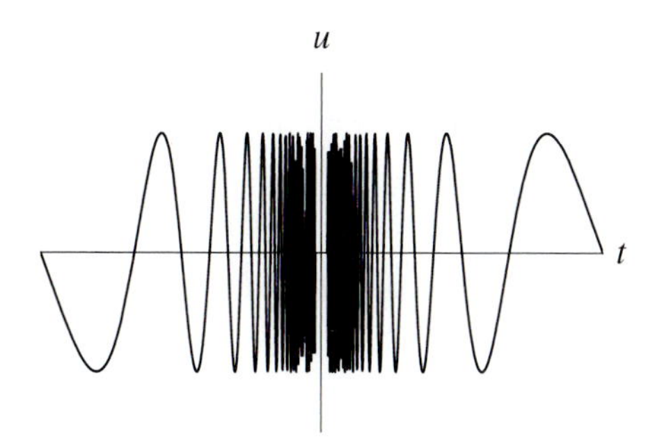

Figure 3.5: Infinite Oscilation at $t = 0$

Theorem 3.2.1. *The standard initial-value problem* (3.2.5) *has one and only one solution in* (α, β).

Proof: The proof of this important existence and uniqueness theorem is beyond the scope of the book. However, if we assume that $p(t)$ and $q(t)$ are constants then we can prove the existence part of this theorem by actually constructing the solution. We do this in Sections 3.6—3.10. □

The following corollary treats the case corresponding to the physical situation in which a system is at rest, is in equilibrium, and has no external forces acting on it. The conclusion can be interpreted as asserting the physically obvious fact that such systems remain permanently in equilibrium.

Corollary 3.2.1. *The standard homogeneous initial-value problem*

$$\begin{aligned} u'' + p(t)u' + q(t)u &= 0 \\ u(t_0) = 0, \quad u'(t_0) &= 0 \end{aligned} \tag{3.2.6}$$

has the unique solution $u(t) = 0$ *in* (α, β).

Proof: This system satisfies the hypothesis of the theorem. Hence, it has one and only one solution. Since $u(t) = 0$ is a solution, it is the only solution. □

•EXERCISES

1. Which of these functions have a discontinuity at $t = 0$? Indicate whether the discontinuity is a jump.
   (a) $f(t) = \csc t$. (b) $f(t) = \ln |t|$.
   (c) $f(t) = |t|$. (d) $f(t) = t^{-2}$.
   (e) $f(t) = \begin{cases} t^{-1}\sin t, & t \neq 0 \\ 0, & t = 0 \end{cases}$ (f) $f(t) = \begin{cases} t^{-1}\sin \pi/t, & t \neq 0 \\ 0, & t = 0 \end{cases}$

2. Which of the following functions are sectionally continuous?
   (a) $f(t) = \ln t, \quad t > 0$ (b) $f(t) = \ln t, \quad t > 1$
   (c) $f(t) = \begin{cases} 1/t, & t \neq 0 \\ 0, & t = 0 \end{cases}$ (d) $f(t) = \operatorname{sgn} t = \begin{cases} 1 & \text{if } t > 0 \\ 0 & \text{if } t = 0 \\ -1 & \text{if } t < 0 \end{cases}$
   (e) $f(t) = \begin{cases} |\sin t|, & t \geq 0 \\ 0, & t < 0 \end{cases}$ (f) $f(t) = \begin{cases} 1, & 0 \leq t \leq 1 \\ 0, & \text{otherwise} \end{cases}$

3. What is the analogue of Theorem 3.2.1 for the first-order, linear, initial-value problem $u' + p(t)u = f(t)$, $u(t_0) = b$?

3.3 Linear Differential Operators

Given any twice differentiable function $u(t)$, the expression

$$L(u) = u'' + p(t)u' + q(t)u \tag{3.3.1}$$

defines a function $g(t) = L(u)$ that may be viewed in an "operator" context: For each allowable input $u(t)$, L produces the output $g(t)$. We describe this diagrammatically by

$$u(t) \to \boxed{L} \to g(t)$$

Thus, $u(t)$ is a solution of (3.2.1) if and only if $L(u) = f(t)$. For example, $u_1(t) = 1 - t$ is a solution of $u'' + 3u' + 2u = -2t - 1$ because

$$u_1(t) = 1 - t \to \boxed{L} \to -2t - 1$$

where $L = D^2 + 3D + 2$ and $D = d/dt$, $D^2 = d^2/dt^2$.

Example 3.3.1. *Find $L(u) = u'' + 3u' + 2u$ for each of the following inputs:*
   (a) $u_1(t) = 1 - t$. (b) $u_2(t) = e^t$.
   (c) $u_3(t) = e^{-t}$. (d) $u_4(t) = k$.

Solution: The calculations are straightforward. The operator is $L = D^2 + 3D + 2$ and

(a) $f_1(t) = L(1-t) = 0 + 3(-1) + 2(1-t) = -2t - 1$.
(b) $f_2(t) = L(e^t) = e^t + 3e^t + 2e^t = 6e^t$.
(c) $f_3(t) = L(e^{-t}) = e^{-t} - 3e^{-t} + 2e^{-t} = 0$.
(d) $f_4(t) = L(k) = 0 + 3 \cdot 0 + 2k = 2k$.

Theorem 3.3.1. *For each pair of constants c_1 and c_2*

$$L(c_1 u_1 + c_2 u_2) = c_1 L(u_1) + c_2 L(u_2) \tag{3.3.2}$$

Proof: The proof is set as Problem 1 in the exercise set at the end of the section. □

The property described in (3.3.2) is shared by many operators. The derivative $D = d/dt$ and the integral $\int$ are two familiar examples. In Chapter 5 we study the Laplace transform, another operator satisfying (3.3.2). Any operator that satisfies (3.3.2) is called *linear*. Note that linearity implies $L(0) = 0$, since we may choose $c_1 = c_2 = 0$. The following corollary can be used as an alternative definition of linearity.

Corollary 3.3.1. *If L is a linear operator, then*

$$\begin{array}{ll} (a) & L(cu) = cL(u) \\ (b) & L(u_1 + u_2) = L(u_1) + L(u_2) \end{array} \tag{3.3.3}$$

Conversely, if L satisfies (a) *and* (b) *for every c, then L is linear.*

Proof: First, (a) follows from (3.3.2) from Theorem 3.3.1 by choosing $c_1 = c$ and $c_2 = 0$. Second, (b) follows by setting $c_1 = c_2 = 1$ in this same equation. Conversely, $L(c_1 u_1 + c_2 u_2) = L(c_1 u_1) + L(c_2 u_2)$ from (b). Now invoke (a) twice to deduce (3.3.2). □

Theorem 3.3.2 (Superposition Principle I). *If $u_1(t)$ and $u_2(t)$ are solutions of a standard* ODE $L(u) = 0$, *then for every c_1 and c_2,*

$$c_1 u_1(t) + c_2 u_2(t)$$

is also a solution of $L(u) = 0$.

Proof: By hypothesis, $L(u_1) = L(u_2) = 0$. By (3.3.3),

$$\begin{aligned} L(c_1 u_1 + c_2 u_2) &= c_1 L(u_1) + c_2 L(u_2) \\ &= c_1 0 + c_2 0 = 0 \end{aligned}$$

as required. □

Theorem 3.3.3 (Superposition Principle II). *If $u_1(t)$ and $u_2(t)$ are solutions of the standard* ODE's $L(u) = f_1(t)$ *and* $L(u) = f_2(t)$, *respectively, then*

$$a\, u_1(t) + b\, u_2(t)$$

is a solution of

$$L(u) = a f_1(t) + b f_2(t) \tag{3.3.4}$$

Proof: The proof is essentially the argument given for Theorem 3.3.2. See also Problem 5. □

•EXERCISES

1. Derive the linearity property, (3.3.2).

2. Define $L_1(u) = u' + p(t)u$. Show that L_1 is a linear operator, that is, verify (3.3.2) for L_1.

3. Define $L_n(u) = u^{(n)} + p_0(t)u^{(n-1)} + \cdots + p_{n-2}(t)u^{(1)} + p_{n-1}(t)u$. Verify that L_n satisfies (3.3.2).

4. Let $L(u) = u'' + 2pu' + qu$, where p and q are constants. Show that $L(e^{\lambda t}) = e^{\lambda t}(\lambda^2 + 2p\lambda + q)$.

5. Suppose that $c_1 + c_2 = 1$ and that L is a linear operator with the properties that $L(u) = L(v) = f(t)$. Prove that $L(c_1 u + c_2 v) = f(t)$.

6. Suppose L is a linear operator with the properties that $L(u) = 0$ and $L(u_p) = f(t)$. Show that $L(cu + u_p) = f(t)$.
7. Suppose L is a linear operator. If $L(u_1) = L(u_2) = f(t)$, show that $L(u_1 - u_2) = 0$.
8. Interpret the following as operators. Which, if either, are linear?

$$\text{(a) } \mathbf{A} \to \boxed{\text{det}} \to \det \mathbf{A}, \quad \text{(b) } \mathbf{x} \to \boxed{\mathbf{A}} \to \mathbf{Ax}$$

3.4 Linear Independence and the Wronskian

The list of functions $u_1(t), u_2(t), \ldots, u_k(t)$ is said to be *linearly dependent* if there exists a list of constants $a_1, a_2, \ldots, a_k$, not all zero, such that for all t in (α, β),

$$a_1 u_1(t) + a_2 u_2(t) + \cdots + a_k u_k(t) = 0 \tag{3.4.1}$$

For example, the list $t - 1, 2 - 2t$ is linearly dependent in $(-\infty, \infty)$ because $a_1 = 2, a_2 = 1$ leads to the identity $2(t-1) + 1(2-2t) = 0$. Likewise, the list $e^t, e^{-t}, \sinh t$ is linearly dependent because, for all t,

$$-\tfrac{1}{2}e^t + \tfrac{1}{2}e^{-t} + \sinh t = 0$$

(Recall that $\sinh t = (e^t - e^{-t})/2$.) A list that is not linearly dependent is called *linearly independent.*

The problem of which lists are linearly independent and which are not can be quite complicated. However, *if the functions in the list are solutions of a linear ODE,* then this issue can be completely resolved by constructing a special "testing" function known as the Wronskian of the solutions.

The *Wronskian* of the solutions $u_1(t)$ and $u_2(t)$ of

$$u'' + p(t)u' + q(t)u = 0 \tag{3.4.2}$$

written $W(t) = W(u_1, u_2)$, is defined by

$$W(t) = \det \begin{bmatrix} u_1(t) & u_2(t) \\ u_1'(t) & u_2'(t) \end{bmatrix} = u_1(t)u_2'(t) - u_1'(t)u_2(t) \tag{3.4.3}$$

Theorem 3.4.1. *The Wronskian $W(t)$ of two solutions of the standard* ODE $u'' + p(t)u' + q(t)u = 0$ *is a solution of*

$$W' + p(t)W = 0 \tag{3.4.4}$$

on the interval (α, β).

Proof: From (3.4.3),

$$\begin{aligned} W'(t) &= u_1 u_2'' + u_1' u_2' - u_1'' u_2 - u_1' u_2' \\ &= u_1 u_2'' - u_1'' u_2 \end{aligned} \tag{3.4.5}$$

Since both $u_1(t)$ and $u_2(t)$ are assumed to be solutions of (3.4.2), we have these equations:

$$u_1'' = -pu_1' - qu_1 \tag{3.4.6}$$

$$u_2'' = -pu_2' - qu_2 \tag{3.4.7}$$

If (3.4.6) and (3.4.7) are substituted into (3.4.5), we obtain

$$\begin{aligned} W'(t) &= u_1 \left(-pu_2' - qu_2\right) - u_2 \left(-pu_1' - qu_1\right) \\ &= -p \left(u_1 u_2' - u_1' u_2\right) = -p(t)W(t) \end{aligned}$$

which is (3.4.4). □

Because $W(t)$ satisfies a first-order linear ODE, we can write an explicit formula for $W(t)$ in terms of $p(t)$. As in Chapter 1, set

$$P(t) = \int p(t)\, dt \tag{3.4.8}$$

Corollary 3.4.1. *Under the hypothesis of Theorem 3.4.1,*

$$W(t) = Ke^{-P(t)} \tag{3.4.9}$$

Proof: By Corollary 1.3.2 and Theorem 3.4.1, all solutions of (3.4.4) are given by (3.4.9). □

Since Corollary 3.4.1 shows that the Wronskian is determined (up to a multiplicative constant) by $p(t)$, we shall often use the shorter statement "the Wronskian of the equation" in place of the more elaborate "the Wronskian of any two solutions of the equation."

Corollary 3.4.2. *Under the hypothesis of Theorem 3.4.1, if* $W(t_0) = 0$ *for any* t_0 *in* (α, β), *then* $W(t) = 0$ *for all* t *in* (α, β).

Proof: Since $p(t)$ is continuous, so is $P(t)$, and therefore, $e^{-P(t)}$ is positive. From (3.4.9), we see that $W(t_0) = 0$ implies $K = 0$, and therefore, $W(t)$ is identically zero in (α, β). □

Example 3.4.1. *Find the Wronskian of the constant coefficient ODE*

$$u'' + au' + bu = 0$$

Solution: Since $p(t) = a$, it follows that $P(t) = \int a\,dt = at$ and, hence, that $W(t) = Ke^{-at}$.

Note: If we had included the arbitrary constant so that $P(t) = at + c$, then

$$W(t) = Ke^{-at-c} = Ke^{-c}e^{-at} = K_1e^{-at}$$

where K_1 is simply another constant. Hence, it is unnecessary to include the arbitrary constant in constructing $P(t)$.

Example 3.4.2. *Find the Wronskian of* $u'' + q(t)u = 0$.

Solution: Here, $p(t) = 0$, so $P(t) = 0$, and hence,

$$W(t) = Ke^0 = K$$

Example 3.4.3. *Find the Wronskian of the Cauchy-Euler equation,*

$$t^2u'' + atu' + bu = 0, \quad t > 0$$

Solution: The definition of $p(t)$ requires that the coefficient of u'' be 1. So we divide the given equation by t^2. Thus, $p(t) = a/t$, and

$$P(t) = \int \frac{a}{t}\, dt = \ln t$$

Hence

$$W(t) = Ke^{-a \ln t} = Ke^{\ln t^{-a}} = Kt^{-a}$$

If $a = -1$ in Example 3.4.3, then $W(t) = Kt$ and $W(0) = 0$, an apparent contradiction of Corollary 3.4.2. It is not a contradiction because $p(t)$ is assumed to be continuous in any interval in which we apply our theorems. In this case $p(t) = 1/t$, and thus, $p(t)$ is clearly not continuous in any interval containing $t = 0$.

The connection between the linear independence of solutions and the Wronskian of these solutions is provided by the next critical theorem.

Theorem 3.4.2. *The solutions* $u_1(t)$ *and* $u_2(t)$ *of*

$$u'' + p(t)u' + q(t)u = 0 \tag{3.4.10}$$

are linearly independent on (α, β) *if and only if* $W(u_1, u_2)$ *does not vanish anywhere in* (α, β).

Proof: Let us suppose that $W(t_0) = W[u_1(t_0), u_2(t_0)] = 0$ for some t_0 in $\alpha < t < \beta$. Let us check for dependency using (3.4.1). The simultaneous equations (in the unknowns a_1, a_2)

$$a_1 u_1(t_0) + a_2 u_2(t_0) = 0 \tag{3.4.11}$$

$$a_1 u_1'(t_0) + a_2 u_2'(t_0) = 0 \tag{3.4.12}$$

have a nontrivial solution because the determinant of the coefficients is $W(t_0) = 0$. Now $u(t) = a_1 u_1(t) + a_2 u_2(t)$ satisfies (3.4.10). By (3.4.11) and (3.4.12), $u(t)$ satisfies the initial conditions $u(t_0) = u'(t_0) = 0$. Thus, by Corollary 3.4.2, $u(t) = 0$ for all t in (α, β). But then $u_1(t), u_2(t)$ are linearly dependent by definition.

Conversely, if $u_1(t)$ and $u_2(t)$ are linearly dependent, then there exist constants a_1, a_2 (not both zero) such that

$$a_1 u_1(t) + a_2 u_2(t) = 0 \tag{3.4.13}$$

for all t in (α, β). So, by differentiating (3.4.13), we deduce

$$a_1 u_1'(t) + a_2 u_2'(t) = 0 \tag{3.4.14}$$

for all t in (α, β). Once again we make the observation that the determinant of the unknowns a_1 and a_2 in (3.4.13) and (3.4.14) is $W(t)$. If $W(t_0) \neq 0$ for any t_0 in (α, β), we conclude that $a_1 = a_2 = 0$, a contradiction. □

Any set of two linearly independent solutions of the homogeneous equation $u'' + p(t)u' + q(t)u = 0$ is called a *basic set.* (Some authors call these sets "fundamental." We use a more restrictive definition for "fundamental.")

Theorem 3.4.3. *The standard homogeneous equation*

$$u'' + p(t)u' + q(t)u = 0 \tag{3.4.15}$$

always has a basic set of solutions.

Proof: Theorem 3.2.1 guarantees a solution of (3.4.15) satisfying

$$u(t_0) = 1, \qquad u'(t_0) = 0 \tag{3.4.16}$$

Let us denote this solution by $u_1(t)$. Likewise let $u_2(t)$ be the solution of (3.4.15) satisfying

$$u(t_0) = 0, \qquad u'(t_0) = 1 \tag{3.4.17}$$

Then,

$$W(t_0) = \det \begin{bmatrix} u_1(t_0) & u_2(t_0) \\ u_1'(t_0) & u_2'(t_0) \end{bmatrix} = \det \begin{bmatrix} 1 & 0 \\ 0 & 1 \end{bmatrix} = 1$$

By Corollary 3.4.2, $W(t) \neq 0$ in $\alpha < t < \beta$. Therefore, by Theorem 3.4.2, $u_1(t)$, $u_2(t)$ is a linearly independent set of solutions of (3.4.15). So $u_1(t)$, $u_2(t)$ is a basic set. □

Theorem 3.4.4. *Suppose* $u_1(t), u_2(t)$ *is a basic set of solutions of*

$$u'' + p(t)u' + q(t)u = 0 \tag{3.4.18}$$

and $u_h(t)$ *is any solution of* (3.4.18). *Then there exist constants* c_1, c_2 *such that*

$$u_h(t) = c_1 u_1(t) + c_2 u_2(t) \tag{3.4.19}$$

Proof: The simultaneous equations in the unknowns, c_1, c_2

$$\begin{aligned} c_1 u_1(t_0) + c_2 u_2(t_0) &= u_h(t_0) \\ c_1 u_1'(t_0) + c_2 u_2'(t_0) &= u_h'(t_0) \end{aligned} \tag{3.4.20}$$

have a solution (and it is unique) if and only if the determinant of their coefficients is nonzero. But this determinant is the Wronskian $W(t)$ of $u_1(t)$ and $u_2(t)$. Because $u_1(t)$, $u_2(t)$ is a basic set, $W(t) \neq 0$ for t in (α, β). Indeed,

$$c_1 = \frac{W[u_h(t_0), u_2(t_0)]}{W(t_0)}, \quad c_2 = \frac{W[u_1(t_0), u_h(t_0)]}{W(t_0)}$$

Using these values for c_1 and c_2 to define $u_q(t) = c_1 u_1(t) + c_2 u_2(t)$, it is easy to verify that the function

$$w(t) = u_h(t) - u_q(t)$$

is a solution of the initial-value problem

$$u'' + p(t)u' + q(t)u = 0, \quad u(t_0) = 0, u'(t_0) = 0$$

By Corollary 3.4.2, $w(t) = 0$, and hence, $u_q(t) = u_h(t)$. □

The point of this theorem is this: The set $c_1 u_1(t) + c_2 u_2(t)$ contains all solutions of $u'' + p(t)u' + q(t)u = 0$. This is one of the reasons we call $u_h(t)$ a general solution of $u'' + p(t)u' + q(t)u = 0$. A second reason, closely related to the first, is that the initial-value problem

$$u'' + p(t)u' + q(t)u = 0, \qquad u(0) = b_1,\ u'(0) = b_2 \tag{3.4.21}$$

always has a solution obtainable from the general solution by specifying the constants c_1 and c_2.

The basic set meeting the initial conditions (3.4.16) and (3.4.17) is called a *fundamental set.* The solution of the standard initial-value problem is greatly simplified by using the fundamental set of solutions. For example, suppose we wish to solve the initial-value problem (3.4.21). If $u_1(t)$, $u_2(t)$ is a fundamental set for $u'' + p(t)u' + q(t)u = 0$, then

$$u(t) = b_1u_1(t) + b_2u_2(t)$$

is the desired solution because, by (3.4.16) and (3.4.17),

$$\begin{aligned} u(t_0) &= b_1u_1(t_0) + b_2u_2(t_0) = b_1 \cdot 1 + b_2 \cdot 0 = b_1 \\ u'(t_0) &= b_1u_1'(t_0) + b_2u_2'(t_0) = b_1 \cdot 0 + b_2 \cdot 1 = b_2 \end{aligned}$$

Example 3.4.4. *Show that the functions*

$$\begin{aligned} u_1(t) &= t^2 - 2t + 2 \\ u_2(t) &= t - 1 \end{aligned}$$

comprise a fundamental set for the equation

$$t(t-2)u'' - 2(t-1)u' + 2u = 0$$

at $t_0 = 1$.

Solution: The verification that these two functions satisfy the differential equation is left for Problem 5 of the exercise set. We need only note that $u_1'(t) = 2(t-1)$ and that $u_2'(t) = 1$. So, to have $u_1(t_0) = 1$, we require that $u_1(t_0) = t_0^2 - 2t_0 + 2 = 1$, and hence, $t_0 = 1$. Thus, $W[u_1(1), u_2(1)] = 1$. Thereforeon $u_1(t)$ and $u_2(t)$ form a fundamental set.

Example 3.4.5. *Show that* $\cosh t, \sinh t$ *is a fundamental set of solutions at* $t_0 = 0$ *of* $u'' - u = 0$.

Solution: It is easy to verify that this pair of functions is a solution and to check that $\sinh 0 = 0$ and $\cosh 0 = 1$. Since $D(\sinh t) = \cosh t$ and $D(\cosh t) = \sinh t$, we see that the derivative condition is also met. Thus, $W(\cosh t, \sinh t) = \cosh^2 t - \sinh^2 t = 1$ and they are a fundamental set.

•EXERCISES

1. Find the Wronskian of each differential equation.
   (a) $u'' + 2u' = 0$. (b) $u'' + 2u = 0$.
   (c) $u'' + tu = 0$. (d) $u'' + tu' = 0$.
   (e) $u'' + 2tu' + tu = 0$. (f) $u'' + atu' + bt^2u = 0$.

2. If u_1, u_2 is a basic set of solutions of $u'' + p(t)u' + q(t)u = 0$, show that the following two sets of solutions are also basic solutions for this equation: $u_1 + u_2$, $u_1 - u_2$ and u_1, $u_1 + u_2$.

3. Show that two solutions of $u'' + p(t)u' + q(t) = 0$ comprise a basic set if and only if they are not proportional, that is, $u_1(t) \neq Ku_2(t)$.

4. One solution of $u'' + 2au' + a^2u = 0$ is e^{-at}. Find a second, linearly independent solution by using the definition of the Wronskian and (3.4.9).

5. Verify that the functions $u_1(t) = t^2 - 2t + 2, u_2(t) = t - 1$ are solutions of $t(t-2)u'' - 2(t-1)u' + 2u = 0$.

6. Verify that $\cosh t, \sinh t, e^t$, and e^{-t} are solutions of $u'' - u = 0$.

7. Verify that $\cos t$, $\sin t$ is a basic set of solutions of $u'' + u = 0$. Is it a fundamental set?

8. Given that u_1, u_2 is a basic set of solutions of $u'' = p(t)u' + q(t)u = 0$, show that $v_1 = \alpha u_1 + \beta u_2$, $v_2 = \gamma u_1 + \delta u_2$ is a fundamental set of this equation if and only if

$$\det \begin{bmatrix} \alpha & \beta \\ \gamma & \delta \end{bmatrix} = \frac{1}{W(u_1, u_2)}$$

3.5 The Nonhomogeneous Equation

Suppose that $u_1(t), u_2(t)$ is a basic set of solutions for the standard complementary equation[2] associated with the nonhomogeneous equation,

$$L(u) = u'' + p(t)u' + q(t)u = f(t) \tag{3.5.1}$$

That is, $L(u_1) = L(u_2) = 0$ and these solutions are linearly independent, so that $W(t) \neq 0$ in the interval (α, b). We now assume that the function $u_p(t)$ has been found that satisfies (3.5.1), so that $L(u_p) = f(t)$. Then, for every choice of c_1 and c_2,

$$u(t) = u_p(t) + c_1u_1(t) + c_2u_2(t) \tag{3.5.2}$$

[2]Recall that the complementary equation has 0 for its forcing function.

also solves (3.5.1) because

$$L(u) = L(u_p) + c_1 L(u_1) + c_2 L(u_2) = f(t) \tag{3.5.3}$$

We call $u(t)$ defined in (3.5.2) a *general solution* of (3.5.1), and $u_p(t)$ a *particular solution.*[3] The reason for this terminology is motivated by the next theorem.

Theorem 3.5.1. *Let $u_p(t)$ be a particular solution of*

$$u'' + p(t)u' + q(t)u = f(t) \tag{3.5.4}$$

and u_1, u_2 be a basic set of solutions of its complementary equation. If $u(t)$ is any solution of (3.5.4), *then there exist constants c_1, c_2 such that*

$$u(t) = u_p(t) + c_1 u_1(t) + c_2 u_2(t) \tag{3.5.5}$$

Proof: The function $u(t) - u_p(t)$ is a solution of the complementary equation of (3.5.4). By Theorem 3.4.4, there exist constants c_1 and c_2 such that $u(t) - u_p(t) = c_1 u_1(t) + c_2 u_2(t)$. □

General solutions are infinite sets of solutions, one solution for each choice of the arbitrary constants that appear in the general solution. The representation of a general solution can take many forms. For example, besides $c_1 e^t + c_2 e^{-t}$, another general solution for $u'' - u = 1$ is $u(t) = -1 + c_1 \cosh t + c_2 \sinh t$. These two sets are the same.

•EXERCISES

1. Given the solution $u_p(t) = 1$ of $t(t-2)u'' - 2(t-1)u' + 2u = 2$ find the general solution of this equation.

2. Use the answer to Problem 1 to find a solution to the following initial-value problems, where $t(t-2)u'' - 2(t-1)u' + 2u = 2$.
(a) $u(1) = 0, \quad u'(1) = 0.$
(b) $u(1) = 1, \quad u'(1) = 0.$
(c) $u(1) = 1, \quad u'(1) = 1.$
(d) $u(1) = -2, \quad u'(1) = 4.$
(e) $u(1) = 4, \quad u'(1) = -10.$

3. Suppose $u_p(t)$ is a solution of $u'' + p(t)u' + q(t)u = f(t)$ with the property that $u_p(t_0) = 0, u_p'(t_0) = 0$. Suppose that $\{u_1(t), u_2(t)\}$ is a

[3] By convention, a particular solution never contains any arbitrary constants.

fundamental set of solutions of its complementary equation. Show that $u(t) = u_p(t) + b_1u_1(t) + b_2u_2(t)$ is the solution with the property $u(t_0) = b_1$, $u'(t_0) = b_2$.

4. Suppose $u_p(t)$ is a solution of $u'' + p(t)u' + q(t)u = f(t)$ and that $u_1(t)$, $u_2(t)$ is a set of fundamental solutions of its complementary equation. Show that the solution $u(t) = u_p(t) - u_p(t_0)\, u_1(t) - u_p'(t_0)\, u_2(t)$ meets the initial conditions $u(t_0) = 0 = u'(t_0)$.

3.6 Constant Coefficient Homogeneous Equations

The most general second-order homogeneous ODE with constant coefficients can be written in the form

$$u'' + pu' + qu = 0 \tag{3.6.1}$$

where p and q are real constants. The initial-value problem for (3.6.1) is

$$u'' + pu' + qu = 0, \qquad u(0) = b_1, \quad u'(0) = b_2 \tag{3.6.2}$$

Experience with linear constant coefficient ODE's shows that exponential functions are major components of the solutions. Motivated by this, we ask whether it is possible to find a constant λ such that $u(t) = e^{\lambda t}$ is a solution of (3.6.1). Since $u'(t) = \lambda e^{\lambda t}$ and $u''(t) = \lambda^2 e^{\lambda t}$, $u(t) = e^{\lambda t}$ is a solution of (3.6.1) if and only if

$$\begin{aligned} u''(t) + pu'(t) + qu(t) &= \lambda^2 e^{\lambda t} + p\lambda e^{\lambda t} + qe^{\lambda t} \\ &= e^{\lambda t}\left(\lambda^2 + p\lambda + q\right) = 0 \end{aligned} \tag{3.6.3}$$

Because $e^{\lambda t} \neq 0$, $u''(t) + pu'(t) + qu(t) = 0$ holds if and only if

$$C(\lambda) = \lambda^2 + p\lambda + q = 0 \tag{3.6.4}$$

We call $C(\lambda)$ the *characteristic polynomial* and $C(\lambda) = 0$ the *characteristic equation* of $u'' + pu' + qu = 0$. Equation (3.6.4) is quadratic with real coefficients and thus has two solutions. These solutions may be real and distinct, real and equal, or both complex. The various cases arise because of the term $\sqrt{p^2 - 4q}$ that appears in the solution of (3.6.4). We treat each case separately.

3.6.1 Real and Unequal Roots

This case occurs if and only if $p^2 - 4q > 0$. Let λ_1 and λ_2 be the two distinct real roots of (3.6.4). Then

$$u_1(t) = e^{\lambda_1 t} \quad \text{and} \quad u_2(t) = e^{\lambda_2 t} \tag{3.6.5}$$

are two solutions of (3.6.1). We compute the Wronskian of these solutions to show that they are linearly independent. (See Section 3.4.) We have

$$u_1'(t) = \lambda_1 e^{\lambda_1 t} \quad \text{and} \quad u_2'(t) = \lambda_2 e^{\lambda_2 t} \tag{3.6.6}$$

and hence,

$$\begin{aligned} W(u_1, u_2) &= \det \begin{bmatrix} u_1 & u_2 \\ u_1' & u_2' \end{bmatrix} \\ &= e^{\lambda_1 t}\lambda_2 e^{\lambda_2 t} - e^{\lambda_2 t}\lambda_1 e^{\lambda_1 t} \\ &= (\lambda_2 - \lambda_1)\, e^{(\lambda_1 - \lambda_2)t} \neq 0 \end{aligned}$$

since $\lambda_1 \neq \lambda_2$ and $e^{kt} \neq 0$ for all k and t. Thus, the general solution of (3.6.1) is given by

$$\begin{aligned} u(t) &= c_1 u_1(t) + c_2 u_2(t) \\ &= c_1 e^{\lambda_1 t} + c_2 e^{\lambda_2 t} \end{aligned} \tag{3.6.7}$$

Obviously, u_1, u_2 is a basic set.

Example 3.6.1. *Find a general solution of $u'' + 5u' + 6u = 0$, and choose the arbitrary constants so that $u(0) = 1$ and $u'(0) = 0$.*

Solution: The characteristic equation is given by (3.6.4)

$$\lambda^2 + 5\lambda + 6 = 0$$

which has solutions $\lambda_1 = -3, \lambda_2 = -2$. Hence, by (3.6.7), a general solution is

$$u(t) = c_1 e^{-3t} + c_2 e^{-2t}$$

The first derivative is

$$u'(t) = -3c_1 e^{-3t} - 2c_2 e^{-2t}$$

In order for $u(t)$ to meet the initial conditions, we set $t = 0$ in both of the preceding equations to obtain the conditions

$$u(0) = c_1 + c_2 = 1, \qquad u'(0) = -3c_1 - 2c_2 = 0$$

A little algebra shows that $c_1 = -2$ and $c_2 = 3$. Hence, the initial-value problem has the unique solution

$$u(t) = -2e^{-3t} + 3e^{-2t}$$

Example 3.6.2. *Find a general solution of $u'' + u' = 0$.*

Solution: The characteristic equation is $\lambda^2 + \lambda = 0$, and this quadratic has solutions $\lambda_1 = -1, \lambda_2 = 0$. Hence, a general solution of $u'' + u' = 0$ is

$$u(t) = c_1 + c_2 e^{-t}$$

Note that the root $\lambda = 0$ corresponds to the constant solution because $e^{0t} = 1$.

3.6.2 Real and Equal Roots

When the roots of $C(\lambda)$ are equal, we have $p^2 - 4q = 0$, and this implies that $q = \frac{1}{4}p^2$. Two facts follow from this observation. First, (3.6.1) now takes the form

$$u'' + pu' + \tfrac{1}{4}p^2 u = 0 \tag{3.6.8}$$

Second, the solutions of the characteristic equation are both $\lambda = -\frac{1}{2}p$, and hence, there is *only one* exponential solution:

$$u_1(t) = e^{-pt/2} \tag{3.6.9}$$

A second linearly independent solution can be discovered by considering trial functions other than $e^{\lambda t}$. Either inspired guesswork or a technique to be presented later on in this section yields another solution:

$$u_2(t) = te^{-pt/2} \tag{3.6.10}$$

That this function is indeed a solution is shown by verifying that

$$u_2''(t) + pu_2'(t) + \tfrac{1}{4}p^2 u_2(t) = 0$$

This is Problem 35 of the exercise set. To show that u_1, u_2 is a basic set, we compute the Wronskian of these functions. To do this conveniently, we express the derivatives of $u_1(t)$ and $u_2(t)$ in terms of $u_1(t)$:

$$u_1'(t) = -\tfrac{1}{2}pe^{-pt/2} = -\tfrac{1}{2}p\,u_1(t) \tag{3.6.11}$$

$$u_2'(t) = e^{-pt/2} - \tfrac{1}{2}p\,te^{-pt/2} = \left(1 - \tfrac{1}{2}p\,t\right)u_1(t) \tag{3.6.12}$$

Now the computation of the Wronskian proceeds easily:

$$W(u_1, u_2) = \left(1 - \tfrac{1}{2}pt\right)u_1^2(t) + \tfrac{1}{2}ptu_1^2(t) = u_1^2(t) \neq 0$$

Thus, a general solution is given by

$$u(t) = (c_1 + c_2t)\,e^{-pt/2} \tag{3.6.13}$$

Example 3.6.3. *Find a general solution of* $u'' + 6u' + 9u = 0$.

Solution: The characteristic equation is

$$\lambda^2 + 6\lambda + 9 = 0$$

with the solutions $\lambda_1 = \lambda_2 = -3$. From (3.6.13), we have

$$u(t) = (c_1 + c_2t)\,e^{-3t}$$

Example 3.6.4. *Find a general solution of* $u'' = 0$.

Solution: The characteristic equation is $\lambda^2 = 0$ with roots $\lambda_1 = \lambda_2 = 0$. From (3.6.13), we have

$$u(t) = c_1 + c_2t$$

a solution more easily obtained by integrating $u'' = 0$ twice.

We now motivate the choice of $te^{-pt/2}$ as the second, linearly independent solution of $L(u) = u'' + pu' + \frac{1}{4}p^2u$ by presenting a special case of an important technique in linear differential equation theory. Let us assume that the second solution is given by

$$u_2(t) = v(t)e^{-pt/2} \tag{3.6.14}$$

for some $v(t)$. Now, differentiating (3.6.14) results in

$$u_2' = \left(-\tfrac{1}{2}pv + v'\right) e^{-pt/2} \tag{3.6.15}$$

The derivative of (3.6.15) yields

$$\begin{aligned} u_2'' &= \left(-\frac{1}{2}pv' + v''\right) e^{-pt/2} - \frac{1}{2}p\left(-\frac{1}{2}pv + v'\right) e^{-pt/2} \\ &= \left(v'' - pv' + \frac{1}{4}p^2 v\right) e^{-pt/2} \end{aligned} \tag{3.6.16}$$

So

$$\begin{aligned} L(u_2) &= \left(\left(v'' - pv' + \frac{1}{4}p^2 v\right) + p\left(-\frac{1}{2}pv + v'\right) + \frac{1}{4}p^2 v\right) e^{-pt/2} \\ &= v'' e^{-pt/2} \end{aligned} \tag{3.6.17}$$

From (3.6.17), we see that $u_2(t)$ is a solution of (3.6.8) if and only if

$$v'' e^{-pt} = 0$$

Since $e^{-pt/2} \neq 0$, we demand that $v'' = 0$; so $v(t) = t$ suffices. Thus, using $v(t) = t$ in (3.6.14) results in

$$u_2(t) = te^{-pt/2} \tag{3.6.18}$$

Of course, $v(t) = c_1 + c_2 t$ also works. This latter choice leads to the solution $u_2(t) = (c_1 + c_2 t)\, e^{-pt/2}$, which is the same as (3.6.13). An alternative method involving the Wronskian is outlined in the Projects at the end of this chapter.

3.6.3 Complex Roots

The roots of $C(\lambda)$ are not real if and only if $p^2 - 4q < 0$. So, if we write these roots as $\lambda_1 = a + ib$ and $\lambda_2 = a - ib$, we have

$$C(\lambda) = (\lambda - \lambda_1)(\lambda - \lambda_2) = \lambda^2 - 2a\lambda + a^2 + b^2 \tag{3.6.19}$$

Note that $p = -2a$ and $q = a^2 + b^2$, and (3.6.1) has the form

$$u'' - 2au' + (a^2 + b^2)u = 0 \tag{3.6.20}$$

A complex-valued solution is therefore

$$u(t) = e^{(a+ib)t} \tag{3.6.21}$$

We can exploit Euler's formula to extract two real solutions from $u(t)$ by taking real and imaginary parts. That is,

$$u_1(t) = \operatorname{Re}\left(e^{(a+ib)t}\right) = e^{at}\cos bt \tag{3.6.22}$$

$$u_2(t) = \operatorname{Im}\left(e^{(a+ib)t}\right) = e^{at}\sin bt \tag{3.6.23}$$

It is not difficult, but it is tedious, to show by substitution that the functions $u_1(t)$ and $u_2(t)$ are indeed solutions of (3.6.20). We show that $u_1(t)$ and $u_2(t)$ is a basic set by observing that the Wronskian of these solutions is not zero. There is some savings in labor in computing the Wronskian by first verifying that $u_1' = au_1 - bu_2$ and $u_2' = au_2 + bu_1$. Then, since $b \neq 0$,

$$\begin{aligned} W(t) &= (au_2 + bu_1)\,u_1 - (au_1 - bu_2)\,u_2 \\ &= b\left(u_1^2 + u_2^2\right) = be^{2at} \neq 0 \end{aligned}$$

Thus, a general solution of (3.6.1), when the solutions of the characteristic equation are not real, is given by

$$\begin{aligned} u(t) &= c_1e^{at}\sin bt + c_2e^{at}\cos bt \\ &= e^{at}\left(c_1\sin bt + c_2\cos bt\right) \end{aligned} \tag{3.6.24}$$

There are alternative forms for the general solution (3.6.24) which have some use in applications. We can choose to write the solution in terms of a sine function or a cosine function. If we choose a cosine function, we use the identity

$$\cos(bt - \theta) = \cos bt\cos\theta + \sin bt\sin\theta$$

Then we write

$$\begin{aligned} c_1\sin bt + c_2\cos bt &= A\cos(bt - \theta) \\ &= A\cos bt\cos\theta + A\sin bt\sin\theta \end{aligned}$$

Equating coefficients of $\sin bt$ and $\cos bt$, respectively, we find that

$$c_1 = A\sin\theta, \qquad c_2 = A\cos\theta \tag{3.6.25}$$

Divide the equations to get

$$\tan\theta = \frac{c_1}{c_2} \tag{3.6.26}$$

Square both equations in (3.6.25) and add:

$$A^2 = c_1^2 + c_2^2 \tag{3.6.27}$$

Finally, we observe that (3.6.24) can take the form

$$u(t) = e^{at} A \cos(bt - \theta) \tag{3.6.28}$$

Of course, we could have selected $A \sin(bt - \theta)$, or $\cos(bt + \theta)$, or $\sin(bt + \theta)$. All of these forms are equivalent and have two arbitrary constants A and θ. A sketch of $u(t)$ with specific values of A, b, and θ is shown in Figure 3.6.

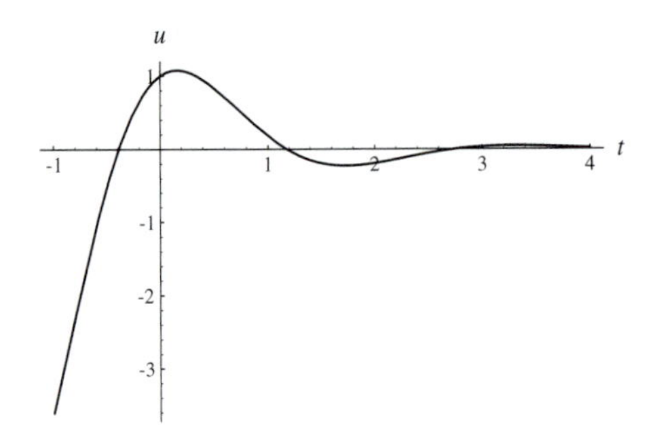

Figure 3.6: The Graph of $u(t) = (\sin 2t + \cos 2t)e^{-t}$

Example 3.6.5. *Find a general solution of $u'' + 2u' + 5u = 0$ and the solution that satisfies the initial conditions $u(0) = u'(0) = 1$. Express the solution in the forms given in* (3.6.24) *and as* (3.6.28).

Solution: The characteristic equation is $\lambda^2 + 2\lambda + 5 = 0$ with nonreal solutions $\lambda_1 = -1 + 2i, \lambda_2 = -1 - 2i$. Therefore, a general solution is given by

$$u(t) = (c_1 \sin 2t + c_2 \cos 2t)\, e^{-t}$$

To determine the solution to the given initial-value problem, we compute

$$u'(t) = ((2c_1 - c_2)\cos 2t - (c_1 + 2c_2)\sin 2t)\, e^{-t}$$

Now set $t = 0$ in $u(t)$ and $u'(t)$ to obtain the equations for c_1 and c_2:

$$\begin{aligned} u(0) &= 1 = c_2 \\ u'(0) &= 1 = 2c_1 - c_2 \end{aligned}$$

Thus, $c_1 = c_2 = 1$, and substituting these values into the general solution leads to this solution of the given initial-value problem:

$$u(t) = (\sin 2t + \cos 2t)\, e^{-t}$$

To put this solution in the form of (3.6.28), use (3.6.26) and (3.6.27). We find $\theta = \tan^{-1} 1 = \pi/4$, and $A = \sqrt{1^2 + 1^2} = \sqrt{2}$. Thus the solution of the initial-value problem has the alternative form

$$u(t) = \sqrt{2}e^{-t}\cos(2t - \pi/4)$$

•EXERCISES

Problems 1—6: Find a general solution of the differential equations.

1. $u'' - u' - 6u = 0.$
2. $u'' - 16u = 0.$
3. $u'' - 4u' - 12u = 0.$
4. $2u'' + 7u' + 5u = 0.$
5. $u'' - 9u' = 0.$
6. $4u'' + u' - 3u = 0.$

Problems 7—10: Find the solution of the initial-value problems.

7. $u'' - 25u = 0,\quad u(0) = 2,\quad u'(0) = 10.$
8. $u'' + u' - 6u = 0,\quad u(0) = 0,\quad u'(0) = -2.$
9. $u'' - 6u' + 5u = 0,\quad u(0) = 10,\quad u'(0) = 0.$
10. $2u'' + 7u' + 5u = 0,\quad u(0) = -2,\quad u'(0) = 4.$

Problems 11—14: Find a general solution of the differential equations.

11. $u'' = 0.$
12. $u'' + 2u' + u = 0.$
13. $u'' + 4u' + 4u = 0.$
14. $u'' + 8u' + 16u = 0.$

Problem 15—17: Find the solution of the initial-value problems.

15. $u'' = 0,\quad u(0) = 2,\quad u'(0) = 4.$
16. $u'' + 4u' + 4u = 0,\quad u(0) = 0,\quad u'(0) = -2.$
17. $u'' + 8u' + 16u = 0,\quad u(0) = 4,\quad u'(0) = 0.$

Problems 18—22: Find a general solution of the differential equations.

18. $u'' + 100u = 0.$
19. $u'' + 4u' + 5u = 0.$
20. $2u'' + 6u' + 5u = 0.$
21. $u'' + 2u' + 10u = 0.$
22. $u'' + 4u' + 40u = 0.$

Problems 23—26: Solve the initial-value problems.

23. $u'' + 2u' + 10u = 0,\quad u(0) = 1,\quad u'(0) = 0.$
24. $u'' + 4u' + 5u = 0,\quad u(0) = 10,\quad u'(0) = 0.$
25. $u'' + u' + u = 0,\quad u(0) = 1,\quad u'(0) = -1.$
26. $u'' + 4u' + 40u = 0,\quad u(0) = -2,\quad u'(0) = 0.$

Problems 27—30: Express the solution of the initial-value problems in the form of (3.6.28).

27. Problem 23.
28. Problem 24.
29. Problem 25.
30. Problem 26.

Problems 31—34: Express the solution of the initial-value problems in the form of $\sin(bt - \theta)$.

31. Problem 23.
32. Problem 24.

33. Problem 25.
34. Problem 26.
35. Verify that $u(t) = te^{-pt/2}$ is a solution of
    $u'' + pu' + \frac{1}{4}p^2u = 0$
36. Verify that
    $e^{at}\cos bt \quad \text{and} \quad e^{at}\sin bt$

are solutions of

$u'' - 2au' + (a^2 + b^2)u = 0$

3.7 Spring-Mass Systems in Free Motion

In this section, we discuss one of the more common applications of second-order ODE's, the free motion of a spring-mass system. We restrict ourselves to systems with *one degree of freedom*, that is, only one independent variable is needed to describe the motion. Systems with more than one independent variable, such as those containing several linked masses and springs, lead to systems of simultaneous ODE's. The analysis of the motion of such systems is taken up in Chapter 4.

Consider the simple spring-mass system shown in Figure 3.7. We make the following simplifying assumptions:

1. The mass M, measured in kilograms, is constrained to move in the vertical direction only.
2. The force due to viscous damping is proportional to the velocity dy/dt. The proportionality constant C, with units of kilograms per second, is the damping coefficient. (For relatively small velocities, this is usually acceptable. However, for large velocities, the damping may be more nearly proportional to the square of the velocity.)
3. The force in the spring is Kh, where h is the distance, measured in meters, from the unstretched position. The spring modulus K, with units of Newtons per meter (N/m), is assumed constant.
4. The masses of the spring and dashpot are negligible compared with the mass M.
5. No external forces act on the system.

Newton's second law states that the sum of the forces acting on a body in any particular direction equals the mass of the body multiplied by the acceleration of the body in that direction. This is written

$$\sum F_y = Ma_y \tag{3.7.1}$$

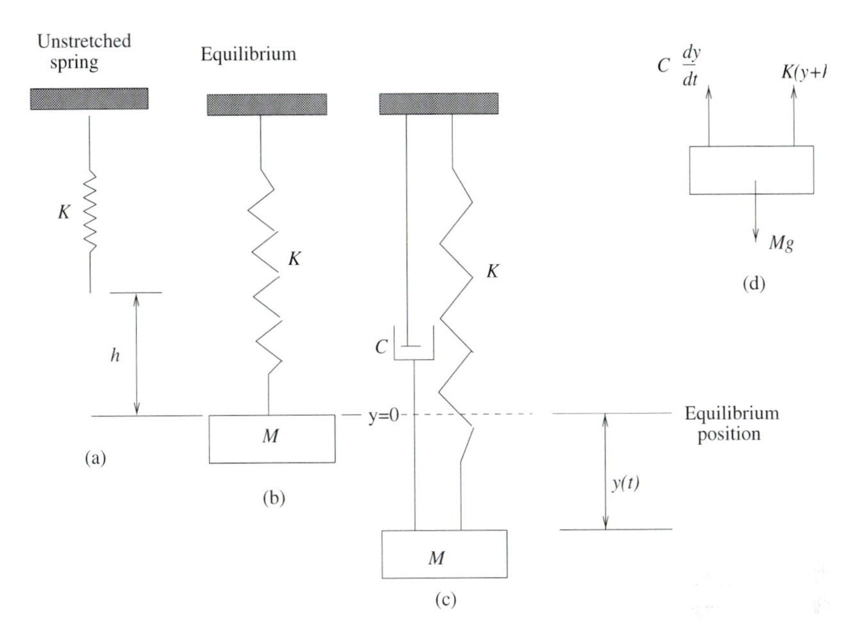

Figure 3.7: A Spring-Mass System

for the y direction. Consider a motionless mass suspended from the unstretched spring as shown in Figure 3.7(b). The spring will deflect a distance h as shown in Figure 3.7(b) where h is found from the relationship

$$Mg = hK \tag{3.7.2}$$

Equation (3.7.2) is the statement that for static equilibrium, the weight must equal the force in the spring. The weight is the mass times the local acceleration of gravity. At this stretched position, we attach a viscous damper – a dashpot – and allow the mass to undergo motion about the unstretched equilibrium position displayed in Figure 3.7(c). A free-body diagram of the mass is shown in Figure 3.7(d). Assume that the positive direction is downward and apply Newton's second law, to obtain

$$Mg - Cy' - K\left(y + h\right) = My'' \tag{3.7.3}$$

We use (3.7.2) in (3.7.3) to find

$$My'' + Cy' + Ky = 0 \tag{3.7.4}$$

From what we have learned about such equations in Section 3.6, we know that the form of the solution will depend on the roots of the characteristic equation associated with (3.7.4) and that the nature of these roots depends on the sign of $C^2 - 4KM$.

3.7.1 Undamped Motion

Over brief time spans and for relatively small C, it is often acceptable to neglect C altogether by assuming that $C = 0$ in (3.7.4). Under these circumstances, (3.7.4) reduces to

$$My'' + Ky = 0 \tag{3.7.5}$$

The characteristic equation is

$$\lambda^2 + K/M = 0 \tag{3.7.6}$$

with roots $\lambda_1 = i\sqrt{K/M}$ and $\lambda_2 = -i\sqrt{K/M}$. Hence, the general solution of (3.7.5) is

$$y(t) = c_1 \cos\sqrt{K/M}\, t + c_2 \sin\sqrt{K/M}\, t \tag{3.7.7}$$

The mass will undergo its first complete cycle as t goes from 0 to $2\pi\sqrt{M/K}$. Thus, one cycle is completed in $2\pi\sqrt{M/K}$ seconds, the *period* of the oscillation. The number of cycles per second, the *frequency*, is the reciprocal of the period, $\sqrt{K/M}/2\pi$. The *angular frequency* ω_0, measured in rad/s, is given by

$$\omega_0 = \sqrt{K/M} \tag{3.7.8}$$

The solution (3.7.7) can now be written as

$$y(t) = c_1 \cos\omega_0 t + c_2 \sin\omega_0 t \tag{3.7.9}$$

For any choice of the constants c_1 and c_2 (other than the trivial selection $c_1 = c_2 = 0$) the solution $y(t)$ is oscillatory. This is the motion of the undamped mass; it is often referred to as *harmonic motion.* It is important to note that the sum of the sine and cosine terms in (3.7.9) can be written as

$$y(t) = A\cos(\omega_0 t - \theta) \tag{3.7.10}$$

where A is the *amplitude* and θ is the *phase* of the oscillation. See the steps leading to (3.6.28). The relationship between the pair of parameters A and θ in (3.7.10) and c_1 and c_2 in (3.7.9) is given by

$$\begin{cases} c_1 = A\cos\theta, & c_2 = A\sin\theta \\ A = \sqrt{c_1^2 + c_2^2}, & \theta = \tan^{-1}(c_2/c_1) \end{cases} \tag{3.7.11}$$

Two conditions – the initial displacement and velocity of the mass – are sufficient to determine the arbitrary constants c_1 and c_2 or A and θ. The conditions are given as specific numerical values in each initial-value problem. Here is an illustrative example.

Example 3.7.1. *A* 4-kg *mass is suspended by a* 100 N/m *spring. The mass is set in motion by giving it an initial downward velocity of* 10 m/s *from its equilibrium position. Find an expression for $y(t)$ if there is no damping.*

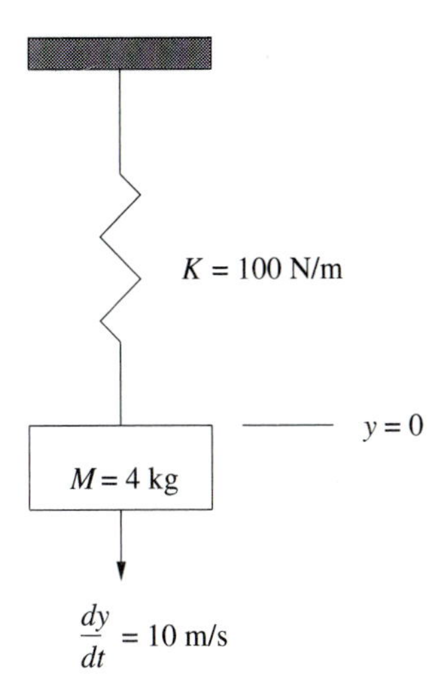

Solution: The ODE describing the motion of the mass is given by (3.7.5),

$$4y'' + 100y = 0$$

The characteristic equation of this differential equation is

$$4\lambda^2 + 100 = 0$$

with solutions $\lambda_1 = 5i$ and $\lambda_2 = -5i$. A general solution is then

$$y(t) = c_1 \cos 5t + c_2 \sin 5t$$

The initial conditions are $y(0) = 0$, $y'(0) = 10$, and these imply

$$\begin{aligned} 0 &= c_1 \cos 0 + c_2 \sin 0 = c_1 \\ 10 &= -5c_1 \sin 0 + 5c_2 \cos 0 = 5c_2 \end{aligned}$$

The values of the parameters are therefore, $c_1 = 0, c_2 = 2$. Hence, the solution is

$$y(t) = 2 \sin 5t$$

•EXERCISES

1. Derive the differential equation describing the motion of a mass M swinging from the end of a string of length L. (Neglect drag.) Replace $\sin\phi$ by ϕ and find a general solution for the resulting differential equation.
2. A spring-mass system has zero damping. Find a general solution for the motion of the mass and determine its frequency of oscillation under the following conditions.
   (a) $M = 2$ kg and $K = 50$ M/m. (b) $M = 1$ kg and $K = 100$ N/m.
   (c) $M = 20$ kg and $K = 5$ N/m. (d) $M = 0.6$ kg and $K = 10$ N/m.
3. Calculate the time necessary for a 2-kg mass to undergo one complete oscillation if there is no damping in the spring-mass system of Figure 3.7 with the following spring constants.
   (a) 32 N/m. (b) 128 N/m.
   (c) 2 N/m. (d) 40 N/m.
4. A 4-kg mass is suspended from a 100-N/m spring. For undamped motion, solve the initial-value problem governing the motion under the following conditions.
   (a) $y(0) = 0$ m, $y'(0) = 10$ m/s.
   (b) $y(0) = 2$ m, $y'(0) = 0$ m/s.
   (c) $y(0) = 2$ m, $y'(0) = -10$ m/s.
   (d) $y(0) = -0.5$ m, $y'(0) = 5$ m/s.

3.7.2 Damped Motion

Let us now include the viscous damping term Cdy/dt in the equation of motion. This is necessary for long time spells because in any real physical situation viscous damping is always present however small, and is necessary for short time spells if the damping coefficient C is not small. The equation describing the motion of the system is (3.7.4), and the roots of its characteristic polynomial are

$$\begin{aligned}\lambda_1 &= -\frac{C}{2M} + \frac{1}{2M}\sqrt{C^2 - 4MK} \\ \lambda_2 &= -\frac{C}{2M} - \frac{1}{2M}\sqrt{C^2 - 4MK}\end{aligned} \qquad (3.7.12)$$

The solution takes three forms, depending on the magnitude of the damping.
Case 1: *Overdamping*
This case arises when $\lambda_1 \neq \lambda_2$ and both are real. Under these circumstances,

$$C^2 - 4KM > 0$$

Case 2: *Critical Damping*
This case arises when $\lambda_1 = \lambda_2$ and both are real. Under these circumstances,

$$C^2 - 4KM = 0$$

Case 3: *Underdamping*
This case arises when λ_1 and λ_2 are not real. Under these circumstances,

$$C^2 - 4KM < 0$$

We handle each case separately and present an example illustrating each.

Overdamping: The solution is a combination of exponentials, which we can write as follows:

$$y(t) = e^{-(C/2M)t}\left(c_1 e^{\Omega t} + c_2 e^{-\Omega t}\right) \tag{3.7.13}$$

where $\Omega = \sqrt{C^2 - 4MK}/2M$. In this overdamped case the two roots are real and are given by

$$\lambda_1 = -\frac{C}{2M} + \Omega, \qquad \lambda_2 = -\frac{C}{2M} - \Omega \tag{3.7.14}$$

However, $C^2 - 4MK < C^2$, and this implies that $\Omega < C/2M$, which in turn implies that $-C/2M \pm \Omega < 0$. Hence, as t tends to infinity, (3.7.13) shows that $y(t)$ approaches zero. Several overdamped motions are shown in Figure 3.8.

Example 3.7.2. *A* 2-kg *mass is suspended by a spring with* $K = 50\,\mathrm{N/m}$ *and a dashpot with* $C = 25\,\mathrm{kg/s}$. *The mass is displaced* 1 m *from its equilibrium position and released from rest (that is, it is given no initial velocity). Calculate the maximum speed attained by the mass.*

Solution: The differential equation describing the motion of the mass as a function of time is

$$2y''(t) + 25y'(t) + 50y(t) = 0$$

A general solution of this equation is obtained by solving its characteristic equation $2\lambda^2 + 25\lambda + 50 = 0$. The solutions of this equation are $\lambda_1 = -10$ and $\lambda_2 = -2.5$. Thus, a general solution is

$$y(t) = c_1 e^{-10t} + c_2 e^{-2.5t}$$

This same equation would result if we used (3.7.13).

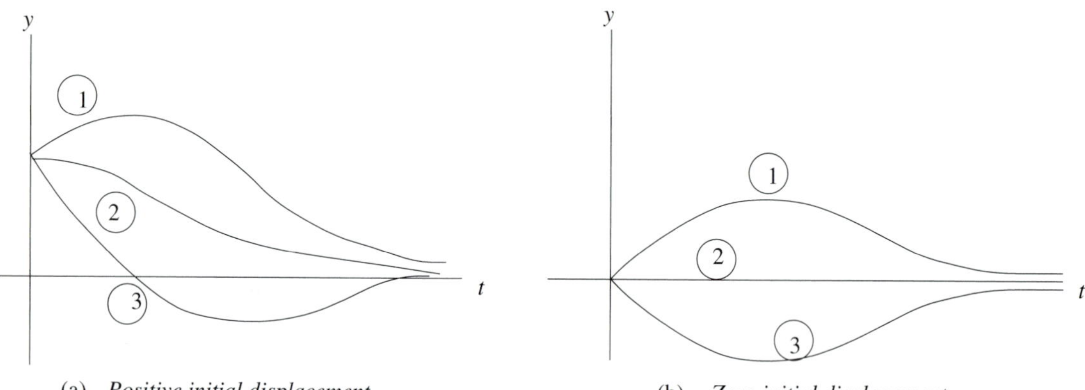

(a) *Positive initial displacement*

(b) *Zero initial displacement*

1. *Positive initial velocity*
2. *Zero initial velocity*
3. *Negative initial velocity*

Figure 3.8: A Variety of Overdamped Motions

The initial conditions are met if we set $t = 0$ in the preceding expression for $y(t)$ and its derivative:

$$y(0) = 1 = c_1 + c_2 \qquad y'(0) = 0 = -10c_1 - 2.5c_2$$

The solution of these simultaneous equations is

$$c_1 = -\tfrac{1}{3} \quad \text{and} \quad c_2 = \tfrac{4}{3}$$

Hence the displacement of the mass is given by

$$y(t) = -\tfrac{1}{3}e^{-10t} + \tfrac{4}{3}e^{-2.5t}$$

The velocity of the mass is given by

$$y'(t) = \tfrac{10}{3}e^{-10t} - \tfrac{10}{3}e^{-2.5t}$$

The function $y'(t)$ is sketched on the next page.

We wish to find the maximum velocity. We invoke a familiar theorem from calculus which asserts that the maximum value of a differentiable function is among the points at which the derivative vanishes or at the endpoints of the interval over which the function is defined. In our case, the function is $y'(t)$ and the derivative is $y''(t)$. The relevant interval is $0 \leq t$. Now, let the time when $y'' = 0$ be designated τ. Then

$$y''(\tau) = -\tfrac{100}{3}e^{-10\tau} + \tfrac{25}{3}e^{-2.5\tau} = 0$$

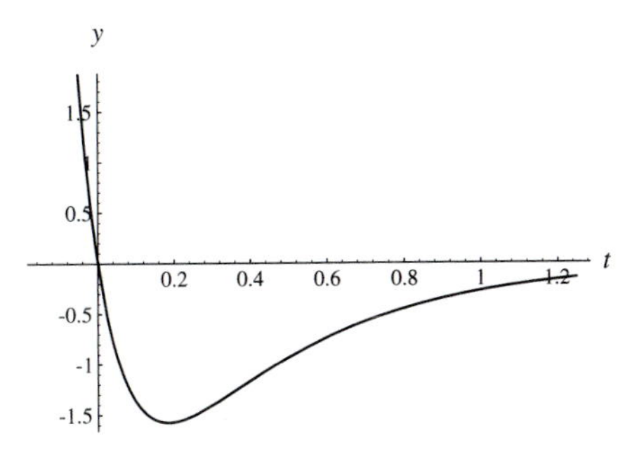

and we find that

$$e^{7.5\tau} = \frac{100}{25} = 4$$

Hence, $\tau = 0.1848$ s is the time at which $y''(t)$ vanishes. At this time $y'(t)$ is given by

$$y'(\tau) = \tfrac{10}{3}e^{-1.848} - \tfrac{10}{3}e^{-0.4621} = -1.575\,\mathrm{m/s}$$

The negative sign shows that the maximum speed is upward, which is expected.

Critical Damping: In this case, $\Omega = 0$ and each of the roots of the characteristic polynomial is equal to $-C/2M$. The solution is then

$$y(t) = e^{-(C/2M)t}\,(c_1 + c_2 t) \tag{3.7.15}$$

As in the overdamped case, $y(t)$ tends to zero as t grows large. (This is a consequence of L'Hôpital's rule applied to te^{-rt}.) A sketch of the solution is not unlike that of the overdamped case (see Figure 3.8.)

Example 3.7.3. *A* 2-kg *mass hangs from a* 50 N/m *spring and a* 20-kg/s *dashpot. Motion occurs as a result of giving the mass an initial velocity of* 20 m/s *at its equilibrium position. Calculate the maximum displacement of the mass.*

Solution: The differential equation describing the motion of the mass is

$$2y'' + 20y' + 50y = 0$$

with characteristic equation $2\lambda^2 + 20\lambda + 50 = 0$. The solutions of this equation are $\lambda_1 = \lambda_2 = -5$. Since these values are equal, the general solution of $2y'' + 20y' + 50y = 0$ is given by

$$y(t) = e^{-5t}\,(c_1 + c_2 t)$$

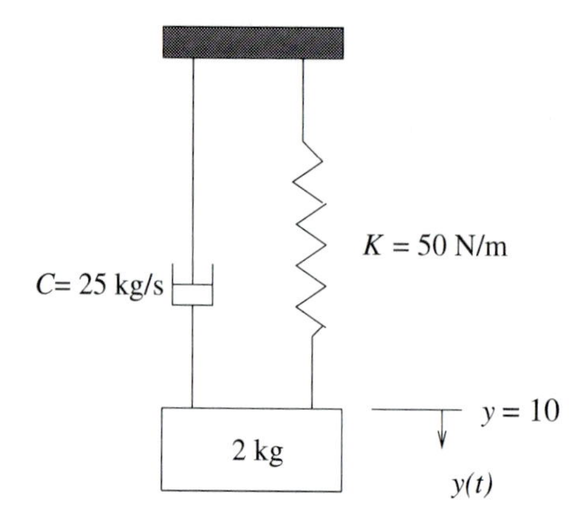

The initial conditions allow us to determine c_1 and c_2:

$$y(0) = 0 = c_1, \qquad y'(0) = 20 = -5c_1 + c_2$$

Hence, $c_1 = 0$ and $c_2 = 20$. The displacement is given, therefore, by

$$y(t) = 20te^{-5t}$$

The maximum displacement occurs either at $t = 0$ or at τ, where $y'(\tau) = 0$. The possibility that $y(0)$ is the maximum dispacement is ruled out by the hypothesis that $y(0) = 0$ is an initial condition. Finally,

$$y'(t) = 20e^{-5\tau} - 100\tau e^{-5\tau} = 0$$

leads to $\tau = \frac{1}{5}$ s. We find the maximum displacement to be

$$y\left(\frac{1}{5}\right) = 4e^{-5(1/5)} = 1.472\,\text{m}$$

as shown in the sketch.

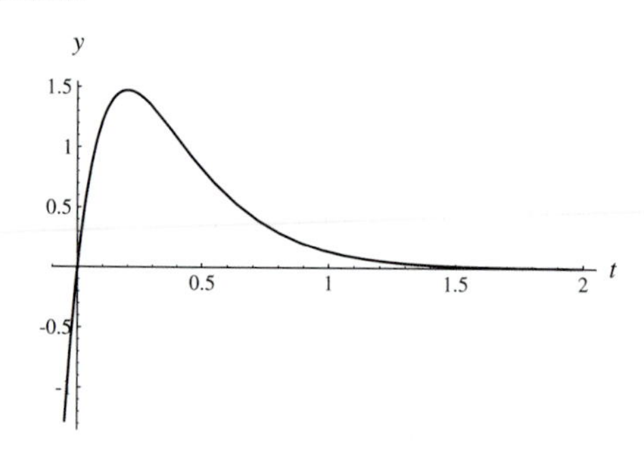

Underdamping: The most interesting of the three cases occurs when $C^2 - 4MK < 0$, that is, when the roots of the characteristic polynomial are not real. The solution can then be written (see (3.6.24))

$$y(t) = e^{-(C/2M)t}\left(c_1 \cos \Omega t + c_2 \sin \Omega t\right) \tag{3.7.16}$$

where we have now defined

$$\Omega = \sqrt{4KM - C^2}/2M$$

Alternatively, we can write

$$y(t) = Ae^{-(C/2M)t} \cos\left(\Omega t - \theta\right) \tag{3.7.17}$$

where $A = \sqrt{c_1^2 + c_2^2}$ and $\tan\theta = c_2/c_1$; if $c_1 = 0$, we take $\theta = \operatorname{sgn}(c_2)\,\pi/2$. We could, of course, elect to write (3.7.17) in terms of $\sin(\Omega t - \theta)$ with $\tan\theta = -c_1/c_2$.

An analysis of (3.7.17) shows that the motion of the mass is oscillatory with decreasing amplitude . The frequency of the oscillation is $\Omega = \sqrt{4MK - C^2}/2M$ and approaches that of the undamped case as $C \to 0$.

A sketch of this motion is given in Figure 3.6 for an initial nonzero displacement and velocity. Note that once again the motion damps out as $t \to \infty$. The ratio of successive maximum amplitudes is a quantity of particular interest. We outline a proof that this ratio is given by

$$\frac{y_n}{y_{n+2}} = e^{\pi C/\Omega M} \tag{3.7.18}$$

in the projects given at the end of the chapter. The ratio is independent of t and the initial conditions. Its logarithm,

$$D = \ln \frac{y_n}{y_{n+2}} = \ln e^{\pi C/\Omega M} = \pi \frac{C}{\Omega M} \tag{3.7.19}$$

is called the *logarithmic decrement.* In view of the definition of Ω, we may write (3.7.19) in the form

$$D = \frac{2\pi C}{\sqrt{4MK - C^2}} \tag{3.7.20}$$

The *critical damping* is defined as the constant

$$C_c = 2\sqrt{MK}$$

Then

$$D = \frac{2\pi C}{\sqrt{C_c^2 - C^2}} \tag{3.7.21}$$

or alternatively,

$$\frac{C}{C_c} = \frac{D}{\sqrt{D^2 + 4\pi^2}} \tag{3.7.22}$$

Equation (3.7.22) allows us to determine the damping. We select a sufficiently large mass so that we have an underdamped system. Then, knowing the spring constant and measuring the ratio of successive maximum amplitudes, we use (3.7.22) to calculate the damping.

Example 3.7.4. *Find the ratio of the damping C to the critical damping C_c if the displacement of a spring-mass system is $y(t) = 10e^{-t/2}\cos 2t$.*

Solution: Comparing the displacement $y(t)$ with (3.7.16), we see that $C/M = 1$, and

$$\Omega = \sqrt{4KM - C^2}/2M = 2$$

From these two equations, we see that $M = C$ and $K = \frac{17}{4}C$. The critical damping is given by $C_c = 2\sqrt{MK} = 2\sqrt{17C^2/4} = \sqrt{17}C$. Hence,

$$\frac{C}{C_c} = \frac{1}{\sqrt{17}} = 0.2425$$

•EXERCISES

1. A damped spring-mass system contains a mass of 4 kg, a spring with $K = 48$ N/m and a dashpot. The mass is displaced 0.5 m from its equilibrium and released from rest. Determine the displacement $y(t)$ if
   (a) $C = 28\,\text{kg/s}$. (b) $C = 32\,\text{kg/s}$.
   (c) $C = 52\,\text{kg/s}$. (d) $C = 60\,\text{kg/s}$.

2. Sketch the displacement (as a function of time) for the first 2 seconds for the spring-mass system of
   (a) Problem 1(b). (b) Problem 1(d).

3. The mass of a damped spring-mass system is given an initial velocity of 40 m/s from the equilibrium position. Find $y(t)$ if
   (a) $M = 4$ kg, $K = 64$ N/m, $C = 40$ kg/s.
   (b) $M = 2$ kg, $K = 24$ N/m, $C = 28$ kg/s.
   (c) $M = 3$ kg, $K = 36$ N/m, $C = 39$ kg/s.
   (d) $M = 2$ kg, $K = 20$ N/m, $C = 20$ kg/s.

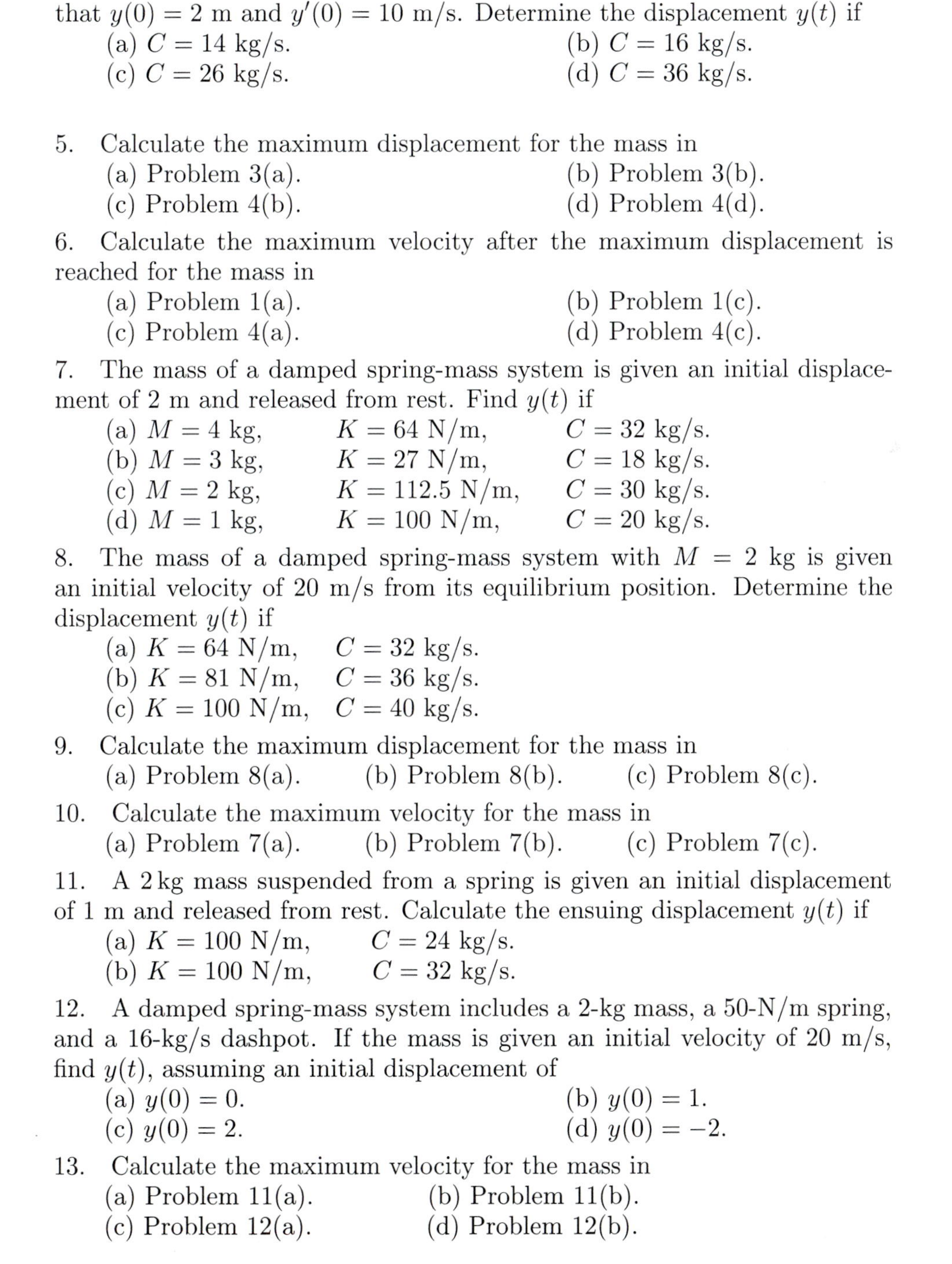

4. The mass of a damped spring-mass system with $M = 2$ kg and a spring with $K = 24$ N/m is subjected to an initial displacement and velocity such that $y(0) = 2$ m and $y'(0) = 10$ m/s. Determine the displacement $y(t)$ if
(a) $C = 14$ kg/s. (b) $C = 16$ kg/s.
(c) $C = 26$ kg/s. (d) $C = 36$ kg/s.

5. Calculate the maximum displacement for the mass in
(a) Problem 3(a). (b) Problem 3(b).
(c) Problem 4(b). (d) Problem 4(d).

6. Calculate the maximum velocity after the maximum displacement is reached for the mass in
(a) Problem 1(a). (b) Problem 1(c).
(c) Problem 4(a). (d) Problem 4(c).

7. The mass of a damped spring-mass system is given an initial displacement of 2 m and released from rest. Find $y(t)$ if
(a) $M = 4$ kg, $K = 64$ N/m, $C = 32$ kg/s.
(b) $M = 3$ kg, $K = 27$ N/m, $C = 18$ kg/s.
(c) $M = 2$ kg, $K = 112.5$ N/m, $C = 30$ kg/s.
(d) $M = 1$ kg, $K = 100$ N/m, $C = 20$ kg/s.

8. The mass of a damped spring-mass system with $M = 2$ kg is given an initial velocity of 20 m/s from its equilibrium position. Determine the displacement $y(t)$ if
(a) $K = 64$ N/m, $C = 32$ kg/s.
(b) $K = 81$ N/m, $C = 36$ kg/s.
(c) $K = 100$ N/m, $C = 40$ kg/s.

9. Calculate the maximum displacement for the mass in
(a) Problem 8(a). (b) Problem 8(b). (c) Problem 8(c).

10. Calculate the maximum velocity for the mass in
(a) Problem 7(a). (b) Problem 7(b). (c) Problem 7(c).

11. A 2 kg mass suspended from a spring is given an initial displacement of 1 m and released from rest. Calculate the ensuing displacement $y(t)$ if
(a) $K = 100$ N/m, $C = 24$ kg/s.
(b) $K = 100$ N/m, $C = 32$ kg/s.

12. A damped spring-mass system includes a 2-kg mass, a 50-N/m spring, and a 16-kg/s dashpot. If the mass is given an initial velocity of 20 m/s, find $y(t)$, assuming an initial displacement of
(a) $y(0) = 0$. (b) $y(0) = 1$.
(c) $y(0) = 2$. (d) $y(0) = -2$.

13. Calculate the maximum velocity for the mass in
(a) Problem 11(a). (b) Problem 11(b).
(c) Problem 12(a). (d) Problem 12(b).

14. Calculate the maximum displacement for the mass in
    Problem 11(a). (b) Problem 11(b).
    (c) Problem 12(a).
15. Find the ratio of the damping to the critical damping if the displacement is as follows:
    (a) $y(t) = 2e^{-t}\sin t$. (b) $y(t) = 4e^{-2t}\sin 4t$.
    (c) $y(t) = e^{-t/2}\cos 5t$.
16. In an undamped spring-mass system, the ratio of successive maximum amplitudes is measured and found to be 1.2. The mass weighs 40 N, and the spring constant is found to be 89.2. Calculate the damping.
17. Calculate the time between successive maximum amplitudes for a spring-mass system in which $M = 30$ kg, $K = 200$ N/m, and $C = 30$ kg/s.

3.8 The Electric Circuit

Consider the circuit in Figure 3.9 containing a resistance R, inductance L, and capacitance C in series.

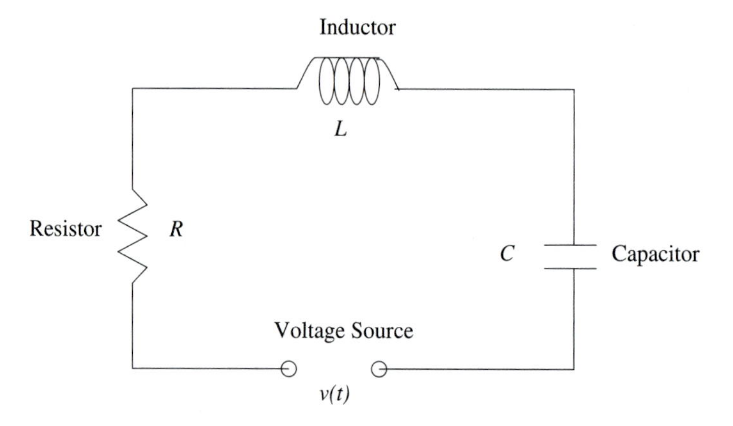

Figure 3.9: An RLC Circuit

This circuit was discussed in Section 1.4.1 where, by omitting selected elements we showed that the equation governing the voltage and current reduced to a first-order ODE. In this section, we study this circuit in its more general context.

One form of the ODE modeling the circuit in Figure 3.9 has q as the dependent variable. Recall that in Section 1.4 we derived

$$v(t) = Lq'' + Rq' + \frac{1}{C}q \tag{3.8.1}$$

If (3.8.1) is differentiated with respect to t and we substitute $i = q'$, where i is the current measured in amperes, we obtain

$$v'(t) = Li'' + Ri' + \frac{1}{C}i \tag{3.8.2}$$

In this section we assume that $v'(t) = 0$, so that (3.8.2) becomes

$$0 = Li'' + Ri' + \frac{1}{C}i \tag{3.8.3}$$

By comparing (3.8.3) with (3.7.4), we see that these equations are identical except for the physical parameters. We can exploit this similarity by interchanging the spring-mass parameters with the circuit parameters as follows:

$$M \to L, \qquad C \to R, \qquad K \to 1/C$$

The various solutions that we have considered for $y(t)$ in Section 3.7 may be taken as solutions for $i(t)$ by using these substitutions.

Thus, the undamped circuit has $R = 0$, and there is no dissipation of electrical energy. The current in this case is given by

$$i(t) = c_1 \cos \omega_0 t + c_2 \sin \omega_0 t \tag{3.8.4}$$

where (see Section 3.7.8),

$$\omega_0 = (LC)^{-1/2} \tag{3.8.5}$$

See (3.7.8). The value of ω_0 is typically quite large for electrical circuits, since both L and C are usually quite small. The damping criteria, in terms of the parameters of the circuit, are:

Case 1:	*Overdamped*	$R^2 - 4L/C > 0.$
Case 2:	*Critically Damped*	$R^2 - 4L/C = 0.$
Case 3:	*Underdamped*	$R^2 - 4L/C < 0.$

Kirchhoff's second law states that the current flowing to a point in a circuit must equal the current flowing away from that point. The law is used to establish the ODE for the parallel electrical circuit shown in Figure 3.10 with the switch closed.

The second law implies that

$$i(t) = i_1 + i_2 + i_3 \tag{3.8.6}$$

We use the observed relationships of current to impressed voltage for the components of the given circuit. (See Section 1.4.1).

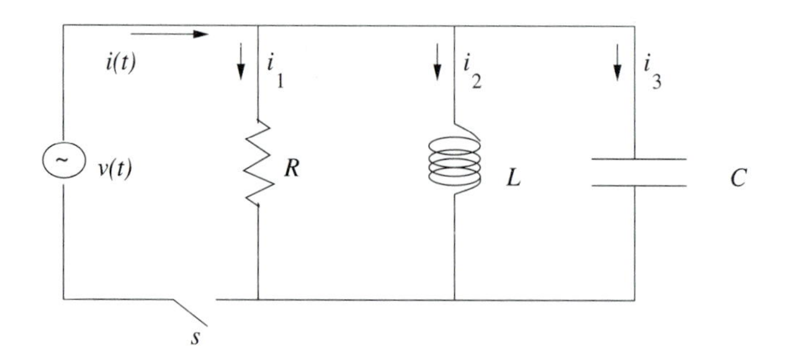

Figure 3.10: A Parallel Circuit

Current through a resistor:	v/R
Current through a capacitor:	Cv'
Current through an inductor:	$\frac{1}{L}\int v\,dt$

Equation (3.8.6) becomes

$$i(t) = \frac{1}{R}v(t) + Cv'(t) + \frac{1}{L}\int v(t)\,dt \tag{3.8.7}$$

If we assume the current source $i(t)$ to be constant, then, differentiating (3.8.7), we obtain

$$Cv''(t) + \frac{1}{R}v'(t) + \frac{1}{L}v(t) = 0 \tag{3.8.8}$$

The analogy with the spring-mass system is made by the "translation"

$$M \to C, \qquad C \to \frac{1}{R}, \qquad K \to \frac{1}{L} \tag{3.8.9}$$

The form of the solution now depends on the sign of the quantity

$$\frac{1}{R^2} - 4\frac{C}{L}$$

The general solutions obtained in Section 3.7 can be used here, making the substitutions indicated in (3.8.9). The specific solutions arising from various initial conditions can be handled similarly. Here are some examples illustrating these points.

Example 3.8.1. *Assume that $v(t) = 0$ and the switch is closed at $t = 0$ in the simple series circuit of Figure 3.9. Determine $i(t)$ if the initial charge is 2 C, $L = 0.05$ H, $C = 2 \times 10^{-4}$ F, and $R = 10\,\Omega$.*

Solution: First we determine whether the circuit is overdamped or underdamped:

$$R^2 - 4\frac{L}{C} = 10^2 - 4 \cdot 0.05/2 \cdot 10^{-4} = -900$$

The circuit is underdamped, and using $\Omega = \sqrt{900}/2 \cdot 0.05 = 300$, we find that the solution for $q(t)$ is, (see (3.7.16),

$$q(t) = e^{-(10/2 \cdot 0.05)t}\,(c_1 \cos 300t + c_2 \sin 300t)$$

The switch is open until $t = 0$, so that $i(0) = 0$. Also, the initial charge is given as $q(0) = 2$. Thus,

$$\begin{aligned} q(0) &= 2 = c_1 \\ q'(0) &= i(0) = 0 = 300c_2 - 100c_1 \end{aligned}$$

Therefore, $c_1 = 2$ and $c_2 = \frac{2}{3}$. The charge on the capacitor is thus

$$q(t) = e^{-100t}\left(2\cos 300t + \tfrac{2}{3}\sin 300t\right)$$

Finally, the current is given by

$$\begin{aligned} i(t) = q'(t) &= -100e^{-100t}(2\cos 300t + \tfrac{2}{3}\sin 300t) \\ &\quad + e^{-100t}(-600\sin 300t + 200\cos 300t) \\ &= -6666.7e^{-100t}\sin 300t \end{aligned}$$

Example 3.8.2. *At* $t = 0$, *the applied voltage in the parallel circuit given in Figure 3.10 is set equal to* 10 V *with* $v' = 0$. *Calculate the resulting voltage* $v(t)$ *if* $L = 0.05\,\mathrm{H}$, $C = 10^{-4}\,\mathrm{F}$, *and* $R = 10\,\Omega$.

Solution: In this case the ODE is overdamped because

$$\frac{1}{R^2} - 4\frac{C}{L} = \frac{1}{10^2} - \frac{4 \cdot 10^{-4}}{0.05} = 0.002$$

Referring to (3.7.13) with $\Omega = \sqrt{0.002}/2 \cdot 10^{-4} = 223.6$, we see that the solution can be written as

$$v(t) = e^{-500t}\left(c_1 e^{223.6t} + c_2 e^{-223.6t}\right)$$

Since we are given $v(0) = 10$ and $v'(0) = 0$, the general solution leads to these conditions on c_1 and c_2:

$$\begin{aligned} 10 &= c_1 + c_2 \\ 0 &= (c_1 + c_2)(-500) + 223.6c_1 - 223.6c_2 \end{aligned}$$

Consequently, $c_1 = 16.18$ and $c_2 = -6.181$, and therefore,

$$v(t) = e^{-500t}\left(16.18e^{223.6t} - 6.181e^{-223.6t}\right)$$

which we can write in the (preferred) form,

$$v(t) = 16.18e^{-276.4t} - 6.181e^{-723.6t}$$

•EXERCISES

1. A series circuit consists of a capacitor of 10^{-4} F and an inductor. A charge of 0.1 coulombs exists on the capacitor at the time $t = 0$ when the switch in the circuit is closed. Find $i(t)$ if there is no applied voltage and L is given as follows:
   (a) $L = 0.02$ H. (b) $L = 0.05$ H.
   (c) $L = 0.01$ H.

2. A series circuit contains a capacitor $C = 2 \cdot 10^{-3}$ F, an inductor $L = 0.01$ H, and a resistor $R = 2\,\Omega$. At time $t = 0$ the current is measured at 10 A. Find $i(t)$ if the charge on the capacitor at $t = 0$ is:
   (a) $q = 0$ C. (b) $q = 0.4$ C. (c) $q = 2$ C.

3. A series circuit consists of a capacitor $C = 1.25 \cdot 10^{-3}$ F, an inductor $L = 0.02$ H and a resistor $R = 8\,\Omega$. If the charge on the capacitor is 2 C and the circuit switch is closed at $t = 0$, determine the maximum current and the time when this maximum current occurs.

4. A capacitor, an inductor, and a resistor are connected in a simple series circuit and the capacitor has a charge of 1.0 C. If the circuit switch is closed at $t = 0$, calculate the current $i(t)$ if
   (a) $C = 10^{-4}$ F, $L = 0.0015$ H, $R = 20\,\Omega$.
   (b) $C = 10^{-4}$ F, $L = 0.01$ H, $R = 20\,\Omega$.
   (c) $C = 10^{-4}$ F, $L = 0.02$ H, $R = 20\,\Omega$.
   (d) $C = 10^{-4}$ F, $L = 0.1$ H, $R = 30\,\Omega$.

5. Calculate the first minimum or maximum current after $t = 0$ and the time at which it occurs for
   (a) Problem 2(a). (c) Problem 2(c)
   (d) Problem 4(a).

6. The parallel circuit of Figure 3.10 has an initial impressed voltage of 100 V with $dv/dt = 0$. Find the current $i_1(t)$ through the resistor if

(a) $R = 20\,\Omega$, $L = 0.015$ H, $C = 2 \cdot 10^{-3}$ F.
(b) $R = 10\,\Omega$, $L = 0.005$ H, $C = 2 \cdot 10^{-4}$ F.
(c) $R = 20\,\Omega$, $L = 0.005$ H, $C = 2 \cdot 10^{-5}$ F.

7. Calculate the maximum magnitude of the current through the capacitor and the time at which it occurs for

(a) Problem 6(a). Problem 6(b). Problem 6(c).

3.9 Undetermined Coefficients

The general, linear, nonhomogeneous, second-order ODE was presented in Section 3.5. Recall proving that a general solution $u(t)$ of this equation is the sum of any of its particular solutions $u_p(t)$ and a general solution of the complementary equation $u_h(t)$. That is,

$$u(t) = u_h(t) + u_p(t)$$

It is important to keep in mind that $u_h(t)$ is a family of functions with two arbitrary constants. Indeed, we have seen earlier in this chapter that $u_h(t)$ is comprised of a rather limited combination of exponential, trigonometric and polynomial functions. We devote our efforts to finding $u_p(t)$ since (at least for the constant coefficient equation) we know $u_h(t)$. There are two common techniques for doing so. In this section we develop the first of these, called *undetermined coefficients.* This method is easy to use and is applicable in many problems in the physical sciences.

Consider the nonhomogeneous ODE

$$u'' + pu' + qu = f(t) \tag{3.9.1}$$

where p and q are real constants; let λ_1 and λ_2 be the roots of the characteristic polynomial

$$\lambda^2 + p\lambda + q = 0 \tag{3.9.2}$$

Suppose

$$f(t) = Ct^k e^{rt} \tag{3.9.3}$$

where k is a nonnegative integer and r is a real or complex number.[4] Then there is a solution of the form

$$u_p(t) = t^s \left(A_0 + A_1 t + \cdots + A_k t^k\right) e^{rt} \tag{3.9.4}$$

for some choice of the constants $A_0, A_1, \ldots, A_k$ and s. The parameter s takes one of the values $0, 1$, or 2, depending on whether r is a root of the characteristic polynomial and the multiplicity of r. In other words there is always a solution which is the product of a polynomial and an exponential. We shall soon see how easy it is to compute the coefficients A_i.

Specifically, the three cases are these:

1. If $\lambda_1 = \lambda_2 = r$, then $s = 2$.
2. If $\lambda_1 = r$ and $\lambda_2 \neq r$, then $s = 1$.
3. If neither root is r, then $s = 0$.

Once the value of s is determined, we can substitute $u_p(t)$ into (3.9.1) to find the values of the constants $A_0, A_1, \ldots, A_k$. These rules are the heart of the method of undetermined coefficients. Although we accept them without proof and use them as guides for guessing trial particular solutions, it is possible to prove that they always result in a solution. We leave the establishment of these rules to the Projects section of this chapter.

Example 3.9.1. *Find a particular solution of $u'' + u = t^2$.*

Solution: For this ODE, the forcing function has the form of (3.9.3), with $r = 0$ and $k = 2$. Since the roots of the characteristic polynomial are $\pm i$, Case 3 applies. So $s = 0$, and we assume a solution of the form

$$u(t) = A_0 + A_1 t + A_2 t^2$$

If $u_p(t)$ is substituted into the original ODE, we obtain

$$2A_2 + A_2 t^2 + A_1 t + A_0 = t^2$$

So

$$A_2 = 1, \qquad A_1 = 0, \qquad 2A_2 + A_0 = 0$$

and hence, $A_0 = -2$, and therefore,

$$u_p(t) = t^2 - 2$$

---

[4]The case where r is complex will be seen to cover the possibility that $f(t) = Ct^k e^{at}\cos(bt)$ and $f(t) = Ct^k e^{at}\sin(bt)$.

Example 3.9.2. *Find a general solution of* $u'' - u = 4e^t$.

Solution: In view of the form of the forcing function, we have $k = 0$ and $r = 1$. However, the roots of the characteristic polynomial are $\lambda_1 = 1$ and $\lambda_2 = -1$. So $s = 1$, and we try a function of the form

$$u_p(t) = t(A_0 e^t)$$

Substitute $u_p(t)$ into the given equation to obtain

$$A_0\left((t+2)e^t - te^t\right) = 2A_0 e^t = 4e^t$$

We find $A_0 = 2$, and therefore,

$$u_p(t) = 2te^t$$

A general solution is given by adding $u_p(t)$ to a general solution of the complementary equation $u'' - u = 0$. Thus,

$$u(t) = 2te^t + c_1 e^t + c_2 e^{-t}$$

Example 3.9.3. *Find a particular solution of* $u'' - 2u' + u = te^t$.

Solution: The characteristic polynomial has roots $\lambda_1 = \lambda_2 = 1$. Since $r = 1$, Case 1 applies with $K = 1$ and $s = 2$. The trial function is

$$u_p(t) = A_1 t^3 e^t$$

After some labor, we find that

$$6A_1 te^t = te^t$$

So $A_1 = 1/6$ leads to the particular solution

$$u_p(t) = \tfrac{1}{6}t^3 e^t$$

Example 3.9.4. *Find a particular solution of* $u'' - 2u' + u = te^t + t$.

Solution: The forcing function is not in the form given by (3.9.3). However, by the second superposition principle, Theorem 3.3.3, if $u_1(t)$ is a solution of

$$u'' - 2u' + u = te^t$$

and $u_2(t)$ is a solution of

$$u'' - 2u' + u = t$$

then $u_1(t) + u_2(t)$ is a solution of

$$u'' - 2u' + u = te^t + t$$

We have already solved $u'' - 2u' + u = te^t$ in the previous example. There we found $u_1(t) = \frac{1}{6}t^3e^t$. So we turn our attention to $u'' - 2u' + u = t$. Since $r = 0$, we try a function of the form $u_2(t) = A_0 + A_1 t$, using $s = 0$. When this function is substituted into $u'' - 2u' + u = t$, we obtain the identity

$$A_0 - 2A_1 + A_1 t = t$$

Thus, $A_0 - 2A_1 = 0$ and $A_1 = 1$. Therefore, $A_0 = 2$, and hence,

$$u_2(t) = 2 + t$$

Finally, we superpose solutions to obtain the particular solution

$$u_p(t) = u_1(t) + u_2(t) = \frac{1}{6}t^3e^t + 2 + t$$

The method of undetermined coefficients can be extended to cover forcing functions of the form

$$(a) \qquad t^k \sin bt\, e^{at} \tag{3.9.5}$$

$$(b) \qquad t^k \cos bt\, e^{at} \tag{3.9.6}$$

One way to do this is to replace these forcing functions by

$$t^k e^{(a+ib)t} \tag{3.9.7}$$

Then the new equation is

$$u'' + pu' + qu = Ct^k e^{(a+ib)t} \tag{3.9.8}$$

and we have the same three cases as we had with (3.9.4), using $r = a + ib$. Actually, Case 1 never occurs for second-order ODE's since complex roots occur in conjugate pairs, so that $\lambda_1 \neq \lambda_2$.

Now if $u_p(t)$ is solution of (3.9.8), we can establish that $\operatorname{Re} u_p(t)$ is a solution of

$$u'' + pu' + qu = t^k \operatorname{Re} e^{(a+ib)t} = t^k \cos bt\, e^{at} \tag{3.9.9}$$

and $\operatorname{Im} u_p(t)$ is a solution of

$$u'' + pu' + qu = t^k \operatorname{Im} e^{(a+ib)t} = t^k \sin bt\, e^{at} \tag{3.9.10}$$

The proof of these facts is set as Problem 33.

If $k = 0$ and $a = 0$ in (3.9.8), we can show that this procedure leads to the following solutions:

(a) If a root is $\lambda_1 = ib$ (so the other root is $\lambda_2 = -ib$), then

$$u_p(t) = At(\cos bt + B \sin bt) \tag{3.9.11}$$

(b) If neither root is r, then

$$u_p(t) = A \cos bt + B \sin bt \tag{3.9.12}$$

The verification of these solutions is set as Problem 34.

Example 3.9.5. *Find a particular solution of* $u'' + 8u = 2 \sin 2t$.

Solution: We begin by replacing the given ODE by (3.9.8), namely,

$$u'' + 8u = 2e^{2it} \tag{1}$$

Then $k = 0, a = 0$, and $b = 2$. Since the roots of the characteristic polynomial are $\lambda_1 = 2\sqrt{2}i$ and $-2\sqrt{2}i$, $s = 0$, and the trial function is $u_p(t) = A_0 e^{2it}$. It is easy to verify that using this choice leads to the identity

$$4A_0 e^{2it} = 2e^{2it}$$

which implies that $A_0 = \frac{1}{2}$. Thus, $u_p(t) = \frac{1}{2}e^{2it}$ is a solution of (1). Hence, a real particular solution is given by the imaginary part. As shown in (3.9.10):

$$\operatorname{Im} u_p(t) = \tfrac{1}{2}\operatorname{Im} e^{2it} = \tfrac{1}{2} \sin 2t$$

and we have

$$u_p(t) = \tfrac{1}{2} \sin 2t$$

The next example shows what happens when the root of the characteristic polynomial agrees with the exponent of the forcing function. Note the change in the form of the solution.

Example 3.9.6. *Find the particular solution of* $u'' + 4u = 2\sin 2t$ *using* (3.9.4), *and then form the general solution.*

Solution: Once again, we consider the modified problem $u'' + 4u = 2e^{2it}$ (see (3.9.8)) with $k = 0$ and $a = 0$. However, $\lambda_1 = 2i = r \neq \lambda_2$; so one of the roots of the characteristic polynomial is the same as the exponent on the forcing function. For this reason, we choose $s = 1$. The trial function is now $u_p(t) = A_0 te^{2it}$. When this function is substituted into the given ODE, we obtain

$$4iA_0e^{2it} = 2e^{2it}$$

We pick $A_0 = -\frac{1}{2}i$. Therefore, $u_p(t) = -\frac{1}{2}ite^{2it}$. We use the imaginary part of $u_p(t)$ because the forcing function is a sine function (see (3.9.10)) and obtain

$$\begin{aligned}\text{Im}u_p(t) &= -\tfrac{1}{2}t\,\text{Im}(ie^{2it}) \\ &= -\tfrac{1}{2}t\,\text{Im}(i\cos 2t - \sin 2t) \\ &= -\frac{1}{2}t\cos 2t\end{aligned}$$

Finally, a general solution of the complementary equation is

$$u_h(t) = c_1\cos 2t + c_2\sin 2t$$

and we construct a general solution of the given ODE:

$$u(t) = -\tfrac{1}{2}t\cos 2t + c_1\sin 2t + c_2\cos 2t$$

Example 3.9.7. *Find a particular solution of* $u'' + u' + u = \cos 2t$ *using* (3.9.11) *or* (3.9.12).

Solution: Since neither solution of the characteristic polynomial is $r = 2i$, we choose (3.9.12) to be the particular solution:

$$u_p(t) = A\cos 2t + B\sin 2t$$

Substitute this function into the given ODE, and equate coefficients of $\cos 2t$ and $\sin 2t$, respectively. There results

$$\begin{aligned}-3A + 2B &= 1 \\ -2A - 3B &= 0\end{aligned}$$

which implies that $A = -\frac{3}{13}$ and $B = \frac{2}{13}$. So the particular solution is

$$u_p(t) = -\tfrac{3}{13}\cos 2t + \tfrac{2}{13}\sin 2t$$

Example 3.9.8. *Find a particular solution of* $u'' - 2u' + 2u = te^t \sin t$.

Solution: We replace the given ODE by

$$u'' - 2u' + 2u = te^t e^{it} = te^{(1+i)t} \tag{1}$$

The characteristic polynomial has roots $\lambda_1 = 1+i, \lambda_2 = 1-i$. Since $r = 1+i$ we have case (2), and we must choose $s = 1$. With $k = 1$ and $a + ib = 1 + i$ we try

$$u_p(t) = t\,(A_0 + A_1 t)\, e^{(1+i)t}$$

After some labor, we find

$$(4iA_1 t + 2iA_0 + 2A_1)\, e^{(1+i)t} = te^{(1+i)t}$$

We easily deduce that $A_0 = \frac{1}{4}$ and $A_1 = -\frac{1}{4}i$. Therefore, a particular solution of (1) is given by

$$\begin{aligned} u_p(t) &= \tfrac{1}{4}t\,(-it + 1)\, e^{(1+i)t} \\ &= \frac{1}{4}te^t\,(-it+1)\, e^{it} \\ &= \tfrac{1}{4}te^t\,(-it+1)\,(\cos t + i \sin t) \\ &= \tfrac{1}{4}te^t\,((\cos t + t\sin t) - i\,(t\cos t - \sin t)) \end{aligned}$$

From (3.9.10), we see that we must use the imaginary part. So

$$u_p(t) = -\tfrac{1}{4}te^t\,(t\cos t - \sin t)$$

is a particular solution of $u'' - 2u' + 2u = t\sin t e^t$.

•EXERCISES

Find a particular solution for each differential equation in Problems 1–14.

1. $u'' + 2u = 2t$.
2. $u'' + u' + 2u = 2t$.
3. $u'' + u = e^{-t}$.
4. $u'' - u = e^t$.
5. $u'' + 10u = 5\sin t$.
6. $u'' + 9u = \cos 3t$.
7. $u'' + 4u' + 4u = e^{-2t}$.
8. $u'' + 9u = t^2 + \sin 3t$.
9. $u'' + u = 4e^t \sin t$.
10. $u'' + u = 4t\sin t$.
11. $u'' - u = t\cos t$.
12. $u'' + u = te^t \sin t$.
13. $u'' + 4u' + 4u = te^t$.
14. $u'' + 9u = t\cos 3t$.

Find a general solution for each differential equation in Problems 15–24.

15. $u'' + u = e^{2t}$.
16. $u'' + 4u = \sin 2t$.
17. $u'' + 9u = t^2$.
18. $u'' - 16u = e^{4t}$.
19. $u'' + 4u' + 4u = t^2 + t + 4$.
20. $u'' + 5u' + 6u = 10 \sin t$.
21. $u'' + 9u = t \cos 2t$.
22. $u'' + 4u = e^t \sin 2t$.
23. $u'' + 5u' + 6u = e^{2t} \sin 2t$.
24. $u'' + 4u = te^t \cos t$.

Problems 25–32: Find the solution for each initial-value problem.

25. $u'' + 4u' + 4u = t^2, \quad u(0) = 0, \quad u'(0) = 1/2$.
26. $u'' + 4u = 2 \sin t, \quad u(0) = 1, \quad u'(0) = 0$.
27. $u'' + 4u' - 5u = 2 - 25t^2, \quad u(0) = 0, \quad u'(0) = 0$.
28. $u'' + 4u = 2 \sin 2t, \quad u(0) = 1, \quad u'(0) = 0$.
29. $u'' + 4u' + 3u = -20 \cos t, \quad u(0) = 0, \quad u'(0) = 0$.
30. $u'' - 16u = 2e^{4t}, \quad u(0) = 0, \quad u'(0) = 0$.
31. $u'' + 4u = 20e^t \cos t, \quad u(0) = 0, \quad u'(0) = 0$.
32. $u'' + 9u = 4te^{2t} \cos 3t, \quad u(0) = 0, \quad u'(0) = 0$.

33. Let $L(u) = u'' + pu' + qu$, where the coefficients are real constants. Suppose $u(t)$ is a solution of $L(u) = t^k e^{(a+ib)t}$. Show that $\operatorname{Re} u(t)$ is a solution of $L(u) = t^k e^{at} \cos bt$ and that $\operatorname{Im} u(t)$ is a solution of $L(u) = t^k e^{at} \sin bt$.

34. Verify that (3.9.11) is a solution of (3.9.8) when $k = a = 0$ and $\lambda_1 = ib, \lambda_2 = -ib$. Also verify that (3.9.12) is a solution of (3.9.8) when $r \neq \lambda_1$ or λ_2.

3.10 The Spring-Mass System: Forced Motion

The spring-mass system shown in Figure 3.11 is acted upon by a force $f(t)$. The ODE describing the motion of the system is found by applying Newton's second law to the mass M. We have

$$f(t) + Mg - K(y + h) - Cy' = My'' \tag{3.10.1}$$

where h is defined as in Figure 3.7, so that $Mg = Kh$. Then (3.10.1) becomes

$$My'' + Cy' + Ky = f(t) \tag{3.10.2}$$

We shall discuss the form of the solution of (3.10.2) for the sinusoidal forcing function

$$f(t) = F_0 \cos \omega t \tag{3.10.3}$$

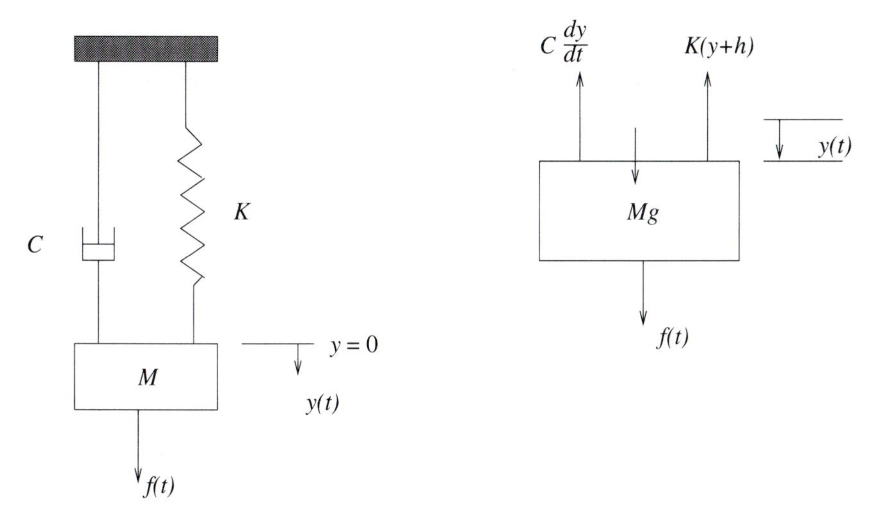

Figure 3.11: A Forced Spring-Mass System

The method of undetermined coefficients (see Section 3.9) is applicable, and after some labor, we find the particular solution (3.9.12), to be

$$y_p(t) = \frac{F_0\left(K - M\omega^2\right)}{\left(K - M\omega^2\right)^2 + \omega^2 C^2}\left(\cos\omega t + \frac{C\omega}{K - M\omega^2}\sin\omega t\right) \tag{3.10.4}$$

As in the free motion problem discussed in Section 3.7, we set

$$\Omega = \sqrt{C^2 - 4MK}/2M \tag{3.10.5}$$

Then the general solution is given by

$$y(t) = e^{-(C/2M)t}\left(c_1 e^{\Omega t} + c_2 e^{-\Omega t}\right) + y_p(t) \tag{3.10.6}$$

We now discuss (3.10.6) in some detail.

3.10.1 Resonance

An interesting and very important phenomenon is observed in the solution, (3.10.6), if we let the damping coefficient C, which is often very small, be zero. The general solution (see (3.7.7)), with $f(t) = F_0 \cos\omega t$, then becomes

$$y(t) = c_1 \cos\omega_0 t + c_2 \sin\omega_0 t + \frac{F_0}{M\left(\omega_0^2 - \omega^2\right)}\cos\omega t \tag{3.10.7}$$

where we have set $\omega_0 = \sqrt{K/M}$. Then $\omega_0/2\pi$ is called the *natural frequency* of the free oscillation. Consider the behavior of $y(t)$ as $\omega \to \omega_0$, that is, as the input frequency approaches the natural frequency. We observe from (3.10.7) that the amplitude

$$\Delta = \frac{F_0}{M\left(\omega_0^2 - \omega^2\right)}$$

of the particular solution becomes unbounded if $\omega = \omega_0$. This condition is known as *resonance.* The amplitude , of course, cannot become unbounded in a physically realistic situation; the damping term usually limits the amplitude , or (3.10.2) no longer is a model for the phenomenon, or the structure fails. The latter must be guarded against in the design of an oscillating system. Soldiers break step on bridges so that resonance will not occur. The spectacular failure of the Tacoma Narrows bridge provided a very impressive example of resonance failure. One must be extremely careful to make the natural frequency of an oscillating system different, if at all possible, from the frequency of any likely forcing function, or provide sufficient damping so that (3.10.4) provides $y_p(t)$.

If $\omega = \omega_0$ (3.10.7) is not a solution of the ODE with zero damping. For in this case, $i\omega_0$ is a solution of the characteristic equation $\lambda^2 + \omega_0^2 = 0$ of the undamped spring-mass system (see (3.9.11)). Again, by the method of undetermined coefficients, the particular solution (3.9.11) takes the form

$$y_p(t) = \frac{F_0}{2M\omega_0} t \sin \omega_0 t \tag{3.10.8}$$

As $t \to \infty$ the amplitude grows without bound. Figure 3.12 shows the behavior of $y_p(t) = t \sin 2\pi t$.

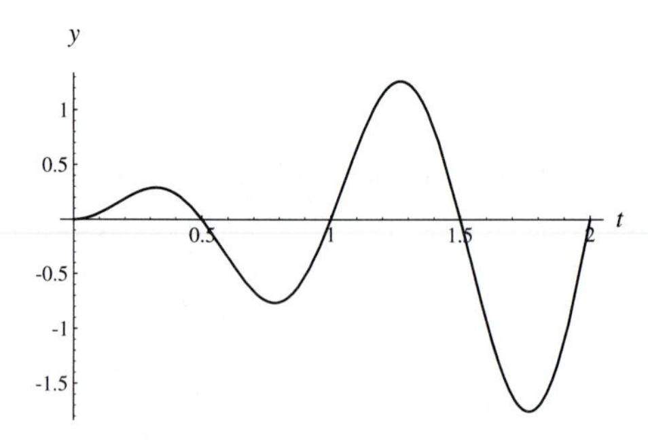

Figure 3.12: Resonance

3.10.2 Near Resonance

Another phenomenon occurs when the forcing frequency is near the natural frequency when damping can be neglected. We simplify our study of this case by setting $y(0) = y'(0) = 0$. The arbitrary constants c_1 and c_2 in (3.10.7) are then

$$c_1 = \frac{F_0}{M\left(\omega_0^2 - \omega^2\right)}, \qquad c_2 = 0 \tag{3.10.9}$$

and (3.10.7) takes the form

$$y(t) = \frac{F_0}{M\left(\omega_0^2 - \omega^2\right)}\left(\cos\omega t - \cos\omega_0 t\right) \tag{3.10.10}$$

With the use of a trigonometric identity, (3.10.10) may be written

$$y(t) = \frac{2F_0}{M\left(\omega_0^2 - \omega^2\right)} \sin\left(\frac{\omega_0 - \omega}{2}t\right)\sin\left(\frac{\omega_0 + \omega}{2}t\right) \tag{3.10.11}$$

A plot of $y(t) = \sin 15t \sin t$ is given in Figure 3.13. The dashed graphs are $\pm\sin t$. Let us write $\epsilon = \frac{1}{2}(\omega_0 - \omega)$, where ϵ is very small because we are assuming that ω is near ω_0. So $\frac{1}{2}(\omega + \omega_0) \approx \omega$. Thus, we have

$$y(t) = \frac{2F_0 \sin\epsilon t}{M\left(\omega_0^2 - \omega^2\right)} \sin\omega t \tag{3.10.12}$$

The wave with larger wavelength appears as a "beat" and can often be heard when two sound waves have approximately the same frequency. Musicians can use this phenomenon to tune their instruments.

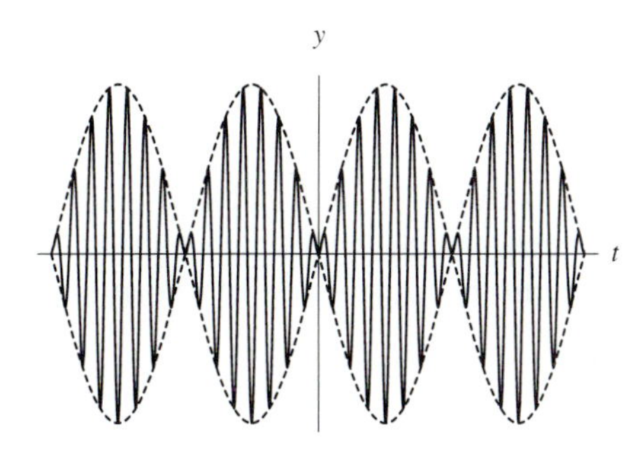

Figure 3.13: The Graph of $y(t) = \sin(15\pi t)\sin(\pi t)$

•EXERCISES

1. Find the solution of $My'' + Cy' + Mg = 0$. Show that this ODE models the motion of a body rising against a constant force of gravity with drag proportional to velocity.

2. In Problem 1, assume that the initial velocity is 100 m/s upward and that $C = 0.4$ kg/s and $M = 2$ kg. How high will the body rise?

3. Calculate the time required for the body in Problem 1 to rise to its maximum height, and compare this with the time it takes to fall back to its original position. Use the values given in Problem 2.

4. A body weighing 100 N is dropped from rest. The drag is assumed to be proportional to the velocity, with the constant of proportionality being 0.5. Approximate the time necessary for the body to attain "terminal" velocity, $0.99V_\infty$, where $V_\infty = \lim_{t\to\infty} V$.

5. Find a general solution to the equation

$$My'' + Ky = F_0 \cos \omega t$$

with $\omega \neq \omega_0$, and verify (3.10.7) by letting $\omega_0 = \sqrt{K/M}$.

6. A 2-kg mass is suspended by a spring with $K = 32$ N/m. A force $f(t) = 0.1 \sin 4t$ is applied to the mass. Calculate the time required for failure to occur if the spring breaks when the amplitude of oscillation exceeds 0.5 m. The motion starts at rest and there is no damping.

7. Solve the following initial-value problems for $y(t)$ if $y(0) = y'(0) = 0$, given the following parameters for an undamped spring-mass system.
   (a) $M = 1$ kg, $K = 25$ N/m, $f(t) = 0.01 \cos 5t$.
   (b) $M = 2$ kg, $K = 32$ N/m, $f(t) = 2 \sin 4t$.
   (c) $M = 1$ kg, $K = 36$ N/m, $f(t) = 2e^{6t}$.
   (e) $M = 3$ kg, $K = 150$ N/m, $f(t) = 0.6 \cos 7t$.
   (f) $M = 2$ kg, $K = 100$ N/m, $f(t) = 4 \sin 7t$.

8. Find the current $i(t)$ for each simple series circuit if $i(0) = q(0) = 0$. See Figure 3.9 and (3.8.3).
   (a) $C = 0.02$ F, $L = 0.5$ H, $R = 0$, $v(t) = 10 \sin 10t$.
   (b) $C = 10^{-4}$ F, $L = 1.0$ H, $R = 0$, $v(t) = 120 \sin 100t$.
   (c) $C = 10^{-3}$ F, $L = 0.1$ H, $R = 0$, $v(t) = 240 \cos 10t$.

9. A 20-N weight is suspended by a frictionless spring $K = 98$ N/m. A force of $f(t) = 2 \cos 7t$ acts on the weight. Calculate the frequency of the "beat", and find the maximum amplitude of the motion, which is assumed to start from rest.

10. Find the solution and show that a "beat" occurs when ω is near ω_0 for the initial-value problem

$$My'' + Ky = F_0 \cos \omega t, \quad y(0) = 1, \quad y'(0) = 10$$

11. A simple series circuit containing a 10^{-3} F capacitor and a 0.1 H inductor has an imposed voltage $v(t) = 120\cos 101t$. Determine the frequency of the "beat" and the maximum current if $q(0) = i(0) = 0$.

12. Show that the amplitude $F_0/M\left(\omega_0^2 - \omega^2\right)$ of the particular solution of (3.10.2) remains unaltered when $C = 0$ and $f(t) = F_0 \sin\omega t$.

13. Show that the particular solution given by (3.10.8) when $\omega = \omega_0$ follows from the appropriate ODE when $f(t) = F_0\cos\omega t$.

3.10.3 Forced Oscillations with Damping

The solution of the equation complementary to (3.10.2) is

$$y_h(t) = e^{-(C/2M)t}\left(C_1 e^{\Omega t} + C_2 e^{-\Omega t}\right) \tag{3.10.13}$$

where, once again we use

$$\Omega = \sqrt{C^2 - 4KM}/2M \tag{3.10.14}$$

This solution includes the factor $e^{-(C/2M)t}$ which is negligibly small after a sufficiently long time. So for large time, the general solution of (3.10.6) tends to the particular solution $y_p(t)$ given in (3.10.4). For this reason, $y_p(t)$ is called the *steady-state* solution. For small t, the "homogeneous" component of the general solution must be considered. We call $y(t) = y_h(t) + y_p(t)$ the *transient* solution.

When damping is included, the amplitude of the particular solution is not unbounded as $\omega \to \omega_0$, but nevertheless, it can still become large. Resonance can be approached for small damping. Hence, when damping is relatively small, it is wise to avoid the condition $\omega \approx \omega_0$ if possible.

We have a special interest in knowing the amplitude of our solutions. To better display the amplitude for the input $F_0\cos\omega t$, write (3.10.4) in the alternative form (see (3.6.28))

$$y_p(t) = \frac{F_0}{\sqrt{M^2\left(\omega_0^2 - \omega^2\right)^2 + \omega^2 C^2}}\cos\left(\omega t - \theta\right) \tag{3.10.15}$$

where we have used $\omega_0^2 = K/M$. The angle θ is called the *phase angle* or *phase lag*. The amplitude of the oscillation is

$$\Delta\left(\omega\right) = \frac{F_0}{\sqrt{M^2\left(\omega_0^2 - \omega^2\right)^2 + \omega^2 C^2}} \tag{3.10.16}$$

We can find the maximum amplitude for any ω by setting $d\Delta/d\omega = 0$. In fact, if we do this, we find that the maximum amplitude occurs when

$$\omega^2 = \omega_0^2 - \frac{C^2}{2M^2} \tag{3.10.17}$$

(See Problem 1.) In view of (3.10.17), we can conclude that for sufficiently large damping, $C^2 > 2M^2\omega_0^2$, there is no value of ω that gives a maximum amplitude . However, if $C^2 < 2M^2\omega_0^2$, then the maximum amplitude occurs when (3.10.17) is met. Substituting this value of ω^2 into (3.10.16) gives the maximum amplitude :

$$\Delta_{max} = \frac{2MF_0}{C\sqrt{4M^2\omega_0^2 - C^2}} \tag{3.10.18}$$

Various amplitudes $\Delta(\omega)$ are sketched in Figure 3.14.

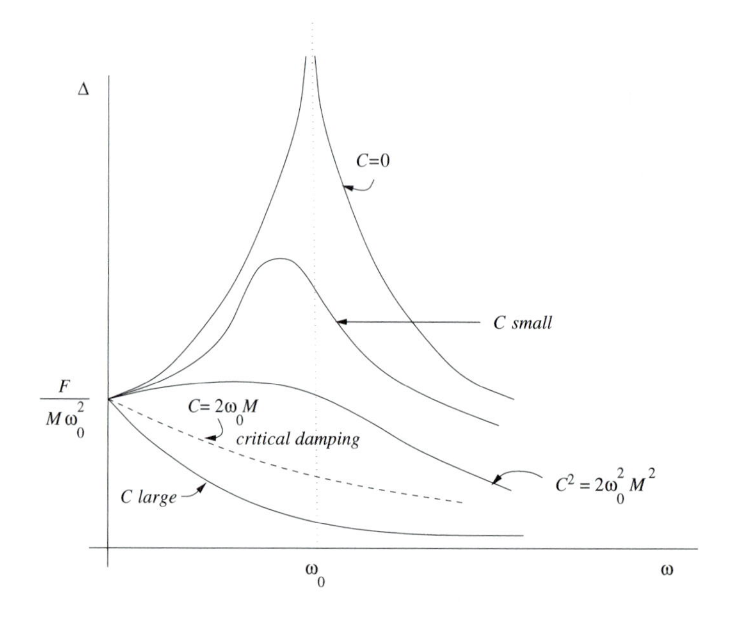

Figure 3.14: Amplitude as a Function of ω

These sketches show that large relative amplitudes can be avoided by sufficient damping or by ensuring that $|\omega - \omega_0|$ is not small.

Example 3.10.1. *The ratio of successive maximum amplitudes for a particular spring-mass system for which $K = 100$ N/m and $M = 4$ kg is found to be 0.8 when the system undergoes free motion. If a forcing function*

$10\cos 4t$ *is imposed on the system, determine the maximum amplitude of the steady state.*

Solution: Damping causes the amplitude indexamplitude of the free motion to decrease with time. The logarithmic decrement is found to be (see(3.7.19))

$$D = \ln \frac{y_n}{y_{n+2}} = \ln \frac{1}{0.8} = 0.223$$

The damping is then calculated from (3.7.22):

$$\begin{aligned} C &= C_c \frac{D}{\sqrt{D^2 + 4\pi^2}} \\ &= 2\sqrt{KM} \frac{D}{\sqrt{D^2 + 4\pi^2}} \\ &= 2\sqrt{400} \frac{0.223}{\sqrt{0.223^2 + 4\pi^2}} = 1.42\,\text{kg/s} \end{aligned}$$

The natural frequency of the undamped system is

$$\omega_0 = \sqrt{K/M} = \sqrt{100/4} = 5\,\text{rad/s}$$

The maximum amplitude (deflection) is given by (3.10.18) and, for this problem, is

$$\Delta_{max} = \frac{2 \cdot 10 \cdot 4}{1.42\sqrt{4 \cdot 4^2 \cdot 5^2 - 1.42^2}} = 1.41\,\text{m}$$

Example 3.10.2. *For the network shown, determine the currents* $i_1(t)$ *and* $i_2(t)$*, assuming that all currents are zero at* $t = 0$.

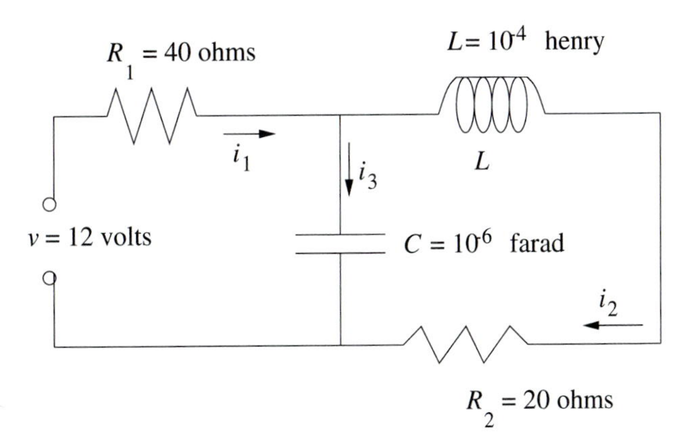

Solution: Using Kirchhoff's first law for the circuit on the left, we find

$$40i_1 + \frac{q}{10^{-6}} = 12 \tag{1}$$

where q is the charge on the capacitor. (See Sections 1.4 and 3.8.) For the circuit around the outside of the network, we have

$$40i_1 + 10^{-4}i_2' + 20i_2 = 12 \tag{2}$$

Kirchhoff's second law requires that

$$i_1 = i_2 + i_3 \tag{3}$$

Using the relationship

$$i_3 = q' \tag{4}$$

and the initial conditions $i_1(0) = i_2(0) = i_3(0) = 0$, we can solve the equations (1)—(4). To do this, substitute (4) and (1) into (3). This results in

$$12 - 10^6 q - 40q' = 40i_2$$

Substituting (1), (3), and (4) into (2) gives, after some algebra,

$$10^{-4}q'' + 22.5q' + 1.5 \cdot 10^6 q = 6$$

The appropriate values for the initial conditions are found from (1) and (4) and are easily seen to be $q(0) = 12 \cdot 10^{-6}$ and $q'(0) = 0$. The general solution for $q(t)$ can be found by the methods already studied. After some labor, we determine that

$$q(t) = e^{-1.12 \cdot 10^5 t}\left(c_1 \cos 48,400t + c_2 \sin 48,200t\right) + 4 \cdot 10^{-6}$$

From the initial conditions, we find that

$$c_1 = 8 \cdot 10^{-6}, \qquad c_2 = 1.851 \cdot 10^{-5}$$

Using (1) we find that

$$i_1(t) = 0.3 - e^{-1.12 \cdot 10^5 t}\left(0.2 \cos 48,400t + 0.463 \sin 48,400t\right)$$

and from (4) and (3),

$$i_2(t) = 0.3 + e^{-1.12 \cdot 10^5 t}\left(-0.2 \cos 48,400t + 2.00 \sin 48,400t\right)$$

Note the high frequency and the rapid decay rate typical of electrical circuits. Also, note that for large $t, i_1 = i_2$, so that $i_3 \to 0$ for large t.

•EXERCISES

1. Show that (3.10.17) follows from (3.10.16) by finding ω such that $d\Delta/d\omega = 0$. Derive (3.10.18).
2. The motion of a 3-kg mass hanging from a spring with $K = 12$ N/m is damped with a dashpot for which $C = 5$ kg/s. (a) Show that (3.10.16) gives the amplitude of the steady-state solution for $f(t) = F_0 \sin \omega t$. (b) Determine the phase lag and the amplitude of the steady-state solution if a force $f(t) = 20 \sin 2t$ acts on the mass. Write the solution in terms of $\sin(\omega t - \theta)$.
3. For Problem 2, let the forcing function be $f(t) = 20 \sin \omega t$. Calculate the maximum amplitude of the steady-state solution and the frequency of the corresponding forcing function.
4. A forcing function $f(t) = 10 \sin 2t$ is imposed on a spring-mass system for which $M = 2$ kg and $K = 8$ N/m. Determine the damping coefficient necessary to limit the amplitude of the motion to a maximum of 2 m.
5. A forcing function $f(t) = 50 \cos \omega t$ is imposed on a spring-mass system for which $M = 4$ kg, $K = 100$ N/m, and $C = 2$ kg/s. Calculate the amplitude of the resulting motion if ω is given as follows:

   (a) 4 rad/s. (b) 4.5 rad/s.
   (c) 5 rad/s. (d) 6 rad/s.

6. Determine the input frequency ω that gives the maximum amplitude for the spring-mass system of Problem 5. For this frequency, what is the maximum amplitude?
7. A constant voltage of 12 V is impressed on a series circuit containing elements $R = 30\,\Omega, L = 10^{-4}$ H, and $C = 10^{-6}$ F. Determine the charge on the capacitor and the current for $q(0) = i(0) = 0$.
8. A series circuit is composed of elements for which $R = 60\,\Omega$, $L = 10^{-4}$ H, and $C = 10^{-5}$ F. Find the steady-state current if a voltage of $120 \cos 120\pi t$ is applied.
9. A circuit is composed of elements for which $R = 80\,\Omega, L = 10^{-4}$ H, and $C = 10^{-6}$ F connected in parallel. The capacitor has an initial charge of 10^{-4} C. There is no current flowing through the capacitor at $t = 0$. What is the current through the resistor at 10^{-4} s?
10. The circuit in Problem 9 is suddenly subjected to a current source of $2 \cos 200t$ at $t = 0$. Find the steady-state voltage across each of the elements.

3.11 The Cauchy-Euler Equation

In the preceding sections we have analyzed linear, constant coefficient ODE's. In this section we study a class of linear second-order ODE's with variable coefficients, the so called *Cauchy-Euler* equations

$$t^2u'' + ptu' + qu = 0 \tag{3.11.1}$$

where p and q are real constants and $t > 0$.

Because the coefficients are polynomials, it is reasonable to search for a solution that is also a polynomial. The simplest choice is surely

$$u(t) = t^\lambda \tag{3.11.2}$$

If this function and its derivatives are substituted into (3.11.1), we obtain

$$(\lambda(\lambda - 1) + p\lambda + q)\, t^\lambda = 0 \tag{3.11.3}$$

Thus t^λ is a solution for $t > 0$ provided that λ is a solution of

$$\lambda^2 + (p-1)\lambda + q = 0 \tag{3.11.4}$$

Equation (3.11.4) is the *characteristic equation* of (3.11.1).

Let λ_1 and λ_2 be the two roots of (3.11.4). If $\lambda_1 \neq \lambda_2$ are real, then

$$u_1(t) = t^{\lambda_1}, \qquad u_2(t) = t^{\lambda_2} \tag{3.11.5}$$

are two linearly independent solutions, and

$$u(t) = c_1 t^{\lambda_1} + c_2 t^{\lambda_2} \tag{3.11.6}$$

is a general solution of the Cauchy-Euler equation in $t > 0$. If the roots are equal, then a second linearly independent solution is not of the form (3.11.5). We can find a second linearly independent solution by the same device we used to find a second solution to the analogous problem for the constant coefficient ODE. The idea is outlined in the Project section. The outcome of this labor is this: When the root $\lambda_1 = \lambda_2$ (and therefore both are real), one solution is $u_1(t) = t^\lambda$, and a second solution is

$$u_2(t) = t^\lambda \ln t \tag{3.11.7}$$

Using (3.11.7), we can write the general solution as

$$u(t) = (c_1 + c_2 \ln t)\, t^\lambda \tag{3.11.8}$$

If the roots of the characteristic polynomial are complex, say, $\lambda_1 = a + ib, \lambda_2 = a - ib$, then we define

$$t^\lambda = t^a t^{ib} = t^a e^{ib \ln t} = t^a \left(\cos\left(b \ln t \right) + i \sin\left(b \ln t \right) \right)$$

for $t > 0$. Two linearly independent solutions are the real and imaginary parts of t^λ:

$$\operatorname{Re} t^\lambda = t^a \cos\left(b \ln t \right) \tag{3.11.9}$$

$$\operatorname{Im} t^\lambda = t^a \sin\left(b \ln t \right) \tag{3.11.10}$$

The proof that these are indeed solutions is simply an exercise in differentiation. The computations are set as Problems 7 and 8.

Example 3.11.1. *Find the solution of*

$$t^2 u'' - 5tu' + 8u = 0, \qquad u(1) = 1, \quad u'(1) = 0$$

Solution: The characteristic polynomial is $\lambda^2 - 6\lambda + 8$, as we see by checking (3.11.4). Its roots $\lambda_1 = 4, \lambda_2 = 2$ lead to the general solution

$$u(t) = c_1 t^4 + c_2 t^2$$

Since

$$u'(t) = 4c_1 t^3 + 2c_2 t$$

we have

$$\begin{aligned} u(1) &= c_1 + c_2 = 1 \\ u'(1) &= 4c_1 + 2c_2 = 0 \end{aligned}$$

These equations have the unique solution $c_1 = -1$ and $c_2 = 2$. Hence, the required solution is

$$u(t) = -t^4 + 2t^2$$

Example 3.11.2. *Find the solution of*

$$t^2 u'' - 3tu' + 4u = 0, \qquad u(1) = 2, \quad u'(1) = 8$$

Solution: The characteristic polynomial is $\lambda^2 - 4\lambda + 4$. Since $\lambda_1 = \lambda_2 = 2$ is a double root, we have the general solution

$$u(t) = \left(c_1 + c_2 \ln t \right) t^2$$

So

$$u'(t) = 2\left(c_1 + c_2 \ln t\right)t + c_2 t$$

and

$$\begin{aligned} u(1) &= c_1 = 2 \\ u'(1) &= 2c_1 + c_2 = 8 \end{aligned}$$

Therefore, the solution is

$$u(t) = 2\left(1 + \ln t^2\right)t^2, \quad t > 0$$

•EXERCISES

1. Find a general solution for each differential equation.
   (a) $t^2u'' + 7tu' + 8u = 0$.
   (b) $t^2u'' + 9tu' + 12u = 0$.
   (c) $t^2u'' - 12u = 24t$.
   (d) $t^2u'' + 2tu' - 12u = 24$.
2. Solve each initial-value problem.
   (a) $t^2u'' + 9tu' + 12u = 0, \quad u(1) = 2, \quad u'(1) = 0.$
   (b) $t^2u'' + 2tu' - 12u = 0, \quad u(1) = 0, \quad u'(1) = 0.$
3. Find the Wronskian of the Cauchy-Euler equation when the roots are real and unequal.
4. Find the Wronskian of the Cauchy-Euler equation when the roots are equal.
5. Show that $u(t) = (c_1 + c_2 \ln t)$ is indeed a general solution of (3.11.1) when the roots are equal.
6. What is the Cauchy-Euler equation of order 1? What is its characteristic equation and its general solution?
7. Show that $t^a \cos(b \ln t)$ is a solution of (3.11.1) when $a + bi$ is a root of the characteristic polynomial.
8. Show that $t^a \sin(b \ln t)$ is a solution of (3.11.1) when $a + bi$ is a root of the characteristic polynomial.
9. The Cauchy-Euler of third-order is $t^3u^{(3)} + p_1t^2u^{(2)} + p_2tu^{(1)} + p_3u = 0$. Show that t^λ is a solution if and only if λ is a root of a polynomial of degree three. Find the coefficients of this cubic polynomial.

3.12 Variation of Parameters

In Section 3.9 we discussed a method of obtaining particular solutions of

$$u'' + pu' + qu = f(t) \tag{3.12.1}$$

when the forcing function $f(t)$ is a member of an important, but limited class, of functions. In this section we present a method applicable to any sectionally continuous forcing function and for equations where $p = p(t)$ and $q = q(t)$ are not necessarily constants. We have already developed this method for the first-order ODE and the first-order system.

Assume that we have already determined two linearly independent solutions of the corresponding homogeneous equation (the complementary equation)

$$u'' + p(t)u' + q(t)u = 0 \tag{3.12.2}$$

Let us denote these solutions by $u_1 = u_1(t)$ and $u_2 = u_2(t)$. We show that there is always a particular solution $u_p = u_p(t)$ of (3.12.1) of the form

$$u_p = v_1 u_1 + v_2 u_2 \tag{3.12.3}$$

for some choice of $v_1 = v_1(t)$ and $v_2 = v_2(t)$.

In order to find v_1 and v_2 so that u_p is a solution of (3.12.1), we compute u_p' and u_p'' and substitute these functions into (3.12.1). So, from (3.12.3)

$$u_p' = v_1 u_1' + v_2 u_2' + v_1' u_1 + v_2' u_2 \tag{3.12.4}$$

Since we have two unknown functions, v_1 and v_2, we need two conditions to determine them. The first condition is picked quite cleverly. We set

$$v_1' u_1 + v_2' u_2 = 0 \tag{3.12.5}$$

so that (3.12.4) simplifies to

$$u_p' = v_1 u_1' + v_2 u_2' \tag{3.12.6}$$

We differentiate (3.12.6) to obtain

$$u_p'' = v_1 u_1'' + v_2 u_2'' + v_1' u_1' + v_2' u_2' \tag{3.12.7}$$

We now substitute (3.12.3), (3.12.6) and, (3.12.7) into (3.12.1) to obtain

$$\begin{aligned} f(t) &= v_1 \left(u_1'' + p(t)u_1' + q(t)u\right) \\ &\quad + v_2 \left(u_2'' + p(t)u_2' + q(t)u_2\right) + v_1' u_1' + v_2' u_2' \end{aligned} \tag{3.12.8}$$

The quantities multiplying v_1 and v_2 are both zero since we are assuming that u_1 and u_2 are solutions of (3.12.2). Thus, (3.12.8) simplifies to

$$f(t) = v_1' u_1' + v_2' u_2' \tag{3.12.9}$$

Equations (3.12.5) and (3.12.9) form a pair of equations for the determination of v_1' and v_2':

$$v_1' u_1 + v_2' u_2 = 0, \qquad v_1' u_1' + v_2' u_2' = 0 \tag{3.12.10}$$

Equations (3.12.10) result from the method of *variation of parameters* for the second-order equation (3.12.1).

Since the determinant of the coefficients of v_1' and v_2' in (3.12.10) is $W(t)$, we have

$$v_1'(t) = -u_2(t)\frac{f(t)}{W(t)}, \qquad v_2'(t) = u_1(t)\frac{f(t)}{W(t)} \tag{3.12.11}$$

where $W(t) = W(u_1, u_2)$. We find $v_1(t)$ and $v_2(t)$ by integrating (3.12.11):

$$v_1(t) = -\int u_2(t)\frac{f(t)}{W(t)}\,dt, \quad v_2(t) = \int u_1(t)\frac{f(t)}{W(t)}\,dt \tag{3.12.12}$$

A particular solution is now given by

$$\begin{aligned} u_p(t) &= u_1(t)v_1(t) + u_2(t)v_2(t) \qquad (3.12.13)\\ &= -u_1(t)\int u_2(t)\frac{f(t)}{W(t)}\,dt + u_2(t)\int u_1(t)\frac{f(t)}{W(t)}\,dt \end{aligned}$$

Finally, a general solution of (3.12.1) is given by

$$u(t) = c_1 u_1(t) + c_2 u_2(t) + u_p(t) \tag{3.12.14}$$

Example 3.12.1. *Use the method of variation of parameters to determine a general solution of* $u'' + u = t^2$.

Solution: The corresponding homogeneous ODE $u''+u = 0$ has the linearly independent solutions $u_1(t) = \sin t$ and $u_2(t) = \cos t$ because $W(\sin t, \cos t) = -1$. We now use (3.12.13) to find a particular solution. We have

$$u_p(t) = \sin t \int t^2 \cos t\,dt - \cos t \int t^2 \sin t\,dt$$

After two integration by parts or a table lookup, we find that $u_p(t) = t^2 - 2$, so

$$u(t) = c_1 \cos t + c_2 \sin t + t^2 - 2$$

Although the particular solution of the ODE given in Example 3.12.1 is the same as the one obtained in Example 3.9.1, not too much should be made of what is, in fact, a coincidence. There are, after all, infinitely many particular solutions of this equation. Here are two others:

$$\sin t + t^2 - 2 \quad \text{and} \quad \cos t + t^2 - 2$$

The reason that no trigonometric terms appeared in the particular solution obtained in Example 3.12.1 was due to our implicit choice of zero for the arbitrary constants of integration when we used (3.12.13). Indeed, had we asked for the solution satisfying the initial conditions $u(0) = 0$ and $u'(0) = 0$, we would have obtained the solution

$$u_p(t) = 2\cos t + t^2 - 2$$

Example 3.12.2. *Use the method of variation of parameters to find a particular solution of*

$$u'' + u = \sec t, \qquad 0 < t < \pi/2.$$

Solution: In view of the work done in Example 3.12.1, we may go directly to

$$\begin{aligned} u_p(t) &= \sin t \int \frac{1}{\cos t} \cos t \, dt - \cos t \int \frac{1}{\cos t} \sin t \, dt \\ &= \int dt + \int \tan t \, dt \\ &= t \sin t + \cos t (\ln \cos t) \end{aligned}$$

since $\int \tan t \, dt = -\ln \cos t$.

•EXERCISES

Use the method of variation of parameters to obtain a particular solution of each ODE in Problems 1–10.

1. $u'' + u = \tan t$.
2. $u'' + 5u' + 4u = te^t$.
3. $u'' + 4u' + 4u = te^{-2t}$.
4. $u'' + u = \csc t$.
5. $u'' - 2u' + u = t^{-1}e^t$.
6. $u'' - 4u' + 4u = e^t$.
7. $t^2u'' + tu' - u = 9$.
8. $t^2u'' + tu' - u = 2t^2$.
9. $t^2u'' - 2tu' - 4u = t$.
10. $tu'' - u' = t(1 + t)$.

11. Show that (3.12.13) may be written in the form

$$u_p(t) = \int_0^t \frac{f(s)}{W(s)} \det \begin{bmatrix} u_1(s) & u_2(s) \\ u_1(t) & u_2(t) \end{bmatrix} ds$$

12. Use the result in Problem 11 to show that the solution of the initial-value problem

$$u'' + p_0(t)u' + p_1(t)u = f(t), \quad u(0) = u'(0) = 0$$

has the solution

$$u(t) \int_0^t f(s) \frac{u_1(s)u_2(t) - u_1(t)u_2(s)}{u_1(s)u_2{}'(s) - u_1'(s)u_2(s)} ds$$

Hint: If $F(t) = \int_{t_0}^t g(t,s)\,ds$, then

$$F'(t) = g(t,s) + \int_0^t \frac{\partial}{\partial t} g(t,s)\,ds$$

13. Use the result of Problem 11 to write a particular solution of

$$u'' + p^2 u = f(t)$$

in the form

$$u_p(t) = \frac{1}{p} \int_0^t f(t) \sin p(t-s)\,ds$$

14. Use the result of Problem 11 to write a particular solution of

$$u'' - p^2 u = f(t)$$

in the form

$$u_p(t) = \frac{1}{p} \int_0^t f(t) \sinh p(t-s)\,ds$$

15. Use the result of Problem 11 to write a particular solution of

$$u'' + 2pu' + p^2 u = f(t)$$

in the form

$$u_p(t) = \frac{1}{p} \int_0^t f(t)(t-s)e^{-p(t-s)}\,ds$$

PROJECTS

1. Changes of Dependent Variable

Consider the second-order equation

$$u'' + p(t)u' + q(t)u = 0 \tag{1}$$

and suppose $u = r(t)y$ for some unspecified function $r = r(t)$. Then

$$\begin{aligned} u &= ry \\ u' &= r'y + ry' \\ u'' &= r''y + 2r'y' + ry'' \end{aligned}$$

(a) Show that substituting these equations into (1) results in the following second-order equation in y:

$$ry'' + (2r' + p(t)r)\, y' + (r'' + p(t)r' + q(t)r)\, y = 0 \tag{2}$$

(b) Use (2) to deduce that the knowledge of one solution of (1) leads to a first-order ODE in y. *Hint*: Let $r(t)$ be the given solution and work first with the equation (with p constant) $u'' + 2pu' + p^2u = 0$ which has the solution $r(t) = e^{-pt}$.

(c) Determine $r(t)$ so that the coefficient of y' in (2) vanishes.

(d) Using the value of $r(t)$ obtained in (c), show that (2) reduces to

$$y'' + I(t)y = 0 \tag{3}$$

where

$$I(t) = q(t) - \tfrac{1}{4}p^2(t) - \tfrac{1}{2}p'(t)$$

Equation (3) is called the *normal form* of (1) and $I(t)$ is called its *invariant*. Suppose that we change variables in (1) by using $u = h(t)w$, resulting in

$$w'' + \hat{p}(t)w' + \hat{q}(t)w = 0$$

Let $\hat{I}(t)$ be the invariant of this ODE.

(e) Show that $I(t) = \hat{I}(t)$.

(f) Show that the normal form of

$$tu'' - 2tu' + \left(a^2t^2 + 2\right) u = 0$$

is $y'' + a^2y = 0$ and use this fact to deduce a general solution of this ODE.

2. The Wronskian

The Wronskian of the solutions $u_1(t)$ and $u_2(t)$ of

$$u'' + p(t)u' + q(t)u = 0 \tag{1}$$

is given by

$$W(t) = u_1(t)u_2'(t) - u_1'(t)u_2(t) = Ke^{-P(t)} \tag{2}$$

where

$$P(t) = \int_{t_0}^{t} p(s)\, ds$$

Suppose we have determined a specific solution $u_1(t)$ of (1). Then (2) is a first-order ODE for the determination of $u_2(t)$.

(a) The equation with equal roots is

$$u'' + pu' + \tfrac{1}{4}p^2 u = 0$$

with one solution $u_1(t) = e^{-pt/2}$. Use (1) with $K = 1$ to find the second linearly independent solution $u_2(t) = te^{-pt/2}$.

3. The Proof of the Logarithmic Decrement Equation

The solution of (3.7.4) in the underdamped case, $C^2 - 4MK < 0$, may be written in the form

$$y(t) = Ae^{-(C/2M)t}\cos(\Omega t - \theta) \tag{1}$$

(a) Show that

$$y'(t) = A\left(\frac{1}{2}\frac{C}{M}\cos(\Omega t - \theta) + \Omega\sin(\Omega t - \theta)\right)e^{-(C/2M)t}$$

and hence that

$$\tan(\Omega t - \theta) = -\frac{1}{2}\frac{C}{M\Omega} \tag{2}$$

is necessary and sufficient for $y'(t) = 0$.

If the solutions of (2) are written t_n, then

$$t_n = \frac{\theta}{\Omega} + \frac{1}{\Omega}\arctan\left(-\frac{1}{2}\frac{C}{M\Omega}\right) + \frac{n\pi}{\Omega} \tag{3}$$

for every integer n. It should be obvious from the nature of the graph of $y(t)$ that its local maxima and minima alternate.

(b) Set $y_n = y(t_n)$ and $y_{n+2} = y(t_{n+2})$ in (1) and verify that

$$\frac{y_n}{y_{n+2}} = \frac{\cos(\Omega t - \theta)}{\cos(\Omega t_{n+2} - \theta)} e^{-(C/2M)(t_n - t_{n+2})}$$

(c) Show that (2) implies that

$$(t_n - t_{n+2})\,\Omega = -2\pi$$

and therefore,

$$\frac{y_n}{y_{n+2}} = \frac{\cos(\Omega t - \theta)}{\cos(\Omega t_{n+2} - \theta)} e^{\pi C/M\Omega}$$

(d) Show that

$$\Omega t_n - \theta = \Omega t_{n+2} - \theta - 2\pi$$

so that

$$\cos(\Omega t_n - \theta) = \cos(\Omega t_{n+2} - \theta)$$

Use this result to show that

$$\frac{y_n}{y_{n+2}} = e^{\pi C/M\Omega}$$

4. The Method of Undetermined Coefficients

We take a general approach to the problem of solving

$$u'' + pu' + qu = t^n e^{rt} \tag{1}$$

using the trial function $u_p(t) = s(t)r^{rt}$, where $s(t)$ is a polynomial of degree n.

(a) Substitute $u_p(t) = s(t)e^{rt}$ into (1) to show that $s(t)e^{rt}$ is a solution of (1) if and only if

$$C(r)s(t) + (2r + p)s'(t) + s''(t) = t^n \tag{2}$$

where $C(\lambda) = \lambda^2 + p\lambda + q$ is the characteristic polynomial of the equation complementary to (1).

(b) Show that $s(t)$ can be found satisfying (2) if

$$C(\lambda) = \lambda^2 + p\lambda + q \neq 0$$

(c) Show that $s(t)$ can be found so that $t\,s(t)$ satisfies (1) if

$$C(\lambda) = \lambda^2 + p\lambda + q = 0 \quad \text{and} \quad C'(\lambda) \neq 0 \tag{3}$$

Show that (3) implies that $\lambda_1 = \lambda$ and $\lambda_2 \neq \lambda$.

(d) Show that

$$u_p(t) = \frac{1}{(n+1)(n+2)} t^{n+2} e^{\lambda t}$$

if $\lambda_1 = \lambda_2 = r$. (*Hint*: In this case, $C(\lambda) = C'(\lambda) = 0$.)

5. The Solution of the Cauchy-Euler Equation for Equal Roots

Let $\lambda_1 = \lambda_2 = \lambda$ be the roots of the characteristic polynomial for the Cauchy-Euler equation (3.11.1).

(a) Show that $2\lambda = 1 - p$.

(b) Show that

$$L\left(vt^\lambda\right) = t^\lambda(t^2v'' + (2\lambda + p)tv' + (\lambda^2 + (p-1)\lambda + q)v)$$

(c) Use (a) to show that vt^λ is a solution of the Cauchy-Euler equation if v is a solution of

$$tv'' + v' = 0 \tag{1}$$

Now show that $v(t) = \ln t$ is a solution of (1) and hence that $t^\lambda \ln t$ is a solution of (3.11.1), when the roots of the characteristic polynomial are equal.

Chapter 4

Higher Order Equations

4.1 Introduction

A solution of a constant coefficient, n^{th}-order ODE is a function whose n^{th} derivative is sectionally continuous in the interval (α, β) and which satisfies

$$u^{(n)} + a_1 u^{(n-1)} + \cdots + a_{n-1} u^{(1)} + a_n u = f(t) \tag{4.1.1}$$

where $f(t)$ is assumed to be sectionally continuous in (α, β). From among all the solutions of (4.1.1), we must often select one that meets n initial conditions specified at $t = t_0$:

$$u(t_0) = b_1, \ \ldots, \ u^{(n-1)}(t_0) = b_n \tag{4.1.2}$$

Equations (4.1.1) and (4.1.2) comprise the nth-order, nonhomogeneous initial-value problem (assuming $f(t) \neq 0$). The *complementary equation* associated with (4.1.1) is the homogeneous equation

$$u^{(n)} + a_1 u^{(n-1)} + \cdots + a_{n-1} u^{(1)} + a_n u = 0 \tag{4.1.3}$$

Not only is this terminology a straightforward extension of that used in the study of first- and second-order linear ODE's, but we can solve these problems in much the same way. Indeed, in Sections 4.2 and 4.3 we do exactly this.

This approach is convenient and effective when the roots of the characteristic polynomial are known. However, when we are confronted with finding particular solutions of the nonhomogeneous equation, or when we must resort to numerical methods for finding solutions, it is often more convenient to replace the nth-order ODE by a system of first-order ODE's. This places the burden of proving theorems and finding solutions on the theory of systems, a theory already developed in Chapter 2. We do this work in Sections 4.4 and 4.5. One happy consequence of converting an nth-order ODE to a system of first-order ODE's is that the method of variation of parameters (Section 4.6) which seems to involve arbitrary and unmotivated decisions in the nth-order ODE, is quite natural in the context of systems.

4.2 The Homogeneous Equation

4.2.1 General Solutions

We learn about the nature of solutions of (4.1.3) by deriving the conditions under which this equation has a solution of the form $u(t) = e^{\lambda t}$. Since $u^{(k)}(t) = \lambda^k e^{\lambda t}$, it follows that $u(t)$ is a solution of (4.1.1) if and only if λ satisfies

$$C(\lambda) = \lambda^n + a_1\lambda^{n-1} + \cdots + a_{n-1}\lambda + a_n = 0 \tag{4.2.1}$$

The polynomial $C(\lambda)$ is called the *characteristic polynomial* and (4.2.1) is called the *characteristic* or *auxiliary equation* of (4.1.3). As is the case for second-order ODE's, each distinct real root of (4.2.1) yields an exponential solution; each complex root results in two solutions with trigonometric terms $t^s e^{at} \sin bt$ and $t^s e^{at} \cos bt$ for some nonnegative integer s.

Every linear combination of solutions of (4.1.3) is also a solution and a general solution of an nth-order ODE is an n-parameter family of solutions such that every set of n initial conditions can be met by some function in the family. The definition of a general solution is provided in Section 4.4. Until then we rely on our experience with second-order ODE's.

Example 4.2.1. *Find a general solution of* $u^{(3)} + 3u^{(2)} + 2u^{(1)} = 0$.

Solution: The characteristic equation is $\lambda^3 + 3\lambda^2 + 2\lambda = 0$ with solutions $r_1 = 0, r_2 = -1$, and $r_3 = -2$. Hence, each of the functions

$$u_1(t) = e^{0t} = 1, \quad u_2(t) = e^{-t}, \quad u_3(t)e^{-2t}$$

is a solution and a general solution is given by the linear combination

$$u(t) = c_1 + c_2 e^{-t} + c_3 e^{-2t}$$

Example 4.2.2. *Find a general solution of $u^{(4)} - u = 0$.*

Solution: The characteristic equation is $\lambda^4 - 1 = 0$ and its solutions are $r_1 = 1, r_2 = -1, r_3 = i, r_4 = -i$. Hence, each of the functions

$$u_1(t) = e^t, \quad u_2(t) = e^{-t}, \quad u_3(t) = \sin t, \quad u_4(t) = \cos t$$

is a solution and a general solution is given by $u(t) = c_1 e^t + c_2 e^{-t} + c_3 \sin t + c_4 \cos t$.

If the root r is real and repeated m times, then r is said to have *multiplicity m*. There are m solutions of (4.1.3) corresponding to a real root of multiplicity m, namely,

$$\begin{aligned} u_1(t) &= e^{\lambda t} \\ u_2(t) &= te^{\lambda t} \\ &\vdots \\ u_m(t) &= t^{m-1}e^{\lambda t} \end{aligned} \tag{4.2.2}$$

If the root r is complex with multiplicity m, (4.2.2) may be used to generate $2m$ real solutions by techniques first introduced in Section 3.6.2. Specifically, let $r = a + ib$, where a and b are real and $b \neq 0$. Then for each integer $j, 1 \leq j \leq k$, the functions

$$\begin{aligned} \operatorname{Re}\left(t^{j-1}e^{(a+ib)t}\right) &= t^{j-1}e^{at}\cos bt \\ \operatorname{Im}\left(t^{j-1}e^{(a+ib)t}\right) &= t^{j-1}e^{at}\sin bt \end{aligned} \tag{4.2.3}$$

provide two linearly independent solutions. Hence, (4.2.3) provides a total of $2m$ linearly independent solutions. Using the root $\bar{r} = a - ib$ in place of r generates no new linearly independent solutions, since

$$\operatorname{Re}\left(t^{j-1}e^{(a-ib)t}\right) = t^{j-1}e^{at}\cos bt$$

and

$$\operatorname{Im}\left(t^{j-1}e^{(a-ib)t}\right) = -t^{j-1}e^{at}\sin bt$$

One final observation: if the root $r = a + ib$ is not repeated, which is the most common case, then the independent solutions associated with $r = a + ib$ are $e^{at}\cos bt$ and $e^{at}\sin bt$ and the general solution is

$$u(t) = e^{at}(c_1 \cos bt + c_2 \sin bt)$$

Example 4.2.3. *Find a general solution of* $u^{(4)} + 2u^{(2)} + u = 0$.

Solution: The characteristic polynomial is $\lambda^4 + 2\lambda^2 + 1 = \left(\lambda^2 + 1\right)^2$, and its roots are $r_1 = i$ and $r_2 = -i$, each with multiplicity 2. Hence, the functions

$$u_1(t) = e^{it}, \quad u_2(t) = te^{it}, \quad u_3(t) = e^{-it}, \quad u_4(t) = te^{-it}$$

are four complex-valued solutions. The corresponding real solutions are obtained by taking the real and imaginary parts of e^{it} and te^{it}:

$$\begin{aligned} \operatorname{Re} e^{it} &= \cos t, & \operatorname{Im} e^{it} &= \sin t \\ \operatorname{Re}\left(te^{it}\right) &= t\cos t & \operatorname{Im}\left(te^{it}\right) &= t\sin t \end{aligned}$$

Hence, a general solution is given by

$$u(t) = (c_1 + c_2 t)\cos t + (c_3 + c_4 t)\sin t$$

Example 4.2.4. *Construct a homogeneous ODE of lowest degree with real coefficients having solutions* $u_1(t) = t$ *and* $u_2(t) = e^{-t}\sin 2t$ *and find its general solution.*

Solution: The solution $u_1(t) = t = te^{0t}$ implies that the characteristic polynomial has roots $r_1 = r_2 = 0$. The solution $u_2(t) = e^{-t}\sin 2t$ implies that $r_3 = -1 + 2i$ and $r_4 = -1 - 2i$ are also roots of the characteristic polynomial. (Complex roots always appear as complex conjugates.) Thus, the characteristic polynomial must be of degree at least 4. One fourth-degree polynomial that suffices is

$$\begin{aligned} C(\lambda) &= (\lambda - r_1)(\lambda - r_2)(\lambda - r_3)(\lambda - r_4) \\ &= \lambda^2(\lambda + 1 - 2i)(\lambda + 1 + 2i) = \lambda^4 + 2\lambda^3 + 5\lambda^2 \end{aligned}$$

Working backwards, we see that a homogeneous ODE with this characteristic polynomial is

$$u^{(4)} + 2u^{(3)} + 5u^{(2)} = 0$$

The assumptions that $u_1(t) = t$ and $u_2(t) = e^{-t}\sin 2t$ are solutions of the equation $u^{(4)} + 2u^{(3)} + 5u^{(2)} = 0$ imply that its general solution may be written in the form $u(t) = c_1 + c_2 t + c_3 e^{-t}\sin 2t + c_4 e^{-t}\cos 2t$

•EXERCISES

1. Construct a homogeneous ODE having real coefficients,the lowest possible order, and the following functions as solutions.

(a) $2,\ e^t$. (b) $\sin t,\ \cos t$. (c) $\sin t,\ \cos t, e^t$.
(d) $t,\ e^t$. (e) $\sinh t,\ \cos t$. (f) $t^2 e^t$.
(g) $te^{3t}\sin 2t$. (h) $\sin t + \cos t$. (i) $\sin t \cos t$.

2. Explain why every linear, constant coefficient ODE that has the solution $e^{at}\sin bt$ must also have $e^{at}\cos bt$ as a solution.

3. Find the characteristic equation and general solution for these equations:
(a) $u^{(3)} - 2u^{(2)} - u^{(1)} + 2u = 0$.
(b) $u^{(3)} - 2u^{(2)} - 3u^{(1)} = 0$.
(c) $10u^{(3)} + u^{(2)} - 7u^{(1)} + 2u = 0$.
(d) $u^{(3)} - 7u^{(2)} + 6u^{(1)} = 0$.
(e) $4u^{(3)} - 13u^{(1)} - 6u = 0$.

4. Repeat Problem 3 for these equations:
(a) $u^{(3)} - 3u^{(2)} + 3u^{(1)} - u = 0$.
(b) $u^{(3)} - 13u^{(2)} + 36u^{(1)} = 0$.
(c) $u^{(4)} - 6u^{(2)} + 8u = 0$.
(d) $u^{(4)} - u^{(3)} - 7u^{(2)} + u^{(1)} + 6u = 0$.
(e) $u^{(3)} + 4u^{(2)} + 5u^{(1)} = 0$.

5. Repeat Problem 3 for these equations:
(a) $2u^{(3)} - u^{(2)} + 32u^{(1)} - 16u = 0$.
(b) $u^{(3)} - u^{(2)} + 4u^{(1)} - 4u = 0$.
(c) $u^{(4)} - 2u^{(3)} - u^{(2)} + 2u^{(1)} = 0$.
(d) $u^{(4)} + 3u^{(3)} + u^{(2)} - 5u^{(1)} = 0$.
(e) $u^{(4)} - u^{(3)} - 12u^{(2)} + 28u^{(1)} - 16u = 0$.

6. Repeat Problem 3 for these equations:
(a) $u^{(5)} - 2u^{(4)} - u^{(1)} + 2u = 0$.
(b) $u^{(4)} - u = 0$.
(c) $u^{(4)} + 4u^{(3)} + 6u^{(2)} + 4u^{(1)} + u = 0$.
(d) $u^{(6)} + 9u^{(4)} + 24u^{(2)} + 16u = 0$.
(e) $u^{(3)} - 3u^{(2)} + 4u = 0$.
(f) $u^{(3)} + 3u^{(2)} + 3u^{(1)} + u = 0$.

MATLAB

7. For each of the following problems, invoke "roots(C)" to find the roots of the characteristic polynomials. Here C is a row vector whose entries are the coefficients of the characteristic polynomial.

(a) Problem 3. (b) Problem 4. (c) Problem 5.

4.2.2 Initial-Value Problems

Having obtained a general solution of the nth-order homogeneous ODE, it is now just a matter of algebra to find a solution of the corresponding initial-value problem. We illustrate this with three examples.

Example 4.2.5. *Find the solution of the initial-value problem*

$$u^{(3)} + 3u^{(2)} + 2u^{(1)} = 0, \qquad u(0) = u^{(1)}(0) = 0, \quad u^{(2)}(0) = 2$$

Solution: A general solution of the given third-order ODE was found in Example 4.2.1, namely, $u(t) = c_1 + c_2e^{-t} + c_3e^{-2t}$. By differentiating $u(t)$ we obtain

$$u^{(1)}(t) = -c_2e^{-t} - 2c_3e^{-2t}, \qquad u^{(2)}(t) = c_2e^{-t} + 4c_3e^{-2t}$$

Now, let $t = 0$ in $u(t)$, $u^{(1)}(t)$, and $u^{(2)}(t)$ and use the initial conditions to obtain the following three equations in three unknowns:

$$\begin{aligned} u(0) &= 0 = c_1 + c_2 + c_3 \\ u^{(1)}(0) &= 0 = -c_2 - 2c_3 \\ u^{(2)}(0) &= 2 = c_2 + 4c_3 \end{aligned}$$

The solution of these equations is $c_1 = 1, c_2 = -2$, and $c_3 = 1$. Replacing the arbitrary constants in the general solution by these values yields the solution that meets the initial conditions:

$$u(t) = 1 - 2e^{-t} + e^{-2t}$$

Example 4.2.6. *Find the solution of the initial-value problem*

$$u^{(4)} - u = 0, \qquad u(0) = u^{(1)}(0) = u^{(2)}(0) = u^{(3)}(0) = 1$$

Solution: A general solution was found in Example 4.2.2:

$$u(t) = c_1e^t + c_2e^{-t} + c_3\cos t + c_4\sin t$$

The derivatives of $u(t)$ are given by

$$\begin{aligned} u^{(1)}(t) &= c_1e^t - c_2e^{-t} - c_3\sin t + c_4\cos t \\ u^{(2)}(t) &= c_1e^t + c_2e^{-t} - c_3\cos t - c_4\sin t \\ u^{(3)}(t) &= c_1e^t - c_2e^{-t} + c_3\sin t - c_4\cos t \end{aligned}$$

Hence,

$$\begin{aligned} u(0) &= 1 = c_1 + c_2 + c_3 \\ u^{(1)}(0) &= 1 = c_1 - c_2 + c_4 \\ u^{(2)}(0) &= 1 = c_1 + c_2 - c_3 \\ u^{(3)}(0) &= 1 = c_1 - c_2 - c_4 \end{aligned}$$

The solution to these equations is $c_1 = 1, c_2 = c_3 = c_4 = 0$. Using these values in the general solution $u(t)$ results in the solution

$$u(t) = e^t$$

Note that although the work leading to the solution was tedious, checking that $u(t) = e^t$ does indeed meet the initial conditions is trivial.

Example 4.2.7. *Find the solution of the initial-value problem*

$$u^{(4)} - u = 0, \qquad u(0) = 1, \quad u^{(1)}(0) = u^{(2)}(0) = u^{(3)}(0) = 0$$

Solution: From the work in Example 4.2.6, we find that

$$\begin{aligned} u(0) &= 1 = c_1 + c_2 + c_3 \\ u^{(1)}(0) &= 0 = c_1 - c_2 + c_4 \\ u^{(2)}(0) &= 0 = c_1 + c_2 - c_3 \\ u^{(3)}(0) &= 0 = c_1 - c_2 - c_4 \end{aligned}$$

Hence, $c_1 = c_2 = \frac{1}{4}, c_3 = \frac{1}{2}, c_4 = 0$, and the desired solution is

$$u(t) = \tfrac{1}{4}\left(e^t + e^{-t} + 2\cos t\right)$$

These three examples illustrate that our intuitive use of the term "general solution" is not unreasonable. We can see from the examples that the crucial issue is whether the simultaneous equations for the determination of the arbitrary constants in the general solution have a solution for every choice of initial conditions. These simultaneous equations, whose unknowns are

$c_1, c_2, \ldots, c_n$, are solved by setting $t = t_0$ (often, $t_0 = 0$) in each of the expressions for $u^{(k-1)}(t), 1 < k < n$, that is,

$$\begin{aligned} c_1u_1(0) + c_2u_2(0) + \cdots + c_nu_n(0) &= b_1 \\ c_1u_1^{(1)}(0) + c_2u_2^{(1)}(0) + \cdots + c_nu_n^{(1)}(0) &= b_2 \\ &\vdots \\ c_1u_1^{(n-1)}(0) + c_2u_2^{(n-1)}(0) + \cdots + c_nu_n^{(n-1)}(0) &= b_n \end{aligned}$$

In matrix form these conditions are represented in this way:

$$\begin{bmatrix} u_1(0) & u_2(0) & \ldots & u_n(0) \\ u_1^{(1)}(0) & u_2^{(1)}(0) & \ldots & u_n^{(1)}(0) \\ \vdots & \vdots & & \vdots \\ u_1^{(n-1)}(0) & u_2^{(n-1)}(0) & \ldots & u_n^{(n-1)}(0) \end{bmatrix} \begin{bmatrix} c_1 \\ c_2 \\ \vdots \\ c_n \end{bmatrix} = \begin{bmatrix} b_1 \\ b_2 \\ \vdots \\ b_n \end{bmatrix} \quad (4.2.4)$$

As in the second-order ODE, we call

$$W(t) = \det \begin{bmatrix} u_1(0) & u_2(0) & \ldots & u_n(0) \\ u_1^{(1)}(0) & u_2^{(1)}(0) & \ldots & u_n^{(1)}(0) \\ \vdots & \vdots & & \vdots \\ u_1^{(n-1)}(0) & u_2^{(n-1)}(0) & \ldots & u_n^{(n-1)}(0) \end{bmatrix} \quad (4.2.5)$$

the *Wronskian* of the list of solutions $u_1(t), u_2(t), \ldots, u_n(t)$. If the Wronskian at $t = 0$ is not zero, that is, if $W(0) \neq 0$, then the coefficient matrix in (4.2.4) is nonsingular and the constants $c_1, c_2, \ldots, c_n$ are uniquely determined. For these values of $c_1, c_2, \ldots, c_n$,

$$u(t) = c_1u_1(t) + c_2u_2(t) + \cdots + c_nu_n(t)$$

which is the solution of (4.1.3) meeting the initial conditions (4.1.2).

Example 4.2.8. *Solve the initial-value problem*

$$u^{(4)} = u \qquad u(0) = 1,\ u^{(1)}(0) = 2,\ u^{(2)}(0) = 3,\ u^{(3)}(0) = 4$$

Solution: From the work done in Example 4.2.6, and by using the formulation given by (4.2.4), the simultaneous equations for c_1, c_2, c_3, c_4 are

$$\begin{bmatrix} 1 & 1 & 1 & 0 \\ 1 & -1 & 0 & 1 \\ 1 & 1 & -1 & 0 \\ 1 & -1 & 0 & -1 \end{bmatrix} \begin{bmatrix} c_1 \\ c_2 \\ c_3 \\ c_4 \end{bmatrix} = \begin{bmatrix} 1 \\ 2 \\ 3 \\ 4 \end{bmatrix}$$

We solve this system by Gaussian elimination on its augmented matrix:

$$\left[\begin{array}{cccc|c} 1 & 1 & 1 & 0 & 1 \\ 1 & -1 & 0 & 1 & 2 \\ 1 & 1 & -1 & 0 & 3 \\ 1 & -1 & 0 & -1 & 4 \end{array}\right] \rightarrow \cdots \rightarrow \left[\begin{array}{cccc|c} 1 & 0 & 0 & 0 & 5/2 \\ 0 & 1 & 0 & 0 & -1/2 \\ 0 & 0 & 1 & 0 & -1 \\ 0 & 0 & 0 & 1 & -1 \end{array}\right]$$

Hence, $c_1 = \frac{5}{2}, c_2 = -\frac{1}{2}, c_3 = c_4 = -1$. The solution of the initial-value problem is, therefore,

$$u(t) = \tfrac{5}{2}e^t - \tfrac{1}{2}e^{-t} - \cos t - \sin t$$

•EXERCISES

1. Find a solution to each of the following initial-value problems.

(a) $u^{(3)} - 4u^{(1)} = 0;$
$$u(0) = 1, \quad u^{(1)}(0) = 0, \quad u^{(2)}(0) = 2.$$

(b) $u^{(3)} - 3u^{(2)} + 2u^{(1)} = 0;$
$$u(0) = 0, \quad u^{(1)}(0) = 2, \quad u^{(2)}(0) = 0.$$

(c) $u^{(3)} - 6u^{(2)} + 11u^{(1)} - 6u = 0;$
$$u(0) = 0, \quad u^{(1)}(0) = 2, \quad u^{(2)}(0) = 0.$$

(d) $u^{(4)} - 2u^{(3)} - u^{(2)} + 2u^{(1)} = 0;$
$$u(0) = 2, \quad u^{(1)}(0) = 0, \quad u^{(2)}(0) = 10, \quad u^{(3)}(0) = 0.$$

(e) $u^{(4)} - 5u^{(3)} + 4u^{(2)} = 0;$
$$u(0) = 2, \quad u^{(1)}(0) = 1, \quad u^{(2)}(0) = 0, \quad u^{(3)}(0) = 0.$$

2. Find a solution to each of the following initial-value problems.

(a) $u^{(3)} - u^{(2)} + 4u^{(1)} - 4u = 0;$
$$u(0) = 2, \quad u^{(1)}(0) = 0, \quad u^{(2)}(0) = 0.$$

(b) $u^{(3)} + u^{(2)} + 4u^{(1)} + 4u = 0;$
$$u(0) = 0, \quad u^{(1)}(0) = 10, \quad u^{(2)}(0) = 0.$$

(c) $u^{(3)} + 2u^{(2)} + 4u^{(1)} + 8u = 0;$
$$u(0) = 0, \quad u^{(1)}(0) = 4, \quad u^{(2)}(0) = 10.$$

(d) $u^{(4)} + 5u^{(2)} + 4u = 0;$
$$u(0) = 4, \quad u^{(1)}(0) = 0, \quad u^{(2)}(0) = 10, \quad u^{(3)}(0) = 0.$$

(e) $u^{(4)} + 3u^{(3)} + u^{(2)} - 5u^{(1)} = 0;$
$$u(0) = 0, \quad u^{(1)}(0) = 10, \quad u^{(2)}(0) = 0, \quad u^{(3)}(0) = 10/3.$$

3. Find a solution to each of the following initial-value problems.

(a) $u^{(4)} - 16u = 0;$

$$u(0) = 4, \quad u^{(1)}(0) = 0, \quad u^{(2)}(0) = 0, \quad u^{(3)}(0) = 0.$$

(b) $u^{(3)} = 0;$

$$u(0) = 2, \quad u^{(1)}(0) = 0, \quad u^{(2)}(0) = 2.$$

(c) $u^{(3)} + 2u^{(2)} + u^{(1)} = 0;$

$$u(0) = 0, \quad u^{(1)}(0) = 1, \quad u^{(2)}(0) = 0.$$

(d) $u^{(3)} - 3u^{(2)} + 3u^{(1)} - u = 0;$

$$u(0) = 1, \quad u^{(1)}(0) = 2, \quad u^{(2)}(0) = 1.$$

4. Find a solution to each of the following initial-value problems.

(a) $u^{(4)} - u = 0;$

$$u(0) = 2, \quad u^{(1)}(0) = 0, \quad u^{(2)}(0) = 10, \quad u^{(3)}(0) = 0.$$

(b) $u^{(4)} - 4u^{(3)} + 6u^{(2)} - 4u^{(1)} + u = 0;$

$$u(0) = 0, \quad u^{(1)}(0) = 1, \quad u^{(2)}(0) = 0, \quad u^{(3)}(0) = 10/3.$$

(c) $u^{(5)} + u^{(3)} = 0;$

$$u(0) = 1, \quad u^{(1)}(0) = 0, \quad u^{(2)}(0) = 2 \quad u^{(3)}(0) = 0, \quad u^{(4)}(0) = 4.$$

4.3 The Nonhomogeneous Equation

Once again we shall see that the analogy between the second-order ODE and the nth-order ODE is sufficiently close that we can extend results valid for the second-order ODE to the equation of nth-order.

Consider the constant coefficient nonhomogeneous ODE

$$u^{(n)} + a_1 u^{(n-1)} + \cdots + a_{n-1} u^{(1)} + a_n u = f(t) \tag{4.3.1}$$

and suppose that

$$f(t) = t^k e^{rt} \tag{4.3.2}$$

Then a natural trial function for a particular solution is

$$u_p(t) = s_k(t) e^{rt} \tag{4.3.3}$$

where $s_k(t)$ is a polynomial of degree k with "undetermined" coefficients. Specifically, suppose

$$s_k(t) = A_0 + A_1 t + \cdots + A_k t^k$$

The assumption that $u_p(t) = s_k(t)e^{rt}$ is a solution of (4.3.1) can be shown to lead to the identity

$$e^{rt}(C(r)s_k(t) + g_{k-1}(t)) = t^k e^{rt} \tag{4.3.4}$$

which is obtained by substituting $u_p(t) = s_k(t)e^{rt}$ and its various derivatives into (4.3.1). The polynomial $C(\lambda)$ is the characteristic polynomial of the equation complementary to (4.3.1), and $g_{k-1}(t)$ is a polynomial of degree at most $k-1$. Since $e^{rt} \neq 0$ (even if r is not real), we may cancel e^{rt} from both sides of (4.3.4) to obtain

$$C(r)s_k(t) + g_{k-1}(t) = t^k \tag{4.3.5}$$

Now if $C(r) \neq 0$, then (4.3.5) determines the values of the constants A_i. But $C(r) \neq 0$ if and only if r is not a root of the characteristic polynomial, that is, if and only if e^{rt} is not a solution of the equation complementary to (4.3.1), namely,

$$u^{(n)} + a_1 u^{(n-1)} + \cdots + a_{n-1}u^{(1)} + a_n u = 0$$

We can easily illustrate the case in which $C(r) \neq 0$. Consider

$$u^{(3)} - u^{(2)} - 2u^{(1)} = te^t \tag{4.3.6}$$

Then $r = 1$, $k = 1$, and $C(\lambda) = \lambda^3 - \lambda^2 - 2\lambda$. Since $C(1) = -2$, r is not a solution of $C(\lambda) = 0$. Thus, a trial function will use $s_2(t) = A_0 + A_1 t$. So,

$$u_p(t) = (A_0 + A_1 t)\, e^t$$

After some labor, we obtain

$$u_p^{(3)}(t) - u_p^{(2)}(t) - 2u_p^{(1)}(t) = (-2A_1 t - 2A_0 - A_1)\, e^t = te^t \tag{4.3.7}$$

This demands that for all t,

$$-2A_1 t - 2A_0 - A_1 = t \tag{4.3.8}$$

This is possible only if $A_1 = -\frac{1}{2}$, and $A_0 = \frac{1}{4}$ and therefore a particular solution of (4.3.6) is

$$u_p(t) = \left(\tfrac{1}{4} - \tfrac{1}{2}t\right) e^t \tag{4.3.9}$$

Suppose, on the other hand, that r is a solution of the characteristic equation $C(\lambda) = 0$. Then, since $C(r) = 0$, (4.3.5) becomes

$$g_{k-1}(t) = t^k \tag{4.3.10}$$

This is contradictory; the degree of $g_{k-1}(t) < k$. As in the second-order case, we try to find a solution of the form

$$u_p(t) = ts_k(t)e^{rt} \tag{4.3.11}$$

The equation

$$u^{(3)} - u^{(2)} - 2u^{(1)} = te^{-t} \tag{4.3.12}$$

provides an illustration. Here $r = -1$, $k = 1$, and $C(-1) = 0$. So we start with the trial function

$$u_p(t) = t\,(A_0 + A_1 t)\,e^{-t} \tag{4.3.13}$$

and find that

$$(6A_1 t + 3A_0 - 8A_1)\,e^{-t} = te^{-t} \tag{4.3.14}$$

So, $A_1 = \frac{1}{6}$, $A_0 = \frac{4}{9}$, is required for (4.3.14) to hold for all t. Therefore a particular solution is given by

$$u_p(t) = t\left(\tfrac{4}{9} + \tfrac{1}{6}t\right)e^{-t} \tag{4.3.15}$$

Example 4.3.1. *Find the solution of the initial-value problem*

$$u^{(3)} - u^{(2)} - 2u^{(1)} = 1, \qquad u(0) = 0 = u^{(2)}(0), \quad u^{(1)}(0) = 1$$

Solution: Since the characteristic equation of the complementary equation is $\lambda^3 - \lambda^2 - 2\lambda = 0$, and its solutions are $r_1 = 0, r_2 = -1, r_3 = 2$, we obtain the following general solution of the complementary equation:

$$u_h(t) = c_1 + c_2 e^{-t} + c_3 e^{2t}$$

Because the forcing function is $1 = e^{0t}, k = 0$ and $r = 0$. But 0 is a solution of the characteristic equation. So our trial function is not a constant (the zero degree polynomial), but rather the first-degree polynomial $u_p(t) = A_1 t$. It should be evident that $A_1 = -\frac{1}{2}$, and so a particular solution is

$$u_p(t) = -\tfrac{1}{2}t$$

and a general solution of the nonhomogeneous equation is

$$u(t) = u_p(t) + u_h(t) = -\tfrac{1}{2}t + c_1 + c_2e^{-t} + c_3e^{2t}$$

The first two derivatives of this solution are

$$\begin{aligned} u^{(1)}(t) &= -\tfrac{1}{2} - c_2e^{-t} + 2c_3e^{2t} \\ u^{(2)}(t) &= c_2e^{-t} + 4c_3e^{2t} \end{aligned}$$

For $u(t)$ to meet the given initial conditions, we must have

$$\begin{aligned} u(0) &= 0 = c_1 + c_2 + c_3 \\ u^{(1)}(0) &= 1 = -\tfrac{1}{2} - c_2 + 2c_3 \\ u^{(2)}(0) &= 0 = c_2 + 4c_3 \end{aligned}$$

After a bit of work, we determine that $c_1 = \frac{3}{4}$, $c_2 = -1$, $c_3 = \frac{1}{4}$, and the desired solution is

$$u(t) = \tfrac{3}{4} - \tfrac{1}{2}t - e^{-t} + \tfrac{1}{4}e^{2t}$$

Example 4.3.2. *Find the solution of the initial-value problem*

$$u^{(3)} + u^{(1)} = \cos 2t, \qquad u(0) = u^{(1)}(0) = 0, \quad u^{(2)}(0) = 1$$

Solution: The characteristic polynomial is $\lambda^3 + \lambda$ with roots $r_1 = 0, r_2 = i$, and $r_3 = -i$. So the complementary equation has the general solution

$$u_h(t) = c_1 + c_2 \sin t + c_3 \cos t$$

As we have seen in Chapter 3, it is often desirable to replace the trigonometric forcing function by an exponential function with the "appropriate" complex exponential. In this case we use $f(t) = e^{2it}$, and since $k = 0$ and $r = 2i$ is not a root of the characteristic polynomial, a trial function for

$$u^{(3} + u^{(1)} = e^{2it} \tag{1}$$

is $u_p(t) = Ae^{2it}$. Its first and third derivatives are

$$u_p^{(1)}(t) = 2iAe^{2it}, \qquad u_p^{(3)}(t) = -8iAe^{2it}$$

In order for $u_p(t)$ to be a solution of (1), we must have

$$-8iAe^{2it} + 2iAe^{2it} = -6ie^{2it} = e^{2it}$$

and therefore, $A = \frac{1}{6}i$. So

$$\operatorname{Re} u_p(t) = \operatorname{Re}\left(\tfrac{1}{6}ie^{2it}\right) = -\tfrac{1}{6}\sin 2t$$

is a particular solution of the given differential equation. Its general solution is

$$u(t) = \operatorname{Re} u_p(t) + u_h(t) = -\tfrac{1}{6}\sin 2t + c_1 + c_2 \sin t + c_3 \cos t$$

The first two derivatives of $u(t)$ are

$$\begin{aligned} u^{(1)}(t) &= -\tfrac{2}{6}\cos 2t + c_2 \cos t - c_3 \sin t \\ u^{(2)}(t) &= \tfrac{4}{6}\sin 2t - c_2 \sin t - c_3 \cos t \end{aligned}$$

In order to satisfy the initial conditions, we evaluate $u(t)$ and these derivatives at $t = 0$ to obtain

$$u(0) = 0 = c_1 + c_3, \quad u^{(1)}(0) = 0 = -\tfrac{2}{6} + c_2, \quad u^{(2)}(0) = 1 = -c_3$$

Hence, $c_1 = 1$, $c_2 = \frac{1}{3}$, and $c_3 = -1$. Using these values, we obtain the solution of the given initial-value problem:

$$u(t) = -\tfrac{1}{6}\sin 2t + 1 + \tfrac{1}{3}\sin t - \cos t$$

Example 4.3.3. *Find a particular solution of* $u^{(4)} + 2u^{(2)} + u = \sin t$.

Solution: Since $\operatorname{Im} e^{it} = \sin t$, we solve the related ODE

$$u^{(4)} + 2u^{(2)} + u = e^{it} \tag{1}$$

For this equation, $k = 0$ and $r = i$. However, $r = i$ is a double root $(m = 2)$ of the characteristic polynomial, and therefore, the appropriate trial function is $u_p(t) = At^2e^{2it}$. (Why isn't Ate^{2it} an effective trial function?) The second and fourth derivatives of $u_p(t)$ are

$$\begin{aligned} u_p^{(2)}(t) &= A\left(-t^2 + 4it + 2\right)e^{it} \\ u_p^{(4)}(t) &= A\left(t^2 - 8it - 12\right)e^{it} \end{aligned}$$

In order for $u_p(t)$ to be a solution of (1), we must have $-8Ae^{it} = e^{it}$, which confirms our choice of At^2e^{it} and leads to $A = -\frac{1}{8}$. Hence, $u_p(t) = -\frac{1}{8}t^2e^{it}$. Since the given forcing function is $\sin t$, we take the imaginary part of $u_p(t)$ to obtain a real solution

$$\operatorname{Im} u_p(t) = \operatorname{Im}\left(-\tfrac{1}{8}t^2e^{it}\right) = -\tfrac{1}{8}t^2 \sin t$$

Example 4.3.4. *Find a general solution of* $u^{(4)} + 2u^{(2)} + u = e^t \cos t$.

Solution: The work in this problem parallels the work in Example 4.3.3. We first note that the solutions of the characteristic equation are $\lambda_1 = \lambda_2 = i$ and $\lambda_3 = \lambda_4 = -i$, and thus, the complementary equation has a general solution

$$u_h(t) = (c_1 + c_2 t)\cos t + (c_3 + c_4 t)\sin t$$

Now for a particular solution. In this case the forcing function is taken as $e^t e^{it} = e^{(1+i)t}$ because its real part is $e^t \cos t$. The trial function for

$$u^{(4)} + 2u^{(2)} + u = e^{(1+i)t}$$

is

$$u_p(t) = Ae^{(1+i)t}$$

since $1 + i$ is not a root of the characteristic polynomial. (Note that we computed the roots of the characteristic polynomial in order to obtain the general solution of the complementary equation so this observation requires no new computations.) But $u_p(t)$ is a particular solution if and only if

$$\begin{aligned} e^{(1+i)t} &= A\left((1+i)^4 + 2(1+i)^2 + 1\right)e^{(1+i)t} \\ &= A\left((1+i)^2 + 1\right)^2 e^{(1+i)t} \end{aligned}$$

is satisfied for all t. After cancelling the exponential function, and invoking the usual simplifications, we find that

$$A = 1/(-3+4i) = -\frac{1}{25}(3+4i)$$

from which it follows that

$$\begin{aligned} u_p(t) &= -\tfrac{1}{25}(3+4i)e^{(1+i)t} = -\tfrac{1}{25}(3+4i)e^t(\cos t + i\sin t) \\ &= \tfrac{1}{25}\left((-3\cos t + 4\sin t) - i(4\cos t + 3\sin t)\right)e^t \end{aligned}$$

A particular solution of the given differential equation is

$$\operatorname{Re} u_p(t) = \tfrac{1}{25}(-3\cos t + 4\sin t)\,e^t$$

Hence, a general solution of the given ODE is

$$u_h(t) + \operatorname{Re} u_p(t) = (c_1 + c_2 t)\cos t + (c_3 + c_4 t)\sin t + \tfrac{1}{25}(-3\cos t + 4\sin t)\,e^t$$

•EXERCISES

1. Verify by substitution that $u(t) = -\frac{1}{8}t^2 \sin t$ is a solution of the ODE of Example 4.3.3.
2. Find a general solution of $u^{(4)} + 2u^{(2)} + u = \cos t$.
3. Use superposition and the solutions to Example 4.3.3 and Problem 2 to find a particular solution of $u^{(4)} + 2u^{(2)} + u = -\sin t + 2\cos t$.
4. Verify by substitution that $u(t) = \frac{1}{25}(-3\cos t + 4\sin t)e^t$ is a solution of $u^{(4)} + 2u^{(2)} + u = e^t \cos t$.

Find the solution of the following initial-value problems. The problem references are to the exercises after Section 4.2.1.

5. Problem 3(a): $u^{(3)} - 2u^{(2)} - u^{(1)} + 2u = 2;$
   $u(0) = 0, \quad u^{(1)}(0) = 0, \quad u^{(2)}(0) = 0.$
6. Problem 3(b): $u^{(3)} - 2u^{(2)} - 3u^{(1)} = 2e^t;$
   $u(0) = 0, \quad u^{(1)}(0) = -1, \quad u^{(2)}(0) = 1.$
7. Problem 3(d): $u^{(3)} - 7u^{(2)} + 6u^{(1)} = 6;$
   $u(0) = 0, \quad u^{(1)}(0) = 0, \quad u^{(2)}(0) = -1.$
8. Problem 3(e): $u^{(3)} - 13u^{(1)} - 6u = 35\cosh t;$
   $u(0) = 0, \quad u^{(1)}(0) = 0, \quad u^{(2)}(0) = 0.$
9. Problem 4(a): $u^{(3)} - 3u^{(2)} + 3u^{(1)} - u = \cos t;$
   $u(0) = 1, \quad u^{(1)}(0) = 1, \quad u^{(2)}(0) = 1.$
10. Problem 4(b): $u^{(3)} - 13u^{(2)} + 36u^{(1)} = 2e^t + 3;$
    $u(0) = \frac{1}{3}, \quad u^{(1)}(0) = \frac{1}{3}, \quad u^{(2)}(0) = \frac{3}{4}.$
11. Problem 4(e): $u^{(3)} + 4u^{(2)} + 5u^{(1)} = \cos t;$
    $u(0) = 1, \quad u^{(1)}(0) = 0, \quad u^{(2)}(0) = 0.$
12. Problem 5(a): $2u^{(3)} - u^{(2)} + 32u^{(1)} - 16u = -t;$
    $u(0) = 0, \quad u^{(1)}(0) = 0, \quad u^{(2)}(0) = 2.$
13. Problem 5(b): $u^{(3)} - u^{(2)} + 4u^{(1)} - 4u = 5e^t;$
    $u(0) = u^{(1)}(0) = u^{(2)}(0) = 0.$
14. Problem 6(b): $u^{(4)} - u = t;$
    $u(0) = 2, \quad u^{(1)}(0) = -1, \quad u^{(2)}(0) = 0.$

4.4 Companion Systems

The theory of the nth-order linear ODE is closely related to the theory of the first-order system $\mathbf{x}' = \mathbf{A}\mathbf{x} + \mathbf{f}(t)$. That such a connection should exist may seem surprising. Actually, the connection is quite natural and can be motivated by physical considerations. Picture a particle whose motion is constrained to lie on a line and is subject to forces that cause the particle to move. Let $u(t)$ measure the distance of the particle from the origin as a function of time. The physics of the problem results in an ODE governing the motion. By this we mean that the distance function satisfies some ODE which we may take, for purposes of illustration, as

$$u^{(3)} - 3u^{(2)} + 2u^{(1)} = \sin t \tag{4.4.1}$$

Initially, the particle is assumed to be at rest at the origin with zero acceleration. This means that $u(0) = u^{(1)}(0) = u^{(2)}(0) = 0$, since $u(t)$ measures distance, $u^{(1)}(t)$ measures velocity, and $u^{(2)}(t)$ measures acceleration. Using the natural variables $v(t)$ for velocity and $a(t)$ for acceleration, we may write (4.4.1) as

$$\begin{aligned} u^{(1)}(t) &= v(t) \\ u^{(2)}(t) &= v'(t) = a(t) \\ u^{(3)}(t) &= a'(t) = 3u^{(2)}(t) - 2u^{(1)}(t) + \sin t \\ &= 3a(t) - 2v(t) + \sin t \end{aligned} \tag{4.4.2}$$

Now set

$$\mathbf{x}(t) = \begin{bmatrix} u(t) \\ v(t) \\ a(t) \end{bmatrix} \tag{4.4.3}$$

Equation (4.4.2) may be written as the system

$$\mathbf{x}'(t) = \begin{bmatrix} u'(t) \\ v'(t) \\ a'(t) \end{bmatrix} = \begin{bmatrix} 0 & 1 & 0 \\ 0 & 0 & 1 \\ 0 & -2 & 3 \end{bmatrix} \begin{bmatrix} u(t) \\ v(t) \\ a(t) \end{bmatrix} + \begin{bmatrix} 0 \\ 0 \\ \sin t \end{bmatrix} \tag{4.4.4}$$

with initial condition

$$\mathbf{x}(0) = \begin{bmatrix} u(0) \\ v(0) \\ a(0) \end{bmatrix} = \begin{bmatrix} 0 \\ 0 \\ 0 \end{bmatrix} = \mathbf{0} \tag{4.4.5}$$

Thus, the simple act of renaming the derivatives of $u(t)$ according to their physical meaning results in the conversion of a third-order ODE into a system of three first-order ODE's. In fact, this conversion can be accomplished for the general n^{th}-order ODE in much the same way, even though a physical interpretation of the components of $\mathbf{x}(t)$ may not be possible. Moreover, the entries in the corresponding coefficient matrix need not be constants, although the constant coefficient ODE is the one most frequently studied in this book. Finally, note that the components of $\mathbf{x}(t)$ are the unknown function $u(t)$ and its successive derivatives.

As another illustration, consider the general third-order ODE

$$u^{(3)} + a_1(t)u^{(2)} + a_2(t)u^{(1)} + a_3(t)u = f(t) \tag{4.4.6}$$

As in the previous illustration, we set

$$\mathbf{x} = \mathbf{x}(t) = \begin{bmatrix} x_1 \\ x_2 \\ x_3 \end{bmatrix} = \begin{bmatrix} u \\ u^{(1)} \\ u^{(2)} \end{bmatrix}$$

It then follows that

$$\mathbf{x}' = \begin{bmatrix} u^{(1)} \\ u^{(2)} \\ u^{(3)} \end{bmatrix} = \begin{bmatrix} x_2 \\ x_3 \\ -a_3x_1 - a_2x_2 - a_1x_3 \end{bmatrix} + \begin{bmatrix} 0 \\ 0 \\ f(t) \end{bmatrix} \tag{4.4.7}$$

Let

$$\mathbf{C}_3(t) = \begin{bmatrix} 0 & 1 & 0 \\ 0 & 0 & 1 \\ -a_3(t) & -a_2(t) & -a_1(t) \end{bmatrix}, \qquad \mathbf{e}_3 = \begin{bmatrix} 0 \\ 0 \\ 1 \end{bmatrix} \tag{4.4.8}$$

Then, using (4.4.8) in (4.4.7), we get the companion system

$$\mathbf{x}' = \mathbf{C}_3(t)\mathbf{x} + f(t)\mathbf{e}_3 \tag{4.4.9}$$

The same approach can be applied to the second-order linear ODE

$$u^{(2)} + a_1(t)u^{(1)} + a_2(t)u = f(t) \tag{4.4.10}$$

Indeed, setting

$$\mathbf{x} = \begin{bmatrix} x_1 \\ x_2 \end{bmatrix} = \begin{bmatrix} u \\ u^{(1)} \end{bmatrix}, \quad \mathbf{C}_2(t) = \begin{bmatrix} 0 & 1 \\ -a_2(t) & -a_1(t) \end{bmatrix}, \quad \mathbf{e}_2 = \begin{bmatrix} 0 \\ 1 \end{bmatrix}$$

results in the companion system of (4.4.10):

$$\mathbf{x}' = \mathbf{C}_2(t)\mathbf{x} + f(t)\mathbf{e}_2 \tag{4.4.11}$$

Equations (4.4.11) and (4.4.9) suggest the general pattern: The system

$$\mathbf{x}' = \mathbf{C}_n(t)\mathbf{x} + f(t)\mathbf{e}_n \tag{4.4.12}$$

where

$$\mathbf{x} = \begin{bmatrix} x_1 \\ x_2 \\ \vdots \\ x_n \end{bmatrix} = \begin{bmatrix} u \\ u^{(1)} \\ \vdots \\ u^{(n-1)} \end{bmatrix}, \quad \mathbf{e}_n = \begin{bmatrix} 0 \\ 0 \\ \vdots \\ 0 \\ 1 \end{bmatrix} \tag{4.4.13}$$

$$\mathbf{C}_n(t) = \begin{bmatrix} 0 & 1 & 0 & \dots & 0 & 0 \\ 0 & 0 & 1 & \dots & 0 & 0 \\ \vdots & \vdots & \vdots & & \vdots & \vdots \\ 0 & 0 & 0 & \dots & 0 & 1 \\ -a_n & -a_{n-1} & -a_{n-2} & \dots & -a_2 & -a_1 \end{bmatrix} \tag{4.4.14}$$

and $a_i = a_i(t), t = 1, 2, \dots, n$ is called the *companion system* of the nth-order linear ODE

$$u^{(n)} + a_1 u^{(n-1)} + \cdots + a_{n-1} u^{(1)} + a_n u = f(t) \tag{4.4.15}$$

The matrix $\mathbf{C}_n(t)$ of this system is called a *companion matrix*, and we call

$$\mathbf{x}' = \mathbf{C}_n(t)\mathbf{x} \tag{4.4.16}$$

the *complementary system* of $\mathbf{x}' = \mathbf{C}_n(t)\mathbf{x} + f(t)\mathbf{e}_n$. Note that the complementary system $\mathbf{x}' = \mathbf{C}_n(t)\mathbf{x}$ is the companion system of the equation complementary to (4.4.15), namely,

$$u^{(n)} + a_1 u^{(n-1)} + \cdots + a_{n-1} u^{(1)} + a_n u = 0 \tag{4.4.17}$$

The connection between the solutions of (4.4.17) and the solutions of its companion system (4.4.16) is given by the following fundamental theorem whose proof is outlined in the Projects.

Theorem 4.4.1. *The function $u(t)$ is a solution of* (4.4.15) *if and only if*

$$\mathbf{x}(t) = \begin{bmatrix} u(t) \\ u^{(1)}(t) \\ \vdots \\ u^{(n-1)}(t) \end{bmatrix} \tag{4.4.18}$$

is a solution of its companion system (4.4.12).

The essence of this theorem is that each solution of (4.4.15) is the first entry of some solution of (4.4.12) and that the successive derivatives of the solutions of (4.4.15) are the successive entries in the corresponding solution of (4.4.12).

Theorem 4.4.1 applies to homogeneous equations. Moreover, if (4.4.15) has constant coefficients and $f(t) = 0$, we know the exact nature of its solutions and therefore Theorem 4.4.1 can be made more specific, see Corollary 4.4.1 to follow. So suppose from now on that

$$u^{(n)} + a_1 u^{(n-1)} + \cdots + a_{n-1} u^{(1)} + a_n u = 0 \tag{4.4.19}$$

has constant coefficients and that r is a solution of the characteristic equation

$$\lambda^n + a_1 \lambda^{n-1} + \cdots + a_{n-1}\lambda + a_n = 0 \tag{4.4.20}$$

Then $u(t) = e^{rt}$ is a solution of (4.4.19). Define $\mathbf{x}(t)$ by

$$\mathbf{x}(t) = \begin{bmatrix} 1 \\ r \\ \vdots \\ r^{n-1} \end{bmatrix} e^{rt} \tag{4.4.21}$$

Note that the kth row of $\mathbf{x}(t)$ is $u^{(k-1)}(t)$. We have the following corollary to Theorem 4.4.1.

Corollary 4.4.1. *Under the hypothesis of Theorem 4.4.1, $\mathbf{x}(t)$ is a solution of the companion system*

$$\mathbf{x}' = \mathbf{C}_n \mathbf{x} \tag{4.4.22}$$

if and only if r is a solution of (4.4.20), i.e., r is an eigenvalue of $\mathbf{C}_n$.

Proof: Equation (4.4.19) has the solution $u(t) = e^{rt}$ if and only if r is a solution of (4.4.20). (We permit r to be nonreal for the purpose of this proof.) Then, for $j = 1, 2, \ldots, n-1$, $u^{(j)}(t) = r^j e^{rt} = r^j u(t)$ and the proof is a consequence of Theorem 4.4.1. □

As a result of this corollary, we have a simple means for writing an invertible solution matrix for the companion system of (4.4.19) if the characteristic polynomial of (4.4.19) has n distinct roots. We state this result as a theorem.

Theorem 4.4.2. *Suppose the characteristic polynomial of*

$$u^{(n)} + a_1 u^{(n-1)} + \cdots + a_{n-1}u^{(1)} + a_n u = 0 \tag{4.4.23}$$

has n distinct real roots $r_1, r_2, \ldots, r_n$. Then

$$\Phi(t) = \begin{bmatrix} \mathbf{x}_1(t) & \mathbf{x}_2(t) & \ldots & \mathbf{x}_n(t) \end{bmatrix} \tag{4.4.24}$$

is a real invertible matrix, and $\mathbf{x}(t) = \Phi(t)\mathbf{c}$ is a general solution of the companion system $\mathbf{x}' = \mathbf{C}_n\mathbf{x}$.

Proof: From the definition of $\mathbf{x}_i(t)$ and Corollary 4.4.1, it follows that $\Phi(t)$ is a solution matrix of $\mathbf{x}' = \mathbf{C}_n(x)\mathbf{x}$. By Corollary 2.5.1, we need to prove that $\Phi(0)$ is invertible. However, since $u_i(0) = \exp(0) = 1$,

$$\Phi(0) = \begin{bmatrix} 1 & 1 & \ldots & 1 \\ r_1 & r_2 & \ldots & r_n \\ \vdots & & & \\ r_1^{n-1} & r_2^{n-1} & \ldots & r_n^{n-1} \end{bmatrix} \tag{4.4.25}$$

Now, $\Phi(0)$ is invertible if and only if its determinant, known as a *Vandermonde* determinant, is not zero. (See Theorem 0.10.4.) It would take us too far afield to construct this argument here, although it is easy to verify for $n \leq 3$. Note that $\Phi(0)$ is always singular if $r_i = r_j$ for any $i \neq j$ because two rows of $\Phi(0)$ are then identical. Finally, by Corollary 2.5.2 and the discussion following that corollary, $\mathbf{x}(t) = \Phi(t)\mathbf{c}$ is a general solution of $\mathbf{x}' = \mathbf{C}_n\mathbf{x}$. □

In the following examples we construct the solution matrix for the companion system by using solutions $u_i(t)$ of the nth-order ODE and Theorem 4.4.2. The purpose of these examples is to illuminate the content of Theorem 4.4.2. They are not meant to suggest that one needs (or ought) to construct solution matrices when the solutions to the nth-order equations are already at hand.

In the event that $r = a + ib$ is a complex root of the characteristic polynomial, we bypass Theorem 4.4.2 and use Theorem 4.4.1 to construct the columns of $\Phi(t)$: The first entry in the column of $\Phi(t)$ corresponding to the root $r = a + ib$ is $e^{at}\cos bt$, and the subsequent entries are its successive derivatives. A second column is generated by $e^{at}\sin bt$ in the same manner. Examples 4.4.3 and 4.4.4 are illustrations of this case.

Example 4.4.1. *Find a solution matrix for the companion system of*

$$u^{(3)} + 3u^{(2)} + 2u^{(1)} = 0$$

Solution: The companion system is, by (4.4.9),

$$\mathbf{x}' = \begin{bmatrix} 0 & 1 & 0 \\ 0 & 0 & 1 \\ 0 & -2 & -3 \end{bmatrix} \mathbf{x}$$

Three linearly independent solutions of $u^{(3)} + 3u^{(2)} + 2u^{(1)} = 0$ were found in Example 4.2.1:

$$u_1(t) = 1, \quad u_2(t) = e^{-t}, \quad u_3(t) = e^{-2t}$$

So three linearly independent solutions of the companion system are

$$\mathbf{x}_1(t) = \begin{bmatrix} 1 \\ 0 \\ 0 \end{bmatrix}, \quad \mathbf{x}_2(t) = \begin{bmatrix} 1 \\ -1 \\ 1 \end{bmatrix} e^{-t}, \quad \mathbf{x}_3(t) = \begin{bmatrix} 1 \\ -2 \\ 4 \end{bmatrix} e^{-2t}$$

We use these solutions for the columns of the solution matrix:

$$\Phi(t) = \begin{bmatrix} 1 & e^{-t} & e^{-2t} \\ 0 & -e^{-t} & -2e^{-2t} \\ 0 & e^{-t} & 4e^{-2t} \end{bmatrix}$$

Note that the second and third rows in $\Phi(t)$ are the first and second derivatives of its first row! We remark in passing that

$$\Phi(t)\mathbf{c} = \begin{bmatrix} 1 & e^{-t} & e^{-2t} \\ 0 & -e^{-t} & -2e^{-2t} \\ 0 & e^{-t} & 4e^{-2t} \end{bmatrix} \mathbf{c} = \begin{bmatrix} c_1 + c_2e^{-t} + c_3e^{-2t} \\ -c_2e^{-t} - 2c_3e^{-2t} \\ c_2e^{-t} + 4c_3e^{-2t} \end{bmatrix}$$

is a general solution of the companion system and that the first row of $\Phi(t)\mathbf{c}$ is a general solution of $u^{(3)} + 3u^{(2)} + 2u^{(1)} = 0$. The second and third rows of $\Phi(t)\mathbf{c}$ are easily seen to be derivatives of its first row.

Example 4.4.2. *Find a solution matrix and the general solution of the companion system of* $u^{(3)} - u^{(2)} - 2u^{(1)} = 0$.

Solution: For this equation, the companion system is

$$\mathbf{x}' = \begin{bmatrix} 0 & 1 & 0 \\ 0 & 0 & 1 \\ 0 & 2 & 1 \end{bmatrix} \mathbf{x}$$

Three linearly independent solutions of $u^{(3)} - u^{(2)} - 2u^{(1)} = 0$ are found to be

$$u_1(t) = 1, \quad u_2(t) = e^{-t}, \quad u_3(t) = e^{2t}$$

So three linearly independent solutions of the companion system are obtained by using (4.4.18):

$$\mathbf{x}_1(t) = \begin{bmatrix} 1 \\ 0 \\ 0 \end{bmatrix}, \quad \mathbf{x}_2(t) = \begin{bmatrix} 1 \\ -1 \\ 1 \end{bmatrix} e^{-t}, \quad \mathbf{x}_3(t) = \begin{bmatrix} 1 \\ 2 \\ 4 \end{bmatrix} e^{2t}$$

(Note that $u^{(k)}(t) = 0$ for $k \geq 1$ whenever $u(t) = 1$ is a solution of the nth-order ODE. This happens if and only if $r = 0$ is a root of the characteristic polynomial.) The columns of the solution matrix are the solutions $\mathbf{x}_1(t), \mathbf{x}_2(t), \mathbf{x}_3(t)$, and $\Phi(0)$ is invertible by Theorem 4.4.2. Thus, an invertible solution matrix is

$$\Phi(t) = [\ \mathbf{x}_1(t) \ \ \mathbf{x}_2(t) \ \ \mathbf{x}_3(t)\ ] = \begin{bmatrix} 1 & e^{-t} & e^{2t} \\ 0 & -e^{-t} & 2e^{2t} \\ 0 & e^{-t} & 4e^{2t} \end{bmatrix}$$

and a general solution is

$$\mathbf{x}(t) = \Phi(t)\mathbf{c} = \begin{bmatrix} 1 & e^{-t} & e^{2t} \\ 0 & -e^{-t} & 2e^{2t} \\ 0 & e^{-t} & 4e^{2t} \end{bmatrix} \begin{bmatrix} c_1 \\ c_2 \\ c_3 \end{bmatrix} = \begin{bmatrix} c_1 + c_2 e^{-t} + c_3 e^{2t} \\ -c_2 e^{-t} + 2c_3 e^{2t} \\ c_2 e^{-t} + 4c_3 e^{2t} \end{bmatrix}$$

Example 4.4.3. *Find a solution matrix for the companion system of*

$$u^{(4)} - u = 0$$

Solution: The companion system is

$$\mathbf{x}' = \begin{bmatrix} 0 & 1 & 0 & 0 \\ 0 & 0 & 1 & 0 \\ 0 & 0 & 0 & 1 \\ 1 & 0 & 0 & 0 \end{bmatrix} \mathbf{x}$$

Four linearly independent solutions of $u^{(4)} - u = 0$ were obtained in Example 4.2.2, namely,

$$u_1(t) = e^t, \quad u_2(t) = e^{-t}, \quad u_3(t) = \cos t, \quad u_4(t) = \sin t$$

Because two of the roots of the characteristic polynomial $\lambda^4 - 1$ are complex, we cannot use Theorem 4.4.2. We use Theorem 4.4.1 instead. By (4.4.18), four solutions of the companion system are

$$\mathbf{x}_1(t) = \begin{bmatrix} 1 \\ 1 \\ 1 \\ 1 \end{bmatrix} e^t, \qquad \mathbf{x}_2(t) = \begin{bmatrix} 1 \\ -1 \\ 1 \\ -1 \end{bmatrix} e^{-t}$$

$$\mathbf{x}_3(t) = \begin{bmatrix} \cos t \\ -\sin t \\ -\cos t \\ \sin t \end{bmatrix}, \qquad \mathbf{x}_4(t) = \begin{bmatrix} \sin t \\ \cos t \\ -\sin t \\ -\cos t \end{bmatrix}$$

We use these four solutions as the columns of the solution matrix $\Phi(t)$:

$$\Phi(t) = \begin{bmatrix} e^t & e^{-t} & \cos t & \sin t \\ e^t & -e^{-t} & -\sin t & \cos t \\ e^t & e^{-t} & -\cos t & -\sin t \\ e^t & -e^{-t} & \sin t & -\cos t \end{bmatrix}$$

Note that $\Phi(0)$ is not of the form (4.4.25) because we do not have four real roots of the characteristic polynomial. However, we can verify that $\Phi(0)$ is invertible:

$$\det \Phi(0) = \det \begin{bmatrix} 1 & 1 & 1 & 0 \\ 1 & -1 & 0 & 1 \\ 1 & 1 & -1 & 0 \\ 1 & -1 & 0 & -1 \end{bmatrix} = \det \begin{bmatrix} 1 & 1 & 1 & 0 \\ 1 & -1 & 0 & 1 \\ 2 & 2 & 0 & 0 \\ 2 & -2 & 0 & 0 \end{bmatrix} = -8$$

As in the previously worked examples, rows 2–4 of $\Phi(t)$ are the successive derivatives of its first row. Note that the first row of $\Phi(t)\mathbf{c}$ is a general solution of $u^{(4)} - u = 0$:

$$u(t) = c_1 e^t + c_2 e^{-t} + c_3 \cos t + c_4 \sin t$$

Example 4.4.4. *Find a solution matrix of the companion system of*

$$u^{(3)} + u^{(1)} = 0$$

Solution: Here the companion matrix is

$$\mathbf{x}' = \begin{bmatrix} 0 & 1 & 0 \\ 0 & 0 & 1 \\ 0 & -1 & 0 \end{bmatrix} \mathbf{x}$$

Three linearly independent solutions of the given ODE are

$$u_1(t) = 1, \quad u_2(t) = \sin t, \quad u_3(t) = \cos t$$

From (4.4.18), we find that

$$\mathbf{x}_1(t) = \begin{bmatrix} 1 \\ 0 \\ 0 \end{bmatrix}, \quad \mathbf{x}_2(t) = \begin{bmatrix} \sin t \\ \cos t \\ -\sin t \end{bmatrix}, \quad \mathbf{x}_3(t) = \begin{bmatrix} \cos t \\ -\sin t \\ -\cos t \end{bmatrix}$$

So

$$\Phi(t) = \begin{bmatrix} 1 & \sin t & \cos t \\ 0 & \cos t & -\sin t \\ 0 & -\sin t & -\cos t \end{bmatrix}$$

is a solution matrix, and $\Phi(0)$ is nonsingular.

Example 4.4.5. *Find the fundamental solution matrix for the companion system of*

$$u'' - 2au' + a^2 u = 0$$

Solution: The companion system for this homogeneous equation is

$$\mathbf{x}' = \begin{bmatrix} 0 & 1 \\ -a^2 & 2a \end{bmatrix}$$

The characteristic equation of this companion system (and the given second-order ODE) is

$$\lambda^2 - 2a\lambda + a^2 = (\lambda - a)^2 = 0$$

Its solutions are $r_1 = r_2 = a$, from which it follows that the given ODE has solutions

$$u_1(t) = e^{at}, \quad u_2(t) = te^{at}$$

Therefore, from (4.4.18), two solutions of the companion system are

$$\mathbf{x}_1(t) = \begin{bmatrix} u_1(t) \\ u_1'(t) \end{bmatrix} = \begin{bmatrix} e^{at} \\ ae^{at} \end{bmatrix}, \quad \mathbf{x}_2(t) = \begin{bmatrix} u_2(t) \\ u_2'(t) \end{bmatrix} = \begin{bmatrix} te^{at} \\ (1 + at)\, e^{at} \end{bmatrix}$$

We use these as columns of a solution matrix. We can find a fundamental matrix as follows,

$$\Psi(t) = \Phi(t)\Phi^{-1}(0) = \begin{bmatrix} e^{at} & te^{at} \\ ae^{at} & (1+at)e^{at} \end{bmatrix} \begin{bmatrix} 1 & 0 \\ a & 1 \end{bmatrix}^{-1}$$

$$= \begin{bmatrix} e^{at} & te^{at} \\ ae^{at} & (1+at)e^{at} \end{bmatrix} \begin{bmatrix} 1 & 0 \\ -a & 1 \end{bmatrix} = e^{at} \begin{bmatrix} 1-at & t \\ -a^2 t & 1+at \end{bmatrix}$$

Example 4.4.6. *Find the fundamental solution matrix of the companion system of $u'' - (a+b)u' + abu = 0$, where $a \neq b$.*

Solution: The characteristic polynomial is $\lambda^2 - (a+b)\lambda + ab$ with roots $r_1 = a$ and $r_2 = b$. Therefore, $u_1(t) = e^{at}$ and $u_2(t) = e^{bt}$ are two solutions. Hence, by (4.4.24) or (4.4.18), two solutions of the companion system are

$$\mathbf{x}_1(t) = \begin{bmatrix} 1 \\ a \end{bmatrix} e^{at}, \quad \mathbf{x}_2(t) = \begin{bmatrix} 1 \\ b \end{bmatrix} e^{bt}$$

It then follows that the fundamental solution matrix is

$$\Psi(t) = \Phi(t)\Phi^{-1}(0) = \begin{bmatrix} e^{at} & e^{bt} \\ ae^{at} & be^{bt} \end{bmatrix} \begin{bmatrix} 1 & 1 \\ a & b \end{bmatrix}^{-1}$$

$$= \frac{1}{b-a} \begin{bmatrix} be^{at} - ae^{bt} & -e^{at} + e^{bt} \\ abe^{at} - abe^{bt} & -ae^{at} + be^{bt} \end{bmatrix}$$

Example 4.4.7. *Find a solution matrix of the companion system of*

$$u'' - 2au' + \left(a^2 + b^2\right) u = 0$$

given that $b \neq 0$.

Solution: This second-order ODE has solutions

$$u_1(t) = e^{at} \cos bt, \quad u_2(t) = e^{at} \sin bt$$

The student is invited (in Problem 17) to show that the fundamental matrix is given by

$$\Psi(t) = \frac{e^{at}}{b} \begin{bmatrix} b\cos bt - a\sin bt & \sin bt \\ -(a^2+b^2)\sin bt & a\sin bt + b\cos bt \end{bmatrix}$$

Note that $\Psi(0) = \mathbf{I}$.

Example 4.4.8. *Find the general solution of the companion system of*

$$u'' - 2u' + u = te^{2t}$$

Solution: A particular solution is easily found by undetermined coefficients, namely, $u_p(t) = (-2+t)e^{2t}$. From (4.4.18), we find that

$$\mathbf{x}_p(t) = \begin{bmatrix} u_p(t) \\ u_p'(t) \end{bmatrix} = \begin{bmatrix} (-2+t)e^{2t} \\ (-3+2t)e^{2t} \end{bmatrix} = \begin{bmatrix} -2+t \\ -3+2t \end{bmatrix} e^{2t}$$

is a particular solution of

$$\mathbf{x}' = \begin{bmatrix} 0 & 1 \\ -1 & 2 \end{bmatrix} \mathbf{x} + \begin{bmatrix} 0 \\ 1 \end{bmatrix} te^{2t}$$

From the results in Example 4.4.5 with $a = 1$, we obtain a general solution

$$\begin{aligned} \mathbf{x}(t) &= \Psi(t)\mathbf{c} + \mathbf{x}_p(t) \\ &= e^t \begin{bmatrix} 1-t & t \\ -t & 1+t \end{bmatrix} \mathbf{c} + \begin{bmatrix} -2+t \\ -3+2t \end{bmatrix} e^{2t} \end{aligned}$$

where $\Psi(t)$ is the fundamental solution matrix for the companion system $\mathbf{x}' = \mathbf{A}\mathbf{x}$. Note that $\Psi(t)\mathbf{c}$ is a general solution of this companion system and $\mathbf{x}_p(t)$ is the particular solution just computed.

•EXERCISES

Problems 1–15 refer to the Exercise set following Section 4.2.1. Find the companion system and a solution matrix for each of these ODE's.

1. Problem 3(b): $u^{(3)} - 2u^{(2)} - 3u^{(1)} = 0.$
2. Problem 3(c): $10u^{(3)} + u^{(2)} + 7u^{(1)} + 2u = 0.$
3. Problem 3(d): $u^{(3)} - 7u^{(2)} + 6u^{(1)} = 0.$
4. Problem 3(e): $4u^{(3)} - 13u^{(1)} - 6u = 0.$
5. Problem 4(a): $u^{(3)} - 3u^{(2)} + 3u^{(1)} - u = 0.$
6. Problem 4(b): $u^{(3)} - 13u^{(2)} + 36u^{(1)} = 0.$
7. Problem 4(d): $u^{(4)} - u^{(3)} - 7u^{(2)} + u^{(1)} + 6u = 0.$
8. Problem 3(a): $u^{(3)} - 2u^{(2)} - u^{(1)} + 2u = 0.$
9. Problem 4(e): $u^{(3)} + 4u^{(2)} + 5u^{(1)} = 0.$
10. Problem 5(a): $2u^{(3)} - u^{(2)} + 32u^{(1)} - 16u = 0.$

11. Problem 5(b): $u^{(3)} - u^{(2)} + 4u^{(1)} - 4u = 0.$
12. Problem 5(c): $u^{(4)} - 2u^{(3)} - u^{(2)} + 2u^{(1)} = 0.$
13. Problem 6(b): $u^{(4)} - u = 0.$
14. Problem 6(e): $u^{(3)} - 3u^{(2)} + 4u = 0.$
15. Problem 6(f): $u^{(3)} + 3u^{(2)} + 3u^{(1)} + u = 0.$
16. Show that the fundamental matrix for $u'' - bu' = 0$ is given by

$$\Psi(t) = \frac{1}{b}\begin{bmatrix} b & -1 + e^{bt} \\ 0 & be^{bt} \end{bmatrix}$$

17. Verify that $\Psi(t)$ given in Examples 4.4.5–4.4.7 are fundamental solution matrices for the companion system of these examples.

4.5 Initial-Value Problems for Homogeneous Companion Systems

In order to solve the initial-value problem

$$u^{(n)} + a_1 u^{(n-1)} + \cdots + a_{n-1}u^{(1)} + a_n u = 0 \tag{4.5.1}$$

$$u(0) = b_1,\ u^{(1)}(0) = b_2,\ \ldots,\ u^{(n-1)}(0) = b_n \tag{4.5.2}$$

it is necessary to obtain not only a general solution of the ODE but also its first $n-1$ derivatives at $t = 0$. These are precisely the components (at $t = 0$) of the solution of the companion system $\mathbf{x}' = \mathbf{C}_n\mathbf{x}$, $\mathbf{x}(0) = \mathbf{b}$, that is,

$$\mathbf{x}(t) = \begin{bmatrix} u(t) \\ u^{(1)}(t) \\ \vdots \\ u^{(n-1)}(t) \end{bmatrix} \tag{4.5.3}$$

such that

$$\mathbf{x}(0) = \begin{bmatrix} u(0) \\ u^{(1)}(0) \\ \vdots \\ u^{(n-1)}(0) \end{bmatrix} = \mathbf{b} = \begin{bmatrix} b_1 \\ b_2 \\ \vdots \\ b_n \end{bmatrix} \tag{4.5.4}$$

It is in this circumstance that the extra work required to find a solution matrix begins to pay dividends. The next theorem, actually a corollary of Theorem 4.4.1, shows how to solve (4.5.1) by use of the companion system.

Theorem 4.5.1. *Let* $\mathbf{x}(t)$ *be the solution of the initial-value problem*

$$\mathbf{x}' = \mathbf{C}_n\mathbf{x}, \quad \mathbf{x}(0) = \mathbf{b} \tag{4.5.5}$$

where $\mathbf{b}$ *is given by* (4.5.2). *Then the first entry of* $\mathbf{x}(t)$ *is the unique solution of the initial-value problem* (4.5.1) *and* (4.5.2).

Proof: By Theorem 4.4.1, the first entry of $\mathbf{x}(t)$ is a solution of (4.5.1). From (4.4.18) of this same theorem, $\mathbf{x}(0) = \mathbf{b}$ which shows that $u(t)$ meets the initial conditions (4.5.2). □

Example 4.5.1. *Find the solution of the initial-value problem*

$$u^{(3)} + 3u^{(2)} + 2u^{(1)} = 0, \quad u(0) = u^{(1)}(0) = 0, \; u^{(2)}(0) = 2$$

Solution: We found a general solution for the companion system of this ODE in Example 4.4.1 (see also Example 4.2.5):

$$\mathbf{x}(t) = \begin{bmatrix} 1 & e^{-t} & e^{-2t} \\ 0 & -e^{-t} & -2e^{-2t} \\ 0 & e^{-t} & 4e^{-2t} \end{bmatrix} \mathbf{c}$$

We find $\mathbf{c}$ by solving

$$\mathbf{x}(0) = \Phi(0)\mathbf{c} = \begin{bmatrix} 1 & 1 & 1 \\ 0 & -1 & -2 \\ 0 & 1 & 4 \end{bmatrix} \mathbf{c} = \mathbf{b} = \begin{bmatrix} 0 \\ 0 \\ 2 \end{bmatrix}$$

This system of equations is solved by row reductions on the augmented system:

$$[\Phi(0)|\, \mathbf{b}] = \left[\begin{array}{ccc|c} 1 & 1 & 1 & 0 \\ 0 & -1 & -2 & 0 \\ 0 & 1 & 4 & 2 \end{array}\right] \to \cdots \to \left[\begin{array}{ccc|c} 1 & 0 & 0 & 1 \\ 0 & 1 & 0 & -2 \\ 0 & 0 & 1 & 1 \end{array}\right]$$

Thus, $c_1 = 1, c_2 = -2$, and $c_3 = 1$. Using these values for the entries of $\mathbf{c}$, we obtain the solution $\mathbf{x}(t)$

$$\mathbf{x}(t) = \Phi(t)\mathbf{c} = \begin{bmatrix} 1 & e^{-t} & e^{-2t} \\ 0 & -e^{-t} & -2e^{-2t} \\ 0 & e^{-t} & 4e^{-2t} \end{bmatrix} \begin{bmatrix} 1 \\ -2 \\ 1 \end{bmatrix} = \begin{bmatrix} 1 - 2e^{-t} + e^{-2t} \\ 2e^{-t} - 2e^{-2t} \\ -2e^{-t} + 4e^{-2t} \end{bmatrix}$$

Therefore, the solution of the given initial-value problem is

$$u(t) = 1 - 2e^{-t} + e^{-2t}$$

Also, the second and third rows of $\mathbf{x}(t)$ provide its derivatives:

$$\begin{aligned} u'(t) &= 2e^{-t} - 2e^{-2t} \\ u''(t) &= -2e^{-t} + 4e^{-2t} \end{aligned}$$

Example 4.5.2. *Find the solution of the initial-value problem*

$$u^{(4)} - u = 0, \qquad u(0) = u^{(1)}(0) = u^{(2)}(0) = 0, \quad u^{(3)}(0) = 4$$

Solution: A general solution for this ODE is implicit in the solution given in Example 4.4.3:

$$\mathbf{x}(t) = \Phi(t)\mathbf{c} = \begin{bmatrix} e^t & e^{-t} & \cos t & \sin t \\ e^t & -e^{-t} & -\sin t & \cos t \\ e^t & e^{-t} & -\cos t & -\sin t \\ e^t & -e^{-t} & \sin t & -\cos t \end{bmatrix} \mathbf{c}$$

The equations that determine $\mathbf{c}$ are given by

$$\mathbf{x}(0) = \Phi(0)\mathbf{c} = \begin{bmatrix} 1 & 1 & 1 & 0 \\ 1 & -1 & 0 & 1 \\ 1 & 1 & -1 & 0 \\ 1 & -1 & 0 & -1 \end{bmatrix} \mathbf{c} = \mathbf{b} = \begin{bmatrix} 0 \\ 0 \\ 0 \\ 4 \end{bmatrix}$$

and we compute as follows:

$$[\Phi(0)|\ \mathbf{b}] = \left[\begin{array}{cccc|c} 1 & 1 & 1 & 0 & 0 \\ 1 & -1 & 0 & 1 & 0 \\ 1 & 1 & -1 & 0 & 0 \\ 1 & -1 & 0 & -1 & 4 \end{array}\right] \to \cdots \to \left[\begin{array}{cccc|c} 1 & 0 & 0 & 0 & 1 \\ 0 & 1 & 0 & 0 & -1 \\ 0 & 0 & 1 & 0 & 0 \\ 0 & 0 & 0 & 1 & -2 \end{array}\right]$$

Hence, $c_1 = 1, c_2 = -1, c_3 = 0$, and $c_4 = -2$. Thus,

$$\mathbf{x}(t) = \Phi(t)\mathbf{c} = \begin{bmatrix} e^t & e^{-t} & \cos t & \sin t \\ e^t & -e^{-t} & -\sin t & \cos t \\ e^t & e^{-t} & -\cos t & -\sin t \\ e^t & -e^{-t} & \sin t & -\cos t \end{bmatrix} \begin{bmatrix} 1 \\ -1 \\ 0 \\ -2 \end{bmatrix}$$

$$= \begin{bmatrix} e^t - e^{-t} - 2\sin t \\ e^t + e^{-t} - 2\cos t \\ e^t - e^{-t} + 2\sin t \\ e^t + e^{-t} + 2\cos t \end{bmatrix}$$

is the unique solution of the companion system associated with the given initial-value problem. So the desired solution is the first entry of $\mathbf{x}(t)$:

$$u(t) = e^t - e^{-t} - 2\sin t$$

Example 4.5.3. *Find the solution of the initial-value problem*

$$u^{(3)} - u^{(2)} - 2u^{(1)} = 0, \qquad u(0) = 1,\ u^{(1)}(0) = 1, \quad u^{(2)}(0) = -1$$

Solution: In Example 4.4.2, we found the general solution matrix

$$\mathbf{x}(t) = \Phi(t)\mathbf{c} = \begin{bmatrix} 1 & e^{-t} & e^{2t} \\ 0 & -e^{-t} & 2e^{2t} \\ 0 & e^{-t} & 4e^{2t} \end{bmatrix} \mathbf{c}$$

We find $\mathbf{c}$ from

$$[\Phi(0)|\ \mathbf{b}] = \left[\begin{array}{ccc|c} 1 & 1 & 1 & 1 \\ 0 & -1 & 2 & 1 \\ 0 & 1 & 4 & -1 \end{array}\right] \to \cdots \to \left[\begin{array}{ccc|c} 1 & 0 & 0 & 2 \\ 0 & 1 & 0 & -1 \\ 0 & 0 & 1 & 0 \end{array}\right]$$

Hence, $c_1 = 2, c_2 = -1$, and $c_3 = 0$, and therefore,

$$\mathbf{x}(t) = \Phi(t)\mathbf{c} = \begin{bmatrix} 1 & e^{-t} & e^{2t} \\ 0 & -e^{-t} & 2e^{2t} \\ 0 & e^{-t} & 4e^{2t} \end{bmatrix} \begin{bmatrix} 2 \\ -1 \\ 0 \end{bmatrix} = \begin{bmatrix} 2 - e^{-t} \\ e^{-t} \\ -e^{-t} \end{bmatrix}$$

and the desired solution is the first row of $\mathbf{x}(t)$:

$$u(t) = 2 - e^{-t}$$

Example 4.5.4. *Find the solution of the initial-value problem*

$$u^{(3)} + u^{(1)} = 0, \qquad u(0) = u^{(1)}(0) = u^{(2)}(0) = 1$$

Solution: A solution matrix for this ODE was found in Example 4.4.4:

$$\Phi(t) = \begin{bmatrix} 1 & \sin t & \cos t \\ 0 & \cos t & -\sin t \\ 0 & -\sin t & -\cos t \end{bmatrix}$$

Hence, a general solution is $\mathbf{x}(t) = \Phi(t)\mathbf{c}$, and $\mathbf{c}$ is found from

$$[\Phi(0)|\, \mathbf{b}] = \left[\begin{array}{ccc|c} 1 & 0 & 1 & 1 \\ 0 & 1 & 0 & 1 \\ 0 & 0 & -1 & 1 \end{array}\right] \to \cdots \to \left[\begin{array}{ccc|c} 1 & 0 & 0 & 2 \\ 0 & 1 & 0 & 1 \\ 0 & 0 & 1 & -1 \end{array}\right]$$

Consequently, $c_1 = 2, c_2 = 1$, and $c_3 = -1$, and

$$\mathbf{x}(t) = \Phi(t)\mathbf{c} = \begin{bmatrix} 1 & \sin t & \cos t \\ 0 & \cos t & -\sin t \\ 0 & -\sin t & -\cos t \end{bmatrix} \begin{bmatrix} 2 \\ 1 \\ -1 \end{bmatrix} = \begin{bmatrix} 2 + \sin t - \cos t \\ \cos t + \sin t \\ -\sin t + \cos t \end{bmatrix}$$

Thus, the solution of the initial-value problem is the first row of $\mathbf{x}(t)$, that is, $u(t) = 2 + \sin t - \cos t$.

Example 4.5.5. *Find the solution of the initial-value problem*

$$u^{(2)} - 2u^{(1)} + u = 0, \qquad u(0) = b_1, \quad u^{(1)}(0) = b_2$$

Solution: The fundamental solution matrix for the companion system of this ODE was found in Example 4.4.5 using $a = 1$. We have, therefore,

$$\mathbf{x}(t) = \Psi(t)\mathbf{c} = e^t \begin{bmatrix} 1-t & t \\ -t & 1+t \end{bmatrix} \mathbf{c}$$

Now we set $t = 0$ and solve

$$\mathbf{x}(0) = \begin{bmatrix} 1 & 0 \\ 0 & 1 \end{bmatrix} \mathbf{c} = \mathbf{b} = \begin{bmatrix} b_1 \\ b_2 \end{bmatrix}$$

So $\mathbf{c} = \mathbf{b}$, and therefore,

$$\mathbf{x}(t) = e^t \begin{bmatrix} 1-t & t \\ -t & 1+t \end{bmatrix} \mathbf{b} = e^t \begin{bmatrix} b_1(1-t) + b_2 t \\ b_2(1+t) - b_1 t \end{bmatrix}$$

and the desired solution is $u(t) = e^t\,(b_1(1-t) + b_2 t)$.

•EXERCISES

Find the solution to each of the following initial-value problems by using the technique of this section. These problems correspond to the similarly numbered problems appearing in the exercise set following Section 4.4.

1. $u^{(3)} - 2u^{(2)} - 3u^{(1)} = 0;$
   $u(0) = 1, \quad u^{(1)}(0) = u^{(2)}(0) = 3.$
2. $10u^{(3)} + u^{(2)} - 7u^{(1)} + 2u = 0;$
   $u(0) = u^{(1)}(0) = 3, \quad u^{(2)}(0) = 0.$
3. $u^{(3)} - 7u^{(2)} + 6u^{(1)} = 0;$
   $u(0) = 1, \quad u^{(1)}(0) = 3, \quad u^{(2)}(0) = 1.$
4. $4u^{(3)} - 13u^{(1)} - 6u = 0;$
   $u(0) = 1, \quad u^{(1)}(0) = 4, \quad u^{(2)}(0) = 0.$
5. $u^{(3)} - 3u^{(2)} + 3u^{(1)} - u = 0;$
   $u(0) = u^{(1)}(0) = 0, \quad u^{(2)}(0) = 2.$
6. $u^{(3)} - 13u^{(2)} + 36u^{(1)} = 0;$
   $u(0) = 1, \quad u^{(1)}(0) = 3, \quad u^{(2)}(0) = 0.$
7. $u^{(4)} - u^{(3)} - 7u^{(2)} + u^{(1)} + 6u = 0;$
   $u(0) = 2, \quad u^{(1)}(0) = 0, \quad u^{(2)}(0) = 0, \quad u^{(3)}(0) = 2.$
8. $u^{(3)} - 2u^{(2)} - u^{(1)} + 2u = 0;$
   $u(0) = 1, \quad u^{(1)}(0) = -1, \quad u^{(2)}(0) = 1.$
9. $u^{(3)} + 4u^{(2)} + 5u^{(1)} = 0;$
   $u(0) = u^{(1)}(0) = u^{(2)}(0) = 1.$
10. $2u^{(3)} - u^{(2)} + 32u^{(1)} - 16u = 0;$
    $u(0) = 0, \quad u^{(1)}(0) = 4, \quad u^{(2)}(0) = 0.$
11. $u^{(3)} - u^{(2)} + 4u^{(1)} - 4u = 0;$
    $u(0) = 1, \quad u^{(1)}(0) = 3, \quad u^{(2)}(0) = 1.$
12. $u^{(4)} - 2u^{(3)} - u^{(2)} + 2u^{(1)} = 0;$
    $u(0) = 2, \quad u^{(1)}(0) = u^{(2)}(0) = u^{(3)}(0) = 1.$
13. $u^{(4)} - u = 0;$
    $u(0) = 1, \quad u^{(1)}(0) = -1, \quad u^{(2)}(0) = 2, \quad u^{(3)}(0) = -1.$
14. $u^{(3)} - 3u^{(2)} + 4u = 0;$
    $u(0) = 0, \quad u^{(1)}(0) = u^{(2)}(0) = -3.$
15. $u^{(3)} + 3u^{(2)} + 3u^{(1)} + u = 0;$
    $u(0) = 1, \quad u^{(1)}(0) = u^{(2)}(0) = 0.$

4.6 Variation of Parameters for Nonhomogeneous Companion Systems

Now we consider the nonhomogeneous equation

$$u^{(n)} + a_1 u^{(n-1)} + \cdots + a_{n-1} u^{(1)} + a_n u = f(t) \tag{4.6.1}$$

Let

$$\mathbf{x}' = \mathbf{C}_n \mathbf{x} + f(t)\mathbf{e}_n \tag{4.6.2}$$

be the companion system of (4.6.1) and recall that $\mathbf{e}_n = [0, 0, \ldots, 1]^T$. Suppose $\Phi(t)$ is an invertible solution matrix of this system. Then, by Corollary 2.7.1, a particular solution of (4.6.2) is given by

$$\mathbf{x}_p(t) = \Phi(t) \int_0^t f(s)\Phi^{-1}(s)\mathbf{e}_n \, ds \tag{4.6.3}$$

$$= \int_0^t f(s)\Phi^{-1}(s)\Phi(t)\mathbf{e}_n \, ds \tag{4.6.4}$$

because the solution matrix $\Phi(t)$ is independent of s. We now apply Theorem 2.7.1 to companion systems.

Theorem 4.6.1. *A general solution of the companion system of* (4.6.1) *is given by*

$$\mathbf{x}(t) = \Phi(t) \int_0^t f(s)\Phi^{-1}(s)\mathbf{e}_n \, ds + \Phi(t)\mathbf{c} \tag{4.6.5}$$

Proof: For companion systems, $\mathbf{f}(s) = f(s)\,\mathbf{e}_n$. Thus, Theorem 4.6.1 is just Theorem 2.7.1. □

Now consider the initial-value problem

$$u^{(n)} + a_1 u^{(n-1)} + \cdots + a_{n-1} u^{(1)} + a_n u = f(t) \tag{4.6.6}$$

$$u(0) = b_1, u^{(1)}(0) = b_2, \ldots, u^{(n-1)}(0) = b_n$$

Define a vector $\mathbf{b}$ whose entries are the initial conditions. That is,

$$\mathbf{b} = \begin{bmatrix} b_1 \\ b_2 \\ \vdots \\ b_n \end{bmatrix} \tag{4.6.7}$$

Corollary 4.6.1. *Under the hypothesis and notation of Theorem 4.6.1, the solution of the initial-value problem* (4.6.6) *is the first row of*

$$\mathbf{x}(t) = \Phi(t) \int_0^t f(s)\Phi^{-1}(s)\mathbf{e}_n\, ds + \Phi(t)\Phi^{-1}(0)\mathbf{b} \tag{4.6.8}$$

Proof: Clearly, (4.6.8) is a special case of (4.6.5), with $\mathbf{c} = \Phi^{-1}(0)\mathbf{b}$. So $\mathbf{x}(t)$, as given in (4.6.8), is a solution of $\mathbf{x}' = \mathbf{C}_n\mathbf{x} + f(t)\mathbf{e}_n$. Now,

$$\begin{aligned} \mathbf{x}(0) &= \int_0^0 f(s)\Phi^{-1}(s)\Phi(t)\mathbf{e}_n\, ds + \Phi(0)\Phi^{-1}(0)\mathbf{b} \\ &= \mathbf{Ib} = \mathbf{b} \end{aligned}$$

This shows that $\mathbf{x}(t)$ meets the initial condition $\mathbf{x}(0) = \mathbf{b}$. Now, the jth row of $\mathbf{x}(0)$ is $u^{(j)}(0)$ — because the jth row of $\mathbf{x}(t)$ is $u^{(j)}(t)$ — and by (4.6.8), the jth row of $\mathbf{x}(0)$ is b_j as well. □

Since this material has already been covered in greater generality in Section 2.7, we illustrate Corollary 4.6.1.

Example 4.6.1. *Find the solution of the initial-value problem*

$$u^{(3)} - u^{(1)} = f(t), \qquad u(0) = u^{(1)}(0) = 1, \quad u^{(2)}(0) = 0$$

Solution: It is easy to verify that three solutions of the complementary companion system

$$\mathbf{x}' = \begin{bmatrix} 0 & 1 & 0 \\ 0 & 0 & 1 \\ 0 & 1 & 0 \end{bmatrix} \mathbf{x}$$

are

$$\mathbf{x}_1(t) = \begin{bmatrix} 1 \\ 0 \\ 0 \end{bmatrix}, \quad \mathbf{x}_2(t) = \begin{bmatrix} 1 \\ -1 \\ 1 \end{bmatrix} e^{-t}, \quad \mathbf{x}_3(t) = \begin{bmatrix} 1 \\ 1 \\ 1 \end{bmatrix} e^{t}$$

Therefore, an invertible solution matrix is given by

$$\Phi(t) = \begin{bmatrix} 1 & e^{-t} & e^t \\ 0 & -e^{-t} & e^t \\ 0 & e^{-t} & e^t \end{bmatrix}$$

We now find $\Phi^{-1}(t)$ by row reductions:

$$\left[\begin{array}{ccc|ccc} 1 & e^{-t} & e^t & 1 & 0 & 0 \\ 0 & -e^{-t} & e^t & 0 & 1 & 0 \\ 0 & e^{-t} & e^t & 0 & 0 & 1 \end{array}\right] \to \cdots \to \left[\begin{array}{ccc|ccc} 1 & 0 & 0 & 1 & -1 & -1 \\ 0 & 1 & 0 & 0 & -e^t/2 & e^t/2 \\ 0 & 0 & 1 & 0 & e^{-t}/2 & e^{-t}/2 \end{array}\right]$$

Hence,

$$\Phi^{-1}(t) = \begin{bmatrix} 1 & -1 & -1 \\ 0 & -e^t/2 & e^t/2 \\ 0 & e^{-t}/2 & e^{-t}/2 \end{bmatrix}$$

and

$$\Phi^{-1}(0) = \begin{bmatrix} 1 & -1 & -1 \\ 0 & -1/2 & 1/2 \\ 0 & 1/2 & 1/2 \end{bmatrix}$$

From these two expressions, we obtain

$$\begin{aligned} \Phi(t)\Phi^{-1}(s)\,\mathbf{e}_3 &= \begin{bmatrix} 1 & e^{-t} & e^t \\ 0 & -e^{-t} & e^t \\ 0 & e^{-t} & e^t \end{bmatrix} \begin{bmatrix} 1 & -1 & -1 \\ 0 & -e^s/2 & e^s/2 \\ 0 & e^{-s}/2 & e^{-s}/2 \end{bmatrix} \begin{bmatrix} 0 \\ 0 \\ 1 \end{bmatrix} \\ &= \frac{1}{2} \begin{bmatrix} 1 & e^{-t} & e^t \\ 0 & -e^{-t} & e^t \\ 0 & e^{-t} & e^t \end{bmatrix} \begin{bmatrix} -2 \\ e^s \\ e^{-s} \end{bmatrix} \\ &= \frac{1}{2} \begin{bmatrix} -2 + e^{s-t} + e^{-(s-t)} \\ -e^{s-t} + e^{-(s-t)} \\ e^{s-t} + e^{-(s-t)} \end{bmatrix} \end{aligned}$$

Also,

$$\begin{aligned} \Phi(t)\Phi^{-1}(0)\mathbf{b} &= \begin{bmatrix} 1 & e^{-t} & e^t \\ 0 & -e^{-t} & e^t \\ 0 & e^{-t} & e^t \end{bmatrix} \begin{bmatrix} 1 & -1 & -1 \\ 0 & -1/2 & 1/2 \\ 0 & 1/2 & 1/2 \end{bmatrix} \begin{bmatrix} 0 \\ 1 \\ 0 \end{bmatrix} \\ &= \begin{bmatrix} 1 & e^{-t} & e^t \\ 0 & -e^{-t} & e^t \\ 0 & e^{-t} & e^t \end{bmatrix} \begin{bmatrix} -1 \\ -1/2 \\ 1/2 \end{bmatrix} \\ &= \frac{1}{2} \begin{bmatrix} -2 - e^{-t} + e^t \\ e^{-t} + e^t \\ -e^{-t} + e^t \end{bmatrix} \end{aligned}$$

From (4.6.8), it follows that the first row of $\mathbf{x}(t)$ is the desired solution:

$$u(t) = \tfrac{1}{2}\int_0^t \left(-2 + e^{s-t} + e^{-(s-t)}\right) f(s)\, ds - 1 - \tfrac{1}{2}e^{-t} + \tfrac{1}{2}e^t$$

•EXERCISES

Find a particular solution to Problems 1–9 using variation of parameters. The first six problems appeared as Problems 1–6 in the exercise set following Section 3.12.

1. $u^{(2)} + u = \tan t.$
2. $u^{(2)} + 5u^{(1)} + 4u = te^t.$
3. $u^{(2)} + 4u^{(1)} + 4u = te^{-2t}.$
4. $u^{(2)} + u = \csc t.$
5. $u^{(2)} - 2u^{(1)} + u = t^{-1}e^t.$
6. $u^{(2)} - 4u^{(1)} + 4u = e^t.$
7. $u^{(3)} + 3u^{(2)} + 3u^{(1)} + u = t^{-3}e^{-t}.$ (See Problem 15, Section 4.4.)
8. $u^{(3)} + u^{(1)} = \sec^2 t.$
9. $u^{(2)} + 2u^{(1)} + u = e^{-t}\ln t.$
10. Find a solution of the initial-value problem
    $u^{(3)} + u^{(1)} = \cosh t, \qquad u(0) = u^{(1)}(0) = 1,\ u^{(2)}(0) = 0.$
    See Example 4.4.4.
11. Find a solution of the initial-value problem
    $u^{(2)} - 2au^{(1)} + a^2u = f(t), \qquad u(0) = b_1,\ u^{(1)}(0) = b_2.$
    See Example 4.4.5.
12. Find a solution of the initial-value problem
    $u^{(2)} - (a+b)u^{(1)} + abu = f(t), \quad a \neq b, \quad u(0) = b_1,\ u^{(1)}(0) = b_2.$
    See Example 4.4.6.
13. Find a solution of the initial-value problem
    $u^{(2)} - 2au^{(1)} + (a^2 + b^2)\, u = f(t), \quad a \neq 0, \quad u(0) = b_1,\ u^{(1)}(0) = b_2.$
    See Example 4.4.7.
14. Find a solution of the initial-value problem
    $u^{(2)} - au^{(1)} = f(t), \quad u(0) = b_1,\ u^{(1)}(0) = b_2.$
15. Find a solution of the initial-value problem
    $u^{(2)} - a^2u = f(t), \quad u(0) = b_1,\ u^{(1)}(0) = b_2.$
16. Find a solution of the initial-value problem
    $u^{(2)} + a^2u = f(t), \quad u(0) = b_1,\ u^{(1)}(0) = b_2.$

PROJECTS

1. Solutions of Companion Systems

(a) Suppose that $u(t)$ is a solution of

$$u^{(n)} + a_1 u^{(n-1)} + \cdots + a_{n-1}u^{(1)} + a_n u = f(t) \tag{1}$$

Define $\mathbf{x}(t)$ and $\mathbf{C}_n$ by

$$\mathbf{x} = \begin{bmatrix} u \\ u^{(1)} \\ \vdots \\ u^{(n-1)} \end{bmatrix}, \quad \mathbf{C}_n = \begin{bmatrix} 0 & 1 & 0 & \cdots & 0 & 0 \\ 0 & 0 & 1 & \cdots & 0 & 0 \\ \vdots & \vdots & \vdots & & \vdots & \vdots \\ 0 & 0 & 0 & \cdots & 0 & 1 \\ -a_n & -a_{n-1} & -a_{n-2} & \cdots & -a_2 & -a_1 \end{bmatrix}$$

Verify that

$$\mathbf{x}' = \mathbf{C}_n\mathbf{x} + \begin{bmatrix} 0 \\ 0 \\ \vdots \\ 0 \\ f(t) \end{bmatrix} + f(t)\mathbf{e}_n$$

(b) Suppose

$$\mathbf{x} = \begin{bmatrix} x_1 \\ x_2 \\ \vdots \\ x_n \end{bmatrix}$$

is a solution of $\mathbf{x}' = \mathbf{C}_n\mathbf{x} + f(t)\mathbf{e}_n$. Equate the first $n-1$ entries of

$$\mathbf{C}_n\mathbf{x} + f(t)\mathbf{e}_n$$

with those of $\mathbf{x}'$ to obtain the following $n-1$ equations:

$$\begin{aligned} x_2 &= x_1^{(1)} \\ &\vdots \\ x_n &= x_{n-1}^{(1)} = x_1^{(n-1)} \end{aligned}$$

Now deduce that $x_1(t)$ is a solution of (1).

2. The Theory of Companion Systems

Recall that the general $n \times n$ companion matrix is

$$\mathbf{C}_n = \begin{bmatrix} 0 & 1 & 0 & \cdots & 0 & 0 \\ 0 & 0 & 1 & \cdots & 0 & 0 \\ \vdots & \vdots & \vdots & & \vdots & \vdots \\ 0 & 0 & 0 & \cdots & 0 & 1 \\ -a_n & -a_{n-1} & -a_{n-2} & \cdots & -a_2 & -a_1 \end{bmatrix}$$

(a) Show that the characteristic equation of the companion matrix $\mathbf{C}_n$ is given by $C(\lambda) = (-1)^n (\lambda^n + a_1 \lambda^{n-1} + \cdots + a_n) = 0$. *Hint*: Use induction and expand $\det(\mathbf{C}_n - \lambda \mathbf{I})$ by cofactors of entries in its first column to obtain

$$\begin{aligned} \det(\mathbf{C}_n - \lambda \mathbf{I}) &= (-\lambda) \det(\mathbf{C}_{n-1} - \lambda \mathbf{I}) + (-\lambda)^{n-1} (-a_n) \det \mathbf{I}_{n-1} \\ &= (-\lambda) \det(\mathbf{C}_{n-1} - \lambda \mathbf{I}) + (-\lambda)^n a_n \end{aligned}$$

(b) If λ is an eigenvalue of $\mathbf{C}_n$, then show that

$$(\lambda, \begin{bmatrix} 1 \\ \lambda \\ \vdots \\ \lambda^{n-1} \end{bmatrix})$$

is an eigenpair of $\mathbf{C}_n$. *Hint*: Verify that

$$\mathbf{C}_n \mathbf{u} = \begin{bmatrix} 0 & 1 & 0 & \cdots & 0 & 0 \\ 0 & 0 & 1 & \cdots & 0 & 0 \\ \vdots & \vdots & \vdots & & \vdots & \vdots \\ 0 & 0 & 0 & \cdots & 0 & 1 \\ -a_n & -a_{n-1} & -a_{n-2} & \cdots & -a_2 & -a_1 \end{bmatrix} \begin{bmatrix} 1 \\ \lambda \\ \vdots \\ \lambda^{n-1} \end{bmatrix} = \lambda \begin{bmatrix} 1 \\ \lambda \\ \vdots \\ \lambda^{n-1} \end{bmatrix}$$

(c) Show: If $(\lambda, \mathbf{v})$ is an eigenpair of $\mathbf{C}_n$, then $\mathbf{v} = c\mathbf{u}$.

Hint: Suppose

$$(\lambda, \begin{bmatrix} v_1 \\ v_2 \\ \vdots \\ v_n \end{bmatrix})$$

is an eigenpair of $\mathbf{C}_n$. Now establish

$$\mathbf{C}_n\mathbf{v} = \begin{bmatrix} 0 & 1 & 0 & \dots & 0 & 0 \\ 0 & 0 & 1 & \dots & 0 & 0 \\ \vdots & \vdots & \vdots & & \vdots & \vdots \\ 0 & 0 & 0 & \dots & 0 & 1 \\ -a_n & -a_{n-1} & -a_{n-2} & \dots & -a_2 & -a_1 \end{bmatrix} \begin{bmatrix} v_1 \\ v_2 \\ \vdots \\ v_n \end{bmatrix}$$

$$= \begin{bmatrix} v_2 \\ v_3 \\ \vdots \\ -a_n v_1 - a_{n-1} v_2 - \dots - a_2 v_{n-1} - a_1 v_n \end{bmatrix}$$

and only if it has n distinct eigenvalues.

3. Fundamental Matrices Are Real

One of the hypotheses of Theorem 4.4.2 was that the roots of the characteristic polynomialare all real. The point of this assumption was to exclude the possibility that the solution matrix

$$\Phi(t) = \begin{bmatrix} u_1 & u_2 & \dots & u_n \\ r_1 u_1 & r_2 u_2 & \dots & r_n u_n \\ \vdots & \vdots & & \vdots \\ r_i^{n-1} u_1 & r_2^{n-1} u_2 & \dots & r_n^{n-1} u_n \end{bmatrix}$$

of $\mathbf{x}' = \mathbf{A}\mathbf{x}$ has complex-valued functions as entries. The following theorems address the issues of how real solution matrices may be obtained from those with nonreal entries and the invertibility of $\Phi(t)$ at $t = 0$. First we need a notational convention. Let

$$u_i = u_i(t) = \exp(r_i t)$$

so that $u_i(0) = 1$ for all $i = 1, 2, \dots, n$.

(a) Show that if $r_1, r_2, \dots, r_n$ are distinct complex numbers, then

$$\mathbf{V}_n = \begin{bmatrix} 1 & 1 & \dots & 1 \\ r_1 & r_2 & \dots & r_n \\ \vdots & \vdots & & \vdots \\ r_1^{n-1} & r_2^{n-1} & \dots & r_n^{n-1} \end{bmatrix}$$

is invertible. *Hint*: Use induction. First let $V_n = \det(\mathbf{V}_n)$. Show that $V_n \neq 0$ for all n. To do this, define a matrix related to $\mathbf{V}_n$:

$$\mathbf{V}_n(r) = \begin{bmatrix} 1 & 1 & \dots & 1 & 1 \\ r_1 & r_2 & \dots & r_n & x \\ \vdots & \vdots & & \vdots & \vdots \\ r_1^{n-1} & r_2^{n-1} & \dots & r_n^{n-1} & x^{n-1} \\ r_1^n & r_2^n & \dots & r_n^n & x^n \end{bmatrix}$$

Prove:

1. $\mathbf{V}_{n+1}(r_{n+1}) = \mathbf{V}_{n+1}$.
2. $\det(\mathbf{V}_{n+1}(r_i)) = 0$ for all $i = 1, 2, \dots, n$.
3. $\det(\mathbf{V}_{n+1}(x)) = V_n x^n + p_{n-1}(x)$, where $p_{n-1}(x)$ is a polynomial of degree $n-1$.
4. $\det(\mathbf{V}_{n+1}(x)) = V_n(x - r_1)(x - r_2)\dots(x - r_n)$.

(b) Suppose $\mathbf{A}$ is a real matrix. Show that if $\Psi(t)$ is a fundamental matrix for the system, $\mathbf{x}' = \mathbf{A}\mathbf{x}$ then $\Psi(t)$ is real for all real t. *Hint*: Write $\Psi(t) = \text{Re}\Psi(t) + i\text{Im}\Psi(t)$, and show that

$$(\text{Im}\Psi(t))' = \mathbf{A}\,\text{Im}\Psi(t)$$

and that $\text{Im}\Psi(0) = \mathbf{O}$. Now show that $\text{Im}\Psi(t)$ is a solution of

$$\mathbf{X}'(t) = \mathbf{A}\mathbf{X}(t), \quad \mathbf{X}(0) = \mathbf{O}$$

(c) Prove: If the characteristic polynomial of

$$u^{(n)} + a_1 u^{(n-1)} + \dots + a_{n-1}u^{(1)} + a_n u \tag{1}$$

has n distinct roots $r_1, r_2, \dots, r_n$, then

$$\Psi(t) = \begin{bmatrix} u_1 & u_2 & \dots & u_n \\ r_1 u_1 & r_2 u_2 & \dots & r_n u_n \\ \vdots & \vdots & & \vdots \\ r_1^{n-1}u_1 & r_2^{n-1}u_2 & \dots & r_n^{n-1}u_n \end{bmatrix} \mathbf{V}_n^{-1}$$

is a real fundamental matrix for the companion system of (1).

(d) Under the hypothesis of (b), prove that $\Phi(t)\mathbf{V}_n^{-1}$ is a real fundamental matrix for the companion system of (1) and that

$$\mathbf{x}(t) = \Phi(t)\mathbf{V}_n^{-1}\mathbf{c}$$

is a general companion system.

Chapter 5

The Laplace Transform

5.1 Introduction

The solution of initial-value problems using the techniques presented thus far can be quite complicated when the forcing function is discontinuous and/or periodic, a situation which frequently arises in practical problems. The Laplace transform is a powerful tool for treating such problems because, among other things, it handles all allowable forcing functions in a uniform manner and proceeds directly to the solution of the initial-value problem without explicitly obtaining the general solution.

5.2 Preliminaries

We assume throughout this chapter that all forcing functions are sectionally continuous for $t > 0$ and that

$$f(0) = f\left(0^{+}\right) = \lim_{t \to 0} f(t) \qquad \text{for} \quad t > 0$$

as shown in Figure 5.1. (See Section 3.2 for a detailed discussion of sectionally continuous functions.)

We shall also assume that $f(t)$ is of *exponential order*, by which we mean that for t sufficiently large,

$$|f(t)| \leq M e^{bt} \tag{5.2.1}$$

for some positive constants M and b, both independent of t. Since functions that are sectionally continuous are bounded in every finite interval, (5.2.1) is

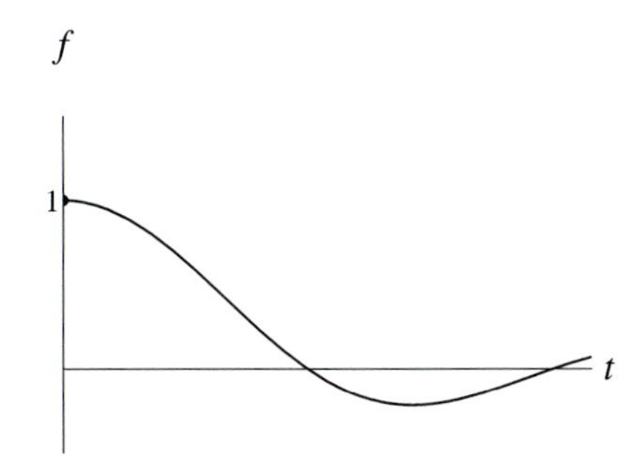

Figure 5.1: The Value of f at $t = 0$

a restriction on the rate of growth of $f(t)$ as $t \to \infty$. Examples of functions *not* considered here are

$$e^{t^2}, \qquad t^{-1/2}, \qquad (t-1)^a, \; a < 0$$

The first grows too rapidly with increasing t and so is not of exponential order; the second fails to meet the condition that $f(0^+)$ exists; the last function is not sectionally continuous in any interval containing $t = 1$.

For convenience we shall call every function that is sectionally continuous and of exponential order *acceptable*. Polynomials, $\sin mt, \cos mt, e^{kt}$, and sums of products of these functions are examples of acceptable functions. In particular, for every choice of constants a and b, $af(t) + bg(t)$ is acceptable if both $f(t)$ and $g(t)$ are acceptable.

The *Laplace transform* of $f(t)$, written $\mathcal{L}[f]$, is the improper integral[1]

$$\mathcal{L}[f] = \int_0^\infty f(t)e^{-st}\,dt \tag{5.2.2}$$

where s is a real variable. Because $\mathcal{L}[f]$ is a function of s and not of the dummy variable of integration t, we often write $F(s)$ rather than the more explicit $\mathcal{L}[f(t)]$. For example, if $f(t) = t$, after integration by parts, there results

$$\int te^{-st}\,dt = C - \frac{1}{s}\left(t + \frac{1}{s}\right)e^{-st} \tag{5.2.3}$$

[1]The integral is termed "improper" because the upper limit of integration is not finite.

By definition of an improper integral and from (5.2.3), for $s > 0$,

$$\begin{aligned}\mathcal{L}[f] = F(s) = \int_0^\infty te^{-st}\,dt &= \lim_{R\to\infty}\left[-\frac{1}{s}\left(t+\frac{1}{s}\right)e^{-st}\Big|_0^R\right] \\ &= \lim_{R\to\infty}\left[-\frac{1}{s}\left(R+\frac{1}{s}\right)e^{-sR}+\frac{1}{s^2}\right] \\ &= \frac{1}{s^2} \qquad (5.2.4)\end{aligned}$$

since $\lim_{R\to\infty} Re^{-sR} = 0$ by l'Hôpital's rule. Note that $F(s)$ is a limit, and therefore, some care needs to be taken to ensure that this limit exists. Also, we must be careful in using the common tools for manipulating integrals. Integration by parts is particularly tricky for functions that have discontinuities.

We conclude this section by proving an existence theorem. As preparation for this theorem, we note that there are two threats to the existence of the Laplace transform. The first is that $f(t)$ may be so badly defined that the definite integral

$$\int_0^R f(t)e^{-st}\,dt$$

does not exist for some R. The function $f(t) = (1-t)^{-1}$ is such an example. The second is that improper integrals may diverge even though $f(t)$ is "well behaved" for all $t > 0$. We avoid both difficulties by requiring that $f(t)$ be acceptable, that is, $f(t)$ must be sectionally continuous and of exponential order. This is the point of the following theorem.

Theorem 5.2.1. *If $f(t)$ is acceptable, then the Laplace transform $F(s)$ of $f(t)$ exists, and moreover,*

1. *$sF(s)$ is bounded as $s \to \infty$, from which it follows that*
2. $\lim_{s\to\infty} F(s) = 0$.

Proof: The existence of the improper integral defining the Laplace transform of $f(t)$ follows by the argument

$$\begin{aligned} 0 \leq \left| \int f(t)e^{-st}\,dt \right| &\leq \int |f(t)|e^{-st}\,dt \\ &\leq \int Me^{bt}e^{-st}\,dt = M\int e^{(b-s)t}\,dt \\ &= M\frac{1}{b-s}e^{(b-s)t} + C \end{aligned} \tag{5.2.5}$$

where C is a constant of integration and t is sufficiently large. From (5.2.2) and (5.2.5), we have

$$|\mathcal{L}[f(t)]| \leq M\left(\frac{1}{s-b} + \lim_{R\to\infty}\frac{1}{b-s}e^{(b-s)R}\right) = M\frac{1}{s-b} \tag{5.2.6}$$

provided that $s > b$. So

$$sF(s) \leq M\frac{s}{s-b} \leq 2M$$

for $s > 2b$. □

We do wish to point out that it is not necessary that $f(t)$ be acceptable for its Laplace transform to exist. For example, it can be shown that $\mathcal{L}\left[t^{-1/2}\right] = \pi/s^{1/2}$ even though $t^{-1/2}$ is not sectionally continuous and therefore not acceptable.

One consequence of Theorem 5.2.1 is that not all elementary functions are Laplace transforms of an acceptable function. For instance, $F(s) = 1$ is not a Laplace transform since both parts of Theorem 5.2.1 are contradicted.

Here are three examples illustrating the definition and our notational conventions.

Example 5.2.1. *Find the Laplace transform of $f(t) = 1$.*

Solution: The computation is routine

$$\mathcal{L}[1] = \int_0^\infty e^{-st}\,dt = \lim_{R\to\infty} -\frac{1}{s}e^{-st}\Big|_0^R = -\frac{1}{s}(0-1) = \frac{1}{s}$$

Example 5.2.2. *Find the Laplace transform of $f(t) = e^{at}$.*

Solution: We compute

$$\mathcal{L}\left[e^{at}\right] = \int_0^\infty e^{at}e^{-st}\,dt = \int_0^\infty e^{(a-s)t}\,dt$$
$$= \lim_{R\to\infty} -\frac{1}{a-s}e^{(a-s)t}\Big|_0^R = \frac{1}{s-a}$$

provided that $s > a$.

From this point on, we do not explicitly refer to the limit process unless some particular argument demands that we do so. Also, as a matter of convenience, we shall ignore any specific reference to the domain over which the transform exists. We tacitly assume that each transform appearing in any specific argument exists on some common interval $0 \leq a \leq s < \infty$.

Example 5.2.3. *Find $\mathcal{L}[\cos bt]$ and $\mathcal{L}[\sin bt]$.*

Solution: By definition,

$$\mathcal{L}[\cos bt] = \int_0^\infty \cos bt\, e^{-st}\,dt$$

By integration by parts twice or a table "lookup,"

$$\int \cos bt\, e^{-st}\,dt = \frac{1}{s^2+b^2}\left(b^2\sin bt - s\cos bt\right)e^{-st}$$

Applying the limits to this expression, we find that

$$\mathcal{L}[\cos bt] = \frac{s}{s^2+b^2}$$

We handle $\sin bt$ in a similar manner and obtain

$$\mathcal{L}[\sin bt] = \frac{b}{s^2+b^2}$$

Here is another approach using the result of Example 5.2.2 with $a = ib$. We have

$$\mathcal{L}\left[e^{ibt}\right] = \frac{1}{s-ib} = \frac{s+ib}{s^2+b^2} = \frac{s}{s^2+b^2} + i\frac{b}{s^2+b^2}$$

Then, taking real and imaginary parts of this expression, we get

$$\text{Re}\mathcal{L}\left[e^{ibt}\right] = \mathcal{L}\left[\text{Re}(e^{ibt})\right] = \mathcal{L}\left[\cos bt\right] = \frac{s}{s^2+b^2}$$

and

$$\text{Im}\mathcal{L}\left[e^{ibt}\right] = \mathcal{L}\left[\text{Im}(e^{ibt})\right] = \mathcal{L}\left[\sin bt\right] = \frac{b}{s^2+b^2}$$

which, when combined with (5.2.7), gives the desired result.

•EXERCISES

Find the Laplace transform of the functions given in Problems 1–15.

1. $2t$.
2. $t-3$.
3. $2-t$.
4. e^{3t}.
5. e^{2t-3}.
6. $\sin bt$.
7. $\cos 4t$.
8. te^{at}.
9. $t\sin 2t$.
10. $2t\cos t$.
11. $2t^2\cos bt$.
12. t^2-3.
13. $\sinh 2t$.
14 $(t-2)^2$.
15. $\cosh 4t$.

Explain why the functions in Problems 16–18 are not transforms of acceptable functions.

16. $s/(s+1)$.
17. $s^a, \quad -1 < a$.
18. $\sec(bs)$.

Show that each limit in Problems 19–23 is 0, where $t \to \infty$ and $s > 0$.

19. $\sin bt\, e^{-st}$.
20. $t\, e^{-st}$.
21. t^2e^{-st}.
22. t^3e^{-st}.
23. $p(t)e^{-st}$ where $p(t)$ is a polynomial.

5.3 General Properties of the Laplace Transform

It is unwieldy to rely solely on the definition to compute the Laplace transform. In this section, we present a number of theorems that make the computations of certain Laplace transforms significantly easier. The first of these results shows that the Laplace transform is a linear operator.

Theorem 5.3.1. *For every pair of scalars a and b and acceptable functions $f(t)$ and $g(t)$,*

$$\mathcal{L}\left[af(t)+bg(t)\right] = a\mathcal{L}\left[f(t)\right] + b\mathcal{L}\left[g(t)\right] \tag{5.3.1}$$

Proof: First of all, $a\,f(t) + b\,g(t)$ is acceptable, so its transform exists. Secondly,

$$\begin{aligned} \mathcal{L}\left[af(t)+bg(t)\right] &= \int_0^\infty (af(t)+bg(t))\,e^{-st}\,dt \\ &= a\int_0^\infty f(t)e^{-st}\,dt + b\int_0^\infty g(t)e^{-st}\,dt \\ &= a\mathcal{L}\left[f(t)\right] + b\mathcal{L}\left[g(t)\right] \end{aligned}$$

Thus the Laplace transform is a linear operator. □

Example 5.3.1. *Find the Laplace transform of $f(t) = 7 - 3e^{2t}$.*

Solution: We have shown in Examples 5.2.1 and 5.2.2 that $\mathcal{L}\left[1\right] = 1/s$ and $\mathcal{L}\left[e^{2t}\right] = 1/(s-2)$. By Theorem 5.3.1,

$$\mathcal{L}\left[7 - 3e^{2t}\right] = 7\mathcal{L}\left[1\right] - 3\mathcal{L}\left[e^{2t}\right] = \frac{7}{s} - \frac{3}{s-2}$$

Example 5.3.2. *Find $\mathcal{L}\left[\cosh bt\right]$ and $\mathcal{L}\left[\sinh bt\right]$.*

Solution: By definition, $\cosh bt = \frac{1}{2}e^{bt} + \frac{1}{2}e^{-bt}$. Hence, by linearity

$$\begin{aligned} \mathcal{L}\left[\cosh bt\right] &= \frac{1}{2}\mathcal{L}\left[e^{bt}\right] + \frac{1}{2}\mathcal{L}\left[e^{-bt}\right] \\ &= \frac{1}{2}\left(\frac{1}{s-b} + \frac{1}{s+b}\right) = \frac{s}{s^2-b^2} \end{aligned}$$

Likewise, $\sinh bt = \frac{1}{2}\left(e^{bt} - e^{-bt}\right)$, so that

$$\mathcal{L}\left[\sinh bt\right] = \frac{1}{2}\left(\frac{1}{s-b} - \frac{1}{s+b}\right) = \frac{b}{s^2-b^2}$$

Theorem 5.3.2. *If* $\mathcal{L}[f(t)] = F(s)$, *then*

$$\mathcal{L}[tf(t)] = -\frac{d}{ds}F(s) \tag{5.3.2}$$

Proof: Since $f(t)$ is acceptable, so is $tf(t)$. Now,

$$\begin{aligned} \frac{d}{ds}F(s) &= \frac{d}{ds}\int_0^\infty f(t)e^{-st}\,dt = \int_0^\infty f(t)\frac{\partial}{\partial s}e^{-st}\,dt \\ &= -\int_0^\infty tf(t)e^{-st}\,dt = -\mathcal{L}[tf(t)] \end{aligned}$$

□

Corollary 5.3.1. *For each positive integer* n,

$$\mathcal{L}[t^n f(t)] = (-1)^n F^{(n)}(s) \tag{5.3.3}$$

Proof: The function $t^n f(t)$ is acceptable. By Theorem 5.3.2,

$$\begin{aligned} \mathcal{L}\left[tt^{n-1}f(t)\right] &= -\frac{d}{ds}\mathcal{L}\left[t^{n-1}f(t)\right] \\ &= (-1)^2\frac{d^2}{ds^2}\mathcal{L}\left[t^{n-2}f(t)\right] \\ &\;\;\vdots \\ &= (-1)^n\frac{d^n}{ds^n}\mathcal{L}[f(t)] \\ &= (-1)^n F^{(n)}(s) \end{aligned}$$

as required. □

Example 5.3.3. *For each nonnegative integer* n, *show that* $\mathcal{L}[t^n] = n!/s^{n+1}$.

Solution: We begin with $\mathcal{L}[1] = 1/s$. Now we can use Corollary 5.3.1:

$$\mathcal{L}[t^n \cdot 1] = (-1)^n \frac{d^n}{ds^n} s^{-1} = \frac{n!}{s^{n+1}}$$

Example 5.3.4. *Find* $\mathcal{L}\left[t^n e^{at}\right]$.

Solution: We use Corollary 5.3.1 and Example 5.2.2 to deduce that

$$\mathcal{L}\left[t^n e^{at}\right] = (-1)^n \frac{d^n}{ds^n} \mathcal{L}\left[e^{at}\right] = (-1)^n \frac{d^n}{ds^n}\left(\frac{1}{s-a}\right) = \frac{n!}{(s-a)^{n+1}}$$

Example 5.3.5. *Find* $\mathcal{L}[t \sin bt]$.

Solution: In Example 5.2.3 we obtained

$$\mathcal{L}[\sin bt] = \frac{b}{s^2 + b^2}$$

Therefore by Theorem 5.3.2,

$$\mathcal{L}[t \sin bt] = -\frac{d}{ds}\left(\frac{b}{s^2+b^2}\right) = \frac{2bs}{(s^2+b^2)^2}$$

We now state and prove a theorem that expresses the Laplace transform of $f'(t)$ in terms of the Laplace transform of $f(t)$. This theorem plays a central role in the applications of the Laplace transform to initial-value problems. Note the appearance of $f(0)$ in the conclusion.

Theorem 5.3.3. *Suppose $f(t)$ is continuous and $f(t)$ and $f'(t)$ are acceptable. Then*

$$\mathcal{L}[f'(t)] = s\mathcal{L}[f(t)] - f(0) \tag{5.3.4}$$

Proof: We use integration by parts with $u = e^{-st}$ and $dv = f'(t)dt$. We have

$$\int_0^R f'(t)e^{-st}\,dt = f(t)e^{-st}\Big|_0^R + s\int_0^R f(t)e^{-st}\,dt$$

which holds because $f(t)$ is continuous. (If $f(t)$ has a jump discontinuity, then the integration-by-parts formula is more complicated than that given here.) Now, let $R \to \infty$ in the preceding displayed equation. Then

$$\begin{aligned}\mathcal{L}\left[f'(t)\right] &= \int_0^\infty f'(t)e^{-st}\,dt = f(t)e^{-st}\Big|_0^\infty + s\mathcal{L}\left[f(t)\right] \\ &= -f(0) + s\mathcal{L}\left[f(t)\right])\end{aligned}$$

since $f(R)e^{-Rs} \to 0$ as $R \to \infty$. □

Example 5.3.6. *Show that* $\mathcal{L}\left[\sin bt + bt\cos bt\right] = 2bs^2/(s^2+b^2)^2$.

Solution: In Example 5.3.5 we derived

$$\mathcal{L}\left[t\sin bt\right] = \frac{2bs}{(s^2+b^2)^2} \tag{1}$$

Now, Theorem 5.3.3 with $f(t) = t\sin bt$ (so that $f(0) = 0$) implies that

$$\mathcal{L}\left[\frac{d}{dt}(t\sin bt)\right] = s\mathcal{L}\left[t\sin bt\right] \tag{2}$$

But $d(t\sin t)/dt = \sin bt + bt\cos bt$. So from (1) and (2) it follows that

$$\begin{aligned}\mathcal{L}\left[\sin bt + bt\cos bt\right] &= \mathcal{L}\left[\frac{d}{dt}(t\sin bt)\right] \\ &= s\mathcal{L}\left[t\sin bt\right] = \frac{2bs^2}{(s^2+b^2)^2}\end{aligned}$$

Under the appropriate hypothesis, Theorem 5.3.3 can be generalized to yield an expression for $\mathcal{L}\left[f^{(n)}(t)\right]$. The following corollary treats the case $n = 2$.

Corollary 5.3.2. *Suppose $f(t)$ and $f'(t)$ are continuous and $f(t), f'(t)$, and $f''(t)$ are acceptable. Then*

$$\mathcal{L}\left[f''(t)\right] = s^2\mathcal{L}\left[f(t)\right] - sf(0) - f'(0) \tag{5.3.5}$$

Proof: We use Theorem 5.3.3 twice:

$$\begin{aligned}\mathcal{L}\left[f''(t)\right] &= s\mathcal{L}\left[f'(t)\right] - f'(0)\\ &= s\left(s\mathcal{L}\left[f(t)\right] - f(0)\right) - f'(0)\\ &= s^2\mathcal{L}\left[f(t)\right] - sf(0) - f'(0)\end{aligned}$$

□

The general case is set as Problem 29 in the exercise set that follows. The result is

$$\mathcal{L}\left[f^{(n)}(t)\right] = s^n\mathcal{L}\left[f(t)\right] - s^{n-1}f(0) - s^{n-2}f^{(1)}(0) - \cdots - f^{(n-1)}(0) \tag{5.3.6}$$

under the appropriate conditions on f and its derivatives.

Theorem 5.3.3 may be used to obtain the transform of the integral of $f(t)$. First, we recall two basic facts about integrals. If we define $g(t)$ by

$$g(t) = \int_0^t f(z)\,dz \tag{5.3.7}$$

then

$$g'(t) = f(t) \quad \text{and} \quad g(0) = \int_0^0 f(z)\,dz = 0 \tag{5.3.8}$$

Corollary 5.3.3. *Let $\mathcal{L}\left[f(t)\right] = F(s)$. Then*

$$\mathcal{L}\left[\int_0^t f(z)\,dz\right] = \frac{1}{s}F(s) \tag{5.3.9}$$

Proof: For $g(t)$ defined by (5.3.7), $g(0) = 0$ and $g'(t) = f(t)$, as we have remarked in (5.3.8). So (5.3.4) yields

$$\mathcal{L}[f(t)] = \mathcal{L}[g'(t)] = s\mathcal{L}[g(t)] - g(0) = s\mathcal{L}[g(t)]$$

Hence, $\mathcal{L}[g(t)] = \mathcal{L}[f(t)]/s$, and in the notation of the corollary,

$$\mathcal{L}\left[\int_0^t f(z)\,dz\right] = \frac{1}{s}F(s)$$

which is the desired formula. □

It is not usually easy to find the transform of a product in terms of the transforms of its factors. Theorem 5.3.2 provides one instance. The next theorem, the *first shifting property*, provides another.

Theorem 5.3.4 (First Shifting Property). *Set* $\mathcal{L}[f(t)] = F(s)$. *For each* a,

$$\mathcal{L}\left[e^{at}f(t)\right] = F(s-a) \tag{5.3.10}$$

Proof: The proof of this theorem follows easily from the properties of integrals. First note that $e^{at}f(t)$ is acceptable. Then

$$\mathcal{L}\left[e^{at}f(t)\right] = \int_0^\infty e^{at}f(t)e^{-st}\,dt = \int_0^\infty f(t)e^{-(s-a)t}\,dt = F(s-a)$$

□

We compute three Laplace transforms by means of this theorem to illustrate its use.

Example 5.3.7. *Find* $\mathcal{L}\left[e^{at}\cos bt\right]$, $\mathcal{L}\left[e^{at}\sin bt\right]$, *and* $\mathcal{L}\left[e^{at}t^n\right]$

Solution: From Theorem 5.3.4 and the fact (Example 5.2.3) that

$$\mathcal{L}[\cos bt] = \frac{s}{s^2+b^2}$$

we deduce

$$\mathcal{L}\left[e^{at}\cos bt\right] = \frac{s-a}{(s-a)^2+b^2}$$

Likewise, from

$$\mathcal{L}[\sin bt] = \frac{b}{s^2+b^2}$$

we deduce

$$\mathcal{L}\left[e^{at}\sin bt\right] = \frac{b}{(s-a)^2+b^2}$$

Finally, it follows from Example 5.3.3 that

$$\mathcal{L}\left[e^{at}t^n\right] = \frac{n!}{(s-a)^{n+1}}$$

•EXERCISES

Use the linearity property and the transforms derived in the examples to find the Laplace transform of each of the functions in Problems 1–6.

1. $3 - e^t$.
2. $4\cos t + 2\sin t$.
3. $2e^{2t} - 3\sin 2t$.
4. $4\cos 2t - 3\sin 2t$.
5. $2 + 5\sin 3t$.
6. $4\cos 5t - 6e^{-2t}$.

Use Theorem 5.3.2 and the transforms derived in the examples to find the Laplace transform of each function in Problems 7–19.

7. $3te^{3t}$.
8. t^2e^{-t}.
9. $3t\cos 4t$.
10. $t\sinh 2t$.
11. $t^3\sin t$.
12. $4t\cosh 2t$
13. $t^2\cos t$.
14. $t(\cos 4t - 2\sin 4t)$.
15. $(t^2+4t+5)\,e^{-2t}$.
16. $t(\sinh 2t + 3\cosh 2t)$.
17. $(t^2 - t)\sin t$.
18. $t^3\sin 2t$.
19. t^4e^{-2t}.

20. Express $\mathcal{L}\left[f^{(4)}(t)\right]$ in terms of $\mathcal{L}[f(t)]$ and the first three derivatives of $f(t)$ at $t=0$ by using Theorem 5.3.3.

21. Use the first shifting property and the transforms derived in the example to find the Laplace transform of each function.

(a) $2\cosh 2t\sin 2t$.
(b) $\cosh 2t\sinh 3t$.
(c) $6\sinh t\cos t$.
(d) $2\sinh 3t\cos 2t$.
(e) $4\sinh 2t\sinh 4t$.
(f) $2\cosh t\cos 2t$.

22. Repeat Problem 21 for the following functions.
    (a) $3te^{3t}$.
    (b) t^2e^{-t}.
    (c) $e^{-2t}\cos 4t$.
    (d) $e^{2t}\sinh 2t$.
    (e) $e^{-t}\sin 2t$.
    (f) $4e^{-2t}\cosh t$.
23. Repeat Problem 21 for the following functions.
    (a) $te^{-t}\sin t$.
    (b) $e^{-2t}(t^2+4t+5)$.
    (c) $e^{-2t}(\cos 4t - 2\sin 4t)$.
    (d) $e^{-2t}(\sinh 2t + 3\cosh 2t)$.
24. Show that $\mathcal{L}[f(t/b)] = bF(bs)$. Use this to derive
$$\mathcal{L}[\cos 4t] = \frac{s}{s^2+16}$$
25. Use the strategy in Example 5.3.6 to derive
$$\mathcal{L}[\sin bt - bt\cos bt] = 2b^3/(s^2+b^2)^{-2}$$
(*Hint*: Consider $2\mathcal{L}[\sin bt] - \mathcal{L}[\sin bt + bt\cos bt]$.)
26. Show that $\mathcal{L}[2\cos bt - bt\sin bt] = 2s^3/(s^2+b^2)^{-2}$.
27. Find the Laplace transform of $e^{at}(\sin bt - bt\cos bt)$. (See Problem 25.)
28. Find the Laplace transform of $a^t\cos bt, a > 1$. (*Hint*: Write $a^t = e^{t(\ln a)}$.)
29. Prove (5.3.6). *Hint*: Use induction.

5.4 Sectionally Continuous Functions

One of the most important uses of the Laplace transform in the theory of ODE's occurs when the forcing function is defined by different formulas in different intervals. We handle this case by introducing the *unit step function*

$$u_a(t) = \begin{cases} 0, & t < a \\ 1, & a \le t \end{cases} \tag{5.4.1}$$

where $a \ge 0$. The graph of this function is given in Figure 5.2.

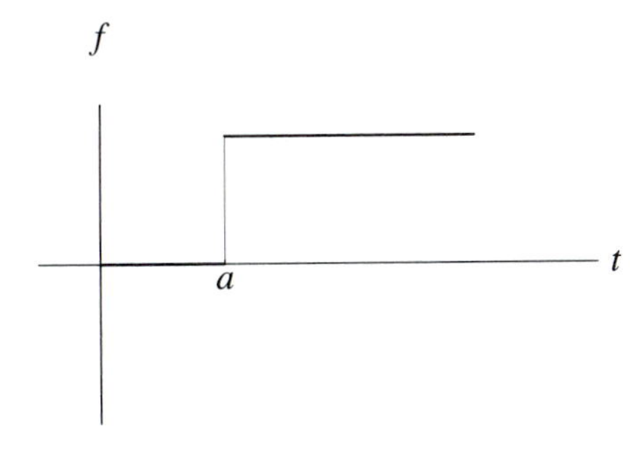

Figure 5.2: The Unit Step at $t = a$

Now suppose that $a < b$. The function $u_a(t) - u_b(t)$ is 1 for all t, $a \le t < b$, and is 0 elsewhere.

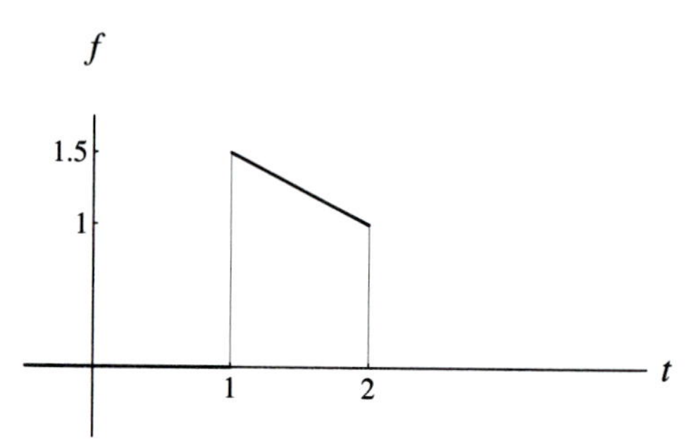

Figure 5.3: The Function $(2 - t/2)\,(u_1(t) - u_2(t))$

Note that $f(t)\,(u_a(t) - u_b(t))$ "turns on" $f(t)$ for $a \le t < b$ and then "turns off" $f(t)$ for $t \ge b$. These features of $f(t)u_a(t)$ and $f(t)\,(u_a(t) - u_b(t))$ enable us to write a single formula for the various sectionally continuous functions that arise in this chapter. For example, let $f(t) = 0$ except in the interval (1,2) where it is given by $f(t) = 2 - t/2$. The graph is portrayed in Figure 5.3. This function can be written by means of the unit step as $(2 - t/2)\,(u_1(t) - u_2(t))$.

Example 5.4.1. *Write $f(t)$ in terms of unit step functions, where*

$$f(t) = \begin{cases} t^2, & 0 \le t < 1 \\ 1, & 1 \le t \end{cases}$$

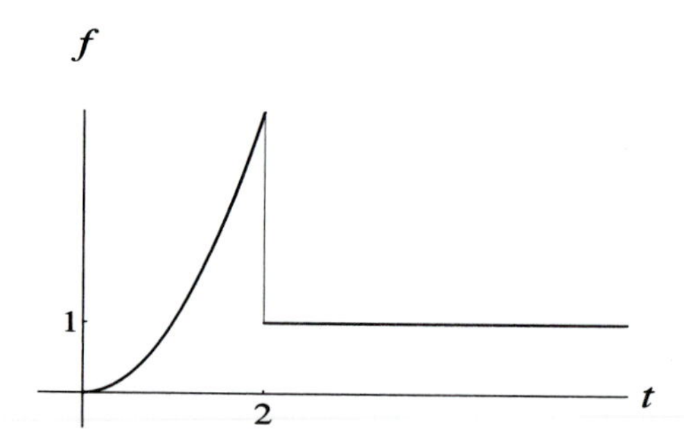

Solution: We turn on t^2 in $0 \le t < 1$ and 1 in $1 \le t$. Thus,

$$f(t) = t^2\,(u_0(t) - u_1(t)) + u_1(t) = t^2 u_0(t) + (1 - t^2)u_1(t)$$

Example 5.4.2. *Define the following function in terms of unit steps:*

$$f(t) = \begin{cases} t, & 0 \le t < 2 \\ 2, & 2 \le t \end{cases}$$

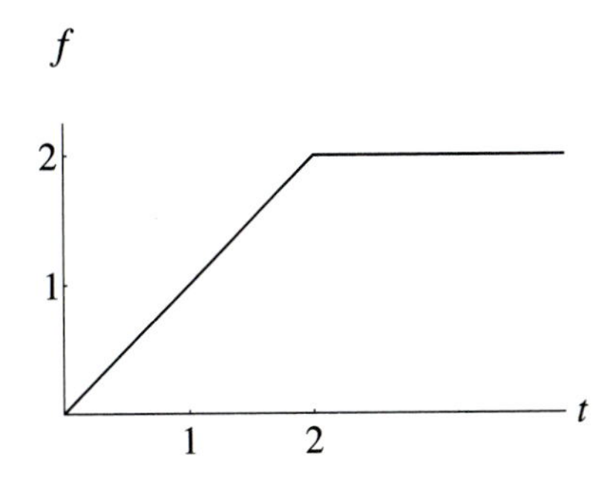

Solution: We turn on t in $0 \le t < 2$ and 2 in $2 \le t$. Thus,

$$f(t) = t\,(u_0(t) - u_2(t)) + 2u_2(t) = tu_0(t) + (2 - t)u_2(t)$$

The next theorem, the *second shifting property*, gives the transform of $u_a(t)f(t)$ in terms of the transform of $f(t+a)$. This is the third instance in which we treat a special case of $\mathcal{L}\,[g(t)f(t)]$.

Theorem 5.4.1 (Second Shifting Property). *Suppose* $\mathcal{L}\,[f(t)] = F(s)$. *Then*

$$\mathcal{L}\,[u_a(t)f(t)] = e^{-as}\mathcal{L}\,[f(t+a)] \tag{5.4.2}$$

Proof: We have, using z as the dummy variable of integration,

$$\mathcal{L}\,[u_a(t)f(t)] = \int_0^\infty u_a(z)f(z)e^{-sz}\,dz = \int_a^\infty f(z)e^{-sz}\,dz$$

Let $t = z - a$. Then, since $dt = dz$,

$$\begin{aligned}
\lim_{R\to\infty} \int_a^R f(z)e^{-sz}\,dz &= \lim_{R\to\infty} \int_0^{R-a} f(t+a)e^{-s(t+a)}\,dt \\
&= e^{-as} \int_0^\infty f(t+a)e^{-st}\,dt = e^{-as}\mathcal{L}\,[f(t+a)]
\end{aligned}$$

□

Example 5.4.3. *Find the Laplace transform of* $f(t) = t^2u_0(t) + u_1(t)\left(1 - t^2\right)$.

Solution: We have,

$$\mathcal{L}\left[f(t)\right] = \mathcal{L}\left[t^2 u_0(t)\right] + \mathcal{L}\left[u_1(t)\left(1 - t^2\right)\right]$$

By the second shifting property,

$$\begin{aligned}\mathcal{L}\left[f(t)\right] &= e^{0s}\mathcal{L}\left[t^2\right] + e^{-s}\mathcal{L}\left[1 - (t+1)^2\right] \\ &= \mathcal{L}\left[t^2\right] - e^{-s}\mathcal{L}\left[t^2 + 2t\right] = \frac{2}{s^3} - e^{-s}\left(\frac{2}{s^3} + \frac{2}{s^2}\right)\end{aligned}$$

Example 5.4.4. *Use the second shifting property* (5.4.2) *to find the Laplace transform of the unit step* $u_a(t)$.

Solution: From the fact that $\mathcal{L}\left[1\right] = 1/s$ and from (5.4.2), we have

$$\mathcal{L}\left[u_a(t)\right] = e^{-as}\mathcal{L}\left[1\right] = e^{-as}/s$$

Example 5.4.5. *Find the transform of*

$$f(t) = \begin{cases} 1, & 0 \le t < 1 \\ t, & 1 \le t < 2 \\ 2, & 2 \le t \end{cases}$$

Solution: From the work in Example 5.4.2,

$$f(t) = u_0(t) + (t-1)u_1(t) + (2-t)u_2(t)$$

Hence

$$\mathcal{L}\left[f(t)\right] = \mathcal{L}\left[u_0(t)\right] + \mathcal{L}\left[u_1(t)(t-1)\right] + \mathcal{L}\left[u_2(t)(2-t)\right]$$

By the second shifting property,

$$\begin{aligned}\mathcal{L}\left[u_1(t)(t-1)\right] &+ \mathcal{L}\left[u_2(t)(2-t)\right] \\ &= e^{-s}\mathcal{L}\left[t + 1 - 1\right] + e^{-2s}\mathcal{L}\left[2 - (t+2)\right] \\ &= e^{-s}\mathcal{L}\left[t\right] - e^{-2s}\mathcal{L}\left[t\right]\end{aligned}$$

Finally, since $\mathcal{L}[u_0(t)] = \mathcal{L}[1] = 1/s$ and $\mathcal{L}[t] = 1/s^2$, we have

$$\mathcal{L}[f(t)] = \frac{1}{s} + \frac{1}{s^2}e^{-s} - \frac{1}{s^2}e^{-2s}$$

Example 5.4.6. *Find the Laplace transform of $f(t) = u_a(t) - u_b(t)$.*

Solution: We have

$$\begin{aligned}\mathcal{L}[f(t)] &= \mathcal{L}[u_a(t)] - \mathcal{L}[u_b(t)] \\ &= \frac{1}{s}e^{-as} - \frac{1}{s}e^{-bs} = \frac{1}{s}\left(e^{-as} - e^{-bs}\right)\end{aligned}$$

Example 5.4.7. *Find the Laplace transform of $f(t)$, shown in the following graph.*

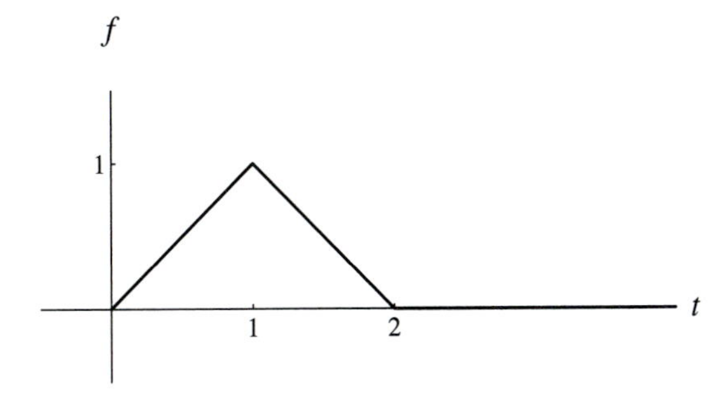

Solution: A study of the graph shows that

$$f(t) = \begin{cases} t, & 0 \le t < 1 \\ 2 - t, & 1 \le t < 2 \\ 0, & 2 \le t \end{cases}$$

Hence,

$$\begin{aligned}f(t) &= t\left(u_0(t) - u_1(t)\right) + (2 - t)\left(u_1(t) - u_2(t)\right) \\ &= tu_0(t) + u_1(t)(2 - 2t) + u_2(t)(t - 2)\end{aligned}$$

By the second shifting property and $\mathcal{L}[tu_0(t)] = e^{0s}\mathcal{L}[t]$,

$$\begin{aligned}\mathcal{L}[f(t)] &= \mathcal{L}[t] + e^{-s}\mathcal{L}[2 - 2(t+1)] + e^{-2s}\mathcal{L}[t + 2 - 2] \\ &= \mathcal{L}[t] - 2e^{-s}\mathcal{L}[t] + e^{-2s}\mathcal{L}[t] = \frac{1}{s^2}\left(1 - 2e^{-s} + e^{-2s}\right)\end{aligned}$$

It is sometimes convenient to use Theorem 5.4.1 with $f(t-a)$ in place of $f(t)$. Then, (5.4.2) becomes, using $\mathcal{L}[f(t)] = F(s)$,

$$\mathcal{L}[u_a(t)f(t-a)] = e^{-as}F(s) \tag{5.4.3}$$

Hence, solving for $F(s)$, we have

$$e^{as}\mathcal{L}[u_a(t)f(t-a)] = F(s) \tag{5.4.4}$$

From the definition of the unit step functions, it is not hard to show that $u_a(t) = u_0(t-a)$. (See Problem 8.) Some tables of Laplace transforms use $u_0(t-a)$ where we use $u_a(t)$.

The Laplace transforms of many of the more common functions are given in Table 5.1 at the end of this chapter. It should be noted here that, as in all tables of integrals, not every expression can be listed. It often requires significant knowledge and technique to use these tables effectively.

•EXERCISES

1. Find the Laplace transform of the following functions.

(a) $u_a(t)$.　　(b) $1 - u_a(t)$.

(c) $2u_1(t) - 1$.　　(d) $tu_a(t)$.

(e) $u_a(t) - u_b(t), \quad a < b$.

2. Use the second shifting property and Table 5.1 to find the Laplace transform of each function. Sketch each function.

(a) $u_2(t)$.　　(b) $u_4(t)\sin \pi t$.

(c) $\frac{1}{2}t - \frac{1}{2}u_4(t)$.　　(d) $u_4(t)(6-t) - u_6(t)(6-t)$.

3. Sketch each of the following functions, write them in terms of unit step functions, and use the second shifting p!g roperty and Table 5.1 to find their Laplace transforms.

(a) $f(t) = \begin{cases} 0, & 0 \le t \\ 1, & 1 \le t < 2 \\ 0, & 2 \le t \end{cases}$　　(b) $f(t) = \begin{cases} 1, & 0 \le t < 4 \\ 2, & 4 \le t \end{cases}$

(c) $f(t) = \begin{cases} 0, & 0 \leq t \\ t, & 1 \leq t < 2 \\ t^2, & 2 \leq t \end{cases}$ (d) $f(t) = \begin{cases} t, & 0 \leq t < 2 \\ 0, & 2 \leq t \end{cases}$

4. Repeat Problems 3 for these functions.

(a) $f(t) = \begin{cases} \sin t, & t < 2\pi \\ 0, & 2\pi \leq t \end{cases}$ (b) $f(t) = \begin{cases} \sin t, & t < \pi \\ \sin 2t, & \pi \leq t \end{cases}$

(c) $f(t) = \begin{cases} 0, & t < 2 \\ 2, & 2 \leq t < 4 \\ 0, & 4 \leq t \end{cases}$ (d) $f(t) = \begin{cases} t, & t < 2 \\ 2, & 2 \leq t < 4 \\ 4 - t, & 4 \leq t \end{cases}$

5. Repeat Problem 3 for these functions.

(a) $f(t) = \begin{cases} 0, & t < 2 \\ t^2 - 4, & 2 \leq t < 4 \\ 0, & 4 \leq t \end{cases}$ (b) $f(t) = \begin{cases} 0, & t < 2 \\ 2t, & 2 \leq t < 4 \\ 4 - t, & 4 \leq t \end{cases}$

(c) $f(t) = \begin{cases} 0, & t < \pi \\ \sin t, & \pi \leq t < 2\pi \\ 0, & 2\pi \leq t \end{cases}$

6. Show that the graph of the function $u_a(t)f(t-a)$ is the graph of the function $f(t)$ translated a units to the right, assuming that $f(t) = 0$ for $t < 0$.

7. Use Problem 6 and (5.4.4) to find the transform of the parabolic arc

$$f(t) = \begin{cases} 0, & t < 2 \\ (t-2)^2, & 2 \leq t \end{cases}$$

Sketch $f(t)$ and$g(t)$ where

$$g(t) = \begin{cases} 0, & t < 0 \\ t^2, & 0 \leq t \end{cases}$$

8. Graph $u_a(t)$ and $u_0(t-a)$ for $a \geq 0$. Note that for all $t < a$, $u_0(t-a) = 0$.

5.5 Laplace Transforms of Periodic Functions

There is one more class of functions whose transforms are important in applications. These functions are periodic and often have discontinuities.

The example portrayed by the graph in Figure 5.4 illustrates many of the features that are shared by most periodic functions.

Clearly, the graph of a periodic function is composed of units, each one a duplicate of every other one. The length of the repetition interval is called

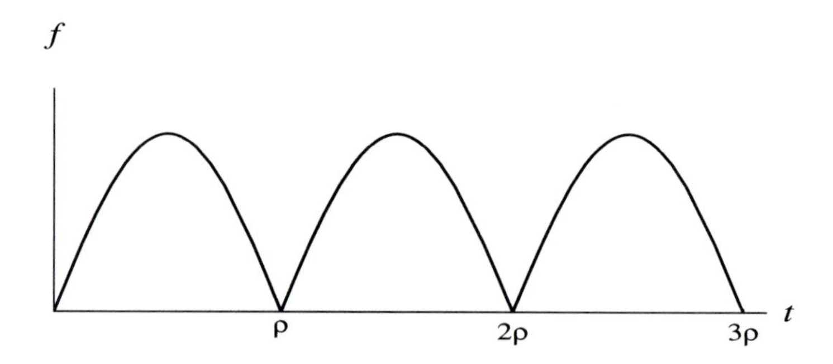

Figure 5.4: A Periodic, Sectionally Continuous Function

the *period* of the function. In the example shown in Figure 5.4, the period is p and the graph is assumed to continue indefinitely. At the jumps, which occur at $t_n = np, n = 1, 2, \ldots$, the function is not defined, although the limits $f\left(t_n^+\right)$ and $f\left(t_n^-\right)$ exist. The periodic nature of these functions can be expressed as

$$f(t) = f(t+p) = f(t+2p) = \cdots = f(t+np) = \ldots \tag{5.5.1}$$

Since the periodic functions we consider are sectionally continuous and bounded in each interval of periodicity, they are bounded for all $t \geq 0$ and hence are acceptable. Indeed, for $f(t)$ of (5.5.1),

$$\begin{aligned} \mathcal{L}\left[f(t)\right] &= \int_0^\infty f(t)e^{-st}\,dt \\ &= \int_0^p f(t)e^{-st}\,dt + \int_p^{2p} f(t)e^{-st}\,dt \\ &\qquad + \cdots + \int_{np}^{(n+1)p} f(t)e^{-st}\,dt + \ldots \end{aligned} \tag{5.5.2}$$

If we make the change of variable $t = z + np$ in the typical integral in the series of integrals and invoke the hypothesis of periodicity, $f(z+np) = f(z)$, we have

$$\begin{aligned} \int_{np}^{(n+1)p} f(t)e^{-st}\,dt &= \int_0^p f(z+np)e^{-s(z+np)}\,dz \\ &= e^{-nsp}\int_0^p f(z)e^{-sz}\,dz \end{aligned} \tag{5.5.3}$$

From (5.5.3), we see that (5.5.2) may be written as

$$\begin{aligned}\mathcal{L}[f(t)] &= \int_0^p f(z)e^{-sz}\,dz + e^{-sp}\int_0^p f(z)e^{-sz}\,dz \\ &\quad + \cdots + e^{-nsp}\int_0^p f(z)e^{-sz}\,dz + \ldots \\ &= (1 + e^{-sp} + \cdots + e^{-nsp} + \ldots)\int_0^p f(z)e^{-sz}\,dz\end{aligned} \tag{5.5.4}$$

Now let $x = e^{-sp}$, and note that $x^n = e^{-nsp}$. Then the series coefficient of the last integral is the geometric series

$$1 + x + \cdots + x^n + \cdots = \frac{1}{1-x} = \frac{1}{1-e^{-sp}} \tag{5.5.5}$$

which converges for all $x = e^{-sp} < 1$ by the ratio test. Because both s and p are positive, it follows that $e^{-sp} < 1$ for all s. Using this result in (5.5.4) is the proof of the following theorem.

Theorem 5.5.1. *If $f(t)$ is acceptable and periodic with period p, then for all $s > 0$,*

$$\mathcal{L}[f(t)] = \frac{1}{1-e^{-sp}}\int_0^p f(t)e^{-st}\,dt \tag{5.5.6}$$

Example 5.5.1. *Find the Laplace transform of the periodic extension of*

$$f(t) = \begin{cases} 1, & 0 < t \le 1 \\ -1, & 1 < t \le 2 \end{cases}$$

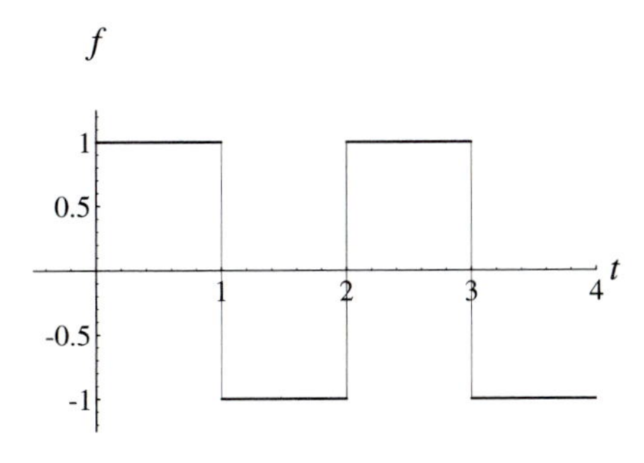

Solution: The given function is the square-wave, periodic with period $p = 2a$ and amplitude 1. In order to use (5.5.6), we first compute

$$\begin{aligned} F(s) = \int_0^2 f(t)e^{-st}\,dt &= \int_0^1 e^{-st}\,dt + \int_1^2 -e^{-st}\,dt \\ &= \frac{1}{s}\left(-e^{-s} + 1 + e^{-2s} - e^{-s}\right) \\ &= \frac{1}{s}\left(e^{-2s} - 2e^{-s} + 1\right) = \frac{1}{s}\left(1 - e^{-s}\right)^2 \end{aligned}$$

Then, since $\left(1 - e^{-2s}\right) = \left(1 - e^{-s}\right)\left(1 + e^{-s}\right)$, Theorem 5.5.1 yields

$$F(s) = \frac{1}{s}\frac{(1 - e^{-s})^2}{1 - e^{-2s}} = \frac{A}{s}\frac{1 - e^{-as}}{1 + e^{-as}}$$

Example 5.5.2. *Find the Laplace transform of the periodic function of period $p = 2$ whose definition in its first period is given by*

$$f(t) = \begin{cases} A, & 0 < t \le 1 \\ 0, & 1 < t \le 2 \end{cases}$$

Solution: Equation 5.5.6 may be used with $p = 2$. Then,

$$F(s) = \frac{1}{1 - e^{-2s}} \int_0^1 Ae^{-st}\,dt = \frac{1}{1 - e^{-2s}}\frac{A}{s}\left(1 - e^{-s}\right) = \frac{A}{s}\frac{1}{1 + e^{-s}}$$

Example 5.5.3. *Assuming that $f(t)$ has period $p = 2\pi$, find the Laplace transform of the periodic extension of*

$$f(t) = \begin{cases} \sin t, & 0 \le t < \pi \\ 0, & \pi \le t < 2\pi \end{cases}$$

Solution: To use Theorem 5.5.1, we need

$$\int_0^{2\pi} f(t)e^{-st}\,dt = \int_0^{\pi} \sin t e^{-st}\,dt = \frac{1}{1 + s^2}\left(1 + e^{-s\pi}\right)$$

Hence, from Theorem 5.5.1,

$$\begin{aligned}\mathcal{L}[f(t)] &= \frac{1+e^{-s\pi}}{1-e^{-2s\pi}}\frac{1}{s^2+1} \\ &= \frac{1}{1-e^{-s\pi}}\frac{1}{s^2+1}\end{aligned}$$

•EXERCISES

In the following problems, the given functions are periodic, the first half of the first period is given, and we assume that the function is zero in the second half of its first period. Sketch two periods for each function. Find the Laplace transform of the periodic extension of these functions.

1. $f(t) = \sin 2t, \quad 0 \le t \le \frac{1}{2}\pi.$
2. $f(t) = 2 - t, \quad 0 \le t < 2.$
3. $f(t) = 1, \quad 0 \le t < 1.$
4. $f(t) = \cos t, \quad 0 \le t < \frac{1}{2}\pi.$
5. $f(t) = t, \quad 0 \le t < 1.$

5.6 The Inverse Laplace Transform

In the solution of ODE's we must reconstruct $f(t)$ given its transform $F(s)$. The operator that yields $f(t)$ from $F(s)$ is called the *inverse Laplace transform* and is denoted by $\mathcal{L}^{-1}$. We write

$$f(t) = \mathcal{L}^{-1}[F(s)] = \mathcal{L}^{-1}[F] \tag{5.6.1}$$

and take as given that $f(t)$ is acceptable. Alternatively (5.6.1) can be written as

$$f(t) = \mathcal{L}^{-1}[\mathcal{L}[f(t]] \tag{5.6.2}$$

Now take the Inverse Laplace transform of both sides of (5.6.1) to obtain

$$F(s) = \mathcal{L}\left[\mathcal{L}^{-1}[F(s)]\right] \tag{5.6.3}$$

Equations (5.6.2) and (5.6.3) show that the Laplace transform and the inverse Laplace transform are inverse operators, thus justifying the name "inverse Laplace transform" for $\mathcal{L}^{-1}$. A table of inverse transforms is given at the end of this chapter.

Here is how the Laplace transform and inverse Laplace transform bear on the theory of constant coefficient linear ODE's.

Consider the initial-value problem

$$y' + y = e^{-t}, \qquad y(0) = -1 \tag{5.6.4}$$

Write $Y(s) = \mathcal{L}[y(t)]$. If we take the Laplace transform of both sides of this ODE, we obtain, for the left-hand side,

$$\begin{aligned} \mathcal{L}[y'] + \mathcal{L}[y] &= sY(s) - y(0) + Y(s) \\ &= sY(s) + 1 + Y(s) = (s+1)Y(s) + 1 \end{aligned}$$

And for the right-hand side, we have

$$\mathcal{L}\left[e^{-t}\right] = \frac{1}{s+1}$$

So equating these two expressions yields

$$(s+1)Y(s) + 1 = \frac{1}{s+1}$$

This latter equation can be solved for $Y(s)$ by using algebra alone:

$$Y(s) = -\frac{1}{s+1} + \frac{1}{(s+1)^2}$$

The inverse Laplace transform is used to reconstruct $y(t)$ from $Y(s)$. Indeed, from Table 5.1, Formula 13, or from Table 5.2, Formula 9, or by the various results obtained so far:

$$y(t) = -e^{-t} + te^{-t} \tag{5.6.5}$$

This is the solution of the given initial-value problem. This example shows how naturally the Laplace transform solves nonhomogeneous initial-value problems by accommodating the initial conditions and the forcing function.

Before we begin a detailed study of the role of the Laplace transform in the theory of linear ODE's, we need to rewrite several of the theorems of the previous section in terms of the inverse Laplace transform. We begin by asserting and proving that the inverse transform, like the Laplace transform, is a linear operator.

Theorem 5.6.1. *For every pair of constants a and b,*

$$\mathcal{L}^{-1}[aF(s) + bG(s)] = a\mathcal{L}^{-1}[F(s)] + b\mathcal{L}^{-1}[G(s)] \tag{5.6.6}$$

Proof: By the linearity of the Laplace transform and from (5.6.3), it follows that

$$\mathcal{L}\left[a\mathcal{L}^{-1}[F(s)] + b\mathcal{L}^{-1}[G(s)]\right] = aF(s) + bG(s)$$

Now take the inverse Laplace transform of both sides of this equation, and use (5.6.2) to derive (5.6.6). □

Theorem 5.6.2. *For* $a \geq 0$,

$$\mathcal{L}^{-1}\left[e^{-as}F(s)\right] = u_a(t)f(t-a) \tag{5.6.7}$$

Proof: From Theorem 5.4.1, using $f(t-a)$ in place of $f(t)$, we obtain

$$\mathcal{L}\left[u_a(t)f(t-a)\right] = e^{-as}\mathcal{L}\left[f(t)\right] \tag{5.6.8}$$

Now take the inverse transform of both sides of (5.6.8). □

Theorem 5.6.3.

$$\mathcal{L}^{-1}\left[\frac{1}{s}F(s)\right] = \int_0^t f(z)\,dz \tag{5.6.9}$$

Proof: This is Corollary 5.3.3 expressed in terms of the inverse Laplace transform. □

Theorem 5.6.4. *For any* a,

$$\mathcal{L}^{-1}\left[F(s+a)\right] = e^{-at}\mathcal{L}^{-1}\left[F(s)\right] \tag{5.6.10}$$

Proof: By taking inverse transforms of the first shifting property,

$$\mathcal{L}\left[e^{at}f(t)\right] = F(s-a) \tag{5.6.11}$$

we obtain

$$e^{at} f(t) = e^{at} \mathcal{L}^{-1} [F(s)] = \mathcal{L}^{-1} [F(s-a)]$$

Now replace $s - a$ by s. □

Theorem 5.6.5. *If $\mathcal{L}^{-1}[F(s)] = f(t)$ and $f(0) = 0$, then*

$$\mathcal{L}^{-1}[sF(s)] = f'(t) \tag{5.6.12}$$

Proof: Simply take the inverse Laplace transform of (5.3.5) of Theorem 5.5.3. □

Example 5.6.1. *Find $f(t)$, given that*

$$\mathcal{L}[f(t)] = \frac{1}{s^2} + \frac{1}{s^2} e^{-s}$$

Solution: From Table 5.1, $\mathcal{L}^{-1}[1/s^2] = t$ and $\mathcal{L}^{-1}[e^{-s}/s^2] = u_1(t)(t-1)$. Hence, using the linearity property, Theorem 5.6.1 and Theorem 5.6.2, we have

$$\begin{aligned} f(t) &= \mathcal{L}^{-1}\left[\frac{1}{s^2} + \frac{1}{s^2} e^{-s}\right] = \mathcal{L}^{-1}\left[\frac{1}{s^2}\right] + \mathcal{L}^{-1}\left[\frac{1}{s^2} e^{-s}\right] \\ &= t + u_1(t)(t-1) \end{aligned}$$

Here is the graph of the solution.

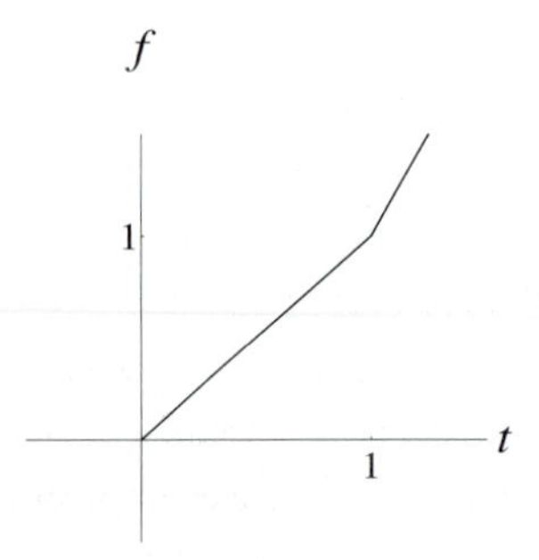

Example 5.6.2. *Find* $f(t)$, *given* $F(s) = \dfrac{1}{s(s^2+1)}$.

Solution: Since

$$\mathcal{L}^{-1}\left[\frac{1}{s^2+1}\right] = \sin t$$

we have, by Theorem 5.6.3,

$$f(t) = \mathcal{L}^{-1}\left[\frac{1}{s\,(s^2+1)}\right] = \int_0^t \sin z\,dz = 1 - \cos t$$

Example 5.6.3. *Find* $f(t)$, *given* $F(s) = \dfrac{1}{s^2+2s+5}$.

Solution: It is often a good idea when searching for the inverse transform of an expression containing a quadratic in s to complete squares. In this case, for example, $s^2+2s+5 = (s+1)^2+4$. The appearance of the term $(s+1)^2$ suggests using Theorem 5.6.4 with $a = 1$. Indeed,

$$f(t) = \mathcal{L}^{-1}\left[\frac{1}{(s+1)^2+4}\right] = e^{-t}\mathcal{L}^{-1}\left[\frac{1}{s^2+4}\right] = \frac{1}{2}e^{-t}\sin 2t$$

Example 5.6.4. *Find* $g(t)$, *given* $G(s) = \dfrac{1}{s\,(s^2+2s+5)}$.

Solution: Set

$$F(s) = \frac{1}{s^2+2s+5}$$

Then $G(s) = \dfrac{1}{s}F(s)$, and this is suggestive of Theorem 5.6.3. From Example 5.6.3,

$$\mathcal{L}^{-1}\left[F(s)\right] = \frac{1}{2}e^{-t}\sin 2t$$

Therefore, by Theorem 5.6.3 and integration by parts,

$$g(t) = \frac{1}{2}\int_0^t e^{-z}\sin 2z\,dz = \frac{1}{5} - \frac{1}{5}\left(\frac{1}{2}\sin 2t + \cos 2t\right)e^{-t}$$

Example 5.6.5. *Find $f(t)$, given $F(s) = \dfrac{4e^{-2s}}{s^2 - 16}$.*

Solution: The factor e^{-2s} suggests using (5.6.7). Following this lead,

$$\frac{4e^{-2s}}{s^2 - 16} = e^{-2s}\left[\frac{4}{s^2 - 16}\right]$$

In Table 5.1, Formula 21 we find $\mathcal{L}[\sinh 4t] = 4/(s^2 - 16)$. So, using the second shifting property, we get

$$f(t) = u_2(t)\sinh 4(t - 2)$$

or

$$f(t) = \begin{cases} 0, & 0 \leq t < 2 \\ \sinh 4(t - 2), & 2 \leq t \end{cases}$$

•EXERCISES

Use the theorems of this section to find the function $f(t)$ whose Laplace transform is given. Use Tables 5.1 and 5.2 as needed.

1. $\dfrac{1}{s}\left(\dfrac{2}{s^2} + \dfrac{1}{s} - 2\right)$.
2. $\dfrac{1}{s^2}\left(\dfrac{3}{s} + 2\right)$.
3. $\dfrac{e^{-as}}{s+1}$.
4. $\dfrac{1}{s^2 + 6s + 10}$.
5. $\dfrac{4s + 3}{s^2 + 4s + 13}$.
6. $\dfrac{4se^{-2s\pi}}{s^2 + 2s + 5}$.
7. $\dfrac{3s + 1}{s^2 - 4s - 5}$.
8. $\dfrac{1}{s(s+1)}$.
9. $\dfrac{1}{s^2 + 2s}$.
10. $\dfrac{e^{-2s}}{s(s+1)^2}$.
11. $\dfrac{1}{s^3 + 4s}$.
12. $\dfrac{1}{s^3 - 9s}$.
13. $\dfrac{1}{s^2(s-2)}$.
14. $\dfrac{2}{s^4 + 4s^2}$.
15. $\dfrac{4}{s^4 + 4s^2}$.
16. $\dfrac{1}{s^2}\dfrac{s - 1}{s^2 + 1}$.
17. $\dfrac{9}{s^4 - 9s^2}$.
18. $\dfrac{1}{(s-1)(s+2)}$.

19. $\dfrac{1}{s}\dfrac{s-1}{s+1}$.

20. $\dfrac{2s}{(s+3)^2}$.

21. $\dfrac{1}{(s-2)(s+1)}$.

22. $\dfrac{2}{s\,(s^2+1)}$.

5.7 Partial Fractions

As we have seen earlier, the Laplace transform of the common functions encountered in engineering practice are often rational functions, the quotient of polynomials.[2] It turns out that it is always possible to find the inverse Laplace transform of rational functions by the tools we will develop. Our technique for doing so is called the *method of partial fractions.* This is one of the methods used in calculus for finding the indefinite integrals of rational functions. In this section we review the method of partial fractions and apply it to obtain the inverse Laplace transform.

Suppose that

$$F(s) = \frac{P(s)}{Q(s)} \tag{5.7.1}$$

where $P(s)$ and $Q(s)$ are polynomials in s. Since we will restrict our attention to rational functions which are acceptable, Theorem 5.2.1 asserts that $F(s) \to 0$ as $s \to \infty$. Hence, the degree of $P(s)$ is less than the degree of $Q(s)$.

We make one reasonable stipulation: We assume that $P(s)$ and $Q(s)$ have no factors in common. In the event that $(s-r)$ is a factor of both of both $P(s)$ and $Q(s)$, we simplify $F(s)$ by cancelling this factor from both numerator and denominator.

5.7.1 Nonrepeated Linear Factors

Suppose that $Q(s)$ is a third-degree polynomial with three distinct real roots, a_1, a_2 and a_3. (The analysis for the nth-degree polynomial proceeds in a similar way.) By the Fundamental Theorem of Algebra, $Q(s)$ has the factored form

$$Q(s) = (s-a_1)\,(s-a_2)\,(s-a_3) \tag{5.7.2}$$

[2]These sections on partial fractions are a review of the material most often covered in calculus. We do not attempt to treat the most general case.

which yields the identity

$$\begin{aligned} F(s) &= \frac{P(s)}{(s-a_1)(s-a_2)(s-a_3)} \\ &= \frac{A}{s-a_1}+\frac{B}{s-a_2}+\frac{C}{s-a_3} \end{aligned} \tag{5.7.3}$$

for some constants A, B, and C. This is called the *partial-fraction decomposition* of $F(s)$. Our aim is to find A, B, and C, given $P(s)$ and $Q(s)$. One way to do this is to multiply (5.7.3) by $(s-a_1)(s-a_2)(s-a_3)$. Then,

$$P(s) = A(s-a_2)(s-a_3)+B(s-a_1)(s-a_3)+C(s-a_1)(s-a_2) \tag{5.7.4}$$

is an identity[3] in s. Setting $s=a_1$ in (5.7.4) leads to A through the equation

$$P(a_1) = A(a_1-a_2)(a_1-a_3) \tag{5.7.5}$$

Similar equations are obtained for B and C by using $s=a_2$ and $s=a_3$, respectively:

$$P(a_2) = B(a_2-a_1)(a_2-a_3) \tag{5.7.6}$$
$$P(a_3) = C(a_3-a_1)(a_3-a_2) \tag{5.7.7}$$

Example 5.7.1. *Find $f(t)$, given $F(s)=\dfrac{s^2+3s-6}{s(s-1)(s-2)}$.*

Solution: We expect a partial-fraction decomposition of $F(s)$ of the form

$$\frac{s^2+3s-6}{s(s-1)(s-2)} = \frac{A}{s}+\frac{B}{s-1}+\frac{C}{s-2}$$

Now clear the denominators by multiplying the expression by $s(s-1)(s-2)$, and thus, obtain the identity

$$s^2+3s-6 = A(s-1)(s-2)+Bs(s-2)+Cs(s-1)$$

To determine A, we substitute $s=0$ in this identity and find $A=-3$. Likewise, substitute $s=1$ to obtain $B=2$. Finally, $C=2$ is found by substituting $s=2$. Thus, using these values, we have

$$F(s) = -\frac{3}{s}+\frac{2}{s-1}+\frac{2}{s-2}$$

[3]We repeatedly use the fact that identical polynomials must have equal coeifficients for like powers of s.

This form for $F(s)$ enables us to compute $f(t)$. Indeed, from Table 5.1 or Table 5.2,

$$f(t) = -3 + 2e^t + 2e^{2t}$$

Example 5.7.2. *Find $f(t)$, given $F(s) = \dfrac{1}{s^3 - s}$.*

Solution: We have

$$F(s) = \frac{1}{s^3 - s} = \frac{1}{s(s-1)(s+1)} = \frac{A}{s} + \frac{B}{s-1} + \frac{C}{s+1}$$

Clear the denominator of $F(s)$ by multiplying by $s\,(s^2 - 1)$. This leads to

$$1 = A(s^2 - 1) + Bs(s+1) + Cs(s-1)$$

Now use $s = 0, s = 1$, and $s = -1$ to determine that $A = -1, B = \frac{1}{2}$, and $C = \frac{1}{2}$. Hence,

$$F(s) = -\frac{1}{s} + \frac{1}{2}\frac{1}{s-1} + \frac{1}{2}\frac{1}{s+1}$$

and therefore, using Table 5.1 or 5.2, we obtain

$$f(t) = -1 + \tfrac{1}{2}e^t + \tfrac{1}{2}e^{-t} = -1 + \cosh t$$

Example 5.7.3. *For $a \neq b$, find $f(t)$ and $g(t)$ given*

(a) $\quad F(s) = \dfrac{1}{(s-a)(s-b)}$

(b) $\quad G(s) = \dfrac{s}{(s-a)(s-b)}$

Solution: (a) We have

$$F(s) = \frac{1}{(s-a)(s-b)} = \frac{A}{s-a} + \frac{B}{s-b}$$

from which it easily follows that $A = -B = 1/(a-b)$. Thus,

$$\frac{1}{(s-a)(s-b)} = \frac{1}{a-b}\left(\frac{1}{s-a} - \frac{1}{s-b}\right)$$

Hence,

$$f(t) = \frac{1}{a-b}\left(e^{at} - e^{bt}\right)$$

(b) In much the same way,

$$G(s) = \frac{s}{(s-a)(s-b)} = \frac{1}{a-b}\left(\frac{a}{s-a} - \frac{b}{s-b}\right)$$

Thus,

$$\mathcal{L}^{-1}\left[G(s)\right] = g(t) = \frac{1}{a-b}\left(ae^{at} - be^{bt}\right)$$

Note that $a \neq b$ is essential. In fact, if $a = b$, then $F(s)$ has a single repeated factor, and $f(t) = te^{at}$ while $g(t) = (1 + at)e^{at}$.

The roots of $Q(s)$ need not be real in order to use partial fractions to determine the inverse Laplace transform.[4] Consider, for example,

$$\frac{2s}{s^2+1} = \frac{1}{s+i} + \frac{1}{s-i} \tag{5.7.8}$$

which leads to

$$\begin{aligned} 2\mathcal{L}^{-1}\left[\frac{s}{s^2+1}\right] &= \mathcal{L}^{-1}\left[\frac{1}{s+i}\right] + \mathcal{L}^{-1}\left[\frac{s}{s-i}\right] \\ &= e^{-it} + e^{it} = 2\cos t \end{aligned} \tag{5.7.9}$$

This is yet another derivation of $\mathcal{L}\left[\cos t\right] = s/(s^2+1)$.

Example 5.7.4. *Find $f(t)$, given $F(s) = \dfrac{8(s+1)}{s\left(s^2+4\right)}$.*

Solution: We have $s^2 + 4 = (s+2i)(s-2i)$, and hence,

$$F(s) = \frac{8(s+1)}{s\left(s^2+4\right)} = \frac{A}{s} + \frac{B}{s+2i} + \frac{C}{s-2i}$$

Thus,

$$8(s+1) = A(s^2+4) + Bs(s-2i) + Cs(s+2i)$$

---

[4]Recall the two Euler formulas, $e^{ibt} + e^{-ibt} = 2\cos bt$ and $e^{ibt} - e^{-ibt} = 2i\sin bt$

Now, $s = 2i$ results in $8(1+2i) = -8C$, and hence, $C = -1-2i$. Likewise, using $s = -2i$ leads to $B = -1+2i$. Finally, $s = 0$ yields $A = 2$. Therefore,

$$F(s) = \frac{2}{s} + (-1-2i)\frac{1}{s+2i} + (-1-2i)\frac{1}{s-2i}$$

and it follows that

$$\begin{aligned} f(t) &= 2 + (-1-2i)e^{-2it} - (1+2i)e^{2it} \\ &= 2 - \left(e^{2it} - e^{-2it}\right) + 2i\left(e^{2it} - e^{-2it}\right) \\ &= 2 - 2\cos 2t - 4\sin 2t \end{aligned}$$

Example 5.7.5. *Find $f(t)$, given $F(s) = \dfrac{s+5}{s\,(s^2+2s+5)}$.*

Solution: Since $s^2+2s+5 = s^2+2s+1+4 = (s+1)^2+4$, we have

$$F(s) = \frac{s+5}{s\,(s^2+2s+5)} = \frac{A}{s} + \frac{B}{s+1+2i} + \frac{C}{s+1-2i}$$

Hence,

$$s+5 = A\left((s+1)^2+4\right) + Bs(s+1-2i) + Cs(s+1+2i)$$

Convenient choices for s are $s = 0$, $s = -1+2i$, and $s = -1-2i$. From $s = 0$, we find $A = 1$. Using $s = -1+2i$ leads to $C = -\frac{1}{2}$, and from $s = -1-2i$, we have $B = -\frac{1}{2}$. Hence,

$$F(s) = \frac{1}{s} - \frac{1}{2}\frac{1}{s+1+2i} - \frac{1}{2}\frac{1}{s+1-2i}$$

Therefore,

$$\begin{aligned} f(t) &= 1 - \tfrac{1}{2}\left(e^{(-1-2i)t} + e^{(-1+2i)t}\right) \\ &= 1 - \tfrac{1}{2}e^{-t}\left(e^{-2it} + e^{2it}\right) = 1 - e^{-t}\cos 2t \end{aligned}$$

Example 5.7.6. *For $a^2 \neq b^2$, find $f(t)$ and $g(t)$ given*

$$\text{(a)} \qquad F(s) = \frac{1}{(s^2+a^2)(s^2+b^2)}$$

$$\text{(b)} \qquad G(s) = \frac{s}{(s^2+a^2)(s^2+b^2)}$$

Solution: In Part (a) we use s^2 in place of s to obtain the following partial fraction decomposition of $F(s)$:

$$F(s) = \frac{1}{b^2-a^2}\left(\frac{1}{s^2+a^2} - \frac{1}{s^2+b^2}\right)$$

and thus, from Table 5.1 or Table 5.2,

$$f(t) = \frac{1}{b^2-a^2}\left(\frac{1}{a}\sin at - \frac{1}{b}\sin bt\right)$$

Part (b) follows from Part (a) by using Theorem 5.6.5:

$$g(t) = \frac{1}{b^2-a^2}(\cos at - \cos bt)$$

The case $a = \pm b$ leads to a repeated root of $Q(s)$. (See Subsection 5.7.3.)

Example 5.7.7. *Find the displacement function $y(t)$ of a forced, frictionless spring-mass system whose Laplace transform is*

$$Y(s) = F_0 \frac{\omega}{M} \frac{1}{(s^2+\omega^2)(s^2+\omega_0^2)}, \qquad \omega \neq \omega_0$$

Solution: (The constants F_0, ω, ω_0, and M are parameters determined from the spring-mass system and the forcing function.) As a direct consequence of Example 5.7.6, using $a^2 = \omega^2$ and $b^2 = \omega_0^2$, we have $a^2 - b^2 = \omega^2 - \omega_0^2$ and therefore,

$$y(t) = \frac{F_0\omega}{M(\omega^2-\omega_0^2)}\left(\frac{1}{\omega_0}\sin\omega_0 t - \frac{1}{\omega}\sin\omega t\right), \qquad \omega \neq \omega_0.$$

•EXERCISES

Use partial fractions or the results obtained in the illustrative examples and theorems to find the inverse transforms of the following functions. Where necessary, refer to Tables 5.1 or 5.2.

1. $\dfrac{1}{s^2+2s}$.

2. $\dfrac{1}{s^2-s}$.

3. $\dfrac{s-1}{s(s+1)}$.

4. $\dfrac{1}{s(s+1)}$.

5. $\dfrac{1}{(s-2)(s+1)}$.

6. $\dfrac{1}{(s-1)(s+2)}$.

7. $\dfrac{e^{-2s}}{s(s+1)}$.

8. $\dfrac{s+2}{s(s-2)(s+1)}$.

9. $\dfrac{e^{-2s}}{s^2(s+1)}$.

10. $\dfrac{1}{s^3+4s}$.

11. $\dfrac{1}{s^4+4s^2}$.

12. $\dfrac{e^{-2s}}{s\left(s^2+1\right)}$.

13. $\dfrac{s+2}{(s-2)\left(s^2+1\right)}$.

14. $\dfrac{1}{s^4-1}$.

15. $\dfrac{1}{s^3-1}$.

16. $\dfrac{s+2}{\left(s^2+4\right)(s+1)}$.

17. $\dfrac{s-1}{\left(s^2+4\right)(s+1)}$.

18. $\dfrac{s^2+1}{\left(s^2+4\right)(s+1)}$.

19. $\dfrac{s-1}{s^2+5s+4}$.

20. $\dfrac{1}{(s-2)\left(s^2+5s+4\right)}$.

21. $\dfrac{s+2}{s(s-2)(s+1)}$.

22. $\dfrac{1}{(s-1)(s-2)(s+1)}$.

23 $\dfrac{s+2}{(s-1)(s-2)(s+1)}$.

24. $\dfrac{s^2+2}{(s-1)(s-2)(s+1)}$.

25. Replace s with s^2 and use the partial fraction decomposition in Example 5.7.3 to find

$$f(t)=\mathcal{L}^{-1}\left[\frac{s^2}{\left(s^2+a^2\right)\left(s^2+b^2\right)}\right], \qquad a^2\neq 0, b^2\neq 0.$$

26. Use the result of Problem 25 and Theorem 5.3.3 to find

$$f(t)=\mathcal{L}^{-1}\left[\frac{s^3}{\left(s^2+a^2\right)\left(s^2+b^2\right)}\right], \qquad a^2\neq 0, b^2\neq 0.$$

27. Do Example 5.7.6 (b) using Theorem 5.6.4.

28. Repeat Problem 25 for the function

$$f(t) = \mathcal{L}^{-1}\left[\frac{s^2}{(s^2-a^2)(s^2-b^2)}\right], \qquad a^2 \neq 0, b^2 \neq 0.$$

29. Repeat Problem 25 for the funtion

$$f(t) = \mathcal{L}^{-1}\left[\frac{s^3}{(s^2-a^2)(s^2-b^2)}\right], \qquad a^2 \neq 0, b^2 \neq 0.$$

5.7.2 Repeated Linear Factors

If $Q(s)$ has a root $s = a$ repeated m times, then $Q(s)$ has the repeated factor, $(s-a)^m$ and the partial expansion of $F(s)$ is complicated by having terms $A_j\,(s-a)^{-j}$ for $j = 1, 2, \ldots, m$. The most common case of repeated roots occurs when $m = 2$, and this is the only case we consider. In this circumstance, the partial-fraction expansion of $F(s)$ is

$$F(s) = \frac{A}{(s-a)^2} + \frac{B}{s-a} + \frac{C}{s-b} + \frac{D}{s-c} + \ldots \tag{5.7.10}$$

If there is more than one repeated root, then we must include terms like the first two terms in (5.7.10) for each repeated root.

Example 5.7.8. *Find $f(t)$, given $F(s) = \dfrac{s}{(s-1)^2}$.*

Solution: From

$$\frac{s}{(s-1)^2} = \frac{A}{(s-1)^2} + \frac{B}{s-1}$$

we obtain $s = A + B(s-1)$. Using $s = 1$, we find $A = 1$. Since using $s = 1$ again does not help, we choose the convenient value $s = 0$ and find $B = 1$. Hence,

$$F(s) = \frac{1}{(s-1)^2} + \frac{1}{s-1}$$

Thus

$$\begin{aligned} f(t) &= \mathcal{L}^{-1}\left[\frac{1}{(s-1)^2}\right] + \mathcal{L}^{-1}\left[\frac{1}{s-1}\right] \\ &= te^t + e^t = (1+t)e^t \end{aligned}$$

Example 5.7.9. *Find $f(t)$, given $F(s) = \dfrac{s^2-1}{(s-2)^2(s+3)}$.*

Solution: We write

$$\frac{s^2-1}{(s-2)^2(s+3)} = \frac{A}{(s-2)^2} + \frac{B}{s-2} + \frac{C}{s+3}$$

so that multiplying by $(s-2)^2(s-3)$, we obtain the identity

$$s^2 - 1 = A(s+3) + B(s-2)(s+3) + C(s-2)^2$$

Now, $s = -3$ leads to $C = \frac{8}{25}$ and $s = 2$ to $A = \frac{3}{5}$. Using $s = 0$ and the values obtained for A and C, we have $B = \frac{17}{25}$. Therefore,

$$\frac{s^2-1}{(s-2)^2(s+3)} = \frac{3}{5}\frac{1}{(s-2)^2} + \frac{17}{25}\frac{1}{s-2} + \frac{8}{25}\frac{1}{s+3}$$

and we have

$$f(t) = \tfrac{3}{5}te^{2t} + \tfrac{17}{25}e^{2t} + \tfrac{8}{25}e^{-3t}$$

Example 5.7.10. *Find $f(t)$, given $F(s) = \dfrac{1}{(s-2)^2(s-1)^2}$.*

Solution: We write

$$\frac{1}{(s-2)^2(s-1)^2} = \frac{A}{(s-2)^2} + \frac{B}{s-2} + \frac{C}{(s-1)^2} + \frac{D}{s-1}$$

and multiply by $(s-2)^2(s-1)^2$ to obtain the identity

$$1 = A(s-1)^2 + B(s-2)(s-1)^2 + C(s-2)^2 + D(s-1)(s-2)^2$$

The choices $s = 2$ and $s = 1$ lead immediately to $A = 1$ and $C = 1$. Using $s = 0$ and the values for A and C, we obtain $2 = B + 2D$. Similarly, we find $2 = 2B + 3D$ by using $s = -1$. Solving these two equations gives $B = -2$ and $D = 2$. Hence,

$$\frac{1}{(s-2)^2(s-1)^2} = \frac{1}{(s-2)^2} + \frac{-2}{s-2} + \frac{1}{(s-1)^2} + \frac{2}{s-1}$$

Therefore,

$$f(t) = te^{2t} - 2e^{2t} + te^t + 2e^t = (t-2)e^{2t} + (t+2)e^t$$

Example 5.7.11. *Find* $f(t) = \mathcal{L}^{-1}\left[\dfrac{9}{(s^2+4)^2(s^2+1)^2}\right]$.

Solution: At first glance, this function is not of the form treated in this subsection. However, since $F(s)$ is a function of s^2, setting $z = s^2$ converts the partial fraction work into an expansion with repeated linear factors in z. That is,

$$\frac{9}{(s^2+4)^2(s^2+1)^2} = \frac{A}{(z+4)^2} + \frac{B}{z+4} + \frac{C}{(z+1)^2} + \frac{D}{z+1}$$

Clearing the denominators by multiplying by $(z+4)^2(z+1)^2$ results in

$$9 = A(z+1)^2 + B(z+4)(z+1)^2 + C(z+4)^2 + D(z+1)(z+4)^2$$

Using $z = -4$ and $z = -1$ leads to $A = C = 1$. Using $z = 0$ results in $-2 = B + 4D$, and using $z = 1$, we have $-2 = 2B + 5D$. Thus, $B = \frac{2}{3}$ and $D = -\frac{2}{3}$. So in terms of s^2, we have

$$\frac{9}{(s^2+4)^2(s^2+1)^2} = \frac{1}{(s^2+4)^2} + \frac{2/3}{s^2+4} + \frac{1}{(s^2+1)^2} - \frac{2/3}{s^2+1}$$

Finally, from Table 5.1 or 5.2,

$$\mathcal{L}[\sin 2t - 2t\cos 2t] = \frac{16}{(s^2+4)^2}$$

so that

$$\begin{aligned} f(t) &= \tfrac{1}{16}(\sin 2t - 2t\cos 2t) + \tfrac{1}{3}\sin 2t + \tfrac{1}{2}(\sin t - t\cos t) - \tfrac{2}{3}\sin t \\ &= \tfrac{19}{48}\sin 2t - \tfrac{1}{6}\sin t - \tfrac{1}{8}t\cos 2t - \tfrac{1}{2}t\cos t \end{aligned}$$

Example 5.7.12. *Find the inverse transform of* $F(s) = \dfrac{9s}{(s^2+4)^2(s^2+1)^2}$.

Solution: Let $sG(s) = F(s)$ define $G(s)$. Then

$$G(s) = \frac{9}{(s^2+4)^2(s^2+1)^2}$$

and we may use the result in Example (5.7.11) to write

$$g(t) = \tfrac{1}{16}\left(\sin 2t - 2t\cos 2t\right) + \tfrac{1}{3}\sin 2t + \tfrac{1}{2}\left(\sin t - t\cos t\right) - \tfrac{2}{3}\sin t$$

Now, since $g(0) = 0$ and $g'(t) = \frac{1}{4}t\sin 2t + \frac{2}{3}\cos 2t + \frac{1}{2}t\sin t - \frac{2}{3}\cos t$, Theorem 5.3.4 yields

$$\begin{aligned} f(t) &= \mathcal{L}^{-1}\left[F(s)\right] = \mathcal{L}^{-1}\left[sG(s)\right] \\ &= g'(t) = \tfrac{1}{4}t\sin 2t + \tfrac{2}{3}\cos 2t + \tfrac{1}{2}t\sin t - \tfrac{2}{3}\cos t \end{aligned}$$

•EXERCISES

Find $f(t)$, given the following Laplace transforms of $f(t)$, using Tables 5.1 or Table 5.2 if needed.

1. $\dfrac{2s}{(s+3)^2}$.
2. $\dfrac{4}{\left(s^2-4\right)^2}$.
3. $\dfrac{4}{s^4+4s^2}$.
4. $\dfrac{s}{(s+1)^2}$.
5. $\dfrac{1}{s^2(s-2)}$.
6. $\dfrac{2}{\left(s^2-1\right)^2\left(s^2+1\right)}$.
7. $\dfrac{e^{-s}}{s(s+2)^2}$.
8. $\dfrac{6}{s^3-9s}$.
9. $\dfrac{s-1}{s^2\left(s^2+1\right)}$.
10. $\dfrac{s^2+1}{(s+1)^2\left(s^2+4\right)}$.
11. $\dfrac{s^2+2s+1}{\left(s^2+2s-3\right)(s-1)}$.
12. $\dfrac{s^2-3s+2}{s^2(s-1)^2\left(s^2-5s+4\right)}$.
13. $\dfrac{5s^2+20}{s(s-1)\left(s^2-5s+4\right)}$.
14. $\dfrac{9}{\left(s^2-2s-3\right)^2}$.
15. $\dfrac{1}{(s-a)^2(s-b)}$.
16. $\dfrac{1}{(s-a)^2(s-b)^2}$.
17. $\dfrac{9}{\left(s^2+4\right)^2\left(s^2+1\right)}$.
18. $\dfrac{9}{\left(s^2+4\right)^2\left(s^2+1\right)^2}$.
19. $\dfrac{2s+3}{\left(s^2+4s+13\right)^2}$.
20. $\dfrac{27s^2}{10\left(s^2+4\right)^2\left(s^2+1\right)^2}$.
21. $\dfrac{9s}{\left(s^2+4\right)^2\left(s^2+1\right)}$.

5.7.3 Repeated Quadratic Factors

Repeated quadratic factors appear less frequently in our applications than do repeated linear factors. The method of partial-fraction expansions is applicable, but generally quite tedious to work by hand. Rather than pursue this method further, we can resort to one of two alternatives:

1. When available, an extensive table of Laplace transforms is the method of choice.

2. Symbolic manipulator computer programs such as Mathematica, Derive, or Maple can be used to obtain the partial fraction expansion or the inverse Laplace transform directly.

5.8 The Convolution Theorem

In view of our efforts to find $f(t)$ given $F(s)$, a natural question is how to express $\mathcal{L}^{-1}[G(s)F(s)]$ in terms of $\mathcal{L}^{-1}[G(s)]$ and $\mathcal{L}^{-1}[F(s)]$. To see how this may be done, define $F(s) = \mathcal{L}[f(t)]$ and $G(s) = \mathcal{L}[g(t)]$. Then, using z as the dummy variable of integration,

$$G(s) = \int_0^\infty e^{-sz} g(z)\, dz \tag{5.8.1}$$

Since $F(s)$ is a "constant" with respect to z, we can pass $F(s)$ through the integral sign in (5.8.1) to obtain

$$F(s)G(s) = \int_0^\infty e^{-sz} F(s) g(z)\, dz \tag{5.8.2}$$

Now, recall (5.4.2), which we write as

$$e^{-sz}F(s) = \mathcal{L}[u_z(t) f(t-z)] \tag{5.8.3}$$

In terms of the definition of the Laplace transform, (5.8.3) can also be written in integral form as

$$\int_0^\infty u_z(t) f(t-z) e^{-st}\, dt = e^{-sz}F(s) \tag{5.8.4}$$

Using this expression for $e^{-sz}F(s)$ in the integrand of (5.8.2), we have

$$\begin{aligned} F(s)G(s) &= \int_0^\infty \int_0^\infty u_z(t) f(t-z) g(z) e^{-st}\, dt\, dz \\ &= \int_0^\infty \int_z^\infty u_z(t) f(t-z) g(z) e^{-st}\, dt\, dz \end{aligned} \tag{5.8.5}$$

since $u_z(t) = 0$ for $t < z$ and $u_z(t) = 1$ for $t \geq z$. The double integral (5.8.5) may be evaluated by integrating first with respect to t from z to ∞ followed by an integration with respect to z from 0 to ∞. On the other hand, we may integrate first with respect to z from 0 to t followed by an integration with respect to t from 0 to ∞. The upshot of this reversal of the order of integration is the following expression for $F(s)G(s)$:

$$F(s)G(s) = \int_0^\infty \left(\int_0^t f(t-z)g(z)\,dz \right) e^{-st}\,dt \tag{5.8.6}$$

But this implies

$$\mathcal{L}\left[\int_0^t f(t-z)g(z)\,dz \right] = F(s)G(s) \tag{5.8.7}$$

or, equivalently,

$$\mathcal{L}^{-1}\left[F(s)G(s)\right] = \int_0^t f(t-z)g(z)\,dz \tag{5.8.8}$$

We call the integral in (5.8.8) a *convolution* and (5.8.8) a *convolution theorem.* We state this result formally as follows.

Theorem 5.8.1. *Given the acceptable functions $f(t)$ and $g(t)$,*

$$\mathcal{L}^{-1}\left[F(s)G(s)\right] = \int_0^t f(t-z)g(z)\,dz \tag{5.8.9}$$

$$= \int_0^t g(t-z)f(z)\,dz \tag{5.8.10}$$

Proof: Equation (5.8.10) follows by interchanging the roles of f and g in (5.8.9). □

It is common to adopt the following notation for the convolution:

$$f * g = \int_0^t f(t-z)g(z)\,dz \tag{5.8.11}$$

Then the convolution theorem may be expressed as

$$\mathcal{L}\left[f * g\right] = F(s)G(s) \tag{5.8.12}$$

Example 5.8.1. *Use the convolution theorem to find $y(t)$, given*

$$Y(s) = \frac{1}{s^2\left(s^2+1\right)}$$

Solution: Let

$$g(t) = \mathcal{L}^{-1}\left[\frac{1}{s^2+1}\right] = \sin t \quad \text{and} \quad f(t) = \mathcal{L}^{-1}\left[\frac{1}{s^2}\right] = t$$

Now invoke the convolution theorem:

$$\begin{aligned} y(t) &= \mathcal{L}^{-1}\left[\frac{1}{s^2\left(s^2+1\right)}\right] \\ &= \int_0^t (t-z)\sin z\,dz = \left(-(t-z)\cos z - \sin z\right)\Big|_0^t \\ &= -\sin t + t \end{aligned}$$

Example 5.8.2. *Use the convolution theorem to find $y(t)$, given*

$$Y(s) = \frac{1}{(s-a)(s-b)}, \quad a \neq b$$

Solution: Let

$$g(t) = \mathcal{L}^{-1}\left[\frac{1}{s-a}\right] = e^{at}$$

and

$$f(t) = \mathcal{L}^{-1}\left[\frac{1}{s-b}\right] = e^{bt}$$

we have

$$\begin{aligned} y(t) &= \int_0^t e^{a(t-z)}e^{bz}\,dz = e^{at}\int_0^t e^{(b-a)z}\,dz \\ &= e^{at}\,\frac{e^{(b-a)t}-1}{b-a} = \frac{1}{b-a}\left(e^{bt}-e^{at}\right) \end{aligned}$$

Note that this avoids a partial fraction decomposition.

Example 5.8.3. *Use the convolution theorem to find $h(t)$, given*

$$H(s) = \frac{4}{(s+1)\left(s^2+1\right)^2}$$

Solution: Let

$$G(s) = \frac{4}{s+1}$$

and

$$F(s) = \frac{1}{(s^2+1)^2}$$

Then

$$g(t) = \mathcal{L}^{-1}[G(s)] = 4e^{-t}$$

and from Table 5.2, Formula 21,

$$f(t) = \mathcal{L}^{-1}[F(s)] = \frac{1}{2}(\sin t - t\cos t)$$

Now we can use (5.8.10):

$$\begin{aligned} h(t) &= \mathcal{L}^{-1}[F(s)G(s)] = 2\int_0^t e^{-(t-z)}(\sin z - z\cos z)\,dz \\ &= 2e^{-t}\int_0^t e^{z}(\sin z - z\cos z)\,dz \end{aligned}$$

Finally, we resort to a table of integrals to deduce that

$$\int_0^t e^{z}(\sin z - z\cos z)\,dz = \tfrac{1}{2}\left[1 - e^{t}\left((t+1)\cos t + (t-2)\sin t\right)\right]$$

So,

$$h(t) = e^{-t} - (t+1)\cos t - (t-2)\sin t$$

Although we have avoided the partial-fraction decomposition, we still must contend with a tedious integration.

Example 5.8.4. *Let* $\mathcal{L}^{-1}[F(s)] = f(t)$ *and* $G(s) = 1/s$. *Use the convolution theorem to derive*

$$\mathcal{L}^{-1}[F(s)/s] = \int_0^t f(z)\,dz$$

Solution: Since $\mathcal{L}^{-1}[1/s] = 1$, we have, from (5.8.9) with $g(t) = 1$,

$$\mathcal{L}^{-1}[F(s)/s] = \int_0^t f(z)\,dz$$

(See Theorem 5.6.3.)

•EXERCISES

1. Establish the following five theorems.
   (a) $f * g = g * f$.
   (b) $f * (g + h) = f * g + f * h$.
   (c) $f * (kg) = k(f * g)$ for any scalar k.
   (d) $1 * f(t) = \int_0^t f(z)\,dz$.
   (e) $1 * f'(t) = f(t) - f(0)$.
2. Compute $f * g$ for the following five functions.
   (a) $f(t) = g(t) = e^{at}$.
   (b) $f(t) = \sin bt, \quad g(t) = e^{-at}$.
   (c) $f(t) = g(t) = \sin bt$.
   (d) $f(t) = t^2, \quad g(t) = e^{-at}$.
   (e) $f(t) = t^2, \quad g(t) = \sin t$.
3. For arbitrary $g(t)$, find $f(t) * g(t)$.
   (a) $f(t) = u_a(t), \quad a > 0$.
   (b) $f(t) = e^{at}$.
4. Use the convolution theorem to find the inverse Laplace transform of the following functions.

   (a) $\dfrac{1}{(s^2 + a^2)^2}$. (b) $\dfrac{s^2}{(s^2 - a^2)^2}$.

   (c) $\dfrac{1}{s - a}\dfrac{1}{s^2 + b^2}, \quad a \neq b$. (d) $\dfrac{1}{s^2 + a^2}\dfrac{1}{s^2 + b^2}, \quad a^2 \neq b^2$.

5.9 The Solution of Initial-Value Problems

We are now in a position to solve initial-value problems in which the forcing function is an acceptable function. The Laplace transform has the advantage over variation of parameters in solving these problems in that the Laplace transform handles sectionally continuous forcing functions in a natural, systematic way. We demonstrate the use of Laplace transforms in the special case of the second-order ODE. The method is applicable to ODE's of any order, provided that they are linear and have constant coefficients. Consider the familiar initial-value problem

$$u'' + pu' + qu = f(t), \qquad u(0) = b_1, \quad u'(0) = b_2 \tag{5.9.1}$$

where we are assuming that $f(t)$ is acceptable. Let $U(s) = \mathcal{L}[u(t)]$. Take the Laplace transform of each side of (5.9.1). By Theorem 5.3.3 and its corollary,

$$\begin{aligned} F(s) &= \mathcal{L}[u'' + pu' + qu] = \mathcal{L}[u''] + p\mathcal{L}[u'] + q\mathcal{L}[u] \\ &= s^2U(s) - su(0) - u'(0) + p[sU(s) - u(0)] + qU(s) \end{aligned}$$

where $F(s) = \mathcal{L}[f(t)]$. Collecting terms, using $b_1 = u(0)$ and $b_2 = u'(0)$ and solving for $U(s)$ leads to

$$U(s) = \frac{(s+p)b_1 + b_2}{s^2 + ps + q} + \frac{F(s)}{s^2 + ps + q} \tag{5.9.2}$$

The algebraic equation (5.9.2) is called the *subsidiary equation* of (5.9.1).

There are two observations to be made about (5.9.2). First, the denominator of the subsidiary equation is the characteristic polynomial of the complementary equation of (5.9.1). Second, the initial conditions are responsible for the first term on the right-hand side, and the forcing function is responsible for the second term.

The initial conditions $u(0) = u'(0) = 0$ simplify (5.9.2) to

$$U(s) = \frac{F(s)}{s^2 + ps + q} \tag{5.9.3}$$

To find $u(t) = \mathcal{L}^{-1}[U(s)]$, the solution of the initial-value problem, we find the inverse transforms of the two terms on the right of (5.9.2).

Example 5.9.1. *Find the solution of the initial-value problem*

$$u'' + u' - 2u = 4e^t + 1, \qquad u(0) = 1, \quad u'(0) = 0$$

Solution: First we compute the transform of the forcing function:

$$\mathcal{L}\left[4e^t + 1\right] = F(s) = \frac{4}{s-1} + \frac{1}{s}$$

Then the subsidiary equation is given by

$$U(s) = \frac{s+1}{s^2 + s - 2} + \left(\frac{4}{s-1} + \frac{1}{s}\right)\left(\frac{1}{s^2 + s - 2}\right)$$

Since $s^2 + s - 2 = (s+2)(s-1)$, the partial-fraction decomposition is

$$U(s) = \frac{A}{s} + \frac{B}{s+2} + \frac{C}{(s-1)^2} + \frac{D}{s-1}$$

The methods in earlier sections lead to $A = -\frac{1}{2}, B = \frac{17}{18}, C = \frac{4}{3}$, and $D = \frac{5}{9}$. (See Problem 15.) Then,

$$u(t) = -\tfrac{1}{2} + \tfrac{17}{18}e^{-2t} + \tfrac{4}{3}te^t + \tfrac{5}{9}e^t$$

Example 5.9.2. *An inductor of* 2 H *and a capacitor of* 0.02 F *is connected in series with an imposed voltage of* $100 \sin \omega t$ *volts. Determine the charge* $q(t)$ *on the capacitor as a function of* ω *if the initial charge on the capacitor and current in the circuit are both zero.*

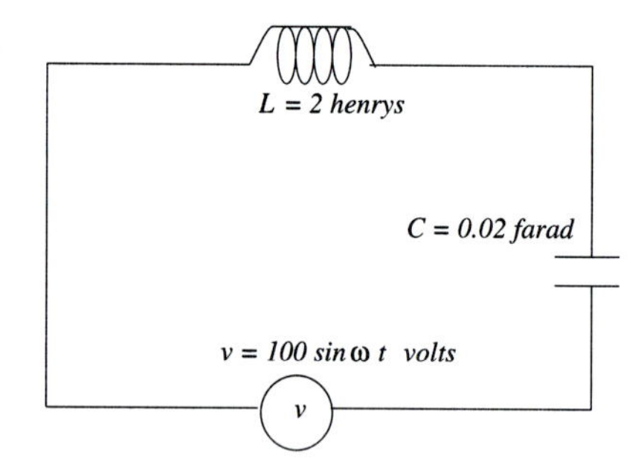

Solution: We use $i = dq/dt$ and Kirchhoff's laws to write the equation governing the charge:

$$2q'' + 50q = 100 \sin \omega t$$

Let $Q(s) = \mathcal{L}[q(t)]$ and note that $q'(0) = i(0) = 0$ and $q(0) = 0$. Hence, as we have seen in (5.9.3),

$$Q(s) = \frac{50\omega}{(s^2 + \omega^2)(s^2 + 25)}$$

From Table 5.1, Formula 27, and a little simplification, we find that

$$q(t) = \frac{10}{25 - \omega^2}(5 \sin \omega t - \sin 5t)$$

valid as long as $\omega \neq 5$.

Note that the amplitude indexamplitude of $q(t)$ tends to ∞ as $\omega \to 5$. If $\omega = 5$

$$Q(s) = \frac{250}{(s^2 + 25)^2}$$

and using Table 5.1 (27), we find

$$q(t) = (\sin 5t - 5t\cos 5t)$$

The case $\omega = 5$ is an illustration of the phenomenon of resonance. The unbounded growth of $q(t)$ as $t \to \infty$ is due to the term $5t\cos 5t$. Similar behavior occurs in all undamped oscillatory systems at resonance.

Example 5.9.3. *Find the motion of a spring-mass system if the mass is displaced* 2 m *and released from rest. Use* $M = 1$ kg, $C = 4$ kg/s, *and* $K = 8$ N/m.

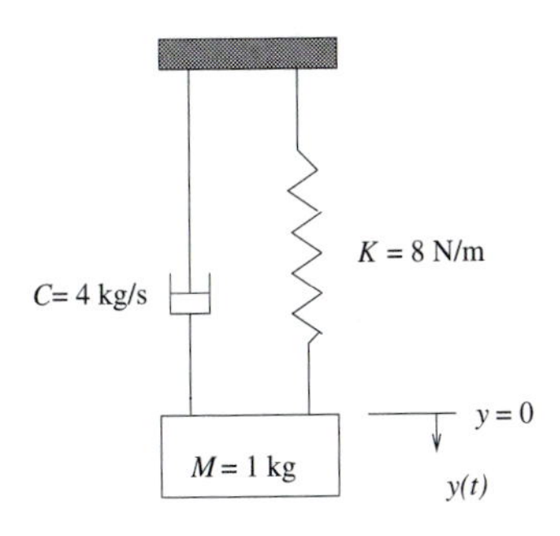

Solution: The initial-value problem describing this damped harmonic motion is

$$y'' + 4y' + 8y = 0, \qquad y(0) = 2, \quad y'(0) = 0$$

The subsidiary equation is found by taking the Laplace transform of both sides of the differential equation. Let $Y(s) = \mathcal{L}\left[y(t)\right]$. Then

$$s^2Y(s) - sy(0) - y'(0) + 4\left(sY(s) - y(0)\right) + 8Y(s) = 0$$

Collecting terms and rearranging leads to

$$Y(s) = 2\frac{s+2+2}{(s+2)^2+4} = 2\frac{s+2}{(s+2)^2+4} + 2\frac{2}{(s+2)^2+4}$$

from which it follows that

$$y(t) = 2e^{-2t}\left(\cos 2t + \sin 2t\right)$$

The Laplace transform can also be used to solve ODE's in which the conditions on the solution are given at more that one point. Such problems are called *boundary-value problems* and are generally much harder to solve than initial-value problems. For one thing, there may not even be solutions! Indeed, it is quite easy to construct such problems. Here is a simple example.

$$u'' - u = 0, \qquad u(0) = 0, \quad u(\pi) = 1$$

The general solution is, of course, $A\cos t + B\sin t$. The two conditions cannot be met by any function in this family. The next example is a case for which there is a solution and the Laplace transform is an aid in finding it.

Example 5.9.4. *A beam of length L and weight w is loaded as shown. The differential equation governing the vertical displacement as a function of x, the distance from the leftmost support, is*

$$y^{(4)} = \frac{w}{EI}$$

If the beam is supported on the right and left, the conditions on the solution are $y(0) = y''(0) = y(L) = y''(L) = 0$. Find the deflection of the beam $y(x)$.

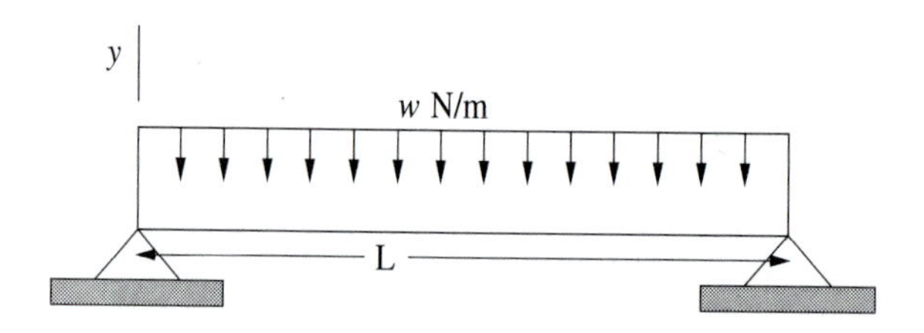

Solution: Write $Y(s) = \mathcal{L}[y(x)]$. The Laplace transform of the given ODE leads to

$$s^4 Y(s) - s^2 y'(0) - y^{(3)}(0) = \frac{w}{EIs}$$

Let $A = y'(0)$ and $B = y^{(3)}(0)$. Then the subsidiary equation is

$$Y(s) = \frac{A}{s^2} + \frac{B}{s^4} + \frac{w}{EIs^5}$$

and hence,

$$y(x) = Ax + \frac{B}{6}x^3 + \frac{1}{24}\frac{w}{EI}x^4$$

The conditions at L can be used to determine A and B, namely

$$y(L) = AL + B\frac{1}{6}L^3 + \frac{1}{24}\frac{w}{EI}L^4 = 0$$

and

$$y''(L) = BL + \frac{1}{2}\frac{w}{EI}L^2 = 0$$

The second equation yields B, and then, using B in the first equation, we obtain A. Specifically,

$$A = \frac{1}{24}\frac{w}{EI}L^3$$

and

$$B = -\frac{1}{2}\frac{w}{EI}L$$

Hence,

$$y(x) = \frac{w}{24EI}(L^3x - 2Lx^3 + x^4)$$

•EXERCISES

Problems 1–14: Determine the solution for each initial-value problem.

1. $u'' + 4u = 0;$ $\quad u(0) = 0, \quad u'(0) = 10.$
2. $u'' - 4u = 0;$ $\quad u(0) = 2, \quad u'(0) = 0.$
3. $u'' + u = 0;$ $\quad u(0) = 0, \quad u'(0) = 2.$
4. $u'' + 4u = 2\cos t;$ $\quad u(0) = 0, \quad u'(0) = 0.$
5. $u'' + 4u = 2\cos 2t;$ $\quad u(0) = 0, \quad u'(0) = 0.$
6. $u'' + u = e^t + 2;$ $\quad u(0) = 0, \quad u'(0) = 0.$
7. $u'' + 5u' + 6u = 0;$ $\quad u(0) = 0, \quad u'(0) = 1.$
8. $u'' + 4u' + 4u = 0;$ $\quad u(0) = 1, \quad u'(0) = 0.$
9. $u'' - 2u' - 8u = 0;$ $\quad u(0) = 1, \quad u'(0) = 0.$
10. $u'' + 5u' + 6u = 0;$ $\quad u(0) = 0, \quad u'(0) = 1.$
11. $u'' + 2u' + u = 2t;$ $\quad u(0) = 0, \quad u'(0) = 0.$
12. $u'' + 4u' + 4u = \sin 2t;$ $\quad u(0) = 0, \quad u'(0) = 0.$
13. $u'' + 4u' + 8u = 10\cos 2t;$ $\quad u(0) = 0, \quad u'(0) = 0.$
14. $u'' + 2u' + 5u = 13e^t \sin t;$ $\quad u(0) = 0, \quad u'(0) = 0.$
15. For the function $U(s)$ of Example 5.9.1, show that
$\mathcal{L}^{-1}[U(s)] = u(t) = -\frac{1}{2} + \frac{17}{18}e^{-2t} + \frac{4}{3}te^t + \frac{5}{9}e^t$

Solve for the displacement $y(t)$ in the figure with Example 5.9.3 if $y(0) = 0$ and $y'(0) = 0$. Use each one of the following friction coefficients: (a) $C = 0$ kg/s (b) $C = 2$ kg/s (c) $C = 24$ kg/s (d) $C = 40$ kg/s with the following forcing functions:

16. $f(t) = 2$ N.
17. $f(t) = 10 \sin 2t$.
18. $f(t) = 10 \sin 6t$.
19. $f(t) = 10(u_0(t) - u_{4\pi}(t))$.
20. $f(t) = 10e^{-0.2t}$.

Use each of these four resistances (a) $R = 0\,\Omega$, (b) $R = 16\,\Omega$, (c) $R = 20\,\Omega$, (d) $R = 25\,\Omega$, in the circuit of Figure 1.2 to calculate the current $i(t)$ if the circuit is quiescent at $t = 0$ (that is, the initial charge on the capacitor is $q(0) = i(0) = 0$). The input voltages are given in Problems 21–25.

21. $v(t) = 10$ V.
22. $v(t) = 10 \sin 10t$.
23. $v(t) = 5 \sin 10t$ V.
24. $v(t) = 10\,[u_0(t) - u_{2\pi}(t)]$.
25. $v(t) = e^{-t}$.

5.10 The Laplace Transform of Systems

The initial-value problem for a system of first-order ODE's is given by

$$\mathbf{x}' = \mathbf{A}\mathbf{x} + \mathbf{f}(t), \qquad \mathbf{x}(0) = \mathbf{b} \tag{5.10.1}$$

Denote by $\mathbf{X}(s)$ and $\mathbf{F}(s)$ the Laplace transforms of $\mathbf{x}(t)$ and $\mathbf{f}(t)$, respectively. Then, from Theorem 5.3.3 and (5.10.1),

$$s\mathbf{X}(s) - \mathbf{x}(0) = \mathbf{A}\mathbf{X}(s) + \mathbf{F}(s) \tag{5.10.2}$$

Therefore,

$$(s\mathbf{I} - \mathbf{A})\,\mathbf{X}(s) = \mathbf{F}(s) + \mathbf{b} \tag{5.10.3}$$

Now set

$$\mathbf{Z}(s) = (s\mathbf{I} - \mathbf{A})^{-1} \tag{5.10.4}$$

Then, as long as s is not an eigenvalue of $\mathbf{A}$, we can solve for $\mathbf{X}(s)$ in (5.10.3) to obtain

$$\mathbf{X}(s) = \mathbf{Z}(s)\,(\mathbf{F}(s) + \mathbf{b}) \tag{5.10.5}$$

The matrix $\mathbf{Z}(s) = (s\mathbf{I} - \mathbf{A})^{-1}$ has an interesting interpretation in terms of the fundamental matrix $\Psi(t)$ associated with (5.10.1). Recall that $\Psi'(t) = \mathbf{A}\Psi(t)$ and $\Psi(0) = \mathbf{I}$, so that the Laplace transform of this matrix differential equation is

$$s\mathcal{L}\,[\Psi(t)] - \Psi(0) = s\mathcal{L}\,[\Psi(t)] - \mathbf{I} = \mathbf{A}\mathcal{L}\,[\Psi(t)]$$

(See Section 2.4)Hence,

$$\mathcal{L}\left[\Psi(t)\right] = (s\mathbf{I} - \mathbf{A})^{-1} \tag{5.10.6}$$

and therefore, from (5.10.4) and (5.10.6), $\mathbf{Z}(s) = \mathcal{L}\left[\Psi(t)\right]$. We state this connection between the fundamental matrix and $(s\mathbf{I} - \mathbf{A})^{-1}$ as a theorem.

Theorem 5.10.1. *The fundamental matrix of the system* $\mathbf{x}' = \mathbf{A}\mathbf{x}$ *is given by*

$$\Psi(t) = \mathcal{L}^{-1}\left[(s\mathbf{I} - \mathbf{A})^{-1}\right] \tag{5.10.7}$$

Proof: The proof is accomplished by taking the inverse transform of $\mathbf{Z}(s) = \mathcal{L}\left[\Psi(t)\right]$. □

Remark: Let $C(s)$ be the characteristic polynomial of $\mathbf{A}$. Then note that

$$\det(s\mathbf{I} - \mathbf{A})^{-1} = \frac{1}{\det(s\mathbf{I} - \mathbf{A})} = (-1)^n \frac{1}{\det(\mathbf{A} - s\mathbf{I})} = (-1)^n \frac{1}{C(s)}$$

One consequence of Theorem 5.10.1 is that we can compute the fundamental matrix and hence a general solution of $\mathbf{x}' = \mathbf{A}\mathbf{x}$, even when the system matrix $\mathbf{A}$ fails to have n linearly independent eigenvectors! Recall that we were unable to do so using the techniques of Chapter 2, except in special cases. The following two examples illustrate how the Laplace transform avoids this obstacle.

Example 5.10.1. *Find the solution to the initial-value problem*

$$\mathbf{x}' = \begin{bmatrix} 1 & 1 \\ -1 & -1 \end{bmatrix}\mathbf{x}, \quad \mathbf{x}(0) = \begin{bmatrix} a \\ b \end{bmatrix}$$

and the fundamental matrix of this system.

Solution: (See Example 2.3.5.) First we find $\mathbf{Z}(s)$ using (5.10.4), and then we use (5.10.5) to find $\mathbf{X}(s)$. Here's the work:

$$\mathbf{Z}(s) = \begin{bmatrix} s-1 & -1 \\ 1 & s+1 \end{bmatrix}^{-1} = s^{-2}\begin{bmatrix} s+1 & 1 \\ -1 & s-1 \end{bmatrix}$$

Since $f(t) = 0$, $\mathbf{F}(s) = \mathbf{0}$ and (5.10.5) becomes

$$\mathbf{X}(s) = \mathbf{Z}(s)\mathbf{b} = s^{-2}\begin{bmatrix} s+1 & 1 \\ -1 & s-1 \end{bmatrix}\begin{bmatrix} a \\ b \end{bmatrix}$$

$$= s^{-2}\begin{bmatrix} as+a+b \\ bs-a-b \end{bmatrix} = s^{-1}\begin{bmatrix} a \\ b \end{bmatrix} + s^{-2}\begin{bmatrix} a+b \\ -a-b \end{bmatrix}$$

The inverse transform of both sides yields the required solution:

$$\mathbf{x}(t) = \begin{bmatrix} a \\ b \end{bmatrix} + t\begin{bmatrix} a+b \\ -a-b \end{bmatrix} = \begin{bmatrix} a+(a+b)t \\ b-(a+b)t \end{bmatrix}$$

Compare this answer with Example 2.3.5. Finally, see also Example 2.4.4:

$$\Psi(t) = \mathcal{L}^{-1}[\mathbf{Z}(s)] = \begin{bmatrix} t+1 & t \\ -t & 1-t \end{bmatrix}$$

Example 5.10.2. *Find the fundamental matrix and the solution of*

$$\mathbf{x}' = \begin{bmatrix} 1 & 1 & 0 \\ 0 & 1 & 1 \\ 0 & 0 & 1 \end{bmatrix}\mathbf{x}, \qquad \mathbf{x}(0) = \begin{bmatrix} 1 \\ 1 \\ 0 \end{bmatrix}$$

Solution: The coefficient matrix of this system has only one linearly independent eigenvector, and hence, the method of Section 2.4 fails. However, using Theorem 5.10.1, we first compute

$$\mathbf{Z}(s) = (s\mathbf{I} - \mathbf{A})^{-1} = \begin{bmatrix} (s-1)^{-1} & (s-1)^{-2} & (s-1)^{-3} \\ 0 & (s-1)^{-1} & (s-1)^{-2} \\ 0 & 0 & (s-1)^{-1} \end{bmatrix}$$

Then, since the inverse transform of each term in $\mathbf{Z}(s)$ is elementary, we find that the fundamental matrix is

$$\Psi(t) = \mathcal{L}^{-1}[\mathbf{Z}(s)] = \begin{bmatrix} e^t & te^t & t^2e^t/2 \\ 0 & e^t & te^t \\ 0 & 0 & e^t \end{bmatrix}$$

Now, since $\mathbf{F}(s) = \mathbf{0}, \mathbf{X}(s) = \mathbf{Z}(s)\mathbf{b}$, and so the transform of the solution is

$$\mathbf{X}(s) = \begin{bmatrix} (s-1)^{-1} & (s-1)^{-2} & (s-1)^{-3} \\ 0 & (s-1)^{-1} & (s-1)^{-2} \\ 0 & 0 & (s-1)^{-1} \end{bmatrix}\begin{bmatrix} 1 \\ 1 \\ 0 \end{bmatrix}$$

$$= \begin{bmatrix} (s-1)^{-1} + (s-1)^{-2} \\ (s-1)^{-1} \\ 0 \end{bmatrix}$$

Thus,

$$\mathbf{x}(t) = \mathcal{L}^{-1}[\mathbf{X}(s)] = \begin{bmatrix} e^t + te^t \\ e^t \\ 0 \end{bmatrix} = e^t \begin{bmatrix} 1+t \\ 1 \\ 0 \end{bmatrix}$$

Example 5.10.3. *Solve the initial-value problem*

$$\mathbf{x}' = \begin{bmatrix} 1 & 0 \\ -1 & 3 \end{bmatrix} \mathbf{x} + \begin{bmatrix} e^{2t} \\ 3 \end{bmatrix}, \qquad \mathbf{x}(0) = \begin{bmatrix} 1 \\ 0 \end{bmatrix}$$

Solution: This is the system given in Example 2.6.5. We compute

$$\mathbf{F}(s) + \mathbf{x}(0) = \begin{bmatrix} (s-2)^{-1} \\ 3s^{-1} \end{bmatrix} + \begin{bmatrix} 1 \\ 0 \end{bmatrix} = \begin{bmatrix} (s-2)^{-1} + 1 \\ 3s^{-1} \end{bmatrix}$$

Also,

$$\mathbf{Z}(s) = \begin{bmatrix} s-1 & 0 \\ 1 & s-3 \end{bmatrix}^{-1} = \frac{1}{(s-1)(s-3)} \begin{bmatrix} s-3 & 0 \\ -1 & s-1 \end{bmatrix}$$

Returning to (5.10.5), we have

$$\mathbf{X}(s) = \mathbf{Z}(s) \begin{bmatrix} (s-2)^{-1} + 1 \\ 3s^{-1} \end{bmatrix} = \frac{1}{s(s-2)} \begin{bmatrix} s \\ 2 \end{bmatrix}$$

Hence, the desired solution is

$$\mathbf{x}(t) = \mathcal{L}^{-1}[\mathbf{X}(s)] = \begin{bmatrix} e^{2t} \\ e^{2t} - 1 \end{bmatrix} = e^{2t} \begin{bmatrix} 1 \\ 1 \end{bmatrix} + \begin{bmatrix} 0 \\ -1 \end{bmatrix}$$

Example 5.10.4. *Find the solution of the system*

$$\mathbf{x}' = \begin{bmatrix} 1 & 1 \\ -1 & 1 \end{bmatrix} + \begin{bmatrix} \cos t \\ -\sin t \end{bmatrix}, \qquad \mathbf{x}(0) = \begin{bmatrix} -1 \\ 0 \end{bmatrix}$$

Solution: As in Example 5.10.3,

$$\mathbf{F}(s) + \mathbf{x}(0) = \frac{1}{s^2+1} \begin{bmatrix} -s^2 + s - 1 \\ -1 \end{bmatrix}$$

Now let $C(s) = \det(s\mathbf{I} - \mathbf{A}) = s^2 - 2s + 2$. Then,

$$\mathbf{Z}(s) = C^{-1}(s) \begin{bmatrix} s-1 & 1 \\ -1 & s-1 \end{bmatrix}$$

and hence,

$$\begin{aligned} \mathbf{X}(s) &= \mathbf{Z}(s)\left(\mathbf{F}(s) + \mathbf{x}(0)\right) \\ &= \frac{C^{-1}(s)}{s^2+1} \begin{bmatrix} s-1 & 1 \\ -1 & s-1 \end{bmatrix} \begin{bmatrix} -s^2+s-1 \\ -1 \end{bmatrix} \\ &= \begin{bmatrix} -\dfrac{s}{s^2+1} \\ \dfrac{1}{s^2+1} \end{bmatrix} \end{aligned}$$

Taking inverse transforms of each entry in $\mathbf{X}(s)$ leads to

$$\mathbf{x}(t) = \begin{bmatrix} -\cos t \\ \sin t \end{bmatrix}$$

Example 5.10.5. *Find the solution of the system*

$$\mathbf{x}' = \begin{bmatrix} 2 & 0 & 1 \\ 0 & 2 & 1 \\ 0 & 0 & -1 \end{bmatrix} \mathbf{x}, \qquad \mathbf{x}(0) = \begin{bmatrix} 1 \\ 1 \\ 1 \end{bmatrix}$$

Solution: From (5.10.5) with $\mathbf{F}(s) = \mathbf{0}$,

$$\mathbf{X}(s) = \begin{bmatrix} s-2 & 0 & -1 \\ 0 & s-2 & -1 \\ 0 & 0 & s+1 \end{bmatrix}^{-1} \begin{bmatrix} 1 \\ 1 \\ 1 \end{bmatrix}$$

Setting $C(s) = \det(s\mathbf{I} - \mathbf{A}) = (s+1)(s-2)^2$, we obtain

$$\begin{aligned} \mathbf{Z}(s) &= \begin{bmatrix} s-2 & 0 & -1 \\ 0 & s-2 & -1 \\ 0 & 0 & s+1 \end{bmatrix}^{-1} \\ &= \frac{s-2}{C(s)} \begin{bmatrix} s+1 & 0 & 1 \\ 0 & s+1 & 1 \\ 0 & 0 & s-2 \end{bmatrix} \end{aligned}$$

so that

$$\begin{aligned}
\mathbf{X}(s) &= \mathbf{Z}(s)\begin{bmatrix} 1 \\ 1 \\ 1 \end{bmatrix} = \frac{s-2}{C(s)}\begin{bmatrix} s+2 \\ s+2 \\ s-2 \end{bmatrix} \\
&= \frac{s^2-4}{C(s)}\begin{bmatrix} 1 \\ 1 \\ 0 \end{bmatrix} + \frac{(s-2)^2}{C(s)}\begin{bmatrix} 0 \\ 0 \\ 1 \end{bmatrix} \\
&= \frac{s+2}{(s+1)(s-2)}\begin{bmatrix} 1 \\ 1 \\ 0 \end{bmatrix} + \frac{1}{s+1}\begin{bmatrix} 0 \\ 0 \\ 1 \end{bmatrix}
\end{aligned}$$

It follows from the partial fraction decomposition

$$\frac{s+2}{(s+1)(s-2)} = \frac{4}{3}\frac{1}{s-2} - \frac{1}{3}\frac{1}{s+1}$$

that the solution of the initial-value problem is

$$\mathbf{x}(t) = \frac{4}{3}\begin{bmatrix} 1 \\ 1 \\ 0 \end{bmatrix} e^{2t} - \frac{1}{3}\begin{bmatrix} 1 \\ 1 \\ 0 \end{bmatrix} e^{-t} + \begin{bmatrix} 0 \\ 0 \\ 1 \end{bmatrix} e^{-t}$$

•EXERCISES

Problems 1–18: Use (5.10.7) to find a fundamental matrix of $\mathbf{x}' = \mathbf{A}\mathbf{x}$.

1. $\mathbf{A} = \begin{bmatrix} 1 & 1 \\ -1 & 1 \end{bmatrix}$.

2. $\mathbf{A} = \begin{bmatrix} 0 & 2 \\ 1 & -1 \end{bmatrix}$.

3. $\mathbf{A} = \begin{bmatrix} 0 & 1 \\ 1 & 0 \end{bmatrix}$.

4. $\mathbf{A} = \begin{bmatrix} 1 & 0 \\ -1 & 3 \end{bmatrix}$.

5. $\mathbf{A} = \begin{bmatrix} 2 & 1 \\ 3 & 0 \end{bmatrix}$.

6. $\mathbf{A} = \begin{bmatrix} 2 & -1 \\ 0 & 1 \end{bmatrix}$.

7. $\mathbf{A} = \begin{bmatrix} 1 & 2 \\ 1 & 2 \end{bmatrix}$.

8. $\mathbf{A} = \begin{bmatrix} 1 & 2 \\ 1 & 0 \end{bmatrix}$.

9. $\mathbf{A} = \begin{bmatrix} 2 & 5 \\ 3 & 0 \end{bmatrix}$.

10. $\mathbf{A} = \begin{bmatrix} 1 & 1 \\ 1 & 1 \end{bmatrix}$.

11. $\mathbf{A} = \begin{bmatrix} 0 & 1 \\ -1 & 2 \end{bmatrix}$.

12. $\mathbf{A} = \begin{bmatrix} 4 & 1 \\ -1 & 2 \end{bmatrix}$.

13. $\mathbf{A} = \begin{bmatrix} 2 & 1 \\ -1 & 2 \end{bmatrix}$.

14. $\mathbf{A} = \begin{bmatrix} 1 & 3 \\ -1 & 3 \end{bmatrix}$.

15. $\mathbf{A} = \begin{bmatrix} 1 & 1 \\ 3 & -1 \end{bmatrix}$.

16. $\mathbf{A} = \begin{bmatrix} 2 & 1 \\ 2 & 1 \end{bmatrix}$.

17. $\mathbf{A} = \begin{bmatrix} 1 & a^2 - 1 \\ 1 & -1 \end{bmatrix}$, $a \neq 0$.

18. $\mathbf{A} = \begin{bmatrix} 2 & 1 \\ 0 & 3 \end{bmatrix}$.

Find a fundamental solution matrix of $\mathbf{x}' = \mathbf{A}\mathbf{x}$ for $\mathbf{A}$ in Problems 19–32.

19. $\mathbf{A} = \begin{bmatrix} 4 & -3 & -2 \\ 2 & -1 & -2 \\ 3 & -3 & -1 \end{bmatrix}$.

20. $\mathbf{A} = \begin{bmatrix} 1 & 1 & 1 \\ 0 & 2 & 1 \\ 0 & 0 & 0 \end{bmatrix}$.

21. $\mathbf{A} = \begin{bmatrix} 1 & 1 & 1 \\ 1 & -1 & -1 \\ 0 & 0 & 0 \end{bmatrix}$.

22. $\mathbf{A} = \begin{bmatrix} 1 & -1 & -1 \\ 0 & 0 & 1 \\ 0 & -2 & -3 \end{bmatrix}$.

23. $\mathbf{A} = \begin{bmatrix} 0 & 1 & 0 \\ 0 & 0 & 1 \\ 1 & -3 & 3 \end{bmatrix}$.

24. $\mathbf{A} = \begin{bmatrix} 1 & 3 & 1 \\ 0 & 1 & -1 \\ 0 & 0 & 1 \end{bmatrix}$.

25. $\mathbf{A} = \begin{bmatrix} 1 & 1 & 1 \\ 1 & 1 & 1 \\ 1 & 1 & 1 \end{bmatrix}$.

26. $\mathbf{A} = \begin{bmatrix} -1 & 0 & 2 \\ 1 & 1 & 1 \\ 2 & 0 & -1 \end{bmatrix}$.

27. $\mathbf{A} = \begin{bmatrix} 2 & 1 & 0 \\ 0 & 2 & 0 \\ 0 & 0 & -1 \end{bmatrix}$.

28. $\mathbf{A} = \begin{bmatrix} 2 & 0 & 0 \\ 0 & 2 & 1 \\ 0 & 0 & -1 \end{bmatrix}$.

29. $\mathbf{A} = \begin{bmatrix} 1 & 1 & 0 & 0 \\ 0 & 1 & 0 & 0 \\ 0 & 0 & 1 & 0 \\ 0 & 0 & 0 & 1 \end{bmatrix}$.

30. $\mathbf{A} = \begin{bmatrix} 1 & 1 & 0 & 0 \\ 0 & 1 & 0 & 0 \\ 0 & 0 & 1 & 1 \\ 0 & 0 & 0 & 1 \end{bmatrix}$.

31. $\mathbf{A} = \begin{bmatrix} 1 & 1 & 0 & 0 \\ 0 & 1 & 1 & 0 \\ 0 & 0 & 1 & 0 \\ 0 & 0 & 0 & 1 \end{bmatrix}$.

32. $\mathbf{A} = \begin{bmatrix} 1 & 1 & 0 & 0 \\ 0 & 1 & 1 & 0 \\ 0 & 0 & 1 & 1 \\ 0 & 0 & 0 & 1 \end{bmatrix}$.

33. Solve the initial-value problems using the method of this section.

(a) $\mathbf{x}' = \begin{bmatrix} 1 & 1 \\ -1 & 1 \end{bmatrix} \mathbf{x} + e^{2t} \begin{bmatrix} 1 \\ 0 \end{bmatrix}$, $\mathbf{x}(0) = \begin{bmatrix} 0 \\ 0 \end{bmatrix}$.

(b) $\mathbf{x}' = \begin{bmatrix} 0 & 2 \\ 1 & -1 \end{bmatrix} \mathbf{x} + \sin t \begin{bmatrix} 1 \\ 1 \end{bmatrix}$, $\mathbf{x}(0) = \begin{bmatrix} 1 \\ 0 \end{bmatrix}$.

(c) $\mathbf{x}' = \begin{bmatrix} 0 & 2 \\ 1 & -1 \end{bmatrix} \mathbf{x} + \sin t \begin{bmatrix} 1 \\ 1 \end{bmatrix}, \quad \mathbf{x}(0) = \begin{bmatrix} 0 \\ 1 \end{bmatrix}.$

(d) $\mathbf{x}' = \begin{bmatrix} 2 & 1 \\ 3 & 0 \end{bmatrix} \mathbf{x} + t \begin{bmatrix} 1 \\ 1 \end{bmatrix}, \quad \mathbf{x}(0) = \begin{bmatrix} 0 \\ 0 \end{bmatrix}.$

(e) $\mathbf{x}' = \begin{bmatrix} 2 & 1 \\ 3 & 0 \end{bmatrix} \mathbf{x} + t \begin{bmatrix} 1 \\ 1 \end{bmatrix}, \quad \mathbf{x}(0) = \begin{bmatrix} 1 \\ 0 \end{bmatrix}.$

34. Solve the initial-value problems using the method of this section.

(a) $\mathbf{x}' = \begin{bmatrix} 0 & 1 \\ -1 & 2 \end{bmatrix} \mathbf{x} + \begin{bmatrix} 1 \\ 1 \end{bmatrix}, \quad \mathbf{x}(0) = \begin{bmatrix} 0 \\ 0 \end{bmatrix}.$

(b) $\mathbf{x}' = \begin{bmatrix} 0 & 1 \\ -1 & 2 \end{bmatrix} \mathbf{x} + e^{-t} \begin{bmatrix} 1 \\ 1 \end{bmatrix}, \quad \mathbf{x}(0) = \begin{bmatrix} 1 \\ 0 \end{bmatrix}.$

(c) $\mathbf{x}' = \begin{bmatrix} 2 & 1 & 0 \\ 0 & 2 & 0 \\ 0 & 0 & -1 \end{bmatrix} \mathbf{x} + \begin{bmatrix} e^t \\ 1 \\ 0 \end{bmatrix}, \quad \mathbf{x}(0) = \begin{bmatrix} 0 \\ 0 \\ 0 \end{bmatrix}.$

(d) $\mathbf{x}' = \begin{bmatrix} 2 & 1 & 0 \\ 0 & 2 & 0 \\ 0 & 0 & -1 \end{bmatrix} \mathbf{x} + \begin{bmatrix} e^t \\ 1 \\ 0 \end{bmatrix}, \quad \mathbf{x}(0) = \begin{bmatrix} 1 \\ 0 \\ 0 \end{bmatrix}.$

(e) $\mathbf{x}' = \begin{bmatrix} 2 & 1 & 0 \\ 0 & 2 & 0 \\ 0 & 0 & -1 \end{bmatrix} \mathbf{x} + \begin{bmatrix} e^t \\ 1 \\ 0 \end{bmatrix}, \quad \mathbf{x}(0) = \begin{bmatrix} 0 \\ 1 \\ 0 \end{bmatrix}.$

35. Repeat Problem 33.

(a) $\mathbf{x}' = \begin{bmatrix} 2 & 1 & 0 \\ 0 & 2 & 0 \\ 0 & 0 & -1 \end{bmatrix} \mathbf{x} + \begin{bmatrix} e^t \\ 1 \\ 0 \end{bmatrix}, \quad \mathbf{x}(0) = \begin{bmatrix} 0 \\ 0 \\ 1 \end{bmatrix}.$

(b) $\mathbf{x}' = \begin{bmatrix} 1 & 1 & 1 \\ 1 & -1 & -1 \\ 0 & 0 & 0 \end{bmatrix} \mathbf{x}+, \quad \mathbf{x}(0) = \begin{bmatrix} 0 \\ 1 \\ 0 \end{bmatrix}.$

(c) $\mathbf{x}' = \begin{bmatrix} -1 & 0 & 2 \\ 1 & 1 & 1 \\ 2 & 0 & -1 \end{bmatrix} \mathbf{x} + \begin{bmatrix} 1 \\ 0 \\ 0 \end{bmatrix}, \quad \mathbf{x}(0) = \begin{bmatrix} 0 \\ 0 \\ 0 \end{bmatrix}.$

5.11 Tables of the Laplace and Inverse Laplace Transforms

This section contains tables of Laplace and inverse Laplace transforms. To use these tables effectively, it is often necessary to perform some preliminary alterations on the form of the function. For example, $\mathcal{L}[t \sinh t]$ does not appear explicitly in Table 5.1. However, Formula 6 in Table 5.1 shows that, with $n = 1$,

$$\mathcal{L}[t \sinh t] = -\frac{d}{ds}\mathcal{L}[\sinh t]$$

and Formula 21 in the same table gives

$$\mathcal{L}[\sinh t] = \frac{1}{s^2 - 1}$$

Therefore,

$$\mathcal{L}[t \sinh t] = -\frac{d}{ds}\left(\frac{1}{s^2 - 1}\right) = \frac{2s}{(s^2 - 1)^2}$$

Likewise, one cannot find

$$\mathcal{L}^{-1}\left[\frac{1}{(s-1)(s^2+4)}\right]$$

in Table 5.2. However, from Formula 1 of Table 5.1,

$$\mathcal{L}^{-1}\left[\frac{1}{(s-1)(s^2+4)}\right] = e^t \mathcal{L}^{-1}\left[\frac{1}{s((s+1)^2+4)}\right]$$

and from Formula 23 of Table 5.1,

$$\mathcal{L}^{-1}\left[\frac{1}{(s+1)^2+4}\right] = \frac{1}{2}e^{-t}\sin 2t$$

Finally, from Formula 7 of this table,

$$\mathcal{L}^{-1}\left[\frac{1}{s((s+1)^2+4)}\right] = \frac{1}{2}\int_0^t e^{-z}\sin 2z\, dz$$

So,

$$\mathcal{L}^{-1}\left[\frac{1}{(s-1)(s^2+4)}\right] = \frac{1}{2}e^t \int_0^t e^{-z}\sin 2z\, dz$$

This same result could be reached by utilizing partial fractions or, most easily, by invoking the convolution theorem, Formula 7 in Table 5.2.

Many of the formulas in Tables 5.1 and 5.2 are the results of the examples, theorems, and problem sets presented in the chapter. Those that cannot be derived from the theorems in the chapter are taken from a more extensive table given in Abramowitz and Stegan, *Handbook of Mathematical Functions*, U.S Department of Commerce, National Bureau of Standards, Applied Mathematical Series, 55, June, 1964. In Table 5.1, [AS] refers to this reference, and a number in parentheses refers to the formula with that number in the table.

Table 5.2 is a rewritten version of Table 5.1 with emphasis given to determining $f(t)$ from its Laplace transform. The references in Table 5.2 are to the corresponding formulas in Table 5.1.

TABLE 5.1 A BRIEF TABLE OF LAPLACE TRANSFORMS

	f(t)	**F(s)**	**Reference**
1.	$e^{at}f(t)$	$F(s-a)$	Theorem 5.3.4
2.	$\cos bt f(t)$	$\text{Re}\,(F(s-ib))$	[AS]
3.	$\sin bt f(t)$	$\text{Im}\,(F(s-ib))$	[AS]
4.	$f\,(t/a)$	$a\,F(as)$	[AS]
5.	$f^{(n)}(t)$	$s^nF(s)-\sum_{k=1}^{n} s^{n-k}f^{(k-1)}(0)$	Theorem 5.3.3
6.	$t^nf(t)$	$(-1)^nF^{(n)}(s)$	Corollary 5.3.1
7.	$\int_0^t f(z)\,dz$	$\frac{1}{s}F(s)$	Corollary 5.3.3
8.	$e^{-bt/a}f\,(t/a)$	$a\,F(as+b)$	(1) and (4)
9.	$f(t)u_a(t)$	$e^{-as}\mathcal{L}\,[f(t+a)]$	Theorem 5.4.1
10.	$\int_0^t f(z)g(t-z)\,dz$	$F(s)G(s)$	Theorem 5.8.1
11.	t^n	$\frac{n!}{s^{n+1}}, \quad n=0,1,\dots$	Example 5.3.3
12.	t^{x-1}	$\frac{\Gamma(x)}{s^x}, x>0$	[AS]
13.	t^ne^{at}	$\frac{n!}{(s-a)^{n+1}}$	(11) and (1)
14.	$t^{x-1}e^{at}$	$\frac{\Gamma}{(s-a)^x}, x>0$	[AS]
15.	$e^{at}-e^{bt}$	$\frac{a-b}{(s-a)(s-b)}, \quad a\neq b$	Example 5.7.3
16.	$ae^{at}-be^{bt}$	$\frac{(a-b)s}{(s-a)(s-b)}, \quad a\neq b$	Example 5.7.3
17.	$u_a(t)$	$\frac{1}{s}e^{-as}$	Example 5.4.4
18.	$\ln t$	$\frac{1}{s}\,(\ln s-0.5772156...)$	[AS]

19. $\sin bt$	$\dfrac{b}{s^2+b^2}$	Example 5.2.3
20. $\cos bt$	$\dfrac{s}{s^2+b^2}$	Example 5.2.3
21. $\sinh bt$	$\dfrac{b}{s^2-b^2}$	(15)
22. $\cosh bt$	$\dfrac{s}{s^2-b^2}$	(15)
23. $e^{at}\sin bt$	$\dfrac{b}{(s-a)^2+b^2}$	(19) and (1)
24. $e^{at}\cos bt$	$\dfrac{s-a}{(s-a)^2+b^2}$	(20) and (1)
25. $1-\cos bt$	$\dfrac{b^2}{s\left(s^2+b^2\right)}$	(20) and (11)
26. $bt-\sin bt$	$\dfrac{b^3}{s^2\left(s^2+b^2\right)}$	Corollary 5.3.3 and (25)
27. $\sin bt-bt\cos bt$	$\dfrac{2b^3}{\left(s^2+b^2\right)^2}$	[AS]
28. $t\sin bt$	$\dfrac{bs}{\left(s^2+b^2\right)^2}$	Example 5.3.5
29. $t\cos bt$	$\dfrac{s^2-b^2}{\left(s^2+b^2\right)^2}$	(20) and (6)
30. $\sin bt+bt\ \cos bt$	$\dfrac{2bs^2}{\left(s^2+b^2\right)^2}$	Example 5.3.6
31. $2\cos bt-bt\ \sin bt$	$\dfrac{s^3}{\left(s^2+b^2\right)^2}$	[AS]
32. $\sin at-\sin bt$	$\dfrac{b^2-a^2}{\left(s^2+a^2\right)\left(s^2+b^2\right)},\ a^2\neq b^2$	Example 5.7.6
33. $\cos at-\cos bt$	$\dfrac{s\left(b^2-a^2\right)}{\left(s^2+a^2\right)\left(s^2+b^2\right)},\ a^2\neq b^2$	Example 5.7.6
34. $\sin bt\ \cosh bt$	$2b\dfrac{b^2+s^2}{s^4+4b^4}$	(3) and (22)

35.	$\cos bt \sinh bt$	$2b\dfrac{s^2-b^2}{s^4+4b^4}$	(2) and (21)
36.	$\sin bt \cosh bt$ $- \cos bt \sinh bt$	$b^3\dfrac{4}{s^4+4b^4}$	(34) and (35)
37.	$\sin bt \cosh bt$ $+ \cos bt \sinh bt$	$b\dfrac{4s^2}{s^4+4b^4}$	(34) and (35)
38.	$\sin bt \sinh bt$	$b^2\dfrac{2s}{s^4+4b^4}$	(3) and (21)
39.	$\sinh bt - \sin bt$	$b^3\dfrac{2}{s^4-b^4}$	(19) and (21)
40.	$\cosh bt - \cos bt$	$b^2\dfrac{2s}{s^4-b^4}$	(20) and (22)

TABLE 5.2 A BRIEF TABLE OF INVERSE LAPLACE TRANSFORMS

	$F(s)$	$f(t)$	**Table 5.1**
1.	$F(s-a)$	$e^{at}\mathcal{L}^{-1}[F(s)]$	(1)
2.	$a\,F(as)$	$f(t/a)$	(4)
3.	$F^{(n)}(s)$	$(-t)^n\mathcal{L}^{-1}[F(s)]$	(6)
4.	$\frac{1}{s}F(s)$	$\int_0^t f(z)\,dz$	(7)
5.	$a\,F(as+b)$	$e^{-bt/a}f(t/a)$	(8)
6.	$e^{-as}F(s)$	$f(t-a)u_a(t)$	(9)
7.	$F(s)G(s)$	$\int_0^t f(z)g(t-z)\,dz$	(10)
8.	$\frac{n!}{s^{n+1}}, n=0,1,\ldots$	t^n	(11)
9.	$\frac{n!}{(s-a)^{n+1}}, n=0,1,\ldots$	$t^n e^{at}$	(13)
10.	$\frac{a-b}{(s-a)(s-b)}$	$e^{at}-e^{bt}$	(15)
11.	$\frac{(a-b)s}{(s-a)(s-b)}$	$ae^{at}-be^{bt}$	(16)
12.	$\frac{1}{s}e^{-as}$	$u_a(t)$	(17)
13.	$\frac{b}{s^2+b^2}$	$\sin bt$	(19)
14.	$\frac{s}{s^2+b^2}$	$\cos bt$	(20)
15.	$\frac{b}{s^2-b^2}$	$\sinh bt$	(21)
16.	$\frac{s}{s^2-b^2}$	$\cosh bt$	(22)
17.	$\frac{b}{(s-a)^2+b^2}$	$e^{at}\sin bt$	(23)

18.	$\dfrac{s-a}{(s-a)^2+b^2}$	$e^{at}\cos bt$	(24)
19.	$\dfrac{b^2}{s\,(s^2+b^2)}$	$1-\cos bt$	(25)
20.	$\dfrac{b^3}{s^2\,(s^2+b^2)}$	$bt-\sin bt$	(26)
21.	$\dfrac{b^3}{(s^2+b^2)^2}$	$\dfrac{1}{2}(\sin bt-bt\cos bt)$	(27)
22.	$\dfrac{bs}{(s^2+b^2)^2}$	$\dfrac{1}{2}t\sin bt$	(28)
23.	$\dfrac{bs^2}{(s^2+b^2)^2}$	$\dfrac{1}{2}(\sin bt+bt\cos bt)$	(30)
24.	$\dfrac{s^3}{(s^2+b^2)^2}$	$2\cos bt-bt\sin bt$	(31)

PROJECTS

1. Trigonometric Shifting Theorems

The first shifting property, Theorem 5.3.4, has a useful extension obtained by allowing c to be complex. Set $c = a + ib$. Then, by Euler's formulas,

$$\text{Re}\left(e^{(a+ib)t}\right) = e^{at}\cos bt \quad \text{and} \quad \text{Im}\left(e^{(a+ib)t}\right) = e^{at}\sin bt$$

(a) Using these results, establish

$$\text{Re}\,\mathcal{L}\left[e^{(a+ib)t}f(t)\right] = \mathcal{L}\left[e^{at}\cos bt f(t)\right]$$

and

$$\text{Im}\,\mathcal{L}\left[e^{(a+ib)t}f(t)\right] = \mathcal{L}\left[e^{at}\sin bt f(t)\right]$$

(b) Use (a) to establish

$$\mathcal{L}\left[\cos bte^{at}f(t)\right] = \text{Re}\left(F(s-a-ib)\right)$$

(c) Use (a) to establish

$$\mathcal{L}\left[\sin bte^{at}f(t)\right] = \text{Im}\left(F(s-a-ib)\right)$$

(d) Use (b) to establish

$$\mathcal{L}\left[\cos bt f(t)\right] = \text{Re}\left(F(s-ib)\right)$$

(e) Use (c) to establish

$$\mathcal{L}\left[\sin bt f(t)\right] = \text{Im}\left(F(s-ib)\right)$$

Use (d) and (e) with the appropriate choice of $f(t)$ to derive the following transforms:

(f) $\mathcal{L}\left[e^{at}\sin bt\right] = \dfrac{b}{(s-a)^2+b^2}$

(g) $\mathcal{L}\left[e^{at}\cos bt\right] = \dfrac{s-a}{(s-a)^2+b^2}$

(h) $\mathcal{L}\left[t\cos bt\right] = \dfrac{s^2-b^2}{\left(s^2+b^2\right)^2}$

(i) $\mathcal{L}\left[t\sin bt\right] = \dfrac{2bs}{\left(s^2+b^2\right)^2}$

2. The Laplace Transform of Series

Suppose that

$$f(t) = \sum_{n=0}^{\infty} a_n t^n$$

converges for all t and $f(t)$ has exponential order. Then

$$F(s) = \mathcal{L}[f] = \sum_{n=0}^{\infty} a_n \mathcal{L}[t^n] = \sum_{n=0}^{\infty} n! a_n s^{-n-1}$$

(a) Use this result to show that

$$\mathcal{L}\left[\frac{e^{-t} - 1}{t}\right] = -\ln\left(1 + \frac{1}{s}\right)$$

(b) The error function $\operatorname{erf}(t)$ is defined by

$$\operatorname{erf}(t) = \frac{2}{\sqrt{\pi}} \int_0^t \exp\left(-x^2\right) dx$$

Show that

$$\mathcal{L}\left[t^{-1/2}\operatorname{erf}\left(\sqrt{t}\right)\right] = \frac{2}{\sqrt{\pi s}} \tan^{-1} \frac{1}{\sqrt{s^2+1}}$$

(*Hint*: Expand $\exp(-x^2)$ in a power series about $x = 0$.)

(c) Show that

$$\mathcal{L}\left[\sin\sqrt{t}\right] = \sqrt{\pi}\frac{e^{-1/4s}}{2s^{3/2}}$$

(d) Show that

$$\mathcal{L}\left[\frac{1}{t}\sin(kt)\right] = \tan^{-1}\frac{k}{s}$$

Chapter 6

Series Methods

6.1 Introduction

The solutions of constant coefficient homogeneous ODE's are combinations of polynomials and exponential functions. By contrast, the Cauchy-Euler equation, a linear ODE with variable coefficients, has solutions that may involve $\ln t$ as well as sums of various nonintegral powers of t. For example, a general solution of $t^2u'' - tu' + u = 0$ is

$$u(t) = (c_1 + c_2 \ln t)\, t, \qquad t > 0 \tag{6.1.1}$$

In fact it often happens that linear ODE's with variable coefficients have solutions which are functions not usually treated in a first course in calculus. Indeed, variable coefficient linear ODE's are frequently a source of new functions. Two important illustrations of this are the Bessel equation (Section 6.7) and the Legendre equation (Section 6.5).

One common method of studying the solutions of variable coefficient linear ODE's is to construct their Taylor series expansions. In this chapter we shall see how to find these expansions from the ODE's they satisfy. We begin with an examination of the definition and properties of Taylor series. The reader is referred to standard calculus texts for a review of this topic.

6.2 Analytic Functions

The set of complex t satisfying the inequality

$$|t - c| \le r, \quad r > 0 \tag{6.2.1}$$

is a disk centered at $t = c$ with radius r, while the set of complex t satisfying

$$|t - c| = r \tag{6.2.2}$$

is the circular boundary of this disk. The set $|t - c| < r$ is called a *neighborhood* of c and is the disk without its bounding circle.

If t and c are restricted to the real numbers, the neighborhood (6.2.1) is an interval of length $2r$ with midpoint c. We are interested in functions all of whose derivatives exist in some neighborhood of $t = c$. Such functions play a central role in mathematics and are described as *analytic* at c. Points at which $f(t)$ fails to be analytic are called *singular points* of $f(t)$. So, for example, the rational function

$$f(t) = \frac{t}{(t^2 + 9)(t - 6)} \tag{6.2.3}$$

is singular at $t = \pm 3i$ and $t = 6$, since $f(t)$ is undefined at these three points.

It is not our intention to develop a theory of analytic functions; such a task is far beyond the reasonable expectations for a first course in ODE's. On the other hand, we do need a number of results from this theory which we shall state without proof. Most of these are reasonable extensions of the analogous results from real variables. In particular, we freely use all the derivative and integration rules from calculus.

•EXERCISES

Determine the largest circle about the origin for which the following functions are necessarily analytic functions of a real variable. (Assume that all constants are real.)

1. e^{-t}.
2. $\cos bt$.
3. $\ln(1 + t)$.
4. $\sqrt{1 + t^2}$.
5. The polynomial $p_n(t) = a_n t^n + A_{n-1} t^{n-1} + \cdots + a_1 t + a_0$.
6. $p_n(e^t)$, where $p_n(t)$ is the polynomial in Problem 5.
7. $\ln(1 + t^2)$.
8. $t^{-x}, \quad x > 0$.
9. $\ln t$.
10. $\tan t$.
11. $\sec t$.
12. $\ln(\sin t)$.
13. $\sinh t$.
14. $\tanh t$.
15. $f^{(n)}(t)$, where $f(t)$ is analytic in $|t| < r$.
16. $\int_0^t f(z)\,dz$, where $f(z)$ is analytic for all z.
17. $f(t) + g(t)$, where $f(t)$ and $g(t)$ are analytic for all t.
18. $f(t)g(t)$, where $f(t)$ and $g(t)$ are analytic for all t.
19. $f(g(t))$, where $f(t)$ and $g(t)$ are analytic for all t.

20. $f(t)/g(t)$, assuming that $f(t)$ and $g(t)$ are analytic for all t, that $g(t) \neq 0$ except at $t_0 \neq 0$, and that $f(t_0) \neq 0$.
21. $\ln(2 - e^t)$.
22. $\sqrt{t^2 + 5t + 4}$.

6.3 Taylor Series of Analytic Functions

Suppose that $f(t)$ is analytic at $t = 0$. The sums

$$\begin{aligned} f_0(t) &= f(0) \\ f_1(t) &= f(0) + f'(0)t \\ &\vdots \\ f_k(t) &= f(0) + f^{(1)}(0)t + \cdots + \frac{1}{k!}f^{(k)}(0)t^k \end{aligned} \tag{6.3.1}$$

are called the *partial sums* of the infinite series

$$f(0) + f^{(1)}(0)t + \cdots + \frac{1}{k!}f^{(k)}(0)t^k + \dots \tag{6.3.2}$$

The infinite series (6.3.2) is called the *Taylor series* of $f(t)$ about $t = 0$. (Series expanded about $t = 0$ are often called *Maclaurin series.*) We say that the Taylor series *converges* at t_0 to $f(t_0)$ if

$$\lim_{k \to \infty} f_k(t_0) = f(t_0) \tag{6.3.3}$$

If this limit does not exist, we say that the Taylor series (6.3.2) *diverges* at t_0. Our first theorem provides a criterion under which (6.3.2) converges to $f(t)$ in a neighborhood of $t_0 = 0$.

Theorem 6.3.1. *Suppose $f(t)$ is analytic at 0 and R is the distance from 0 to the nearest singular point of $f(t)$. Then the Taylor series of $f(t)$ about $t = 0$ converges to $f(t)$ in the interval $|t| < R$ and diverges in $|t| > R$.*

Remarks:

(1) R is called the *radius of convergence* of the Taylor series.

(2) If the real singularity of $f(t)$ nearest to 0 is at t_0, then $R \leq |t_0|$. We cannot be sure that $R = |t_0|$ because $f(t)$ may have a nonreal singularity closer to 0 than to t_0. The rational function (6.2.3), has a real singularity at $t = 6$, but the radius of convergence is $R = 3$ because of the complex singularities at $t = \pm 3i$. See also Problem 4 in the previous exercise set.

One way of finding R without finding the singularities of $f(t)$ is to use the ratio test for absolute convergence of a series. We review this test now.

The ratio test for the series of complex numbers

$$a_0 + a_1 + \cdots + a_k + \ldots \tag{6.3.4}$$

is this: The series, (6.3.4), converges if

$$\lim_{k\to\infty} \left| \frac{a_{k+1}}{a_k} \right| < 1 \tag{6.3.5}$$

and diverges if

$$\lim_{k\to\infty} \left| \frac{a_{k+1}}{a_k} \right| > 1 \tag{6.3.6}$$

So, if the ratio test is to be applied to the power series

$$\sum_{k=0}^{\infty} b_k t^k = b_0 + b_1 t + \cdots + b_k t^k + \cdots \tag{6.3.7}$$

we will have convergence for all t for which

$$\lim_{k\to\infty} \left| \frac{b_{k+1} t^{k+1}}{b_k t^k} \right| = |t| \lim_{k\to\infty} \left| \frac{b_{k+1}}{b_k} \right| < 1 \tag{6.3.8}$$

For convenience, define r by the following limit, if it exists:

$$r = \lim_{k\to\infty} \left| \frac{b_k}{b_{k+1}} \right| \tag{6.3.9}$$

Theorem 6.3.2. *If r exists and is positive, then* (6.3.7) *converges for all t, $|t| < r$, and diverges for all t, $|t| > r$. If $r = \infty$,* (6.3.7) *converges for all t. Thus, $r = R$ is the radius of convergence of* (6.3.7).

Proof: Since we are assuming that $r > 0$, the ratio test and (6.3.8) imply that (6.3.7) converges for all t, $|t| < r$ and diverges for all t, $|t| > r$. □

Note: There are series for which r does not exist and r is not $+\infty$. For these series this theorem is quiet and other means are necessary to find R. This will not be a problem for any of the series with which we shall deal.

We now address an important issue raised implicitly by Theorem 6.3.2. Suppose we pick any set of coefficients $b_0, b_1, \ldots$ as the parameters in (6.3.7). Suppose that this series converges to some function, say $S(t)$, in $|t| < r$. What is the relationship between the coefficients $b_0, b_1, \ldots$ and the function $S(t)$? This question is similar to but not quite the same as this: If the coefficients in (6.3.7) are obtained by selecting $b_k = f^{(k)}(0)/k!$, to what function does (6.3.7) converge? The next theorem provides the answer to both these questions.

Theorem 6.3.3. *The series*

$$\sum_{k=0}^{\infty} b_k t^k$$

converges in some interval $|t| < r$ to the function $f(t)$, if and only if $f(t)$ is analytic for all t, $|t| < r$, and

$$b_k = \frac{1}{k!} f^{(k)}(0) \tag{6.3.10}$$

Corollary 6.3.1. *If for some $r > 0$*

$$\sum_{k=0}^{\infty} b_k t^k = 0 \tag{6.3.11}$$

for all t in $|t| < r$, then $b_k = 0$ for all k.

Proof: This is a trivial consequence of Theorem 6.3.3 because the Taylor coefficients for $f(t) = 0$ are all zero. □

When no center is explicitly mentioned for a Taylor series expansion, we assume that the center is $c = 0$. Also, it is convenient to abbreviate the kth derivative operator by D^k. For example, $D = d/dt$ and $D^2 = d^2/dt^2$.

Example 6.3.1. *Find the Taylor series and its radius of convergence for*

$$f(t) = (1-t)^{-a}$$

Solution: We first find the derivatives of $(1-t)^{-a}$ at $t=0$:

$$D^k(1-t)^{-a}\Big|_{t=0} = a(a+1)(a+2)\cdots(a+k-1)$$

Hence,

$$(1-t)^{-a} = 1 + at + \frac{1}{2!}a(a+1)t^2 + \frac{1}{3!}a(a+1)(a+2)t^3 + \dots$$

by (6.3.10). Since $(1-t)^{-a}$ is singular[1] at $t=1$, we know that $R \le 1$. We can show that $R=1$ by the ratio test. For convenience, set

$$b_k = \frac{1}{k!}a(a+1)(a+2)\cdots(a+k-1)$$

Then, since $(k+1)! = (k+1)k!$, Theorem 6.3.2 and (6.3.9) provide

$$\begin{aligned} R &= \lim_{k\to\infty}\left|\frac{b_k}{b_{k+1}}\right| \\ &= \lim_{k\to\infty}\frac{k+1}{a+k} = 1 \end{aligned}$$

Example 6.3.2. *Find the Taylor series of $f(t) = e^{at}$ about $t=0$.*

Solution: Since $f^{(k)}(t) = a^k e^{at}$, it follows that $f^{(k)}(0) = a^k$, and by (6.3.10),

$$e^{at} = 1 + at + \frac{1}{2!}a^2t^2 + \cdots + \frac{1}{k!}a^kt^k + \dots$$

The ratio test shows that this series converges for all t and all a:

$$R = \lim_{k\to\infty}\left|\frac{\frac{a^k}{k!}}{\frac{a^{k+1}}{(k+1)!}}\right| = \lim_{k\to\infty}\left|\frac{(k+1)!}{ak!}\right| = \lim_{k\to\infty}\frac{|k+1|}{|a|} = \infty$$

---

[1] Unless a is a negative integer in which case $(1-t)^{-a}$ is a polynomial and polynomials have no singularities.

Example 6.3.3. *Find the Taylor series expansion of* $\cos t$ *about* $t = 0$, *and show that this series converges for all* t.

Solution: The derivatives of $\cos t$ are alternatively $\sin t$ and $\cos t$ with "appropriate" signs. That is, for $k = 0, 1, 2, \ldots$

$$D^{2k} \cos t = (-1)^k \cos t, \quad D^{2k+1} \cos t = (-1)^{k+1} \sin t$$

So, at $t = 0$, all the odd order derivatives are zero, while $b_{2k} = (-1)^k/(2k)!$. By (6.3.10), we have

$$\cos t = 1 - \frac{1}{2!}t^2 + \frac{1}{4!}t^4 + \cdots + (-1)^n \frac{1}{(2n)!}t^{2n} + \ldots$$

Since we do not know whether $\cos t$ has any nonreal singularities, we cannot use Theorem 6.3.1 to determine R. The ratio test saves the day. We define

$$b_k = (-1)^k \frac{1}{(2k)!}, \quad k = 0, 1, 2, \cdots$$

Then, from (6.3.5), we find that

$$\lim_{k\to\infty} \left| \frac{b_k t^{2k}}{b_{k+1} t^{2k+2}} \right| = \frac{1}{|t^2|} \lim_{k\to\infty} \left| \frac{\frac{1}{(2k)!}}{\frac{1}{(2k+2)!}} \right| = \frac{1}{|t^2|} \lim_{k\to\infty} (2k+1)(2k+2) = \infty$$

Thus, the Taylor series for $\cos t$ converges for all t, and therefore, from Theorem 6.3.1, $\cos t$ has no singularities in the complex plane.

The expansions for $\cos t$ can also be derived by using Euler's formula and the expansion for e^{at} in Example 6.3.2 with $a = i$ as follows:

$$e^{it} = \sum_{k=0}^{\infty} \frac{1}{k!} i^k t^k \tag{1}$$

So, using $i^{2k} = (-1)^k$ and $i^{2k+1} = (-1)^k i$ in (1) leads to

$$\begin{aligned} \operatorname{Re} e^{it} &= \cos t = 1 - \frac{1}{2!}t^2 + \frac{1}{4!}t^4 + \ldots \\ \operatorname{Im} e^{it} &= \sin t = t - \frac{1}{3!}t^3 + \frac{1}{5!}t^5 + \ldots \end{aligned}$$

Example 6.3.4. *Find the Taylor series of $f(t) = \ln(1 + at)$ about $t = 0$.*

Solution: We have $f(0) = 0$. Also, since $f'(t) = a(1 + at)^{-1}$, we have

$$f^{(k)}(t) = (-1)^{k-1}(k-1)!a^k(1-t)^{-k}, \qquad \text{for} \quad k = 1, 2 \ldots$$

Thus, $f^{(k)}(0) = (-1)^{k-1}(k-1)!\, a^k$ and therefore, (6.3.10) leads to

$$\ln(1 + at) = at - \frac{1}{2}a^2t^2 + \frac{1}{3}a^3t^3 + \cdots + (-1)^{n-1}\frac{1}{n}a^nt^n + \ldots$$

The ratio test shows that $R = |1/a|$. For $a = 1$, we obtain the series

$$\ln(1 + t) = t - \frac{1}{2}t^2 + \frac{1}{3}t^3 + \cdots + (-1)^{n-1}\frac{1}{n}t^n + \ldots$$

Let $a = 1$ in Example 6.3.1. The result is called the *geometric series.* Using the variable x rather than t, we have

$$\frac{1}{1-x} = 1 + x + x^2 + x^3 + \ldots \tag{6.3.12}$$

and this series converges for all $|x| < 1$ and diverges for all $|x| > 1$. In fact, the geometric series does not converge for any $|x| \geq 1$. It sometimes happens that this series can be used to find the Taylor series of certain functions by a simple change of variable. Here is one useful example.

Example 6.3.5. *Find the Taylor series expansion of*

$$f(t) = \left(1 - a^2t^2\right)^{-1}$$

about $t = 0$, and show that $R = |1/a|$.

Solution: Using $x = (at)^2$ in (6.3.12), we have

$$\frac{1}{1 - a^2t^2} = 1 + a^2t^2 + a^4t^4 + \ldots$$

This series converges for $|t| < |1/a|$. Since a singularity occurs when $t = 1/a$, R cannot be greater than $|1/a|$. So $R = |1/a|$.

The special cases $a = 1$ and $a = i$ in Example 6.3.5 lead to the following important expansions:

$$\frac{1}{1-t^2} = 1 + t^2 + t^4 + \cdots + t^{2k} + \ldots \tag{6.3.13}$$

$$\frac{1}{1+t^2} = 1 - t^2 + t^4 + \cdots + (-1)^k t^{2k} + \ldots \tag{6.3.14}$$

Theorem 6.3.4. *Suppose $f(t)$ and $g(t)$ have the Taylor series expansions*

$$f(t) = \sum_{k=0}^{\infty} b_k t^k \quad \text{and} \quad g(t) = \sum_{k=0}^{\infty} c_k t^k \tag{6.3.15}$$

convergent in $|t| < r$. Then $f(t) + g(t)$ has the Taylor series

$$f(t) + g(t) = \sum_{k=0}^{\infty} (b_k + c_k)\, t^k \tag{6.3.16}$$

also convergent in $|t| < r$.

Proof: The coefficients of the Taylor series expansion of $f(t) + g(t)$ are $b_k + c_k$ because

$$D^k (f(t) + g(t))_{t=0} = f^{(k)}(0) + g^{(k)}(0) = k!\, b_k + k!\, c_k$$

Finally, since we are implicitly assuming that there are no singularities of $f(t)$ or $g(t)$ in $|t| < r$, there can be no singularities of $f(t) + g(t)$ in $|t| < r$. □

Note this: Because some of the singularities of $f(t)$ and $g(t)$ can cancel, the radius of convergence of $f(t) + g(t)$ may be larger than r. An example of this phenomenon is given by $f(t) = (t-1)/(t-2)$ and $g(t) = 1/(t-1)$:

$$f(t) + g(t) = \frac{1}{(t-1)(t-2)} + \frac{1}{t-1} = \frac{1}{t-2}$$

in which $R = 2$. Although this example is trivial, it does illustrate the point. More sophisticated examples are not too difficult to produce, but the point of this note should be clear from the simple case shown above.

Corollary 6.3.2. *Two convergent Taylor series represent the same function if and only if their coefficients are identical.*

Proof: Suppose

$$f(t) = \sum_{k=0}^{\infty} b_k t^k \qquad \text{and} \qquad g(t) = \sum_{k=0}^{\infty} c_k t^k$$

Then by Theorem 6.3.3 and Corollary 6.3.1,

$$f(t) - g(t) = \sum_{k=0}^{\infty} (b_k - c_k)\, t^k = 0$$

if and only if $b_k = c_k$ for all k. □

Theorem 6.3.5. *If $f(t)$ has the Taylor series expansion*

$$f(t) = \sum_{k=0}^{\infty} b_k t^k, \quad |t| < r \tag{6.3.17}$$

then $F(t) = \int_0^t f(z)\,dz$ has the Taylor series expansion

$$F(t) = \sum_{k=0}^{\infty} b_k \int_0^t z^k\,dz = \sum_{k=0}^{\infty} \frac{1}{k+1} b_k t^{k+1} \tag{6.3.18}$$

and $f'(t)$ has the Taylor series expansion

$$f'(t) = \sum_{k=0}^{\infty} b_k D t^k = \sum_{k=1}^{\infty} k b_k t^{k-1} \tag{6.3.19}$$

Both expansions converge in $|t| < r$.

Proof: Since $F(t) = \int_0^t f(z)\,dz$, we have

$$F(0) = 0 \qquad \text{and} \qquad F^{(k+1)}(t) = f^{(k)}(t)$$

Thus,

$$\frac{1}{(k+1)!}F^{(k+1)}(0) = \frac{1}{k+1}\left(\frac{1}{k!}f^{(k)}(0)\right) = \frac{1}{k+1}b_k$$

Next, the $(k-1)^{st}$ derivative of $f'(t)$ is given by $f^{(k)}(t)$, so that the coefficient of t^{k-1} is

$$\frac{1}{(k-1)!}f^{(k)}(0) = k\frac{1}{k!}f^{(k)}(0) = kb_k$$

Moreover, if $f(t)$ is analytic in $|t| < r$, then so are $F(t)$ and $f'(t)$. □

Example 6.3.6. *Use Theorem 6.3.5 to find the Taylor series expansion and its radius of convergence of* $f(t) = \arctan t$.

Solution: We start with expansion (6.3.14), namely,

$$\frac{1}{1+t^2} = 1 - t^2 + t^4 + \cdots + (-1)^n t^{2n} + \ldots \tag{1}$$

valid in $|t| < 1$. Now we use (6.3.18) to conclude that

$$\int_0^t \frac{1}{1+s^2}\,ds = \arctan t = t - \frac{1}{3}t^3 + \frac{1}{5}t^5 + \cdots + \frac{(-1)^n}{2n+1}t^{2n+1} + \ldots$$

Since the series (1) converges for $|t| < 1$, so does the series for $\arctan t$. The ratio test establishes that $R = 1$. In fact, the series for $\arctan t$ actually converges for $t = 1$ and, at this value, yields the following well known series for π:

$$\frac{\pi}{4} = 1 - \frac{1}{3} + \frac{1}{5} + \cdots + \frac{(-1)^n}{2n+1} + \ldots$$

It is convenient to have available a table of Taylor series. Such tables provide both a reference and a starting point for obtaining many Taylor series expansions. We provide such a table at the end of the chapter.

It is not always possible to find an explicit expression for the coefficients (in terms of n) in the Taylor series expansion of a function. In such circumstances, we settle for the values of the first few coefficients. For this type of computation, it is convenient to introduce an abbreviation for the "tail end" of a power series. Let us write

$$\mathrm{O}(t^n) = c_n t^n + c_{n+1}t^{n+1} + \ldots \tag{6.3.20}$$

We read $O(t^n)$ as "order" t^n or "big O" of t^n. So for instance, we can write

$$e^t = 1 + t + O(t^2)$$

or

$$e^t = 1 + t + \tfrac{1}{2}t^2 + O(t^3)$$

as we wish.

Example 6.3.7. *Find the first three terms of the series expansion about $t = 0$ of the function*

$$f(t) = \frac{t}{e^t - 1}$$

Solution: Technically, $f(0)$ is undefined. However, if we set $f(0) = 1$, which is the value of $\lim_{t\to 0} f(t)$, we will obtain a Taylor series for this function which converges to it for all t because $f(t)$ has no other singularities! We are content to find only the first three coefficients. First we define b_0, b_1,and b_2 by this series:

$$f(t) = \frac{t}{e^t - 1} = b_0 + b_1 t + b_2 t^2 + O(t^3)$$

We see from the expansion for e^t, (see table), that

$$e^t - 1 = t + \frac{t^2}{2!} + \frac{t^3}{3!} + O(t^4)$$

Now, multiply both sides of the expression for $f(t)$ by the expansion for $e^t - 1$ to obtain the identity

$$t = (t + \frac{t^2}{2!} + \frac{t^3}{3!} + O(t^4))(b_0 + b_1 t + b_2 t^2 + O(t^3))$$

Multiply the two series and simplify to obtain

$$t = b_0 t + (2b_1 + \frac{b_0}{2})t + (b_2 + \frac{b_1}{2} + \frac{b_0}{6})t^2 + O(t^3)$$

For this latter expression to be an identity, the coefficients of all the powers of t on the right-hand side must agree with the coefficients of all the powers of t on the left-hand side. Equating these coefficients leads to $b_0 = 1$, $2b_1 + b_0/2 = 0$, and $b_2 + b_1/2 + b_0/6 = 0$, from which we learn that $b_1 = -1/2$, and $b_2 = 1/12$. From all this we conclude that

$$\frac{t}{e^t - 1} = 1 - \frac{t}{2} + \frac{t^2}{12} + O(t^3)$$

The series yields the same value for $t/(e^t-1)$ at $t=0$ as $\lim_{t\to 0}\left(t/(e^t-1)\right)$.

Example 6.3.8. *Find the first three coefficients in the series expansion about $t=0$ of the function*

$$C(t)=\int_0^t \cos(\tfrac{1}{2}\pi z^2)\,dz$$

Solution: We expand the integrand using the expansion of $\cos t$ with $\frac{1}{2}\pi z^2$ in place of t (see table or Example 6.3.3):

$$\begin{aligned}
C(t) &= \int_0^t \cos(\frac{1}{2}\pi z^2)\,dz \\
&= \int_0^t \left(1-\frac{1}{2!}(\frac{1}{2}\pi z^2)^2+\frac{1}{4!}(\frac{1}{2}\pi z^2)^4+\mathrm{O}(z^{12})\right)dz \\
&= \int_0^t dz-\frac{1}{2!}\frac{\pi^2}{4}\int_0^t z^4\,dz+\frac{1}{4!}\frac{\pi^4}{16}\int_0^t z^8\,dz+\int_0^t \mathrm{O}(z^{12})\,dz \\
&= t-\frac{1}{2!}\frac{\pi^2}{4}\frac{1}{5}t^5+\frac{1}{4!}\frac{\pi^4}{9}\frac{1}{16}t^9+\mathrm{O}(t^{13}) \\
&= t-\frac{\pi^2}{40}t^5+\frac{\pi^4}{3456}t^9+\mathrm{O}(t^{13})
\end{aligned}$$

a series that converges extremely rapidly for $|t|<1$. For example, using only the first two terms, $C(0.1)\approx 0.1-0.0000024674=0.0999975326$. All of these digits are correct!

•EXERCISES

Determine the radius of convergence for each series found in Problems 1–6. (The index of summation, k, runs from 0 to ∞.)

1. $\sum t^k$.
2. $\sum \frac{1}{k!}t^k$.
3. $\sum 2^k(k+1)(k+2)t^k$.
4. $\sum 2^k t^k$.
5. $\sum \frac{1}{k!}(t-2)^k$.
6. $\sum \frac{(-1)^k}{(2k)!}(t-1)^k$.

Find the first four coefficients of the Taylor series expansion for each function in Problems 7–32 about $t=0$. Find its radius of convergence.

7. $\dfrac{t}{1-t}$

8. e^{2t}.

9. $\sin^2 t$.

10. $\cos t \sin t$.

11. $\ln(1-t)$.

12. $\dfrac{1}{2+t}$.

13. $\sinh 2t$.

14. $\cosh \frac{1}{2}t$.

15. $\dfrac{1}{1+t} - \dfrac{1}{1-t}$.

16. $(1+2t)^{1/3}$.

17. $\dfrac{1}{t^2+3t+2}$.

18. $\dfrac{1}{t^2-3t-4}$.

19. e^{2t+1}.

20. $\exp\left(-t^2\right)$.

21. $\sin t^2$.

22. $\cos^2 t$.

23. $\ln \dfrac{t+1}{2}$.

24. $\ln \dfrac{4-t^2}{4}$.

25. $\ln \dfrac{2-t}{t+1}$.

26. $\dfrac{e^t}{t+4}$.

27. $e^{-t} \sin t$.

28. $\dfrac{1}{t^4+4t^2+4}$.

29. $\left(1-t^2/2\right)^{1/2}$.

30. $\dfrac{1}{t^2-9}$.

31. $\dfrac{1}{t^2+5t+4}$.

32. $\dfrac{1}{t^2-4t+3}$.

Find the first four coefficients of the Taylor series expansion for each integral in Problems 33–39 by expanding the integrand about 0 and then integrating the resulting series term by term. Use the table of series expansions.

33. $\displaystyle\int_0^t \frac{ds}{1+s}$.

34. $\displaystyle\int_0^t \frac{ds}{4-s^2}$.

35. $\displaystyle\int_0^t \frac{sds}{1+s^2}$.

36. $\displaystyle\int_0^t \sin^2 s\, ds$.

37. $\displaystyle\int_0^t \ln(1+s)\, ds$.

38. $\displaystyle\int_0^t \exp\left(-s^2\right)\, ds$.

39. $\displaystyle\int_0^t \sin s \cos s\, ds$.

Find the first 4 coefficients of the Taylor series expansion about $t=1$ for each function by replacing t by $z+1$ and expanding the resulting function about $z=0$.

40. $\dfrac{1}{t(t-2)}$.

41. $\dfrac{1}{t^2-4}$.

42. $\dfrac{1}{t}$.

43. $\ln t$.

Find the coefficients b_0, b_1 and b_2 of the Taylor series expansion about $t = 0$ for the following functions. Use the technique illustrated in Example 6.3.7.

44. $f(t) = \tan t$.

45. $f(t) = \sec t$.

46. $f(t) = \dfrac{2e^{t/2}}{e^t + 1}$.

6.4 Power Series Solutions of Linear Differential Equations

In this section we show how to use the theory of Taylor series to find solutions of linear homogeneous ODE's where the coefficients are functions of t. Consider the linear ODE

$$u'' + p(t)u' + q(t)u = 0 \tag{6.4.1}$$

If $p(t)$ and $q(t)$ are analytic at $t = t_0$, then $t = t_0$ is called an *ordinary point* of (6.4.1). Otherwise, $t = t_0$ is a *singular point* of (6.4.1). Thus, $t = 0$ is a singular point of (6.4.1) if it is a singular point of either $p(t)$ or $q(t)$. Note the two different uses of the word "singular" in this chapter. We avoid confusion by always referring to points that are not ordinary points as "singular points of the differential equation." Our first theorem (stated without proof) tells us that there always exists two solutions represented by Taylor series if $t = 0$ is an ordinary point of the ODE.

Theorem 6.4.1. *If $t = 0$ is an ordinary point of (6.4.1), then there exists a pair of basic solutions*

$$u_1(t) = \sum_{k=0}^{\infty} a_k t^k, \qquad u_2(t) = \sum_{k=0}^{\infty} b_k t^k \tag{6.4.2}$$

Both series converge in a disk $|t| < r$, where r is at least as large as the distance from the origin to the nearest singular point of the set of singular points of the functions $p(t)$ and $q(t)$.

Corollary 6.4.1. *If $p(t)$ and $q(t)$ are polynomials, then (6.4.2) converge for all t.*

Proof: Polynomials are everywhere analytic. In particular, we see that the constant coefficient linear, homogeneous ODE has solutions that are analytic everywhere. (All solutions of linear, homogeneous ODE's, are sums of terms of the form $t^k e^{at}$, and such functions have no singularities.) □

Note that Theorem 6.4.1 asserts that the Taylor series expansions of the solutions converge in an interval at least as large as the distance to the nearest singularity of the ODE. The interval of convergence may, however, be significantly larger in specific problems. Many examples of this phenomenon will appear later on in this chapter. A simple illustration is provided by the Cauchy-Euler equation,

$$u'' - 5t^{-1}u' + 8t^{-2}u = 0$$

Since $t = 0$ is a singular point of the equation, Theorem 6.4.1 is silent on whether there are analytic solutions (about $t = 0$) of this equation. A general solution of this equation is easily obtained by the methods of Chapter 3. (See Section 3.11.) The solution is

$$u(t) = c_1 t^2 + c_2 t^4$$

and $u(t)$ is analytic for all t.

The hypothesis in Theorem 6.4.1, that $t = 0$ is an ordinary point of the ODE, is a conclusion relevant to the Taylor series expanded about $t = 0$. For expansions about $t = a$, make the change of variable $\tau = t - a$ in the ODE. Then Theorem 6.4.1 is applicable at $\tau = 0$, which corresponds to $t = a$ in the original equation.

The existence of convergent Taylor series solutions of

$$u'' + p(t)u' + q(t)u = 0 \tag{6.4.3}$$

for analytic coefficients $p(t)$ and $q(t)$ is guaranteed by Theorem 6.4.1. There still remains the issue of determining the coefficients of these series. In this section we show how this can be done for many of the ODE's in which $t = 0$ is an ordinary point of the equation.

Suppose that a series representation for the solution $u(t)$ of (6.4.1) exists and is given by

$$u(t) = \sum_{k=0} b_k t^k \tag{6.4.4}$$

For convenience, we omit the ∞ upper limit on the summation sign. Theorem 6.3.5 allows

$$u'(t) = \sum_{k=1} k b_k t^{k-1} \tag{6.4.5}$$

$$u''(t) = \sum_{k=2} k(k-1) b_k t^{k-2} \tag{6.4.6}$$

The steps from now on are best explained by example. We choose an illustration that exhibits few complications and which can be solved by elementary means. This provides an easy confirmation of our solution method.

Consider the initial-value problem

$$u'' + 9u = 0, \qquad u(0) = 1, \quad u'(0) = 0 \tag{6.4.7}$$

When the series for $u(t)$ and $u''(t)$ given by (6.4.4) and (6.4.6) are substituted into (6.4.7), we obtain

$$\sum_{k=2} k(k-1) b_k t^{k-2} + 9 \sum_{k=0} b_k t^k = 0 \tag{6.4.8}$$

At this point we make use of a property of summations.

If the starting value of the index of summation k is lowered by s, each appearance of k in the summand is replaced by $s + k$.

This property is easily proved by expanding both sides of

$$\sum_{k=s} P(k) b_k t^k = \sum_{k=0} P(k+s) b_{k+s} t^{k+s} \tag{6.4.9}$$

We apply this fact to the first term on the left-hand side of (6.4.8):

$$\sum_{k=2} k(k-1) b_k t^{k-2} = \sum_{k=0} (k+2)(k+1) b_{k+2} t^k \tag{6.4.10}$$

Now use (6.4.10) in (6.4.8) to obtain

$$\begin{aligned} u'' + 9u &= \sum_{k=0} (k+2)(k+1) b_{k+2} t^k + 9 \sum_{k=0} b_k t^k \\ &= \sum_{k=0} \left((k+2)(k+1) b_{k+2} + 9 b_k\right) t^k = 0 \end{aligned}$$

By Corollary 6.3.1, a power series is identically zero if and only if all its coefficients are zero. So we conclude that

$$(k+2)(k+1)b_{k+2} + 9b_k = 0, \quad \text{for all } k \geq 0 \tag{6.4.11}$$

Solving for b_{k+2} results in

$$b_{k+2} = -\frac{9}{(k+2)(k+1)} b_k \tag{6.4.12}$$

Equation (6.4.12) is called a *recurrence relation.* In this particular example,

Table 6.1: *Recurrence Table for* $b_{n+2} = -\dfrac{9}{(n+2)(n+1)} b_n$

Row number	Relationship between k and n	recurrence n even
1	$n = 0$	$b_2 = \dfrac{-9}{1 \cdot 2} b_0$
2	$n = 2$	$b_4 = \dfrac{-9}{3 \cdot 4} b_2$
$\vdots$	$\vdots$	$\vdots$
k	$n = 2k - 2$	$b_{2k} = \dfrac{-9}{(2k-1)2k} b_{2k-2}$

the even-subscripted coefficients are related to b_0 while the odd-subscripted coefficients are related to b_1. Our objective is to find the expression relating b_{2k} to b_0. To do this conveniently, we replace k by n in (6.4.12) and construct a table with k rows. (See Table 6.1.) Let us study this table in some detail. The first column, headed "Row number," counts the number of rows in the table. By convention we shall always have k rows in our tables. The second column shows how n varies according to the row number. This variation is determined by the fact that the subscripts in (6.4.12) jump by 2. The last column results from substituting $2k - 2$ for k in (6.4.12).

Now multiply the k equations in the third column to obtain

$$b_2 b_4 \cdots b_{2k-2} b_{2k} = \frac{(-1)^k 9^k}{1 \cdot 2 \cdots (2k-1)(2k)} b_0 b_2 \cdots b_{2k-2} \tag{6.4.13}$$

The factors $b_2, b_4, \ldots, b_{2k-2}$ are common to both sides of (6.4.13) and, when cancelled, leave b_{2k} as a function of b_0:

$$b_{2k} = (-1)^k \frac{9^k}{(2k)!} b_0 \tag{6.4.14}$$

The initial condition $u(0) = 1$ implies that $b_0 = 1$ because

$$u(0) = \sum_{k=0} b_k t^k \bigg|_{t=0} = b_0 = 1$$

Similarly, the initial condition $u'(0) = 0$ implies that $b_1 = 0$. Since $b_1 = 0$, the recurrence relation (6.4.12) implies that all odd-subscripted coefficients are zero. Hence, setting $b_0 = 1$ and all odd subscripts zero and using (6.4.14), we have

$$u(t) = 1 + \sum_{k=1} b_{2k} t^{2k} = 1 + \sum_{k=1} \frac{(-1)^k}{(2k)!} 9^k t^{2k} \tag{6.4.15}$$

The given ODE has constant coefficients and is therefore amenable to solution by the methods of Chapter 3. We find the solution $u(t) = \cos 3t$. Since the solution is unique, we must have

$$\cos 3t = 1 + \sum_{k=1} \frac{(-1)^k}{(2k)!} 9^k t^{2k} \tag{6.4.16}$$

Alternatively, (6.4.16) can be derived by direct reference to the expansion of $\cos 3t$ in Taylor series about the origin.

With the exception of the Cauchy-Euler equation and certain other rather special ODE's, none of the solution methods developed prior to this chapter would succeed in yielding a solution for ODE's with variable coefficients. We now illustrate how the "power series" method can be used in many such cases.

Consider this linear equation with variable coefficients:

$$u'' - 2tu' + u = 0$$

No elementary technique seems likely to solve this equation, so we try to obtain a Taylor series expansion for its solution. As in the case of the previous illustration, we substitute the power series for $u(t)$ and its derivatives (6.4.4)—(6.4.6) into $u'' - 2tu' + u = 0$ to obtain the identity

$$\sum_{k=2} k(k-1) b_k t^{k-2} - 2 \sum_{k=1} k b_k t^k + \sum_{k=0} b_k t^k = 0 \tag{6.4.17}$$

where, in the second summation, we have multiplied t^{k-1} by t to obtain t^k. In order to combine these terms into one sum, we must have the indices of each of the summations start at the same value. We use (6.4.9) with $s = 2$ in the first summation:

$$\sum_{k=2} k(k-1)b_k t^{k-2} = \sum_{k=0} (k+2)(k+1)b_{k+2}t^k \tag{6.4.18}$$

We do not use (6.4.3) in the second summation, but simply write

$$\sum_{k=1} kb_k t^k = \sum_{k=0} kb_k t^k \tag{6.4.19}$$

because the value of the summand in the summation on the right-hand side of (6.4.19) is zero when $k = 0$. Now, using (6.4.18) and (6.4.19) in (6.4.17), there results

$$\sum_{k=0} \left((k+2)(k+1)b_{k+2} - (2k-1)b_k\right) t^k = 0 \tag{6.4.20}$$

from which we deduce

$$b_{k+2} = \frac{(2k-1)}{(k+1)(k+2)} b_k, \qquad k \geq 0 \tag{6.4.21}$$

We will find two solutions for this second-order ODE. The subscripts differ by two, so we obtain the even-subscripted coefficients b_{2k} as a function of b_0 by starting with $k = 0$ and solving the resulting recurrence relationship. Table 6.2 presents the relationship for the even-subscripted coefficients.

Multiplying the equations in column three leads to

$$b_{2k} = \frac{(-1) \cdot 3 \cdot 7 \cdots (4k-5)}{(2k)!} b_0 \tag{6.4.22}$$

We find a series expansion containing only even powers of t for the first solution by setting $b_0 = 1$ and $b_1 = 0$. This is our first solution:

$$u_1(t) = \sum_{k=0} b_{2k} t^{2k} = 1 - \sum_{k=1} \frac{3 \cdot 7 \cdots (4k-5)}{(2k)!} t^{2k} \tag{6.4.23}$$

Likewise, the odd-subscripted coefficients b_{2k+1} are functions of b_1 and are obtained by starting with $k = 1$. Table 6.3 shows the recurrence relationship for the odd-subscripted coefficients.

Table 6.2: *Recurrence Table for* $b_{n+2} = -\dfrac{2n-1}{(n+2)(n+1)} b_n$

Row number	Relationship between k and n	recurrence n even
1	$n = 0$	$b_2 = \dfrac{-1}{1 \cdot 2} b_0$
2	$n = 2$	$b_4 = \dfrac{3}{3 \cdot 4} b_2$
$\vdots$	$\vdots$	$\vdots$
k	$n = 2k-2$	$b_{2k} = \dfrac{4k-5}{(2k-1)2k} b_{2k-2}$

The product of the functions in column three leads to

$$b_{2k+1} = \frac{1 \cdot 5 \cdot 9 \cdots (4k-3)}{(2k+1)!} b_1 \tag{6.4.24}$$

In this case we set $b_1 = 1$ and $b_0 = 0$ to obtain a second solution containing only odd powers of t:

$$\begin{aligned} u_2(t) &= \sum_{k=0} b_{2k+1} t^{2k+1} \\ &= t + \sum_{k=1} \frac{1 \cdot 5 \cdot 9 \cdots (4k-3)}{(2k+1)!} t^{2k+1} \end{aligned} \tag{6.4.25}$$

The functions $u_1(t)$ and $u_2(t)$ make up a fundamental set of solutions because their Wronskian at $t = 0$ is 1:

$$W(0) = \det \begin{bmatrix} u_1(0) & u_2(0) \\ u_1'(0) & u_2'(0) \end{bmatrix} = \det \begin{bmatrix} 1 & 0 \\ 0 & 1 \end{bmatrix} = 1$$

The general solution is then

$$\begin{aligned} u(t) &= c_1 u_1(t) + c_2 u_2(t) \\ &= c_1 \left(1 - \sum_{k=1} \frac{3 \cdot 7 \cdots (4k-5)}{(2k)!} t^{2k} \right) \\ &\quad + c_2 \left(t + \sum_{k=1} \frac{1 \cdot 5 \cdot 9 \cdots (4k-3)}{(2k+1)!} t^{2k+1} \right) \end{aligned}$$

Table 6.3: *Recurrence Table for* $b_{n+2} = -\dfrac{2n-1}{(n+2)(n+1)} b_n$

Row number	Relationship between k and n	recurrence n odd
1	$n = 1$	$b_3 = \dfrac{1}{2 \cdot 3} b_1$
2	$n = 3$	$b_5 = \dfrac{5}{4 \cdot 5} b_3$
$\vdots$	$\vdots$	$\vdots$
k	$n = 2k-1$	$b_{2k+1} = \dfrac{4k-3}{(2k+1)2k} b_{2k-1}$

•EXERCISES

In Problems 1–6, list all the singular points of each ODE and determine the distance from the origin to the nearest of these singular points.

1. $\cos t u'' + u = 0.$
2. $(t^2 - 1)\, u'' + u = 0.$
3. $t\,(t^2 + 4)\, u'' + tu' = 0.$
4. $u'' + tu = 0.$
5. $(t+1)u'' + (t-1)u' + u = 0.$
6. $u'' + (t^2 - 1)\, u = 0.$

Find the first four terms in the Taylor series representation of a general solution for the ODE's in Problems 7–20. What is the radius of the largest disk in which the series must converge?

7. $u'' + tu' = 0.$
8. $u'' + 4u' = 0.$
9. $u'' + 4u = 0.$
10. $u'' + 6u' + 5u = 0.$
11. $u'' + tu = 0.$
12. $u'' - 4u = 0.$
13. $u'' + u' + 4u = 0.$
14. $u'' - u' - 6u = 0.$
15. $u'' + t^2 u = 0.$
16. $(1-t)u'' + u = 0.$
17. $(t^2 - 1)\, u'' - 4u = 0.$
18. $u'' + 3tu' + 3u = 0.$
19. $(t^2 + 1)\, u'' - 2u = 0.$
20. $(1 - t^2)\, u'' - 12tu' - 18u = 0.$

Problems 21—24: Find the first four terms in the Taylor series solution for the following initial-value problems,

21. $tu' + \sin tu = 0, \qquad u(0) = 1, \quad u'(0) = 0.$
22. $(4 - t^2)\, u'' + 2u = 0, \qquad u(0) = 0, \quad u'(0) = 1.$
23. $u'' + (1-t)u = 0, \qquad u(0) = 1, \quad u'(0) = 0.$
24. $u'' - t^2 u' + \sin tu = 0, \qquad u(0) = 0, \quad u'(0) = 1.$

6.5 Legendre's Equation

An ODE often encountered when modeling physical phenomena in spherical coordinates is the *Legendre equation*,

$$(1 - t^2)u'' - 2tu' + \lambda(\lambda + 1)u = 0 \tag{6.5.1}$$

The parameter λ is frequently a positive integer, but in any case λ is a real, nonnegative constant. In order to determine the interval of convergence of the Taylor series representation of the solutions of (6.5.1), we divide the Legendre equation by $1 - t^2$. Then, by definition,

$$p(t) = -\frac{2t}{1 - t^2}, \qquad q(t) = \frac{\lambda(\lambda + 1)}{1 - t^2}$$

The origin is therefore an ordinary point of the Legendre equation, and the nearest singularities are $t = \pm 1$. For this reason, we can expect to find a Taylor series expansion (about $t = 0$) for each of the two linearly independent solutions of (6.5.1). Moreover, the radius of convergence of each series is at least 1. Let

$$u(t) = \sum_{k=0} b_k t^k \tag{6.5.2}$$

be a solution of (6.5.1) and set $\alpha = \lambda(\lambda + 1)$. Then substituting their Taylor series expansions for $u(t), u'(t)$, and $u''(t)$ into (6.5.1) results in

$$\left(1 - t^2\right) \sum_{k=2} k(k-1) b_k t^{k-2} - 2t \sum_{k=1} k b_k t^{k-1} + \alpha \sum_{k=0} b_k t^k = 0$$

After multiplying the first series by $1 - t^2$ and the second by $-2t$ and collecting terms, we obtain

$$\sum_{k=2} k(k-1) b_k t^{k-2} - \sum_{k=0} (k(k-1) + 2k - \alpha) b_k t^k = 0$$

Next, here's what we get by lowering the index of summation in the first sum:

$$\sum_{k=2} k(k-1) b_k t^{k-2} = \sum_{k=0} (k+2)(k+1) b_{k+2} t^k$$

When substituted into the previous equation, this results in

$$\sum_{k=0} \left((k+2)(k+1) b_{k+2} - (k^2 + k - \alpha) b_k\right) t^k = 0$$

Now, using $\alpha = \lambda(\lambda+1)$, the last summand becomes

$$k^2 + k - \alpha = k^2 + k - \lambda^2 - \lambda = (k-\lambda)(k+\lambda+1) \tag{6.5.3}$$

and thus, for all k,

$$(k+2)(k+1)b_{k+2} - ((k-\lambda)(k+\lambda+1))\, b_k = 0 \tag{6.5.4}$$

Equation (6.5.4) leads to the recurrence relation

$$b_{k+2} = \frac{(k-\lambda)(k+\lambda+1)}{(k+2)(k+1)} b_k \tag{6.5.5}$$

We can use the recurrence table method illustrated in Section 6.4 to obtain formulas of all the even-subscripted coefficients in terms of b_0 and all odd subscripted coefficients in terms of b_1. However, the resulting formulas are rather complicated. For this reason, we restrict ourselves to the computation of the first few coefficients. By setting $b_0 = 1$ and $b_1 = 0$, we obtain a solution involving only even powers of t. The relationship between the first three coefficients is obtained from (6.5.5) and is given by

$$b_0 = 1, \quad b_2 = -\frac{1}{2}\lambda(\lambda+1)\, b_0, \quad b_4 = \frac{1}{12}(2-\lambda)(3+\lambda)\, b_2$$

So, denoting the solution by $u_1(t)$, we have

$$\begin{aligned} u_1(t) &= 1 - \frac{1}{2!}\lambda(\lambda+1)\, t^2 \\ &\quad + \frac{1}{4!}\lambda(\lambda-2)(\lambda+1)(\lambda+3)\, t^4 + \mathrm{O}(t^5) \end{aligned} \tag{6.5.6}$$

Likewise, setting $b_0 = 0$ and $b_1 = 1$ leads to a Taylor series for a second solution involving only odd powers of t. Again from (6.5.5), we learn that the first three coefficients are related as follows:

$$b_1 = 1, \quad b_3 = \frac{1}{6}(1-\lambda)(2+\lambda)\, b_1, \quad b_5 = \frac{1}{20}(3-\lambda)(4+\lambda)\, b_3$$

So,

$$\begin{aligned} u_2(t) &= t - \frac{1}{3!}(\lambda-1)(\lambda+2)\, t^3 \\ &\quad + \frac{1}{5!}(\lambda-1)(\lambda-3)(\lambda+2)(\lambda+4)\, t^5 + \mathrm{O}(t^7) \end{aligned} \tag{6.5.7}$$

The solutions $u_1(t), u_2(t)$ are linearly independent and constitute a basic set. Note that they are functions of λ as well as t.

The choice of λ as a positive integer is particularly interesting. If we write $\lambda = 2m$, then (6.5.6) implies that $u_1(t)$ is a polynomial of degree $2m$ because all terms with subscript greater than $2m$ are zero. Here are a few examples:

$$\begin{aligned} \lambda = 0: &\quad u_1(t) = 1 \\ \lambda = 2: &\quad u_1(t) = 1 - 3t^2 \\ \lambda = 4: &\quad u_1(t) = 1 - 10t^2 + \frac{35}{3}t^4 \end{aligned} \tag{6.5.8}$$

For λ an odd integer, write $\lambda = 2m + 1$, and note that (6.5.7) results in a polynomial solution of degree $2m + 1$. Here are some examples for various choices of λ:

$$\begin{aligned} \lambda = 1: &\quad u_2(t) = t \\ \lambda = 3: &\quad u_2(t) = t - \frac{5}{3}t^3 \\ \lambda = 5: &\quad u_2(t) = t - \frac{14}{3}t^3 + \frac{21}{5}t^5 \end{aligned} \tag{6.5.9}$$

These polynomials illustrate the fact that at least one solution of the Legendre equation is a polynomial when the parameter λ is a nonnegative integer. Since multiples of solutions of homogeneous ODE's are themselves solutions, we can "normalize" each polynomial so that its value is 1 at $t = 1$. The resulting polynomials are called *Legendre polynomials.* The first six Legendre polynomials are easy to write down by normalizing the polynomial solutions given in (6.5.8) and (6.5.9):

$$\begin{aligned} P_0(t) &= 1 \\ P_1(t) &= t \\ P_2(t) &= \tfrac{3}{2}t^2 - \tfrac{1}{2} \\ P_3(t) &= \tfrac{5}{2}t^3 - \tfrac{3}{2}t \\ P_4(t) &= \tfrac{35}{8}t^4 - \tfrac{30}{8}t^2 + \tfrac{3}{8} \\ P_5(t) &= \tfrac{63}{8}t^5 - \tfrac{70}{8}t^3 + \tfrac{15}{8}t \end{aligned}$$

In general, although it is assuredly not obvious,

$$P_n(t) = 2^{-n} \sum_{k=0}^{N} \frac{(-1)^k (2n-2k)!}{k!(n-k)!(n-2k)!} t^{n-2k} \tag{6.5.10}$$

where $N = n/2$ if n is even and $N = (n-1)/2$ if n is odd. A sketch of $P_{11}(t)$ is given in Figure 6.1. The oscillatory phenomenon is typical of the Legendre polynomials.

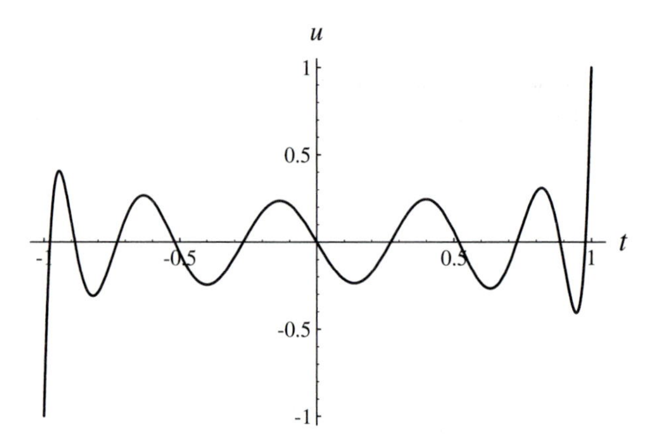

Figure 6.1: The Legendre Polynomial $P_{11}(t)$

It is possible to show, but we do not do so here, that a second linearly independent solution of the Legendre equation must have singularities at $t = \pm 1$. When suitably normalized, these solutions are called *Legendre functions of the second kind.* For the sake of completeness, here are the first five Legendre functions of the second kind, corresponding to $\lambda = 0, 1, 2, 3, 4$:

$$
\begin{aligned}
Q_0(t) &= \frac{1}{2} \ln \frac{1+t}{1-t} \\
Q_1(t) &= P_1(t)Q_0(t) - 1 \\
Q_2(t) &= P_2(t)Q_0(t) - \tfrac{3}{2}t \\
Q_3(t) &= P_3(t)Q_0(t) - \tfrac{5}{2}t^2 + \tfrac{2}{3} \\
Q_4(t) &= P_4(t)Q_0(t) - \tfrac{35}{8}t^3 + \tfrac{55}{24}t
\end{aligned}
$$

A general solution of the Legendre equation with nonnegative integer $\lambda = n$ is then

$$u(t) = c_1 P_n(t) + c_2 Q_n(t) \tag{6.5.11}$$

Note that solutions that exist at $t = 1$ or $t = -1$ must have $c_2 = 0$ since $Q_0(t)$ is singular at these points. Here is an example.

Example 6.5.1. *Find the solution of the differential equation*

$$(1 - t^2)u'' - 2tu' + 12u = 0$$

that exists at $t = 1$ *and for which* $u'(0) = 4$.

Solution: The given ODE is Legendre's equation with $\lambda(\lambda+1)=12$. The nonnegative integer solution of this relationship is $\lambda = 3$ and the fact that the solution exists at $t = 1$ forces $c_2 = 0$ in the general solution (6.5.11). Hence,

$$u(t) = c_1 P_3(t) = c_1\left(\tfrac{5}{2}t^3 - \tfrac{3}{2}t\right)$$

from which we deduce that

$$u'(0) = c_1 P_3'(0) = -\tfrac{3}{2}c_1 = 4$$

Therefore, $c_1 = -\frac{8}{3}$, and the required solution is

$$u(t) = -\tfrac{20}{3}t^3 + 4t$$

Note that we cannot arbitrarily specify the solution at $t = 0$. We are forced to take $u(0) = 0$.

•EXERCISES

1. Verify by substitution that the Legendre equation (for appropriate values of λ) is satisfied by the polynomials $P_2(t)$ and $P_3(t)$.
2. Find $P_6(t)$ and $P_7(t)$.
3. Find a general solution of these ODE's in terms of $P_n(t)$ and $Q_n(t)$. (*Hint*: Use undetermined coefficients to find a particular solution for these nonhomogeneous equations.)
   (a) $(1-t^2)u'' - 2tu' + 6u = 6$.
   (b) $(1-t^2)u'' - 2tu' + 20u = 36t$.
   (c) $(1-t^2)u'' - 2tu' + 30u = 12t^2$.
4. Find the first three nonzero terms in the Taylor series expansion (about $t = 0$) of the solution of each of the following initial-value problems.
   (a) $(1-t^2)u'' - 2tu' + 3u = 0$; $\quad u(0) = 1, \quad u'(0) = 0$.
   (b) $(1-t^2)u'' - 2tu' + 20u = 18t$; $\quad u(0) = 0, \quad u'(0) = 1$.
   (c) $9(1-t^2)u'' - 18tu' + 4u = 0$; $\quad u(0) = 0, \quad u'(0) = 1$.
   (d) $(1-t^2)u'' - 2tu' + 20u = 0$; $\quad u(0) = 1, \quad u'(1)$ finite.
   (e) $(1-t^2)u'' - 2tu' + 20u = 14t^2$; $\quad u(0) = 3, \quad u'(1)$ finite.
5. Show that $Q_0(t)$ is a solution of Legendre's equation with $\lambda = 0$.

6.6 Three Important Examples

In this section we will present three examples that find application in several areas of applied physics. First we find polynomial solutions for the *Hermite equation*

$$u'' - 2tu' + 2qu = 0 \tag{6.6.1}$$

where q is a constant. When the Taylor series for u, u', and u'' are substituted into the Hermite equation we obtain

$$\sum_{k=0} \left((k+2)(k+1)b_{k+2} - 2(k-q)b_k\right) t^k = 0 \tag{6.6.2}$$

from which it follows that

$$b_{k+2} = \frac{2(k-q)}{(k+1)(k+2)} b_k, \qquad k \geq 0 \tag{6.6.3}$$

As usual, the even-subscripted coefficients are functions of $u(0) = b_0$, and the odd subscripted coefficients are functions of $u'(0) = b_1$. Tables 6.4 and 6.5 present these relations.

Table 6.4: *Recurrence Table for* $b_{n+2} = \frac{2(n-q)}{(n+1)(n+2)} b_n$

Row number	Relationship between k and n	recurrence n even
1	$n = 0$	$b_2 = 2\frac{-q}{1 \cdot 2} b_0$
2	$n = 2$	$b_4 = 2\frac{2-q}{3 \cdot 4} b_2$
$\vdots$	$\vdots$	$\vdots$
k	$n = 2k - 2$	$b_{2k} = 2\frac{2k-2-q}{2k(2k-1)} b_{2k-2}$

Multiplying the equations in column three leads to

$$b_{2k} = -2^k \frac{q(2-q)\cdots(2k-2-q)}{(2k)!} b_0 \tag{6.6.4}$$

Now b_0 and b_1 are arbitrary. By taking $b_0 = 1$ we insure that the solution satisfies the initial-condition $u(0) = 1$. By taking $b_1 = 0$ we insure that the solution meets the condition $u'(0) = 0$. In fact, choosing $b_1 = 0$ also implies that $u(t)$ is an even function since all the coefficients of odd powers of t are forced to be zero. All this leads to the solution

$$u_1(t) = 1 - \sum_{k=1} 2^k \frac{q(2-q)\cdots(2k-2-q)}{(2k)!} t^{2k} \tag{6.6.5}$$

Now we pick $b_0 = 0$ and $b_1 = 1$ and construct Table 6.5.

Table 6.5: *Recurrence Table for* $b_{n+2} = \dfrac{2(n-q)}{(n+1)(n+2)} b_n$

Row number	Relationship between k and n	recurrence n odd
1	$n = 1$	$b_3 = 2\dfrac{1-q}{2 \cdot 3} b_0$
2	$n = 3$	$b_5 = 2\dfrac{3-q}{4 \cdot 5} b_2$
$\vdots$	$\vdots$	$\vdots$
k	$n = 2k-1$	$b_{2k+1} = 2\dfrac{2k-1-q}{2k(2k+1)} b_{2k-1}$

The product of the functions in column three leads to

$$b_{2k+1} = 2^k \frac{(1-q)(3-q)\cdots(2k-1-q)}{(2k+1)!} b_0 \tag{6.6.6}$$

and, by selecting $b_1 = 1$ and $b_0 = 0$, a second solution is

$$u_2(t) = t + \sum_{k=1} 2^k \frac{(1-q)(3-q)\cdots(2k-1-q)}{(2k+1)!} t^{2k+1} \tag{6.6.7}$$

If q is a nonnegative integer then either $u_1(t)$ or $u_2(t)$ is a polynomial. Specifically, if q is even, the solution $u_1(t)$ simplifies to these polynomials:

$$\begin{aligned} q = 0: &\qquad u_1(t) = 1 \\ q = 2: &\qquad u_1(t) = 1 - 2t^2 \\ q = 4: &\qquad u_1(t) = 1 - 4t^2 + \frac{4}{3}t^4 \end{aligned} \tag{6.6.8}$$

For q odd, $u_2(t)$ is given by

$$\begin{aligned}
q=1: &\qquad u_2(t) = t \\
q=3: &\qquad u_2(t) = t - \frac{2}{3}t^3 \\
q=5: &\qquad u_2(t) = t - \frac{4}{3}t^3 + \frac{4}{15}t^5
\end{aligned} \tag{6.6.9}$$

The *Hermite polynomials* $H_n(t)$ are the multiples of these polynomials for which the coefficients of the highest power of t is 2^n. Here are the first four Hermite polynomials:

$$\begin{aligned}
H_0(t) &= 1 \\
H_1(t) &= 2t \\
H_2(t) &= -2 + 4t^2 \\
H_3(t) &= -12t + 8t^3
\end{aligned}$$

In general, setting $q = n$, we find that

$$H_n(t) = n! \sum_{k=0}^{N} \frac{(-1)^k 2^{n-2k}}{k!(n-2k)!} t^{n-2k} \tag{6.6.10}$$

where $N = n/2$ if n is even and $N = (n-1)/2$ if n is odd.

In the second example, we find the polynomial solutions to the *Laguerre equation* for each nonnegative integer q, namely,

$$tu'' + (1-t)u' + qu = 0 \tag{6.6.11}$$

where q is a constant. Although the Laguerre equation has a singular point at the origin, and hence, Theorem 6.4.1 is silent on the nature of its solutions, it turns out that there is always one solution. We begin as usual:

$$t \sum_{k=2} k(k-1)b_k t^{k-2} + (1-t) \sum_{k=1} k b_k t^{k-1} + q \sum_{k=1} b_k t^k = 0 \tag{6.6.12}$$

which can be simplified to yield

$$\sum_{k=1} \left((k+1)^2 b_{k+1} + (q-k) b_k \right) t^k = 0 \tag{6.6.13}$$

Hence,

$$(k+1)^2 b_{k+1} + (q-k) b_k = 0 \tag{6.6.14}$$

For this example, only a single recurrence can be constructed, and thus, we obtain only one solution of (6.6.11). A second solution will be found in Section 6.8.

The product of the functions in column three of Table 6.6 leads to

$$b_k = (-1)^k \frac{q(q-1)\cdots(q-k+1)}{k!k!} b_0 \tag{6.6.15}$$

and hence, with $b_0 = 1$, to the solution

$$u_1(t) = 1 + \sum_{k=1} (-1)^k \frac{q(q-1)\cdots(q-k+1)}{k!\,k!} t^k \tag{6.6.16}$$

Table 6.6: *Recurrence Table for* $b_n = \dfrac{q-n-1}{n^2} b_{n-1}$

Row number	Relationship between k and n	recurrence
1	$n = 1$	$b_1 = -qb_0$
2	$n = 2$	$b_2 = -2^{-2}(2q-1)b_1$
$\vdots$	$\vdots$	$\vdots$
k	$n = k$	$b_k = \dfrac{q-k-1}{k^2} b_{k-1}$

Since q is a nonnegative integer, we see that $u_1(t)$ is a polynomial of degree q. A simple expression for these polynomials is obtained by noting that the binomial coefficients $\binom{q}{k}$, $1 \le k \le q$, may be written in terms of factorials:

$$\binom{q}{k} = \frac{q!}{k!(q-k)!} = \frac{q(q-1)\cdots(q-k+1)}{k!} \tag{6.6.17}$$

And therefore, the *Laguerre polynomials* are given by

$$L_q(t) = 1 + \sum_{k=1}^{q} \frac{(-1)^k}{k!} \binom{q}{k} t^k \tag{6.6.18}$$

Table 6.7: *Recurrence Table for* $b_n = \dfrac{-1}{(n)(n-p)}b_{n-1}$

Row number	Relationship between k and n	recurrence
1	$n=1$	$b_1 = \dfrac{-1}{1-p}b_0$
2	$n=2$	$b_2 = \dfrac{-1}{2(2-p)}b_1$
$\vdots$	$\vdots$	$\vdots$
k	$n=k$	$b_k = \dfrac{-1}{k(k-p)}b_{k-1}$

All solutions that are not multiples of $L_q(t)$ are singular at $t=0$. We shall see why this is the case in Section 6.8, Example 6.8.1.

In our third example we find a Taylor series solution for the *Bessel-Clifford equation*

$$tu'' + (1-p)u' + u = 0 \tag{6.6.19}$$

where $p \neq 1,\ 2\ \ldots$ is a constant. Applying the techniques of this chapter to the Bessel-Clifford equation leads to the identity

$$\sum_{k=0} ((k+1)(k+1-p)b_{k+1} + b_k)t^k = 0 \tag{6.6.20}$$

from which we obtain

$$b_{k+1} = \frac{-1}{(k+1)(k+1-p)}b_k, \qquad k \geq 0$$

Table 6.7 is a recurrence table for the relationship between b_k and b_0, providing one solution of (6.6.19). We find a second solution in Section 6.8. With $b_0 = 1$, the product of the functions in column three results in

$$b_k = \frac{(-1)^k}{k!(1-p)(2-p)\cdots(k-p)}, \qquad k \geq 1 \tag{6.6.21}$$

So

$$u_1(t) = 1 + \sum_{k=1} \frac{(-1)^k}{k!(1-p)(2-p)\cdots(k-p)} t^k \tag{6.6.22}$$

valid for all t provided that $p \neq 1, 2, \ldots$. In Example 6.9.1 we obtain a solution for every p.

We remark in passing the recurrence relationship shows that every b_n is a function of b_0; we do not have the possibility of two linearly independent analytic solutions. Indeed, every solution linearly independent of $u_1(t)$ must be singular at 0 — we will demonstrate this fact for $p = 0$ in Section 6.8. Note also that although the Clifford-Bessel equation has a singular point at the origin, there is an analytic solution (at the origin) for all p except p a positive integer.

•EXERCISES

1. Find
   (a) $H_4(t)$. (b) $H_5(t)$.
2. Find three nonzero terms in the Taylor series representation of a general solution of each of the following Hermite equations.
   (a) $u'' - 2tu' + 6u = 0$.
   (b) $u'' - 2tu' + 10u = 0$.
   (c) $u'' - 2tu' + 4u = 0$.
3. Find three nonzero terms in the Taylor series representation of the solution of each initial-value problem.
   (a) $u'' - 2tu' + 6u = 0$, $u(0) = 2$ $u'(0) = 10$.
   (b) $u'' - 2tu' + 10u = 0$, $u(0) = 1$ $u'(0) = 0$.
   (c) $u'' - 2tu' + 10u = 8t$, $u(0) = 1$ $u'(0) = 0$.
4. Problems (a) and (b) involve the Laguerre equation.
   (a) Find R for the series of (6.6.16) by the ratio test.
   (b) Find the Laguerre polynomials $L_0(t), L_1(t), \ldots, L_5(t)$.
5. Given that a general solution of Laguerre's equation is
   $u(t) = c_1 L_q(t) + c_2 u_2(t)$
   solve these problems knowing that $u_2(t)$ is singular at $t = 0$.
   (a) $tu'' + (1-t)u' + 3u = 0$, $u(0) =$ finite, $u(1) = 1$.
   (b) $tu'' + (1-t)u' + 4u = 0$, $u(0) =$ finite, $u(2) = 2$.
   (c) $tu'' + (1-t)u' + 4u = 3t$, $u(0) =$ finite, $u(1) = 4$.

6. Problems (a), (b) (c) concern the Bessel-Clifford equation.

(a) Show that the series of (6.6.22) converges for all t.

(b) Set $p = 0$ in (6.6.19), and derive the solution

$$u(t) = \sum_{k=0} (-1)^k \frac{1}{k!\,k!} t^k$$

(c) Set $p = 1$ in (6.6.19), and derive the solution

$$u(t) = \sum_{k=0} (-1)^k \frac{k}{k!\,k!} t^k$$

7. Solve Problems (a), (b), and (c) using the fact that all solutions of the Bessel-Clifford equation which are not multiples of $u_1(t)$ have singularities at $t = 0$.

(a) $tu'' + 4u' + u = 0, \qquad u(0) = \text{finite}, \quad u(0) = 1.$

(b) $tu'' + 4u' + u = 0, \qquad u(0) = \text{finite}, \quad u(0) = -2.$

(c) $tu'' + 4u' + u = 0, \qquad u(0) = \text{finite}, \quad u(0) = 2.$

6.7 Bessel's Equation

The *Bessel ODE*

$$t^2u'' + tu' + (t^2 - n^2)u = 0 \tag{6.7.1}$$

is one of the most important ODE's in mathematical physics. The properties of its solutions have been studied in literally thousands of research papers. It often appears in the process of solving partial differential equations in cylindrical coordinates.

The parameter n is real and nonnegative and, as we shall see, affects the nature of the solution in a dramatic way. Since the Bessel equation has a singularity at $t = 0$, we cannot expect a Taylor series expansion about $t = 0$ to represent its solutions. As it turns out, however, if n is an integer, then a series representation about $t = 0$ for one solution exists and converges for all t. In this case, a second solution may be obtained by the methods we will introduce in Sections 6.8 and 6.9.

Suppose $n = m$, m a nonnegative integer. The assumption that $u_1(t)$ is a solution expandable in a Taylor series about $t = 0$ leads to

$$-m^2b_0 + \left(1 - m^2\right) b_1 t + \sum_{k=2} \left((k^2 - m^2)b_k + b_{k-2}\right) t^k = 0 \tag{6.7.2}$$

The first two terms external to the summation yield, according to Theorem

6.3.1,

$$m^2 b_0 = 0 \tag{6.7.3}$$

$$\left(1 - m^2\right) b_1 = 0 \tag{6.7.4}$$

The remaining coefficients satisfy

$$\left(k^2 - m^2\right) b_k + b_{k-2} = 0, \qquad k \geq 2 \tag{6.7.5}$$

Equations (6.7.3), (6.7.4) and (6.7.5) imply that all $b_k = 0$ for $k < m$. When $k = m$, the coefficient of b_k is zero and (6.7.5) becomes the identity $0 = 0$. Therefore, b_m is arbitrary. Note that $b_{m+1} = b_{m+3} = \cdots = 0$ is also a consequence of (6.7.3), (6.7.4), and (6.7.5).

Let us choose $b_m = 1$ and determine the coefficients $b_{m+2}, b_{m+4}, \ldots$, by using (6.7.5) with $k = m+2, m+4, \ldots$. We have

$$b_k = \frac{-1}{k^2 - m^2} b_{k-2} = \frac{-1}{(k-m)(k+m)} b_{k-2} \tag{6.7.6}$$

and the recurrence table, Table 6.8, results.

Table 6.8: *Recurrence Table for* $b_n = \dfrac{-1}{(n-m)(n+m)} b_{n-2}$

Row number	Relationship between k and n	recurrence
1	$n = m + 2$	$b_3 = \dfrac{-1}{2(2m+2)} b_m$
2	$n = m + 4$	$b_5 = \dfrac{-1}{4(2m+4)} b_{m+2}$
$\vdots$	$\vdots$	$\vdots$
k	$n = m + 2k$	$b_{m+2k} = \dfrac{-1}{2k(2m+2k)} b_{m+2k-2}$

From this table, we deduce that for $m \geq 0$, $k \geq 1$, and $b_m = 1$,

$$b_{m+2k} = 2^{-2k} \frac{(-1)^k}{k!(m+1)(m+2)\cdots(m+k)} \tag{6.7.7}$$

Since $b_k = 0$ for all $k < m$, one solution of Bessel's equation is

$$u_1(t) = \sum_{k=0} 2^{-2k} \frac{(-1)^k}{k!(m+1)(m+2)\cdots(m+k)} t^{2k+m} \tag{6.7.8}$$

The *Bessel function of the first kind of order* n, written $J_n(t)$, is a multiple of $u_1(t)$. (It is customary to use n for the order of the Bessel function.) Specifically, for each nonnegative integer n,

$$J_n(t) = 2^{-n} \frac{1}{n!} u_1(t) = \sum_{k=0} 2^{-2k-n} \frac{(-1)^k}{k!(n+k)!} t^{2k+n} \tag{6.7.9}$$

The first two Bessel functions are

$$J_0(t) = \sum_{k=0} 2^{-2k} \frac{(-1)^k}{k!\,k!} t^{2k} \tag{6.7.10}$$

and

$$J_1(t) = \sum_{k=0} 2^{-2k-1} \frac{(-1)^k}{k!(k+1)!} t^{2k+1} \tag{6.7.11}$$

The graph of $J_0(t)$ shows the damped oscillation of this function. Note that the distance between x-axis intercepts is not constant.

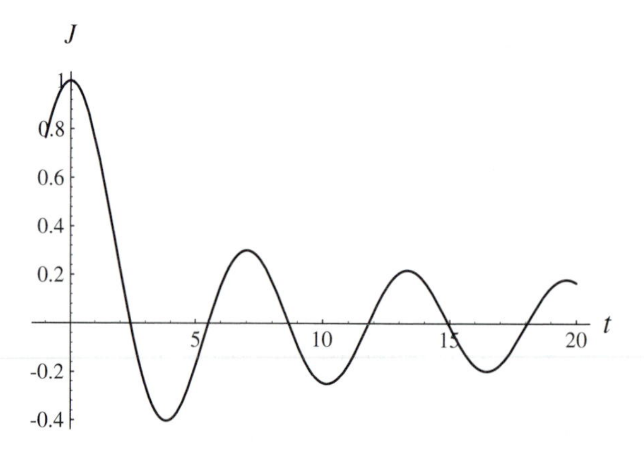

Figure 6.2: The Bessel function $J_0(t)$

A second linearly independent solution (suitably normalized) is called the *Bessel function of the second kind* of order n and is written $Y_n(t)$. It is beyond the scope of this book to derive the expression for $Y_n(t)$. However, it may be of some interest to note that $Y_n(t)$ is given by

$$Y_n(t) = \frac{2}{\pi} J_n(t) \left(\ln\left(\frac{t}{2}\right) + \gamma \right) + R(t) + u(t) \tag{6.7.12}$$

where $R(t)$ is a rational function, $\gamma \approx 0.577215665$ is the *Euler constant*, and $u(t)$ is a power series that converges for all t. Note that $Y_n(t)$ is singular at $t = 0$ because of the term $\ln(t/2)$ in (6.7.12). Consequently, we seldom require $Y_n(t)$ — most of our needs will be for solutions that are analytic at the origin.

•EXERCISES

1. Verify the conclusions after (6.7.5) for $k = m = 6$ by showing that $b_i = 0$ for all $i \leq 5$ and that $b_{2k+1} = 0$ for all $k \geq 0$.
2. Find four nonzero terms in the series expansion valid about $t = 0$:
   (a) $J_0(t)$. (b) $J_1(t)$. (c) $J_2(t)$.
3. Write the general solution using $J_n(t)$ and $Y_n(t)$ for these equations.
   (a) $t^2u'' + tu' + (t^2 - 4)\, u = 0$.
   (b) $t^2u'' + tu' + (t^2 - 9)\, u = 0$.
   (c) $t^2u'' + tu' + (t^2 - 16)\, u = 0$.
4. Let $I_n(t)$ be an analytic solution of the *modified Bessel equation* $t^2u''(t) + tu'(t) - (t^2 + n^2)u = 0$
   (a) Find a Taylor series expansion for $I_n(t)$ around $t = 0$.
   (b) Use the result in Problem (a) to show that $I_n(t) = i^{-n} J_n(it)$.

6.8 The Wronskian Method

All of the ODE's we have studied, and a great many more that have singular points at $t = 0$, have the form

$$t^2u'' + tp(t)u' + q(t)u = 0 \tag{6.8.1}$$

where $p(t)$ and $q(t)$ are analytic at the origin. That is, $p(t)$ and $q(t)$ necessarily have convergent Taylor series in some interval about 0. Any linear, second-order homogeneous ODE with this property is said to have a *regular singular point* at 0. Suppose $u_1(t)$ and $u_2(t)$ are two linearly independent solutions of (6.8.1). Then the Wronskian of these two solutions is

$$W(t) = W(u_1, u_2) = u_1(t)u_2'(t) - u_1'(t)u_2(t) \tag{6.8.2}$$

By Corollary 3.4.1, with $K = 1$,

$$W(t) = e^{-P(t)} \tag{6.8.3}$$

where

$$P(t) = \int \frac{p(t)}{t}\, dt \tag{6.8.4}$$

It follows from the quotient rule and (6.8.3) that

$$\left(\frac{u_2(t)}{u_1(t)}\right)' = \frac{u_1(t)u_2'(t) - u_1'(t)u_2(t)}{u_1^2(t)} = \frac{1}{u_1^2(t)} e^{-P(t)} \tag{6.8.5}$$

which leads to

$$u_2(t) = u_1(t) \int \frac{1}{u_1^2(t)} e^{-P(t)}\, dt \tag{6.8.6}$$

Equation (6.8.6) provides an expression for the second solution $u_2(t)$ in terms of the first solution $u_1(t)$ and the coefficient function $p(t)$.

It is often the case that the integral defining $P(t)$ and the integral on the right-hand side of (6.8.6) cannot be evaluated in closed form. However, we can obtain a number of terms of the Taylor series representation of $u_2(t)$ by using a few terms in the series representation of $p(t)$ and $u_1(t)$. We can demonstrate the main ideas of this procedure by using the Bessel-Clifford equation of Section 6.6 with $p = 0$. In the standard form given by (6.8.1), the Bessel-Clifford equation is

$$t^2u'' + tu' + tu = 0 \tag{6.8.7}$$

So $p(t) = 1$ in (6.8.1) and since $\int t^{-1}\, dt = \ln t + C$,

$$e^{-P(t)} = e^{-\ln t + C} = t^{-1}e^C = Kt^{-1} \tag{6.8.8}$$

where we have written K in place of e^C. Since we need only one new solution, we set $K = 1$. One solution of (6.8.7) is given in (6.6.22) with $p = 0$:

$$u_1(t) = 1 + \sum_{k=1} (-1)^k \frac{1}{k!\,k!} t^k = 1 - t + \tfrac{1}{4}t^2 + \mathrm{O}(t^3) \tag{6.8.9}$$

Consequently,

$$\begin{aligned}
\frac{1}{u_1^2(t)} &= \left(1 - \left(t - \tfrac{1}{4}t^2 + \mathrm{O}(t^3)\right)\right)^{-2} \\
&= 1 + 2\left(t - \tfrac{1}{4}t^2 + \mathrm{O}(t^3)\right) + 3\left(t - \tfrac{1}{4}t^2 + \mathrm{O}(t^3)\right)^2 + \mathrm{O}(t^3) \\
&= 1 + 2t + \tfrac{5}{2}t^2 + \mathrm{O}(t^3)
\end{aligned} \tag{6.8.10}$$

where we have used the series expansion of $(1-x)^{-2}$ with $t - \frac{1}{4}t^2 + O(t^3)$ in place of x. (To find $(1-x)^{-2}$, simply square $(1-x)^{-1}$). Substituting (6.8.8), (6.8.9), and (6.8.10) into (6.8.6), we obtain

$$u_2(t) = u_1(t) \int \left(1 + 2t + \tfrac{5}{2}t^2 + O(t^3)\right) \frac{1}{t}\, dt \tag{6.8.11}$$

Since,

$$\int \left(1 + 2t + \tfrac{5}{2}t^2 + O(t^3)\right) \frac{1}{t}\, dt = \ln t + 2t + \tfrac{5}{4}t^2 + O(t^3)$$

we have

$$\begin{aligned} u_2(t) &= u_1(t)\left(\ln t + 2t + \tfrac{5}{4}t^2 + O(t^3)\right) \\ &= u_1(t)\ln t + \left(1 - t + \tfrac{1}{4}t^2 + O(t^3)\right)\left(2t + \tfrac{5}{4}t^2 + O(t^3)\right) \\ &= u_1(t)\ln t + 2t - \tfrac{3}{4}t^2 + O(t^3) \end{aligned}$$

From this expression for $u_2(t)$, we can write the general solution of the Bessel-Clifford equation with $p = 0$ as

$$u(t) = c_1 u_1(t) + c_2\left(u_1(t)\ln t + 2t - \tfrac{3}{4}t^2 + O(t^3)\right) \tag{6.8.12}$$

Note that $u_2(t)$ is singular at $t = 0$, so if a solution is required which exists at $t = 0$, we are forced to choose $c_2 = 0$.

Example 6.8.1. *Find an approximate solution up to $O(t^3)$ for the Laguerre equation with $q = 1$.*

Solution: The Laguerre equation with $q = 1$ is given by (6.6.11). We multiply this equation by t so that it takes the form

$$t^2 u'' + t(1-t)u' + tu = 0$$

Thus, $p(t) = 1 - t$, and therefore, $P(t) = \ln t - t + C$. Once again, we choose $C = 0$ for the constant of integration. Hence, $e^{-P(t)} = e^{t - \ln t} = t^{-1}e^t$. By setting $q = 1$ in (6.6.16), we obtain the polynomial solution $L_1(t) = 1 - t$. Therefore, $u_1^2(t) = (1-t)^2$, and we have (see Problem 9)

$$\begin{aligned} u_2(t) &= (1-t)\int \frac{e^t}{t(1-t)^2}\, dt \\ &= (1-t)\int e^t\left(\frac{1}{t} + \frac{1}{1-t} + \frac{1}{(1-t)^2}\right) dt \\ &= (1-t)\ln t + 3t - \tfrac{1}{4}t^2 + O(t^3) \end{aligned}$$

Example 6.8.2. *Find an expression for a second, linearly independent solution of Bessel's equation of order* 0.

Solution: The Bessel equation of order 0, written in the form (6.8.1), is

$$t^2u'' + tu' + t^2u = 0$$

For this equation, $p(t) = 1$ and $u_1(t) = J_0(t)$. Using (6.8.6), we have,

$$u_2(t) = J_0(t) \int \frac{1}{tJ_0^2(t)}\, dt$$

Then, from (6.7.10),

$$\begin{aligned} J_0(t) &= 1 - \frac{1}{4}t^2 + \frac{1}{64}t^4 + \mathrm{O}(t^6) \\ J_0(t)^{-2} &= 1 + \frac{1}{2}t^2 + \frac{5}{32}t^4 + \mathrm{O}(t^6)) \end{aligned}$$

Therefore,

$$\begin{aligned} u_2(t) &= J_0(t) \int \frac{1}{t}\left(1 + \tfrac{1}{2}t^2 + \tfrac{5}{32}t^4 + \mathrm{O}\left(t^6\right)\right) dt \\ &= J_0(t)\left(\ln t + \tfrac{1}{4}t^2 + \tfrac{5}{128}t^4 + \mathrm{O}(t^6)\right) \\ &= J_0(t)\ln t + \left(1 - \tfrac{1}{4}t^2 + \tfrac{1}{64}t^4 + \mathrm{O}(t^6)\right)\left(\tfrac{1}{4}t^2 + \tfrac{5}{128}t^4 + \mathrm{O}(t^6)\right) \\ &= J_0(t)\ln t + \tfrac{1}{4}t^2 - \tfrac{3}{128}t^4 + \mathrm{O}(t^6) \end{aligned}$$

The Wronskian method is not restricted to equations with a singular point at 0. A beautiful example of its use at an ordinary point occurs in the Legendre equation that follows.

Example 6.8.3. *Find a second, linearly independent solution of Legendre's equation for* $\lambda = 0$.

Solution: The relevant ODE (see (6.5.1)) may be written in the form

$$t^2u'' - \frac{2t^3}{1-t^2}u' = 0$$

So $p(t) = -2t^2/\left(1-t^2\right)$, and hence, $P(t) = \ln\left(1-t^2\right)$ as long as $|t| < 1$. So

$$e^{-P(t)} = \exp\left(-\ln\left(1-t^2\right)\right) = \frac{1}{1-t^2}$$

We know that $P_0(t) = 1$ so that (6.8.6) leads to

$$\begin{aligned} u_2(t) &= \int \frac{1}{u_1^2(t)} \frac{1}{1-t^2}\, dt = \int \frac{1}{1-t^2}\, dt \\ &= \frac{1}{2} \int \left(\frac{1}{1+t} + \frac{1}{1-t} \right) dt = \frac{1}{2} \ln \frac{1+t}{1-t} = Q_0(t) \end{aligned}$$

Note that $u_2(t)$ is not singular at $t = 0$. This is not surprising, since $t = 0$ is an ordinary point of Legendre's equation. (An alternative method of solving Legendre's equation with $\lambda = 0$ is to observe that this equation is linear and of first order in u'.)

•EXERCISES

The next eight functions are analytic solutions of an equation with a regular singular point at $t = 0$. Show that

$$u_2(t) = u_1(t) \ln t + a_{-2} t^{-2} + a_{-1} t^{-1} + a_0 + a_1 t + a_2 t^2 + O(t^3)$$

is the form for every solution independent of $u_1(t)$. Find explicit values for a_{-2}, a_{-1}, a_0, a_1, and a_2. Here $I_n(t)$ are modified Bessel functions (see Problem 4 in the previous exercise set), and $L_n(t)$ are the Laguerre polynomials (see (6.6.18)).

1. $I_0(t)$.
2. $I_1(t)$.
3. $J_1(t)$.
4. $L_1(t)$.
5. $L_2(t)$.
6. $J_2(t)$.
7. $I_2(t)$.

8. Use for $u_1(t)$, the solution of the Bessel-Clifford equation $tu'' + u = 0$.
9. Use the first two terms in the power series expansion of $e^t, (1-t)^{-1}$, and $(1-t)^{-2}$ to find the first two terms in a power series expansion of

$$\int \frac{e^t}{t(1-t)^2}\, dt = \int e^t \left(\frac{1}{t} + \frac{1}{1-t} + \frac{1}{(1-t)^2} \right) dt$$

valid to two terms, and thereby verify the result of Example 6.8.1.

10. In Example 6.8.1 show that

$$u_2(t) = (1-t)u_1(t) + 3t - \tfrac{1}{4}t^2 - \tfrac{1}{36}t^3 + O(t^4)$$

11. Use one more term in (6.8.9) to show that

$$u_1^{-2}(t) = 1 + 2t + \tfrac{5}{2}t^2 + \tfrac{23}{9}t^3 + O(t^4)$$

and hence that

$$u_2(t) = u_1(t) \ln t + 2t - \tfrac{3}{4}t^2 + \tfrac{11}{108}t^3 + O(t^4)$$

6.9 The Frobenius Method

There are second-order linear ODE'e that appear in physical applications which do not have two linearly independent analytic solutions about $t = 0$. The Bessel equation (6.8.1) is perhaps the most thoroughly studied example; it is surely the most important. Here is another somewhat simpler example:

$$t^2u'' + \frac{3}{2}tu' - \frac{1}{2}u = 0 \tag{6.9.1}$$

This is a Cauchy-Euler equation (see Section 3.11). We readily verify that $u_1(t) = t^{-1}$ and $u_2(t) = t^{1/2}$ are linearly independent solutions valid for all $t > 0$. Neither of these functions is analytic at the origin, and therefore neither has a Taylor series expansion about the origin. However, in 1847, F. Georg Frobenius (1849—1917) showed that the class of ODE's with a regular singular point at the origin has solutions that can be represented by a slight generalization of a Taylor series. These series have the form

$$t^r \sum_{k=0} b_k t^k = \sum_{k=0} b_k t^{k+r}, \quad b_0 \neq 0 \tag{6.9.2}$$

where the series $\sum b_k t^k$ converges in some neighborhood of $t = 0$.

The Cauchy-Euler equation and the Bessel equation are two examples of equations that belong to this class. The series (6.9.2) is known as a *Frobenius series*, and the method for obtaining r and the coefficients b_k is commonly known as the *Frobenius method* in his honor. We restrict our attention to the most common example to which the Frobenius method is applicable. Specifically, we study the following ODE, which has a regular singular point at $t = 0$:

$$t^2u'' + tp(t)u' + q(t)u = 0 \tag{6.9.3}$$

Here $p(t)$ and $q(t)$ are most often low degree polynomials.

We begin by assuming that there is a solution of (6.9.3) which has an expansion that converges in $0 < |t| < R$ of the form

$$u(t) = \sum_{k=0} b_k t^{k+r} \tag{6.9.4}$$

Then,

$$u'(t) = \sum_{k=0} (k+r) b_k t^{k+r-1} \tag{6.9.5}$$

and

$$u''(t) = \sum_{k=0} (k+r)(k+r-1)b_k t^{k+r-2} \tag{6.9.6}$$

Suppose

$$p(t) = p_0 + p_1 t + \cdots + p_n t^n + \dots \tag{6.9.7}$$
$$q(t) = q_0 + q_1 t + \cdots + q_n t^n + \dots \tag{6.9.8}$$

If these expansions are substituted into (6.9.3), we obtain

$$\begin{aligned} t^2u'' &+ tp(t)u' + q(t) \\ &= (r(r-1) + p_0 r + q_0)b_0 \\ &\qquad + c_1 t + \cdots + c_n t^n + \cdots = 0 \end{aligned} \tag{6.9.9}$$

The generic coefficient c_n $(n \geq 1)$ in (6.9.9) depends on n and all the earlier coefficients. A general formula for c_n is quite complicated and not particularly useful for our purposes. For these reasons, we elect not to derive it. Indeed, as we shall see, it is unnecessary to have such a formula for the specific examples we wish to study. However, we do wish to draw one general conclusion from (6.9.9), namely,

$$F(r) = r(r-1) + p_0 r + q_0 \tag{6.9.10}$$

Equation (6.9.10) is called the *indicial equation*. Its two roots provide the values of r that are used in (6.9.4).

At this point we abandon the general technique and resort to explanation by example.

Example 6.9.1. *Find a Frobenius series solution for the Bessel-Clifford equation* (6.6.19).

Solution: The Bessel-Clifford equation is written in the form of (6.9.3), that is,

$$t^2u'' + (1-p)tu' + tu = 0$$

where p is a constant, so that $p_0 = 1 - p$ and $q_0 = 0$. For this equation the indicial equation is

$$r(r-1) + (1-p)r = r(r-p) = 0$$

with the two roots $r = 0$ and p. Using the root $r = 0$ results in the series solution already obtained in (6.6.22), so we try the second root, $r = p$. Using

$r = p$, we have

$$
\begin{aligned}
tu(t) &= \sum_{k=0} b_k t^{k+p+1} \\
(1-p)tu'(t) &= \sum_{k=0} (1-p)(k+p)b_k t^{k+p} \\
t^2u''(t) &= \sum_{k=0} (k+p)(k+p-1)b_k t^{k+p}
\end{aligned}
$$

By adding these expressions and simplifying, we obtain

$$
t^2u'' + (1-p)tu' + tu = \sum_{k=0} k(k+p)b_k t^{k+p} + \sum_{k=0} b_k t^{k+p+1}
$$

The first term in the first summation is zero, so by adjusting the index of summation in the second sum and combining sums, we obtain

$$
\sum_{k=1} (k(k+p)b_k + b_{k-1})t^{k+p} = 0
$$

From this identity it follows that

$$
k(k+p)b_k + b_{k-1} = 0, \qquad k \geq 1
$$

This recurrence relationship can be solved by the technique introduced earlier in the chapter. Indeed, we find that

$$
b_k = \frac{(-1)^k}{k!(1+p)(2+p)\cdots(k+p)} b_0, \qquad k = 1, 2, \ldots
$$

So, setting $b_0 = 1$, we obtain the Frobenius series representation for the solution:

$$
u_2(t) = t^p \left(1 + \sum_{k=1} \frac{(-1)^k}{k!(1+p)(2+p)\cdots(k+p)} t^k \right)
$$

There are a number of remarks that should be made regarding $u_2(t)$ in the preceding example.

1. If $p = 0$, this solution is exactly the one obtained in (6.6.22).

2. If p is a nonnegative integer, the Frobenius series solution reduces to a Taylor series. Indeed, the cases excluded in Section 6.6 (p a positive integer) are now covered.

3. If p is not an integer $u_2(t)$ is singular at $t = 0$ and, with $u_1(t)$ as given in (6.6.22), forms a linearly independent set of solutions of the Bessel-Clifford equation for $t > 0$.

Example 6.9.2. *Find a Frobenius series solution of Bessel's equation*

$$t^2u'' + tu' + (t^2 - \lambda^2)u = 0$$

Solution: One solution of this equation was already obtained in Section 6.8 in the special case where λ is an integer. So in this example we assume that λ is not an integer.

Since $p(t) = 1$ and $q(t) = -\lambda^2 + t^2$, we have $p_0 = 1$ and $q_0 = -\lambda^2$. This leads to the indicial equation

$$r(r-1) + r - \lambda^2 = r^2 - \lambda^2 = 0$$

whose solutions are $r = \pm\lambda$. We pick $r = \lambda$. So using (6.9.4), (6.9.5), and (6.9.6), we find the three series are

$$\begin{aligned}
\left(t^2 - \lambda^2\right) u(t) &= \sum_{k=0} b_k t^{k+\lambda+2} - \sum_{k=0} \lambda^2 b_k t^{k+\lambda} \\
tu'(t) &= \sum_{k=0} (k+\lambda)\, b_k t^{k+\lambda} \\
t^2u''(t) &= \sum_{k=0} (k+\lambda)(k+\lambda-1)\, b_k t^{k+\lambda}
\end{aligned}$$

We combine these summations to obtain

$$\sum_{k=0} k\,(k+2\lambda)\, b_k t^{k+\lambda} - \sum_{k=0} b_k t^{k+\lambda+2} = 0$$

We then step up the index by 2 in the second summation to obtain

$$(1+2\lambda)\, b_1 t^{1+\lambda} - \sum_{k=2} \left(k\,(k+2\lambda)\, b_k + b_{k-2}\right) t^{k+\lambda} = 0$$

So $(1+2\lambda)\, b_1 = 0$, and

$$k\,(k+2\lambda)\, b_k + b_{k-2} = 0, \qquad k \geq 2 \tag{1}$$

One solution of these recurrence relations is obtained by setting $b_0 = 1$ and $b_1 = 0$. Then, because we are assuming that λ is not an integer, and $b_1 = 0$, (1) implies that all odd-subscripted coefficients are zero and that

$$b_k = \frac{-1}{k(2\lambda + k)} b_{k-2}, \qquad k = 2, 3, \ldots$$

Therefore, the familiar recurrence table gives, with $b_0 = 1$,

$$b_{2k} = \frac{(-1)^k 2^{-2k}}{k!(1+\lambda)(2+\lambda)\cdots(k+\lambda)}$$

and consequently,

$$u(t) = t^\lambda + \sum_{k=1} \frac{(-1)^k 2^{-2k}}{k!(1+\lambda)(2+\lambda)\cdots(k+\lambda)} t^{2k+\lambda}$$

Note that this series converges for all t. (Why?)

A more detailed study of the Frobenius method is beyond the scope of this text. For further information see see Potter and Goldberg, *Mathematical Methods*, Second Edition, Great Lakes Press, 1995.

•EXERCISES

Find the indicial equation for the differential equations in Problems 1–10, and by using the root that is not an integer, find the first three coefficients of the Frobenius series solution.

1. $2tu'' + (1-t)u' + u = 0.$
2. $16tu'' + 3\left(1 + t^{-1}\right)u = 0.$
3. $2t(1-t)u'' + u' - u = 0.$
4. $2tu'' + (1+4t)u' + u = 0.$
5. $4t^2(1-t)u'' - tu' + (1-t)u = 0.$
6. $2t^2u'' - tu' + (t-5)u = 0.$
7. $2t^2u'' + t(t-1)u' + u = 0.$
8. $2tu'' + (1-t)u' - u = 0.$
9. $3tu'' + 2(1-t)u' - 4u = 0.$
10. $3t^2u'' - tu' - 4u = 0.$
11. Find the indicial equation for the Cauchy-Euler equation
$$t^2u'' + ptu' + qu = 0$$
12. Show that the roots of the indicial equation are equal for the Laguerre equation
$$tu'' + (1-t)u' + qu = 0$$

A BRIEF TABLE OF TAYLOR SERIES EXPANSIONS

1. $\dfrac{1}{1-x} = 1 + x + x^2 + \cdots + x^n + \ldots \qquad -1 < x < 1$

2. $\ln(1+x) = x - \dfrac{x^2}{2} + \cdots + (-1)^{n-1}\dfrac{x^{2n-1}}{2n-1} + \ldots \qquad -1 \leq x < 1$

3. $\sin x = x - \dfrac{x^3}{3!} + \dfrac{x^5}{5!} + \cdots + (-1)^{n-1}\dfrac{x^{2n-1}}{(2n-1)!} + \ldots \qquad -\infty < x < \infty$

4. $\cos x = 1 - \dfrac{x^2}{2!} + \dfrac{x^4}{4!} + \cdots + (-1)^{n-1}\dfrac{x^{2n}}{(2n)!} + \ldots \qquad -\infty < x < \infty$

5. $e^x = 1 + x + \dfrac{x^2}{2!} + \dfrac{x^3}{3!} + \cdots + \dfrac{x^n}{n!} + \ldots \qquad -\infty < x < \infty$

6. $\sinh t = x + \dfrac{x^3}{3!} + \dfrac{x^5}{5!} + \cdots + \dfrac{x^{2n-1}}{(2n-1)!} + \ldots \qquad -\infty < x < \infty$

7. $\cosh t = 1 + \dfrac{x^2}{2!} + \dfrac{x^4}{4!} + \cdots + \dfrac{x^{2n}}{(2n)!} + \ldots \qquad -\infty < x < \infty$

8. $\arctan x = x - \dfrac{x^3}{3} + \dfrac{x^5}{5} + \cdots + (-1)^{n-1}\dfrac{x^{2n-1}}{2n-1} + \ldots \qquad -1 < x \leq 1$

9. $(1+x)^a$

$$= 1 + ax + \frac{1}{2!}a(a-1)x^2$$

$$+\frac{1}{3!}a(a-1)(a-2)x^3 + \cdots + \qquad -1 < x < 1$$

PROJECTS

1. Taylor Series for First-Order Equations

Let $u(t) = \sum_{n=0}^{\infty} a_n t^n$ be the Taylor series of a solution of

$$tu' + \lambda u = f(t) \tag{1}$$

where $f(t) = \sum_{n=0}^{\infty} f_n t^n$
(a) Show that

$$u(t) = \sum_{n=0}^{\infty} \frac{f_n}{n+\lambda} t^n$$

(b) Show that

$$\sum_{n=0}^{\infty} \frac{f_n}{n+\lambda} t^n = t^{-\lambda} \sum_{n=0}^{\infty} \frac{f_n}{n+\lambda} t^{n+\lambda-1} = t^{-p} \int_0^t \sum_{n=0}^{\infty} f_n x^{n+\lambda}\, dx$$
$$= t^{-\lambda} \int_0^t x^{\lambda} \sum_{n=0}^{\infty} f_n x^n\, dx = t^{-\lambda} \int_0^t x^{\lambda} f(x)\, dx$$

(c) Substitute

$$u(t) = t^{-\lambda} \int_0^t x^{\lambda} f(x)\, dx$$

into (1) to show that $u(t)$ is indeed a solution.

2. Legendre Polynomials

(a) Define $P_n(t)$ as the coefficients of the expansion

$$\left(1 - 2tz + z^2\right)^{1/2} = \sum_{n=0}^{\infty} P_n(t) z^n \tag{1}$$

By computing the various partial derivatives of $\left(1 - 2tz + z^2\right)^{1/2}$ with respect to t verify that $P_n(t)$ satisfies the Legendre equation.
(b) Explain why $P_n(t)$ is a polynomial of degree n.
(c) Show that $P_n(-t) = (-1)^n P_n(t)$. (*Hint*: Replace t by $-t$ and z by $-z$ in (1).)
(d) Show that $P_n'(-t) = (-1)^{n+1} P_n'(t)$.
(e) Show that $P_{2n}(0) = \dfrac{(-1)^n (2n)!}{2^{2n} (n!)^2}$.
(e) Show that $P_{2n+1}(0) = 0$.

3. Hermite Polynomials

(a) Define $P_n(t)$ as the coefficients of the expansion

$$\exp\left(2tz - z^2\right) = \sum_{n=0}^{\infty} \frac{1}{n!} H_n(t) z^n \tag{1}$$

By computing the various partial derivatives of $\exp(2tz - z^2)$ with respect to t, verify that $H_n(t)$ satisfies the Hermite equation.

(b) Explain why $H_n(t)$ is a polynomial of degree n.

(c) Show that $H_n(-t) = (-1)^n H_n(t)$.
(d) Show that $H_{2n}(0) = (-1)^n 2^{2n}$.
(e) Show that $H_{2n+1}(0) = 0$.

4. The Gamma Function

The Gamma function $\Gamma(x)$ plays a fundamental role in many areas of advanced mathematics.

$$\Gamma(s+1) = \int_0^{\infty} e^{-t} t^s \, dt, \qquad s > -1$$

(a) Show that $\Gamma(1) = 1$.
(b) Use integration by parts to show that $\Gamma(s+1) = s\Gamma(s)$.
(c) If s is a positive integer, show that $\Gamma(s) = s!$.
(d) Suppose $r > 0$. Then $\mathcal{L}[t^r] = \int_0^{\infty} e^{-st} t^r \, dt$. Use this to show that

$$\mathcal{L}\left[t^{r-1}\right] = \frac{\Gamma(r)}{s^r}$$

(e) Show that $\Gamma(1/2) = 2\int_{-\infty}^{\infty} e^{-x^2} \, dx = \sqrt{\pi}$.

(f) Use (b) to show that

$$x(x+h)(x+2h)\cdots(x+(n-1)h) = h^n \frac{\Gamma(n + x/h)}{\Gamma(x/h)}$$

(g) Thus, show that

$$1 \cdot 3 \cdot 5 \cdots (2n-1) = 2^n \frac{\Gamma(n+1/2)}{\Gamma(1/2)} = \frac{2^n}{\sqrt{\pi}} \Gamma(n+1/2)$$

(h) Since $1 \cdot 3 \cdot 5 \cdots (2n-1) = (2n)!/(2^n n!)$, show that

$$\Gamma(n+1/2) = \frac{(2n)!}{2^n n!}\sqrt{\pi}, \text{ for } n = 1, 2, \ldots$$

Chapter 7

Numerical Methods

7.1 Introduction

With the exception of a few special cases, the only ODE's which have "closed form" solutions are those that are linear with constant coefficients.[1] Unfortunately, we often encounter ODE's that are nonlinear, or if linear, have variable coefficients. To solve these equations, we resort to approximation methods. One such method provides solutions in the form of power series, an idea we explored in Chapter 6. Another widely used method involves Fourier series, a subject we explore in Chapter 8.

In this chapter we present a variety of approximation techniques known collectively as numerical methods. When applied to initial-value problems, all these methods share one feature: They use the values of the solution and its derivatives at a finite set of points to determine an approximate value of the solution at a new point.

7.2 Direction Fields

For each t_1 and u_1, the first-order ODE

$$u' = f(t, u) \tag{7.2.1}$$

defines $f(t_1, u_1)$, the slope of the solution function passing through the point (t_1, u_1). For example, if $u' = x^2 - u$, then $f(t, u) = x^2 - u$ and at $(1, 2)$,

[1]By "closed form" we mean a finite composition of the rational and elementary transcendental functions.

the slope of the tangent line to the solution curve $u(t)$ is $f(1,2) = -1$. We can easily verify this fact since the solution of this (initial-value) problem is $u(t) = 2e^{t-1} + t - 1$. Indeed, the equation of the tangent line to the solution function at this point is given by

$$u = u_1 + f(t_1, u_1)(t - t_1) \tag{7.2.2}$$

The important point to keep in mind is that we do not need the solution to find the slope of the tangent line! This idea can be easily exploited. Suppose that for each of many points (t_i, u_i), we plot portions of the tangent line, (7.2.2), in the interval $t_i - h \leq t \leq t_i + h$, where h is small. This plot is called a *direction field* of (7.2.1). In Figure 7.1, we exhibit a direction field for the ODE

$$u' = -\frac{1 + u^2}{t\,u} \tag{7.2.3}$$

The function

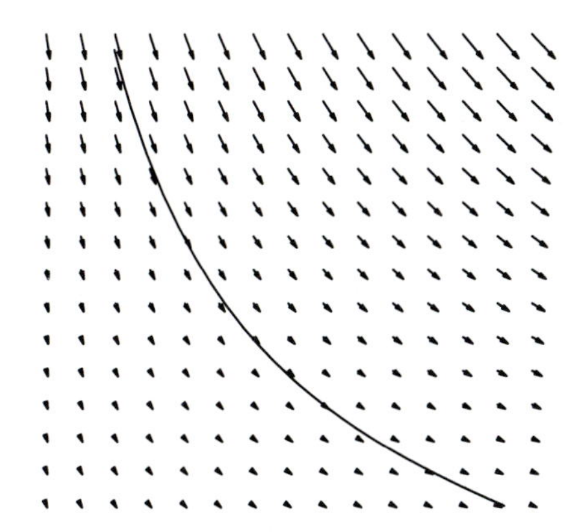

Figure 7.1: The Direction Field for $tuu' = -1 - u^2$

$$u(t) = \frac{\sqrt{45 - t^2}}{t} \tag{7.2.4}$$

is a solution of (7.2.3) that passes through the point (3,2), that is, the function $u(t) = \sqrt{45 - t^2}/t$ solves the initial-value problem

$$u' = -\frac{1 + u^2}{t\,u}, \qquad u(3) = 2$$

The graph of the solution is superimposed on the vector field displayed in Figure 7.1. Notice how the arrows "point the way" for the solution almost like children's "dot to dot". Here is another example. In Figure 7.2, we show a direction field for the linear ODE

$$u' = u + t \tag{7.2.5}$$

and superimpose on the direction field the graphs of two solutions of $u' = u + t$, namely, $u(t) = 2e^t - t - 1$ and $u(t) = -e^t - t - 1$.

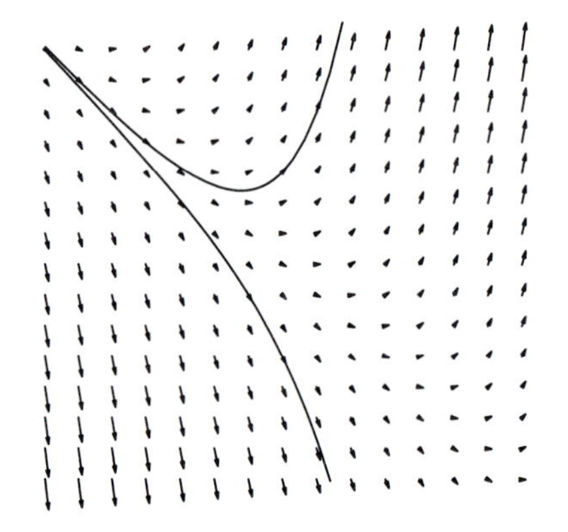

Figure 7.2: The Direction Field for $u' = t + u$

These two illustrations are meant to show how the direction field can give insight into the family of functions that solve (7.2.1).

•EXERCISES

MATLAB
Use MATLAB to construct the direction fields for the following initial-value problems. Choose (t_i, u_i) on a grid with spacing 0.5 on both axes.

1. $u' + 2tu = 0,\qquad u(0) = -2.$
2. $u' = 2u - 1,\qquad u(0) = 2.$
3. $u' - u = 0,\qquad u(0) = 2.$
4. $(u')^2 + 2u = 0,\qquad u(0) = 2.$
5. $u' - u^2 = 4,\qquad u(0) = 1.$
6. $tuu' = -1 - u^2,\qquad u(0) = 2.$

7.3 Notational Conventions

Consider the first-order initial-value problem

$$u' = f(t, u), \qquad u(t_0) = u_0 \tag{7.3.1}$$

Suppose the interval $t_0 \le t \le t_n$ is divided into n equal subintervals of size $h = (t_n - t_0)/n$. Then the points $t_k = t_0 + k(t_n - t_0)/n$, $k = 0, 1, 2, \ldots, n$, mark the beginning and the end of each of these subintervals.

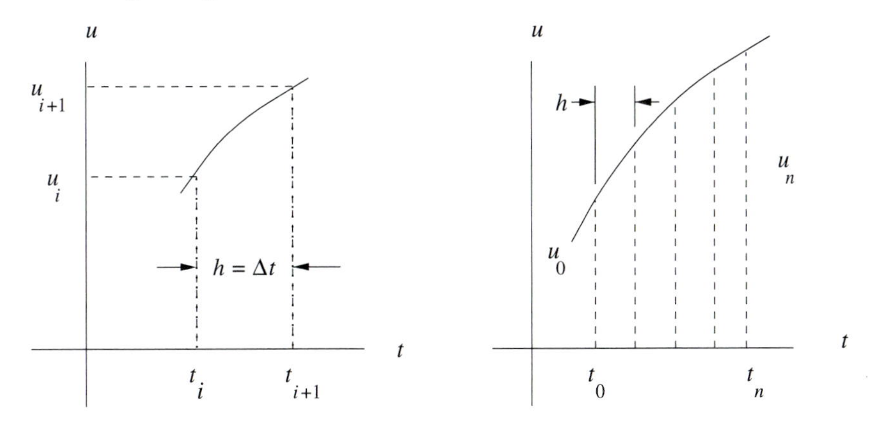

Figure 7.3: Subdividing the Interval $t_0 \le t \le t_n$ in Parts of Length h

By definition, $u(t)$ is the solution of (7.3.1) that takes on the value u_0 when $t = t_0$. The graph of $u(t)$ passes through the point (t_0, u_0) with slope $u'(t_0) = f(t_0, u_0)$. For example, consider the initial-value problem

$$u' = 4t - 2tu, \qquad u(0) = 1 \tag{7.3.2}$$

where $f(t, u) = 4t - 2tu$. The initial condition $u(0) = 1$ implies that $u'(0) = 0$. So we know that the graph of every solution of (7.3.3) passes through the point (0,1) with slope 0.

The "stepsize" $(t_n - t_0)/n$ is often denoted by h, and we adopt this convention now. So suppose that $h = 0.2$. If $n = 5$, then $t_0 = 0$, $t_1 = 0.2$, $t_2 = 0.4$, $t_3 = 0.6$, $t_4 = 0.8$, and $t_5 = 1.0$. How can we estimate the values of $u(t_i)$ at these values of t? The answer to this question constitutes the main issue for numerical analysis.

The stepsize h is usually taken as some fixed, convenient constant. (Many of the more advanced methods allow h to vary according to some criteria designed to use smaller stepsizes when the solution function is changing rapidly. These schemes are sometimes called adaptive methods.) The idea behind this can be illustrated in an elementary way. Suppose we compute approximate solutions using a stepsizes of h and $h/2$. Compare the two solutions. If the difference between the two solutions is smaller than some tolerance, we accept the solution (either one!). If not, then we do the computation again using $h/4$ and compare the new solution with the one obtained using $h/2$. We continue this process of halving the stepsize until either the tolerance condition is met or the stepsize is smaller than a preset limit. If the latter is the case, we are probably wise to switch to a different method!

The stepsize affects the accuracy of the approximation method in two ways. Loosely speaking, the smaller the stepsize, the smaller is the error that arises from the approximation method. This error is known as the *truncation error*. However, a small h implies a large n, which means that we will use more computation steps. The more computations, the greater is the error due to the rounding of numbers in the computer. This error is known as *round-off error*. One way to compare different numerical methods is to compare round-off and truncation errors. Which method to use and how to best trade off truncation against round-off errors is the art of numerical analysis.

One more notational convention: It is customary and convenient to write

$$\Delta t = h = t_{k+1} - t_k \tag{7.3.3}$$

$$u_k = u(t_k) \tag{7.3.4}$$

$$u'_k = u'(t_k) \tag{7.3.5}$$

so that

$$\Delta u_k = u_{k+1} - u_k \tag{7.3.6}$$

We now examine a few of the methods that have been developed for the solution of the initial-value problem. We made the choice of which method to examine based on a compromise between ease of understanding and practical value. We wish to make it clear that none of the methods we describe are as powerful as many of those used in practice or as hard to comprehend.

7.4 The Euler Method

The simplest, least accurate but most instructive numerical method is known as *Euler's method*. This method approximates the derivative by the difference quotient, that is,

$$u' \approx \frac{\Delta u}{\Delta t} \tag{7.4.1}$$

Then, setting $\Delta t = h$ and $\Delta u = u_{k+1} - u_k$, we see that (7.4.1) becomes

$$u_{k+1} - u_k \approx h\,u' = h\,f(t_k, u_k) \tag{7.4.2}$$

and it follows that

$$u_{k+1} \approx u_k + h\, f(t_k, u_k) \tag{7.4.3}$$

Now set

$$a_k = u'_k = f(t_k, u_k) \tag{7.4.4}$$

So the equations defining the Euler method are these:

$$u_{k+1} = u_k + h\, a_k, \qquad \text{for} \quad k = 0,\, 1,\, \ldots,\, n \tag{7.4.5}$$

Note that (7.4.5) estimates the value of $u_{k+1} = u(t_{k+1})$ by using the value $u_k = u(t_k)$. Geometrically, we have essentially replaced the graph of $u(t)$ by the graph of its tangent line between t_k and t_{k+1}. Equation 7.4.5 introduces truncation errors. We determine the size of these errors as a function of h in Section 7.6. For the time being, we shall simply accept the fact that the truncation error is proportional to h^2. In effect, halving the stepsize leads to a decrease in the truncation error proportional to 1/4. We do not examine the effect of round-off errors here.

Example 7.4.1. *Use the Euler method with $h = 0.2$ and $h = 0.1$ to find an approximate solution of the following initial-value problem:*

$$u' + 2t\, u = 4t, \qquad u(0) = 1$$

Compare these solutions with the exact solution.

Solution: We can find the exact solution by elementary methods. Indeed, the solution is $u(t) = 2 - e^{-t^2}$, as is easily verified. We obtain an approximate solution using $h = 0.2$ as follows. First of all, $f(t, u) = 4t - 2tu$, $t_0 = 0$ and $u(0) = 1$. Then $a_0 = f(0, 1) = 0$. From these data at $t = 0$, we determine an approximation to $u(0.2)$:

$$u(0.2) \approx u(t_1) = u_1 = u_0 + h\, a_0 = 1 + 0.2(0) = 1$$

We have an estimate for $u(0.2)$, so the next step is to compute a_1, which will lead to an estimate for $u(0.4)$. By definition,

$$a_1 = f(0.2, 1) = 4 \cdot 0.2 - 2 \cdot 0.2 \cdot 1 = 0.4$$

and therefore,

$$u(0.4) \approx u_2 = u_1 + h\, a_1 = 1 + 0.2 \cdot 0.4 = 1.08$$

Once again, we move a step further by determining a_2 from

$$a_2 = f(0.4, 1.08) = 4 \cdot 0.4 - 2 \cdot 0.4 \cdot 1.08 = 0.736$$

Tables 7.1 and 7.2 exhibit the approximate values of $u(t_k)$, the values of $2 - e^{-t^2}$, and the error in using the Euler method with $h = 0.2$ and $h = 0.1$, respectively. All the computed values are approximate, and the exact solution values were computed retaining six places of accuracy. We display only three places and the values at $t = 0.2k$, $k = 0, 1, 2, 3, 4, 5$.

Table 7.1: *Euler's method applied to*

$$u' + 2t\,u = 4t, \quad u(0) = 1; \quad h = 0.2$$

t_k	u_k	$u(t) = 2 - e^{-t^2}$	$\lvert u_k - u(t)\rvert$
0.0	1.000	1.000	0.000
0.2	1.000	1.039	0.039
0.4	1.080	1.148	0.068
0.6	1.227	1.302	0.075
0.8	1.413	1.472	0.059
1.0	1.601	1.632	0.031

Table 7.2: *Euler's method applied to*

$$u' + 2t\,u = 4t, \quad u(0) = 1; \quad h = 0.1$$

t_k	u_k	$u(t) = 2 - e^{-t^2}$	$\lvert u_k - u(t)\rvert$
0.0	1.000	1.000	0.000
0.2	1.020	1.039	0.019
0.4	1.116	1.148	0.032
0.6	1.268	1.302	0.034
0.8	1.446	1.473	0.027
1.0	1.618	1.632	0.014

Example 7.4.2. *Use Euler's method with $h = 0.1$ and $-1 \leq t \leq 0$ to approximate the solution of $u' = u^2$, $u(-1) = 1$. Compare the approximate solution with the of the exact solution.*

Solution: The exact solution of this initial-value problem is $u(t) = -t^{-1}$. This can easily be verified. The various calculations necessary to effect the Euler method are presented in Table 7.3. We calculate with six places of accuracy but display only three.

Table 7.3: *The error in Euler's method*

$$u' = u^2, \quad u(-1) = 1; \quad h = 0.1$$

t_k	u_k	$u(t) = -t^{-1}$	$\lvert u_k + t^{-1}\rvert$
-1.0	1.000	1.000	0.000
-0.9	1.100	1.111	0.011
-0.8	1.221	1.250	0.029
-0.7	1.375	1.429	0.054
-0.6	1.558	1.667	0.109
-0.5	1.801	2.000	0.199
-0.4	2.125	2.500	0.338
-0.3	2.577	3.333	0.756
-0.2	3.241	5.000	1.759
-0.1	4.291	10.00	5.709

Note that the accuracy decreases rapidly as we near $t = 0$ because of the large size of $u'(t) = t^{-2}$ near $t = 0$. The tangent line does a poor job of "hugging" this curve for small values of t.

•EXERCISES

MATLAB

Problems 1—6: Find an approximation to the solution by using Euler's method with $h = 0.1$ on the interval $[0, 1]$. Find the exact solutions and compare these to the corresponding approximate solutions.

1. $u' + 2tu = 0, \qquad u(0) = -2.$
2. $u' = 2u - 1, \qquad u(0) = 2.$
3. $u' - u = 0, \qquad u(0) = 2.$
4. $(u')^2 + 2u = 0, \qquad u(0) = 2.$
5. $u' - u^2 = 4, \qquad u(0) = 1.$
6. $tuu' = -1 - u^2, \qquad u(0) = 2.$

Problems 7–10: Find the approximate solutions on $[0, 1]$ using $h = 0.1$.

7. $(u')^2 + 2u = 0, \qquad u(0) = 2.$
8. $u' - 2u^2 = 4, \qquad u(0) = 1.$
9. $u' - \sin u = 2e^t, \qquad u(0) = 0.$
10. $u' = 2\sin u\, e^{t/2}, \qquad u(0) = 0.$

7.5 Heun's Method

In this section we study a slight variation of Euler's method called *Heun's method.* In Euler's method,

$$u_{k+1} = u_k + h\, a_k \tag{7.5.1}$$

where $a_k = f(t_k, u_k)$. Thus, the increment added to u_k to get u_{k+1} is $hf(t_k, u_k)$. Instead of using only information at t_k, we can use some information at t_{k+1}. Here's what we do: Set

$$a_k = f(t_k, u_k) \tag{7.5.2}$$

$$b_k = f(t_{k+1}, u_k + h\, a_k) \tag{7.5.3}$$

Then the equations defining Heun's method are

$$u_{k+1} = u_k + \tfrac{1}{2}h(a_k + b_k) \tag{7.5.4}$$

for $k = 0, 1 \ldots, n-1$. Compare (7.5.4) with (7.4.5). In (7.5.4) the term $\frac{1}{2}(a_k + b_k)$ is the average of two estimated slopes, one at t_k and the other at t_{k+1}. Now the slope at t_{k+1} is $f(t_{k+1}, u_k + h\, a_k)$. However, $u_k + h\, a_k$ is itself an estimate, since the value that ought to be the second variable in f is u_{k+1}, the very quantity we are trying to estimate! In Section 7.7, we show that the truncation error in Heun's method is proportional to h^3, a significant improvement over Euler's method.

Example 7.5.1. *Use Heun's method with $h = 0.1$ to find an approximate solution of the initial-value problem given in Example 7.4.1, that is,*

$$u' + 2t\, u = 4t, \qquad u(0) = 1$$

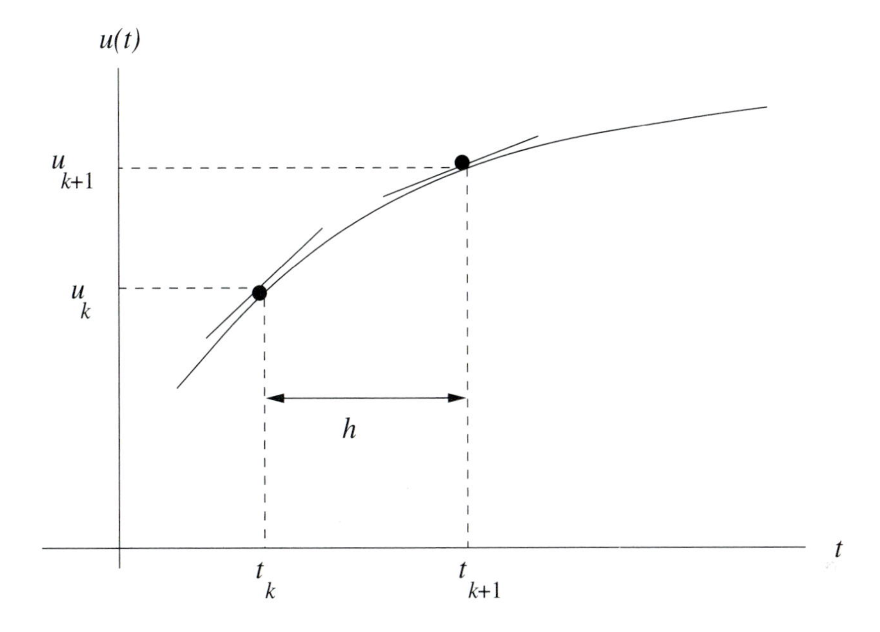

Figure 7.4: The Two Slopes Used in the Heun Method

Solution: We need to compute a_k and b_k. First of all, $a_0 = u'(0) = 0$, and $u(0.1) \approx 1 + 0.1 \cdot 0 = 1$. Thus, $b_0 = f(0.1, 1) = 4 \cdot 0.1 - 2 \cdot 0.1 \cdot 1 = 0.2$. The average of the slopes at $t_0 = 0$ and $t_1 = 0.1$ is

$$\tfrac{1}{2}h(a_0 + b_0) = \tfrac{1}{2} \cdot 0.1(0 + 0.2) = 0.01$$

We now use this value of the slope to approximate u at $t_1 = 0.1$:

$$u_1 = u_0 + \tfrac{1}{2}h(a_0 + b_0) = 1 + 0.01 = 1.01$$

We repeat these steps using the values at $t_1 = 0.1$ in place of the values at t_0. So, $a_1 = f(0.1, 1.01) = 4 \cdot 0.1 - 2 \cdot 0.1 \cdot 1.01 = 0.198$. Hence,

$$u(0.2) \approx u_1 + h\, a_1 = 1.01 + 0.1 \cdot 0.198 = 1.0298$$

by Euler's method. Thus,

$$b_1 = f(0.2, u(0.2)) = 4 \cdot 0.2 - 2 \cdot 1.0298 \cdot 0.2 = 0.38808$$

The average of the values of a_1 and b_1 along with h and u_1 is used to obtain a better approximation to $u(0.2)$:

$$u_2 = u_1 + \frac{1}{2}h(a_1 + b_1) = 1.01 + 0.1 \cdot 0.29304 = 1.03930$$

We leave the remaining computational details to the exercise set that follows. Table 7.4 gives the values of the approximate solution obtained by Heun's method. Contrast the accuracy of Euler's and Heun's method by comparing Table 7.2 with Table 7.4.

Table 7.4: *Heun's method applied to*

$$u' + 2t\,u = 4t, \quad u(0) = 1; \quad h = 0.2$$

t_k	u_k	$u(t) = 2 - e^{-t^2}$	$\|u_k - u(t)\|$
0.0	1.000	1.000	0.000
0.2	1.039	1.039	0.000
0.4	1.148	1.148	0.000
0.6	1.302	1.302	0.000
0.8	1.472	1.472	0.000
1.0	1.631	1.632	0.001

•EXERCISES

MATLAB

Problems 1–10 are the same initial-value problems as in the previous exercise set. This time use Heun's method with $h = 0.1$. Compare the approximate solutions obtained by Heun's method with the approximate solutions obtained with Euler's method. In Problems 1—6 compare the Heun approximation with the exact solutions.

1. $u' + 2tu = 0, \qquad u(0) = -2.$
2. $u' = 2u - 1, \qquad u(0) = 2.$
3. $u' - u = 0, \qquad u(0) = 2.$
4. $(u')^2 + 2u = 0, \qquad u(0) = 2.$
5. $u' - u^2 = 4, \qquad u(0) = 1.$
6. $tuu' = -1 - u^2, \qquad u(0) = 2.$
7. $(u')^2 + 2u = 0, \qquad u(0) = 2.$
8. $u' - 2u^2 = 4, \qquad u(0) = 1.$
9. $u' - \sin u = 2e^t, \qquad u(0) = 0.$
10. $u' = 2\sin u\, e^{t/2}, \qquad u(0) = 0.$

7.6 Taylor Series Methods

Taylor series methods are based on the Taylor series expansion of $u(t+h)$ about t. The relevant theorem is this:

$$u(t+h) = u(t) + hu'(t) + \cdots + \frac{1}{n!}h^n u^n(t) + \mathrm{O}(h^{n+1}) \tag{7.6.1}$$

provided that $u(t)$ has a continuous n^{th}-derivative in $[0, h]$. The truncation error in using

$$u(t+h) = u(t) + hu'(t) + \frac{1}{2}h^2 u''(t) + \cdots + \frac{1}{n!}h^n u^n(t) \tag{7.6.2}$$

is proportional to h^{n+1}, and the resulting method is called an nth-order method.

For example, the approximate value of $u(t+h)$ as given in Euler's method is obtained from (7.6.2) by choosing $n = 1$. From (7.6.1),

$$u(t+h) = u(t) + hu'(t) + \mathrm{O}(h^2) \tag{7.6.3}$$

we learn that Euler's method is first order and has error proportional to h^2.

In order to use the Taylor series method of order two, we need $u''(t)$, which we obtain by differentiating $u' = f(t, u)$. By the chain rule,

$$u'' = f_t(t, u) + f_u(t, u)\, f \tag{7.6.4}$$

where f_t and f_u are the partial derivatives of f with respect to t and u, respectively. For example, if $f(t, u) = tu^2$, then $f_t = u^2$ and $f_u = 2tu$. From (7.6.2) with $n = 2$ and (7.6.4), we have

$$\begin{aligned} u(t+h) &= u(t) + h\,u'(t) + \frac{1}{2}h^2 u''(t) && (7.6.5) \\ &= u(t) + hu'(t) + \frac{1}{2}h^2(f_t(t, u) + f_u(t, u)\, u') && (7.6.6) \end{aligned}$$

and in terms of the notation introduced earlier,

$$u_{k+1} = u_k + h\, u'_k + \tfrac{1}{2}h^2(f_t(t_k, u_k) + f_u(t_k, u_k)\, u'_k) \tag{7.6.7}$$

Example 7.6.1. *Determine the specific form of* (7.6.7) *for the initial-value problem*

$$u' = 4t - 2tu, \qquad u(0) = 1$$

and use this method to find a second-order approximation to its solution on $0 \le t \le 1$. *Use* $h = 0.2$.

Solution: We begin by recognizing that $f(t,u) = 4t - 2tu$, so that $f_t = 4 - 2u$ and $f_u = -2t$. Now use (7.6.7) to deduce

$$u_{k+1} = u_k + hu'_k + \tfrac{1}{2}h^2(4 - 2u_k - 2tu'_k)$$

We start the approximation by using the initial data, $t = 0$, $u_0 = 1$ and $u'_0 = f(0,1) = 0$. Our first computation yields an approximation to $u(0.2)$:

$$\begin{aligned} u(0.2) \approx u_1 &= u_0 + hu'_0 + \tfrac{1}{2}h^2(4 - 2u_0 - 2tu_0) \\ &= 1 + 0.2 \cdot 0 + \tfrac{1}{2}(0.2)^2(4 - 2 \cdot 1 - 2 \cdot 0.2 \cdot 0) = 1.04 \end{aligned}$$

Also, we need to compute an approximation to $u'(0.2)$:

$$u'(0.2) \approx f(0.2, 1.04) = u'_1 = 4 \cdot 0.2 - 2 \cdot 0.2 \cdot 1.04 = 0.384$$

With these data, we can proceed to approximate $u(0.4)$:

$$\begin{aligned} u(0.4) \approx u_2 &= u_2 + hu'_1 + \tfrac{1}{2}h^2(4 - 2u_1 - 2tu_1) \\ &= 1.04 + 0.2 \cdot 0.384 + \tfrac{1}{2}(0.2)^2(4 - 2 \cdot 1.04 - 2 \cdot 1.02 \cdot 0.384) \\ &= 1.15212 \end{aligned}$$

It is interesting to compare the methods of Heun and Taylor on the same problem. We do so in Table 7.5.

Table 7.5: *Comparing Heun's and Taylor's methods for*

$$u' + 2t\,u = 4t, \quad u(0) = 1; \quad h = 0.2$$

t_k	Heun	Taylor	$u(t) = 2 - e^{-t^2}$
0.0	1.000	1.000	1.000
0.2	1.039	1.040	1.039
0.4	1.148	1.152	1.148
0.6	1.302	1.311	1.302
0.8	1.472	1.484	1.472
1.0	1.631	1.632	1.632

Table 7.5 supports the assertion made earlier that the Heun method is second order, as is seen by comparing the columns headed Heun and Taylor.

In the next section we shall see that the Heun method may also be viewed as one of a variety of methods known collectively as Runge-Kutta methods.

The third-order Taylor method requires the computation of $u'''(t)$. Since $u''' = d(u'')/dt$, we have, invoking the product rule and (7.6.4),

$$\begin{aligned} u''' &= \frac{d}{dt}(f_t + f_u\, f) \\ &= \frac{d}{dt} f_t + \frac{d}{dt}(f_u\, f) \end{aligned} \tag{7.6.9}$$

Now we apply the chain rule to the first summand:

$$\frac{d}{dt} f_t = f_{tt} + f_{tu}\, f \tag{7.6.10}$$

Next we apply the product and the chain rule to the second summand:

$$\frac{d}{dt}(f_u\, f) = f\, \frac{d}{dt} f_u + f_u\, \frac{d}{dt} f = f\, f_{ut} + f^2 f_{uu} + f_u\, f_t + f_u^2 f \tag{7.6.11}$$

Therefore, noting that $f_{tu} = f_{tu}$ we can write u''' in terms of various derivatives of f:

$$\begin{aligned} u''' &= f\, f_{ut} + f^2 f_{uu} + f_u\, f_t + f_u^2 f + f_{tt} + f_{tu} f && (7.6.12) \\ &= f_{tt} + 2 f_{tu} f + f_{uu} f^2 + f_u f_t + f_u^2 f && (7.6.13) \end{aligned}$$

Higher-order Taylor methods lead to an exponential growth in terms and so to a significant increase in function evaluations. These methods have been replaced by more efficient and accurate schemes.

•EXERCISES

MATLAB

Using Taylor's method of order two, find approximate solutions of the following initial-value problems. Compare these approximations with those obtained by Heun's method in the previous exercise set.

1. $u' + 2tu = 0, \qquad u(0) = -2.$
2. $u' = 2u - 1, \qquad u(0) = 2.$
3. $u' - u = 0, \qquad u(0) = 2.$
4. $(u')^2 + 2u = 0, \qquad u(0) = 2.$
5. $u' - u^2 = 4, \qquad u(0) = 1.$
6. $tuu' = -1 - u^2, \qquad u(0) = 2.$

7. $(u')^2 + 2u = 0,$ $u(0) = 2.$
8. $u' - 2u^2 = 4,$ $u(0) = 1.$
9. $u' - \sin u = 2e^t,$ $u(0) = 0.$
10. $u' = 2\sin u\, e^{t/2},$ $u(0) = 0.$

7.7 Runge-Kutta Methods

One of the most popular and widely used numerical methods for the solution of initial-value problems is called the *Runge-Kutta* method. In fact like the Taylor method, the Runge-Kutta method is a family of methods of various orders. We begin by examining the simplest such method, the Runge-Kutta method of second order.

Actually, the second-order Runge-Kutta method is a generalization of Heun's method and is most easily understood by reviewing the ideas behind that method. Recall the approximation scheme defining Heun's method:

$$\begin{aligned} u_{k+1} &= u_k + \tfrac{1}{2}h\left(f(t_k, u_k) + f(t_{k+1}, u_k + hu'_k)\right) \\ &= u_k + \tfrac{1}{2}hf(t_k, u_k) + \tfrac{1}{2}hf(t_{k+1}, u_k + hu'_k) \end{aligned} \tag{7.7.1}$$

In place of the average of the slopes appearing in the second summand, we use an arbitrary combination and demand that it be second order. Here's what we mean: Consider the expression,

$$\begin{aligned} u_{k+1} = {} & u_k + c_1hf(t_k, u_k) \\ & + c_2hf(t_k + \alpha h, u_k + \beta u'_k) \end{aligned} \tag{7.7.2}$$

The last term in this sum can itself be expanded and yields

$$\begin{aligned} f(t_k + \alpha h, u_k + \beta u'_k) = {} & f(t_k, u_k) + \alpha hf_t(t_k, u_k) \\ & + \beta hf_u(t_k, u_k)\, f(t_k, u_k) + kh^2 \end{aligned} \tag{7.7.3}$$

(See the Projects section at the end of the chapter.) We now substitute this result into the previous displayed equation and simplify.

$$\begin{aligned} u_{k+1} = {} & u_k + (c_1 + c_2)hf(t_k, u_k) \\ + {} & \alpha c_2h^2 f_t(t_k, u_k) + \beta c_2h^2 f_u(t_k, u_k)\, f(t_k, u_k) + kh^3 \end{aligned} \tag{7.7.4}$$

On the other hand, we have seen in (7.6.7) that every second order method must have the form

$$u_{k+1} = u_k + hu'_k + \frac{1}{2}h^2\left(f_t(t_k, u_k) + f_u(t_k, u_k)u'_k\right) \tag{7.7.5}$$

Comparing these two expressions leads us to conclude that the Runge-Kutta method will be second order if we choose the parameters α, β, c_1, and c_2 so that (7.7.4) reduces to (7.7.5). Thus,

$$\begin{aligned} c_1 + c_2 &= 1 \\ \alpha c_2 &= \beta c_2 = \frac{1}{2} \end{aligned} \tag{7.7.6}$$

We leave it to the reader to verify that choosing $\alpha = \beta = 1$ leads to $c_1 = c_2 = \frac{1}{2}$ and that this is the Heun method.

Runge-Kutta methods of the fourth order are based on a similar idea. We take combinations of the values of $f(t, u)$ at various points in the interval $t_k \leq t \leq t_{k+1}$ and adjust the parameters so that our approximation to u_{k+1} agrees with the Taylor expansion (7.6.2) up to and including the term involving h^4. The details of this work are quite complicated and beyond the scope of this book. But it is easy to write down the conclusions, and we do so now. Define

$$u_{k+1} = u_k + \frac{1}{6}h(a_k + 2b_k + 2c_k + d_k) \tag{7.7.7}$$

where

$$\begin{aligned} a_k &= f(t_k, u_k) \\ b_k &= f(t_k + \tfrac{1}{2}h, u_k) + \tfrac{1}{2}ha_k \\ c_k &= f(t_k + \tfrac{1}{2}h, u_k) + \tfrac{1}{2}hb_k \\ d_k &= f(t_k + h, u_k) + hc_k \end{aligned} \tag{7.7.8}$$

for $k = 0, 1, \ldots, n-1$. From this set of relationships, we can see that a_k is the approximate slope at the leftmost node — the point (t_k, u_k) — and d_k is the approximate slope at the rightmost node. The numbers b_k and c_k are approximate slopes at the center of the interval. The two center slopes are weighted twice as much as the end slopes. The factor $\frac{1}{6}$ makes the second term on the right-hand side of (7.7.9) a weighted average of these slopes.

Although we do not need derivatives of f in this Runge-Kutta scheme, we do need to evaluate f four times for each step. The truncation error is proportional to h^5. Up until recently, this was the method of choice for most users of numerical differential equation solvers. Runge-Kutta methods of order 5 and 6 are now commonly used.

The power of the fourth-order Runge-Kutta method is best illustrated by solving our familiar initial-value problem $u' + 2tu = 4t$, $u(0) = 1$. The

Table 7.6: *Runge-Kutta fourth-order method applied to*

$$2t\,u = 4t, \quad u(0) = 1, \quad h = 0.2$$

t_k	u_k	$u(t) = 2 - e^{-t^2}$	$\lvert u_k - u(t)\rvert$
0.0	1.000	1.000	0.000
0.2	1.039	1.039	0.000
0.4	1.148	1.148	0.000
0.6	1.302	1.302	0.000
0.8	1.473	1.473	0.000
1.0	1.632	1.632	0.000

following table was computed by using MATLAB's version of this Runge-Kutta scheme.[2]

•EXERCISES

MATLAB

Problems 1—10 are the same initial-value problems that we have used in each of the exercise sets. This time use the fourth-order Runge-Kutta method to obtain the approximate solutions.

1. $u' + 2tu = 0, \qquad u(0) = -2.$
2. $u' = 2u - 1, \qquad u(0) = 2.$
3. $u' - u = 0, \qquad u(0) = 2.$
4. $(u')^2 + 2u = 0, \qquad u(0) = 2.$
5. $u' - u^2 = 4, \qquad u(0) = 1.$
6. $tuu' = -1 - u^2, \qquad u(0) = 2.$
7. $(u')^2 + 2u = 0, \qquad u(0) = 2.$
8. $u' - 2u^2 = 4, \qquad u(0) = 1.$
9. $u' - \sin u = 2e^t, \qquad u(0) = 0.$
10. $u' = 2\sin u\, e^{t/2}, \qquad u(0) = 0.$
11. Show that the choices $\alpha = \beta = 1$ in (7.7.6) imply that $c_1 = c_2 = \frac{1}{2}$, and then the Runge-Kutta method is precisely Heun's method.

[2]MATLAB labels this method "ode45."

7.8 Multivariable Methods

Consider the initial-value problem for the system

$$\mathbf{x}' = \mathbf{f}(t, \mathbf{x}), \qquad \mathbf{x}(0) = \mathbf{x}_0 \tag{7.8.1}$$

where, as usual, we use the abbreviations $\mathbf{x} = \mathbf{x}(t)$ and $\mathbf{x}' = \mathbf{x}'(t)$, and

$$\mathbf{x}(t) = \begin{bmatrix} x_1(t) \\ x_2(t) \\ \vdots \\ x_n(t) \end{bmatrix}, \qquad \mathbf{x}'(t) = \begin{bmatrix} x_1'(t) \\ x_2'(t) \\ \vdots \\ x_n'(t) \end{bmatrix} \tag{7.8.2}$$

The vector of initial values is given by

$$\mathbf{x}(0) = \begin{bmatrix} x_1(0) \\ x_2(0) \\ \vdots \\ x_n(0) \end{bmatrix} \tag{7.8.3}$$

and $\mathbf{f}$ is given by

$$\mathbf{f}(t, \mathbf{x}) = \begin{bmatrix} f_1(t, \mathbf{x}) \\ f_2(t, \mathbf{x}) \\ \vdots \\ f_n(t, \mathbf{x}) \end{bmatrix} \tag{7.8.4}$$

Suppose that $\mathbf{x}(t)$ is the exact solution of (7.8.1) and $\mathbf{x}_k = \mathbf{x}_k(t)$ is an approximate solution. We measure the error at $t_k = kh$ by using $\mathbf{x}_k(t)$ in place of $\mathbf{x}(t)$ by the norm

$$\|\mathbf{x}_k - \mathbf{x}(t_k)\| = \sqrt{\sum_{k=1}^{n} (x_k - x(t_k))^2} \tag{7.8.5}$$

Each of the methods discussed in the previous sections has a counterpart in vector form that is remarkably similar to the scalar form. As in the scalar case, the Euler method is easiest to understand, and we present this first. The resemblance between the scalar and multivariable Euler method is strikingly close. Indeed, if we allow vectors to contain a single entry, then the multivariable method subsumes the scalar method as a special case.

In the Euler method the approximation $\mathbf{x}_k$ is given by

$$\mathbf{x}_{k+1} = \mathbf{x}_k + h\mathbf{f}(t_k, \mathbf{x}_k) \tag{7.8.6}$$

for $k = 0, 1, \ldots, n-1$. The following example illustrates the use of this approximation scheme. The reader should pay particular attention to the parallel between the scalar (one variable) and the vector (many variable) schemes.

Example 7.8.1. *Use the multivariable Euler method with $h = 0.25$ and $0 \leq t \leq 1$ to find an approximate solution of the initial-value problem*

$$\mathbf{x}' = \begin{bmatrix} 0 & -1 \\ 1 & 0 \end{bmatrix} \mathbf{x}, \qquad \mathbf{x}(0) = \begin{bmatrix} 1 \\ 0 \end{bmatrix}$$

Solution: For this system, (7.8.6) becomes

$$\mathbf{x} = \begin{bmatrix} x_1 \\ x_2 \end{bmatrix}, \qquad \mathbf{x}' = \mathbf{f}(t, \mathbf{x}) = \begin{bmatrix} -x_2 \\ x_1 \end{bmatrix}$$

where

$$\mathbf{f}(t, \mathbf{x}) = \begin{bmatrix} 0 & -1 \\ 1 & 0 \end{bmatrix} \begin{bmatrix} x_1 \\ x_2 \end{bmatrix} = \begin{bmatrix} -x_2 \\ x_1 \end{bmatrix}$$

Equation (7.8.6) provides the approximate values of the solution. To find $\mathbf{x}_1$ we use $\mathbf{x}_0$ and $\mathbf{f}(0, \mathbf{x}_0)$:

$$\begin{aligned} \mathbf{x}_1 &= \mathbf{x}_0 + h\mathbf{f}(0, \mathbf{x}_0) \\ &= \begin{bmatrix} 1 \\ 0 \end{bmatrix} + 0.25 \begin{bmatrix} 0 & -1 \\ 1 & 0 \end{bmatrix} \begin{bmatrix} 1 \\ 0 \end{bmatrix} \\ &= \begin{bmatrix} 1 \\ 0.25 \end{bmatrix} \end{aligned}$$

The remaining values are obtained in a like manner, and the results are recorded in Table 7.7. The exact solution can be obtained by the methods used in Chapter 2, and this solution is found to be

$$\mathbf{x}(t) = \begin{bmatrix} \cos t \\ \sin t \end{bmatrix}$$

Table 7.7 shows a comparison between the approximate solution and the exact solution.

Table 7.7: *Multivariable Euler's method applied to*

$$\mathbf{x}' = \begin{bmatrix} 0 & -1 \\ 1 & 0 \end{bmatrix} \mathbf{x}, \qquad \mathbf{x}(0) = \begin{bmatrix} 1 \\ 0 \end{bmatrix}; \quad h = 0.25$$

t_k	$\mathbf{x}_k$	$\mathbf{x}(t) = \begin{bmatrix} \cos t \\ \sin t \end{bmatrix}$	$\lVert\mathbf{x}_k - \mathbf{x}(t)\rVert$
0.00	$\begin{bmatrix} 1.000 \\ 0.000 \end{bmatrix}$	$\begin{bmatrix} 1.000 \\ 0.000 \end{bmatrix}$	0.000
0.25	$\begin{bmatrix} 1.000 \\ 0.250 \end{bmatrix}$	$\begin{bmatrix} 0.969 \\ 0.247 \end{bmatrix}$	0.031
0.50	$\begin{bmatrix} 0.938 \\ 0.5000 \end{bmatrix}$	$\begin{bmatrix} 0.878 \\ 0.479 \end{bmatrix}$	0.064
0.75	$\begin{bmatrix} 0.813 \\ 0.734 \end{bmatrix}$	$\begin{bmatrix} 0.732 \\ 0.682 \end{bmatrix}$	0.096
1.00	$\begin{bmatrix} 0.629 \\ 0.938 \end{bmatrix}$	$\begin{bmatrix} 0.540302 \\ 0.841470 \end{bmatrix}$	0.132

Here are the equations defining the multivariable Heun method. For $k = 0, 1, \ldots, n-1$

$$\begin{aligned} \mathbf{x}_{k+1} &= \mathbf{x}_k + \tfrac{1}{2}h(\mathbf{a}_k + \mathbf{b}_k) && (7.8.7) \\ \mathbf{a}_k &= \mathbf{f}(t_k, \mathbf{x}_k) && (7.8.8) \\ \mathbf{b}_k &= \mathbf{f}(t_{k+1}, \mathbf{x}_k + h\mathbf{a}_k) \end{aligned}$$

Again, it is worth noting that the multivariable Heun method reduces to the scalar Heun method if the vectors contain only one entry.

Example 7.8.2. *Use the multivariable Heun method to solve the initial-value problem given in Example 7.8.1 and compare the results with Euler's method:*

$$\mathbf{x}' = \begin{bmatrix} 0 & -1 \\ 1 & 0 \end{bmatrix} \mathbf{x}, \qquad \mathbf{x}(0) = \begin{bmatrix} 1 \\ 0 \end{bmatrix}$$

Solution: Here we use (7.8.7) and (7.8.8) and we summarize the results of the computation in Table 7.8.

Table 7.8: *Comparison between Euler's and Heun's methods for*

$$\mathbf{x}' = \begin{bmatrix} 0 & -1 \\ 1 & 0 \end{bmatrix}\mathbf{x}, \qquad \mathbf{x}(0) = \begin{bmatrix} 1 \\ 0 \end{bmatrix}; \quad h = 0.25$$

t_k	Euler	$\mathbf{x}(t) = \begin{bmatrix} \cos t \\ \sin t \end{bmatrix}$	Heun
0.00	$\begin{bmatrix} 1.000 \\ 0.000 \end{bmatrix}$	$\begin{bmatrix} 1.000 \\ 0.000 \end{bmatrix}$	$\begin{bmatrix} 1.000 \\ 0.000 \end{bmatrix}$
0.25	$\begin{bmatrix} 1.000 \\ 0.250 \end{bmatrix}$	$\begin{bmatrix} 0.969 \\ 0.247 \end{bmatrix}$	$\begin{bmatrix} 0.969 \\ 0.250 \end{bmatrix}$
0.50	$\begin{bmatrix} 0.938 \\ 0.500 \end{bmatrix}$	$\begin{bmatrix} 0.878 \\ 0.479 \end{bmatrix}$	$\begin{bmatrix} 0.876 \\ 0.484 \end{bmatrix}$
0.75	$\begin{bmatrix} 0.813 \\ 0.734 \end{bmatrix}$	$\begin{bmatrix} 0.732 \\ 0.682 \end{bmatrix}$	$\begin{bmatrix} 0.728 \\ 0.688 \end{bmatrix}$
1.00	$\begin{bmatrix} 0.629 \\ 0.938 \end{bmatrix}$	$\begin{bmatrix} 0.540 \\ 0.841 \end{bmatrix}$	$\begin{bmatrix} 0.533 \\ 0.849 \end{bmatrix}$

•EXERCISES

MATLAB

For each initial-value problem, obtain an approximate solution at $t = 0.1$, 0.2, and $t = 0.3$ using the multivariable Euler and Heun methods. Use $h = 0.1$.

1. $\mathbf{x}' = \begin{bmatrix} 0 & -1 \\ 1 & 0 \end{bmatrix}\mathbf{x}, \qquad \mathbf{x}(0) = \begin{bmatrix} 1 \\ 1 \end{bmatrix}$

2. $\mathbf{x}' = \begin{bmatrix} t & -1 \\ 0 & 0 \end{bmatrix}\mathbf{x}, \qquad \mathbf{x}(0) = \begin{bmatrix} 1 \\ 0 \end{bmatrix}$

3. $\mathbf{x}' = \begin{bmatrix} 0 & -1 \\ 1 & 0 \end{bmatrix} \mathbf{x} + \begin{bmatrix} 1 \\ t \end{bmatrix}, \qquad \mathbf{x}(0) = \begin{bmatrix} 0 \\ 0 \end{bmatrix}$

4. $\mathbf{x}' = \begin{bmatrix} 1 & -1 \\ 0 & t \end{bmatrix} \mathbf{x} + \begin{bmatrix} 1 \\ 1 \end{bmatrix}, \qquad \mathbf{x}(0) = \begin{bmatrix} 0 \\ 0 \end{bmatrix}$

5. $\mathbf{x}' = \begin{bmatrix} 0 & 1 & 0 \\ 0 & 0 & 1 \\ 2 & -3 & 1 \end{bmatrix} \mathbf{x} + \begin{bmatrix} 0 \\ 1 \\ t \end{bmatrix}, \qquad \mathbf{x}(0) = \begin{bmatrix} 0 \\ 0 \\ 0 \end{bmatrix}$

7.9 Higher-Order Equations

In Chapter 4 we saw how linear n^{th}-order ODE's can be transformed into systems of n first-order equations. The same idea works for nonlinear equations, as we now shall see.

We explain the idea by converting a general second-order, nonlinear, initial-value problem into an initial-value problem involving a system of two first-order equations. Consider

$$u'' = f(t, u, u'), \qquad u(0) = a, \quad u'(0) = b \tag{7.9.1}$$

Define $v = u'$. Then $v' = u'' = f(t, u, u')$, so that (7.9.1) may be written as

$$\begin{aligned} u' &= v \\ v' &= f(t, u, u') \qquad u(0) = a, \quad v(0) = b \end{aligned} \tag{7.9.2}$$

We put (7.9.2) in vector notation by setting

$$\mathbf{x}' = \begin{bmatrix} u' \\ v' \end{bmatrix}, \qquad \mathbf{x} = \begin{bmatrix} u \\ v \end{bmatrix}, \qquad \mathbf{x}(0) = \begin{bmatrix} a \\ b \end{bmatrix}$$

Then (7.9.1) is transformed into the system

$$\mathbf{x}'(t) = \begin{bmatrix} v(t) \\ f(t, \mathbf{x}) \end{bmatrix}, \qquad \mathbf{x}(0) = \begin{bmatrix} a \\ b \end{bmatrix} \tag{7.9.3}$$

We illustrate this in the next example by using a linear, second-order initial-value problem whose solution can be written in closed form.

Example 7.9.1. *Find an approximate solution of*

$$u'' + 16u = 24 \sin 2t, \qquad u(0) = u'(0) = 0$$

using the multivariable Euler method on the system equivalent to this initial-value problem. Use $h = 0.1$, and solve in the interval $0 \le t \le 0.5$.

Solution: The equivalent system is

$$\mathbf{x}'(t) = \begin{bmatrix} v(t) \\ 24\sin 2t - 16u \end{bmatrix}, \qquad \mathbf{x}(0) = \begin{bmatrix} 0 \\ 0 \end{bmatrix}$$

Because this equation is linear, we can also put the system in this form:

$$\mathbf{x}'(t) = \begin{bmatrix} 0 & 1 \\ -16 & 0 \end{bmatrix} \begin{bmatrix} u(t) \\ v(t) \end{bmatrix} + \begin{bmatrix} 0 \\ 24\sin 2t \end{bmatrix}, \qquad \mathbf{x}(0) = \begin{bmatrix} 0 \\ 0 \end{bmatrix}$$

At each stage of the computation we obtain a vector whose first component

Table 7.9: *Multivariable Euler's method applied to*

$u'' + 16u = 24\sin 2t, \quad u(0) = u'(0) = 0; \quad h = 0.1$

t_k	u_k	$2\sin 2t - 4\sin 4t$	$\lvert u_k - u(t)\rvert$
0.0	0.000	0.000	0.000
0.1	0.000	0.008	0.008
0.2	0.000	0.061	0.061
0.3	0.048	0.197	0.149
0.4	0.189	0.435	0.246
0.5	0.458	0.774	0.316

is an approximation to the solution we desire and whose second component is $u'(t)$. In Table 7.9, we display only the approximation to u, the exact solution $u(t) = 2\sin 2t - 4\sin 4t$, and the error, $|u_k(t) - u(t)|$.

Example 7.9.2. *Use the multivariable Heun method on the initial-value problem in Example 7.9.1:*

$$u'' + 16u = 24\sin 2t, \qquad u(0) = u'(0) = 0$$

Compare the results of the Heun and Euler methods.

Solution: As we should expect, the Heun method yields much improved accuracy. This is seen in the next example. The results of this computation are shown in Table 7.10.

Table 7.10: *Multivariable Heun method applied to*

$$u'' + 16u = 24\sin 2t, \quad u(0) = u'(0) = 0; \quad h = 0.1$$

t_k	Euler	$2\sin 2t - 4\sin 4t$	Heun
0.0	0.000	0.000	0.000
0.1	0.000	0.008	0.000
0.2	0.000	0.061	0.048
0.3	0.048	0.197	0.183
0.4	0.189	0.435	0.428
0.5	0.458	0.774	0.781

•EXERCISES

MATLAB

1. Verify Table 7.9.

2. Verify Table 7.10.

3. Use the multivariable Heun method with $h = 0.1$ to approximate the solution of $u'' + 16u^2 = 24\sin 2t$, $u(0) = u'(0) = 0$ at $t = 0.1, 0.2$ and 0.3.

4. Use the multivariable Heun method with $h = 0.1$ at $t = 0.1$, 0.2 and 0.3 to approximate the solution of

$$u'' + 0.2(u')^2 + 2u^2 = 4e^{-t}\sin t, \qquad u(0) = u'(0) = 0$$

Chapter 8

Boundary-Value Problems

8.1 Introduction

The vibrations of the strings of a violin, the motion of the membrane of a drum, the oscillation of the earth, the movement of water waves and acoustic waves, the passage of electricity in a transmission line, the diffusion of heat in a conductor, gravitation, magnetic and electric potentials, and the flow of an incompressible fluid are some of the many phenomena from physics and engineering whose behavior is modeled by partial differential equations, PDQ's. Although not primarily a topic of this book, we include its study here because under certain rather common circumstances, the solution of these PDE's can be made to depend on the study of related boundary-value problems in ODE's.

8.1.1 Wave Motion

One of the first phenomena modeled by a PDE was the motion of a tightly stretched, flexible string. The problem is to describe the position of the string shown in Figure 8.1 as a function of time.

To do this, we seek an equation whose solution $u(x, t)$ gives the deflection of the string for any position x and any time t. Let us make the following assumptions about this "ideal" string:

1. The string offers no resistance to bending so that no shearing force exists on a surface normal to the string.
2. The tension is so large that the string's weight is negligible.

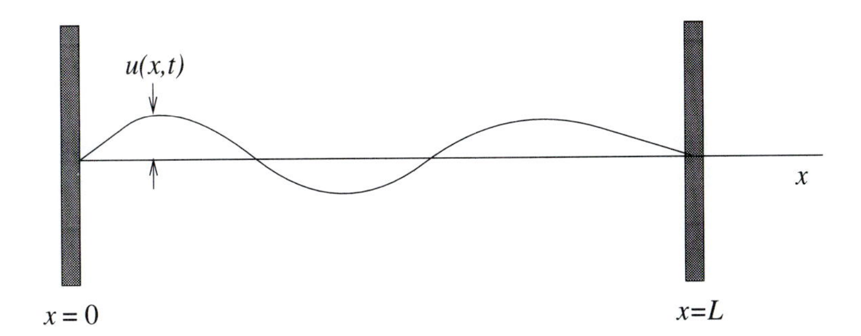

Figure 8.1: A Deformed, Flexible String at Instant t

3. Every element moves normal to the x-axis.
4. The slope of the deflection curve is small.
5. The string's mass per unit length is constant.
6. The effects of friction are negligible.

Under these assumptions, we find that $u = u(x,t)$ satisfies

$$u_{tt} = a^2 u_{xx} \tag{8.1.1}$$

where the subscripts denote partial differentiation with respect to x and t, respectively.[1]

Equation (8.1.1) is the *one-dimensional wave equation.* The constant a is the *wave speed,* a positive parameter related to the mass and tension in the string. We study (8.1.1) in Sections 8.2 and 8.3.

8.1.2 Diffusion

A metal bar that has an uneven distribution of temperature experiences a warming of the cooler sections at the expense of a cooling of the warmer ones through the mechanism of heat transfer. This passage of heat is known as conduction. The transfer of heat from the hot end of a spoon to its handle is a familiar and sometimes painful example of diffusion. Under the appropriate assumption, the first law of thermodynamics leads to the *diffusion equation*

$$T_t = k(T_{xx} + T_{yy} + T_{zz}) \tag{8.1.2}$$

[1]That is, $u_x = \partial u/\partial x, u_t = \partial u/\partial t, u_{xx} = \partial^2 u/\partial x^2$, and $u_{tt} = \partial^2 u/\partial t^2$.

where $T = T(x, y, z, t)$ is the temperature at time t at the point (x, y, z). The parameter k is a physically measured constant, the *thermal diffusivity.* Two special cases of the diffusion equation are of interest to us. The first occurs when we try to predict the temperature in a long slender rod with insulated sides. Then T is a function of t and one space variable, say, x, and (8.1.2) reduces to

$$T_t = kT_{xx} \tag{8.1.3}$$

Equation (8.1.3) is the *one-dimensional heat equation* and we study this PDE in Section 8.5. The second arises when T does not vary with time and this case is studied in Section 8.6 and introduced next.

8.1.3 Laplace's Equation

If we assume that T does not vary in time, then $T_t = 0$ and (8.1.2) reduces to *Laplace's equation*

$$T_{xx} + T_{yy} + T_{zz} = 0 \tag{8.1.4}$$

Steady-state heat conduction is one problem modeled by Laplace's equation. Other problems involving Laplace's equation arise in the study of certain gravitational, magnetic, and electric potential applications. We consider these problems in Sections 8.6 and 8.7.

8.2 Separation of Variables

We now present a powerful technique called *separation of variables* that reduces the solution of many important PDE's to the solution of boundary-value problems for ODE's. We illustrate this method by studying the following problem involving wave motion.

Consider a taut string of length L that is held fixed at its ends, which we take as $x = 0$ and $x = L$ and which has an initial displacement $f(x)$ and no initial velocity. (See Figure 8.1.) For each $t > 0$ and each x, $0 \le x \le L$, the displacement $u = u(x, t)$ satisfies the wave equation

$$u_{tt} = a^2 u_{xx} \tag{8.2.1}$$

Since the string is assumed fixed at the endpoints $x = 0$ and $x = L$,

$$u(0, t) = u(L, t) = 0 \tag{8.2.2}$$

Because the string is at rest, the initial vertical velocity normal to the string is zero at each x. In terms of u, this means that

$$u_t(x,0) = 0 \tag{8.2.3}$$

The last condition on u arises from the fact that at $t = 0$ the string's position is given by $f(x)$, that is,

$$u(x,0) = f(x) \tag{8.2.4}$$

The fundamental assumption of the method of separation of variables is this: There is a solution of (8.2.1) that is the product of a function of t and a function of x. Put more specifically, there are functions $T = T(t)$ and $X = X(x)$ such that

$$u(x,t) = T(t)\,X(x) \tag{8.2.5}$$

is a solution of (8.2.1). For this to be the case, it is necessary that

$$u_{tt} = XT'' = a^2 u_{xx} = a^2 X''T \tag{8.2.6}$$

where the primes denote differentiation with respect to the corresponding independent variable. Now write (8.2.6) as

$$\frac{1}{a^2}\frac{T''}{T} = \frac{X''}{X} \tag{8.2.7}$$

The left-hand side of (8.2.7) is a function of t alone, while the right-hand side varies only with x. This is possible only if both the left- and right-hand sides are constant. This constant value, which we call the *separation constant*, will be denoted by μ. Then

$$\frac{1}{a^2}\frac{T''}{T} = \mu = \frac{X''}{X} \tag{8.2.8}$$

Equation (8.2.8) is, in fact, two equations, namely,

$$T'' - \mu a^2 T = 0 \tag{8.2.9}$$

$$X'' - \mu X = 0 \tag{8.2.10}$$

We assume that μ is real, so either $\mu \geq 0$ or $\mu < 0$. It is not difficult to convince oneself that $\mu \geq 0$ is a physical impossibility, so if this is accepted, then $\mu < 0$. (If μ were nonnegative, the solutions of (8.2.9) would involve exponential functions, which cannot describe the motion of a string fixed at its ends.) For notational convenience, we set $b^2 = -\mu$. Then the general solutions of (8.2.9) and (8.2.10) are

$$T(t) = A\cos(bat) + B\sin(bat) \tag{8.2.11}$$

and

$$X(t) = C\cos bx + D\sin bx \tag{8.2.12}$$

From these two equations, we use (8.2.5) and obtain

$$\begin{aligned} u(x,t) &= T(t)X(x) \\ &= (A\cos bat + B\sin bat)(C\cos bx + D\sin bx) \end{aligned} \tag{8.2.13}$$

Now, since we are assuming that $u(0,t) = 0$, (8.2.13) reduces to

$$(A\cos bat + B\sin bat)C = 0 \tag{8.2.14}$$

which must hold for all t. If $C \neq 0$, then (8.2.14) requires that both A and B be zero. But then $u(x,t) = 0$, and this is a solution, albeit quite uninteresting. So we shall take $C = 0$. Then (8.2.13) simplifies to

$$u(x,t) = D\sin bx(A\cos bat + B\sin bat) \tag{8.2.15}$$

The next step is to use (8.2.3), which requires that $u_t(x,0) = 0$. We find

$$\begin{aligned} u_t(x,0) &= D\sin bx(Aab\sin 0 + Bab\cos 0) \\ &= DBab\sin bx \\ &= 0 \end{aligned} \tag{8.2.16}$$

For the same reason that we did not pick $C = 0$, we assume that $D \neq 0$ and $b \neq 0$ and $a \neq 0$. This leaves $B = 0$ and that choice leads to

$$u(x,t) = E\cos bat\sin bx \tag{8.2.17}$$

where we have used E for DA. The condition $u(L,t) = 0$ is particularly restrictive. For then $u(L,t) = E\cos bat\sin bL = 0$ implies $\sin bL = 0$. (Why is the other possibility, $E = 0$, undesirable?) Now $\sin bL = 0$ only for bL an integer multiple of π. That is, $b = n\pi/L$, where $n = 0, \pm 1, \pm 2, \ldots$. We denote by $u_n(x,t)$ that solution of (8.2.1) obtained by using $b = n\pi/L$ in (8.2.17). So for each positive[2] integer

$$u_n(x,t) = a_n\cos\frac{n\pi a}{L}t\,\sin\frac{n\pi}{L}x \tag{8.2.18}$$

[2]Choosing $n = 0$ is useless, and choosing $n < 0$ offers nothing that is not obtained by forcing n to be positive.

where we have used a_n in place of E to indicate that a different constant is possible for each choice of n. We can significantly enlarge our set of solutions by observing that linear combinations of (8.2.18), satisfying (8.2.2) and (8.2.3), are also solutions. Indeed, the expressions

$$u(x,t) = \sum_{n=1}^{M} u_n = \sum_{n=1}^{M} a_n \cos \frac{n\pi a}{L} t \sin \frac{n\pi}{L} x \tag{8.2.19}$$

are solutions for every $M = 1, 2, \ldots$.

Up to this point we have constructed solutions of (8.2.1) that meet the conditions in (8.2.2) and (8.2.3). There remains only the problem of selecting from these solutions the ones that satisfy $u(x,0) = f(x)$; and this is where the real difficulty lies. For,

$$\begin{aligned} u(x,0) = f(x) &= \sum_{n=1}^{M} a_n \cos 0 \sin \frac{n\pi}{L} x \\ &= \sum_{n=1}^{M} a_n \sin \frac{n\pi}{L} x \end{aligned} \tag{8.2.20}$$

But we have no physical reason to restrict $f(x)$ to be of the form (8.2.20). The implication of this fact is that our method will work only for a limited class of functions $f(x)$. To enlarge the class of acceptable $f(x)$, we take the dramatic step of letting $M \to \infty$, that is,

$$f(x) = \sum_{n=1}^{\infty} a_n \sin \frac{n\pi}{L} x \tag{8.2.21}$$

We are now confronted with questions that literally changed the nature of mathematics:[3] Under what conditions on $f(x)$ can we find constants $a_1, a_2, \ldots,$ so that the infinite series (8.2.21), converges to $f(x)$? And how are the constants $a_1, a_2, \ldots$ determined from $f(x)$? (These are the same questions we asked about Taylor series, only, as we shall soon see, the answers are far more difficult to come by.)

•EXERCISES

1. Find the four boundary conditions satisfied by a taut string with fixed ends and with an initial velocity, $g(t)$ from its equilibrium position.
2. Find the four boundary conditions satisfied by a taut string initially at

[3]The modern theory of integration is an outgrowth of the attempt to answer the question of which functions have series expansions in terms of trigonometric functions.

rest with left end fixed but with the right end moving harmonically with amplitude A.
3. If $\mu \geq 0$, find the solutions $T(t)$ and $X(x)$ each with two arbitrary constants. Show that at least one of the boundary conditions must be violated by these solutions.
4. If $u(x, 0) = 0$, $u_t(x, 0) = g(x)$, and $u(0, t) = u(L, 0) = 0$, find a solution similar in form to (8.2.21).
5. If $u(x, 0) = f(x)$, $u_t(x, 0) = g(x)$, and $u(0, t) = u(L, 0) = 0$, find a solution similar in form to (8.2.19).

8.3 Fourier Series Expansions

Equation 8.2.21 and similar problems ultimately reduce to this question: Given $f(x)$ in $-\rho \leq x \leq \rho$, written $[-\rho, \rho]$, under what circumstances can we find scalars $a_0, a_1, \ldots$, and $b_1, b_2, \ldots$, such that

$$f(x) = \frac{1}{2}a_0 + \sum_{n=1}^{\infty} a_n \cos\frac{n\pi}{\rho}x + b_n \sin\frac{n\pi}{\rho}x \tag{8.3.1}$$

Although we do not explore this issue here (see Section 8.4), there are good reasons for defining a_n and b_n by the equations

$$a_n = \frac{1}{\rho}\int_{-\rho}^{\rho} f(x)\cos\frac{n\pi}{\rho}x\,dx \tag{8.3.2}$$

$$b_n = \frac{1}{\rho}\int_{-\rho}^{\rho} f(x)\sin\frac{n\pi}{\rho}x\,dx \tag{8.3.3}$$

for $n = 0, 1, \ldots$. (Note that $b_0 = 0$ and the constant term in (8.3.1) is taken as $\frac{1}{2}a_0$.)

The constants $a_0, a_1, \ldots$ and $b_1, b_2, \ldots$ are called the *Fourier coefficients* of $f(x)$ relevant to the interval $[-\rho, \rho]$. Suppose that $\{a_n\}$ and $\{b_n\}$ are the Fourier coefficients of $f(x)$, then we call (8.3.1) the *Fourier series* expansion of $f(x)$ in $[-\rho, \rho]$. Conditions on $f(x)$ which ensure that (8.3.1) converges to $f(x)$ are known as *Fourier theorems.* Theorem 8.3.1 to follow, is presented without proof as an example of a Fourier theorem whose hypotheses are met by most of the functions encountered in physical applications. We need to recall the notation and definitions of sectionally continuous functions. (See Chapter 3.)

At each point x_0 in $(-\rho, \rho)$ we define

$$f(x_0^-) = \lim_{x \to x_0} f(x) \qquad x < x_0$$

and

$$f(x_0^+) = \lim_{x \to x_0} f(x), \qquad x > x_0$$

known respectively as the limits from *below* and *above*. We say that f has a *jump* discontinuity at $x = x_0$ if the limit from below, $f(x_0^-)$, and the limit from above, $f(x_0^+)$, both exist and are unequal. We define $f(x_0)$ as the average of the limits from below and above:

$$f(x_0) = \tfrac{1}{2}(f(x_0^+) + f(x_0^-)) \tag{8.3.4}$$

In addition to the conditions in the interior of the interval $[-\rho, \rho]$, we need to discuss the behavior of f at the endpoints $x = -\rho$ and $x = \rho$. We require that $f(-\rho^+)$ and $f(\rho^-)$ exist and define, if necessary, the values of f by

$$f(-\rho) = f(\rho) = \tfrac{1}{2}(f(-\rho^+) - f(\rho^+)) \tag{8.3.5}$$

Two examples are illustrated in Figure 8.2.

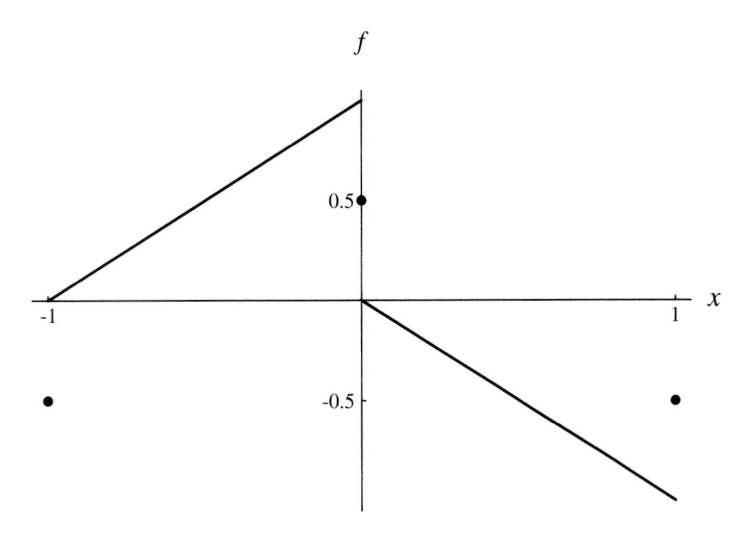

Figure 8.2: The Values at Jumps and at End Points

When f has only a finite number of jump singularities or is continuous in $-\rho < x < \rho$ and meets the conditions previously specified for the endpoints, we say that f is *sectionally continuous*. Now we can state the crucial Fourier theorem.

Theorem 8.3.1. *Suppose $f(x)$ and $f'(x)$ are sectionally continuous in $[-\rho, \rho]$, and suppose a_n and b_n are the Fourier coefficients of $f(x)$ relevant to this interval. Then the Fourier series (8.3.1), converges to $f(x)$ at every point in $[-\rho, \rho]$.*

Proof: This proof can be found in textbooks devoted to the theory of Fourier series but is too complicated to develop here. We accept the theorem without proof. □

Example 8.3.1. *Find the Fourier series for*

$$f(x) = \begin{cases} 0 & -\pi = x \\ x & -\pi < x < \pi \\ 0 & \pi = x \end{cases}$$

Solution: Since $f(x)$ is continuous in $-\pi < x < \pi$ and is defined at $-\pi$ and π so that (8.3.5) is true, we need not modify f's definition. Using (8.3.2) with $n > 0$, we have

$$a_n = \frac{1}{\pi}\int_{-\pi}^{\pi} x\cos nx\,dx = \frac{1}{\pi}\frac{1}{n}(x\sin nx + \frac{1}{n}\cos nx)\Big|_{-\pi}^{\pi} = 0$$

And, using (8.3.3), we obtain

$$\begin{aligned} b_n &= \frac{1}{\pi}\int_{-\pi}^{\pi} x\sin nx\,dx = \frac{1}{\pi}\frac{1}{n}(-x\cos nx + \frac{1}{n}\sin nx)\Big|_{-\pi}^{\pi} \\ &= -\frac{1}{n}\left[(-1)^n + (-1)^n\right] \\ &= \frac{2}{n}(-1)^{n+1} \end{aligned}$$

where we have used $\cos n\pi = (-1)^n$.

The computation of a_0 must be handled separately since the general formula for a_n breaks down in this case. (Why?) We have

$$a_0 = \frac{1}{\pi}\int_{-\pi}^{\pi} x\,dx = \frac{1}{\pi}\frac{1}{2}x^2\Big|_{-\pi}^{\pi} = 0$$

Hence, by the Fourier theorem,

$$x = 2\sum_{n=1}^{\infty}(-1)^{n+1}\frac{1}{n}\sin nx, \qquad \text{for all } x, \quad -\pi \le x \le \pi$$

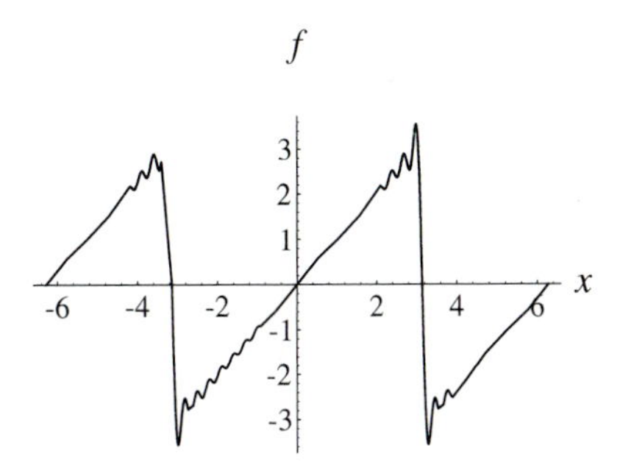

Figure 8.3: The Graph of the Fourier Series of $f(x) = x$

The graph of 20 terms of this Fourier series is shown in Figure 8.3.

Example 8.3.2. *Find the Fourier series for the unit step function*

$$f(x) = \begin{cases} -1 & -\pi < x < 0 \\ 1 & 0 < x < \pi \end{cases}$$

Solution: This function is not defined at $\pm\pi$ or at 0. We extend its definition by defining $f(\pm\pi) = f(0) = 0$. Now $f(x)$ meets the conditions stipulated for the Fourier theorem and all we need do is compute the Fourier coefficients. First, for $n \geq 1$, we have

$$\begin{aligned} a_n &= -\frac{1}{\pi}\int_{-\pi}^{0} \cos nx\, dx + \frac{1}{\pi}\int_{0}^{\pi} \cos nx\, dx \\ &= -\frac{1}{\pi}\frac{1}{n}\sin nx\Big|_{-\pi}^{0} + \frac{1}{\pi}\frac{1}{n}\sin nx\Big|_{0}^{\pi} \\ &= 0 \end{aligned}$$

and

$$\begin{aligned} b_n &= -\frac{1}{\pi}\int_{-\pi}^{0} \sin nx\, dx + \frac{1}{\pi}\int_{0}^{\pi} \sin nx\, dx \\ &= \frac{1}{\pi}\frac{1}{n}\cos nx\Big|_{-\pi}^{0} - \frac{1}{\pi}\frac{1}{n}\cos nx\Big|_{0}^{\pi} \\ &= \frac{1}{\pi}\frac{1}{n}(1 - (-1)^n - (-1)^n + 1) \\ &= \frac{2}{\pi}\frac{1}{n}(1 - (-1)^n) \end{aligned}$$

Finally, the special case

$$a_0 = -\frac{1}{\pi}\int_{-\pi}^{0} dx + \frac{1}{\pi}\int_{0}^{\pi}, dx = \frac{1}{\pi}\left(x\Big|_{-\pi}^{0} + x\Big|_{0}^{\pi}\right) = -1 + 1 = 0$$

So by Theorem 8.3.1,

$$f(x) = \frac{2}{\pi}\sum_{n=1}^{\infty}\frac{1}{n}(1-(-1)^n)\sin nx$$

When n is even, $(1-(-1)^n) = 0$, and when n is odd, this term is 2. Hence,

$$f(x) = \frac{4}{\pi}\sum_{n=1}^{\infty}\frac{1}{2n-1}\sin(2n-1)x$$

is a more compact form for the solution. The graph of the sum of 30 terms of this Fourier series is shown in Figure 8.4.

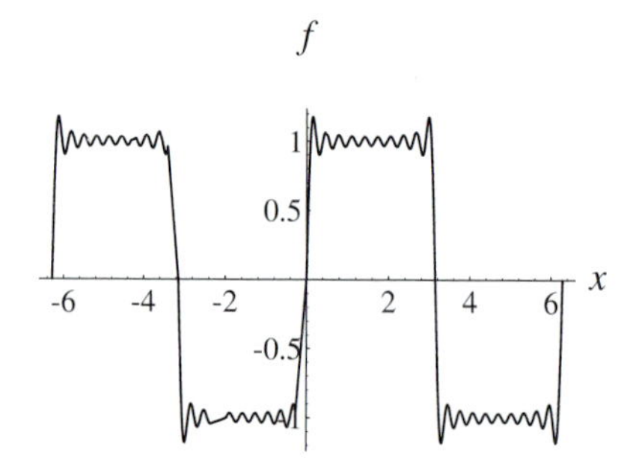

Figure 8.4: The Graph of the Fourier Series of a Unit Step

Example 8.3.3. *Find the Fourier series for*

$$f(x) = \begin{cases} 0 & -\pi \le x < 0 \\ \sin x & 0 \le x \le \pi \end{cases}$$

Solution: Since $f(x)$ is continuous in $[-\pi, \pi]$ and $f(\pm\pi) = 0$, we may proceed directly to the computation of the Fourier coefficients. We have

$$a_0 = \frac{1}{\pi}\int_0^{\pi}\sin x\,dx = -\frac{1}{\pi}\cos x\Big|_0^{\pi} = \frac{2}{\pi}$$

Next,

$$\begin{aligned} a_n &= \frac{1}{\pi}\int_0^\pi \sin x \cos nx\, dx = \frac{1}{2\pi}\int_0^\pi [\sin(x+nx)+\sin(x-nx)]\, dx \\ &= \frac{1}{2\pi}\left[-\frac{\cos(n+1)x}{n+1}+\frac{\cos(n-1)x}{n-1}\right]_0^\pi \\ &= \frac{1}{2\pi}\left[\frac{(-1)^n}{n+1}-\frac{(-1)^n}{n-1}\right]+\frac{1}{2\pi}\left[\frac{1}{n+1}-\frac{1}{n-1}\right]=\frac{1+(-1)^n}{\pi(1-n^2)} \end{aligned}$$

Clearly, this integration is invalid if $n = 1$, so we treat this case separately:

$$a_1 = \frac{1}{\pi}\int_0^\pi \sin x \cos x\, dx = \frac{1}{2\pi}\left(\sin x\right)^2\Big|_0^\pi = 0$$

The computation of b_n is similar. Once again, we must treat the case $n = 1$ separately. We have

$$\begin{aligned} b_n &= \frac{1}{\pi}\int_0^\pi \sin x \sin nx\, dx = \frac{1}{2\pi}\int_0^\pi [-\cos(x+nx)+\cos(x-nx)]\, dx \\ &= -\frac{1}{2\pi}\left[\frac{\sin(n+1)x}{n+1}+\frac{\sin(n-1)x}{n-1}\right]_0^\pi = 0 \end{aligned}$$

and

$$b_1 = \frac{1}{\pi}\int_0^\pi \sin x \sin x\, dx = \frac{1}{2\pi}\left[x-\frac{1}{2}\sin 2x\right]_0^\pi = \frac{1}{2}$$

Therefore,

$$f(x) = \frac{1}{\pi}+\frac{1}{2}\sin x-\frac{2}{\pi}\sum_{n=1}^{\infty}\frac{1}{4n^2-1}\cos 2nx$$

The graph of the sum of three terms of this series is shown in Figure 8.5.

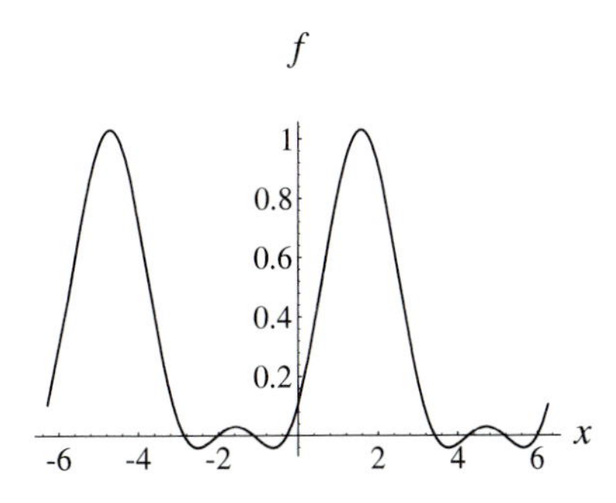

Figure 8.5: The Fourier Series of the Function in Example 8.3.3

•EXERCISES

Find the Fourier series expansions of the functions $f(x)$ given in Problems 1–7. Explain why Theorem 8.3.1 is applicable and determine the value of the Fourier series at each point of discontinuity of $f(x)$ in $[-\rho, \rho]$ and at $x = \pm\rho$.

1. $f(x) = |x|, \quad -1 \leq x \leq 1.$

2. $f(x) = \begin{cases} 1 + x & -1 \leq x < 0 \\ 1 - x & 0 \leq x \leq 1 \end{cases}$

3. $f(x) = x^2, \quad -1 \leq x \leq 1.$

4. $f(x) = \cos \frac{1}{2}x, \quad -\pi \leq x \leq \pi.$

5. $f(x) = \begin{cases} 0 & -1 \leq x < 0 \\ x & 0 \leq x \leq 1 \end{cases}$ see Problem 1.

6. $f(x) = \begin{cases} 0 & -1 \leq x < 0 \\ x^2 & 0 \leq x \leq 1 \end{cases}$ see Problem 3.

7. $f(x) = \begin{cases} -\sin x & -\pi \leq x < 0 \\ 0 & 0 \leq x \leq \pi \end{cases}$ see Example 8.3.3.

8. Use the expansion for $f(x)$ given in Example 8.3.3 to deduce

$$\sin\frac{\pi}{2} = \frac{1}{\pi} + \frac{1}{2}\sin\frac{\pi}{2} - \frac{2}{\pi}\sum_{n=1}^{\infty}\frac{2}{4n^2 - 1}$$

Simplify this expression, and conclude

$$\frac{\pi}{2} = 1 - \sum_{n=1}^{\infty}\frac{4}{4n^2 - 1}$$

8.3.1 Fourier Sine Series

It often happens in applications of the wave equation that $f(x)$ is defined in $0 \leq x \leq \rho$. (See Section 8.2.) The Fourier series expansion of f relevant to this interval will generally contain cosine as well as sine terms. In some circumstances (see Section 8.4), we will require expansions that converge to f in this interval but involve only sine terms, that is, the coefficients of the cosine terms must all be zero. One way to accomplish this is to extend the definition of $f(x)$ to the symmetric interval $[-\rho, \rho]$ in such a way that $f(x) = -f(-x)$. Functions with the property that $f(x) = -f(-x)$ are called *odd* functions.[4] The crucial fact about odd functions is that their Fourier cosine coefficients a_n are all zero. Here's how we exploit this fact: Assume that $f(x)$ is defined and is sectionally continuous in the interval $[0, \rho]$. Define $f_o(x)$ on $[-\rho, \rho]$ so that

1. $f_o(x) = f(x), \qquad \text{for} \quad 0 < x < \rho$
2. $f_o(x) = -f(-x) \qquad \text{for} \quad -\rho < x < 0$
3. $f_o(0) = f_o(\pm\rho) = 0$

Item 1 defines f_o for x in $0 < x < \rho$ in such a manner that f_o is f at these x. Item 2 makes f_o an odd function. Item 3 is required so that the Fourier series will converge to $f_o(x)$ at $x = 0$, $x = -\rho$, and $x = \rho$. Finally, if f is sectionally continuous on $0 \le x \le \rho$, then f_o satisfies the hypothesis of Theorem 8.3.1 in $[-\rho, \rho]$. This leads us to the following variation of Theorem 8.3.1.

Theorem 8.3.2. *If $f(x)$ is sectionally continuous on $[0, \rho]$, and $f_o(x)$ is as just defined, then the Fourier expansion of $f_o(x)$ converges to $f(x)$ on $0 < x < \rho$, converges to zero at $x = 0$, $x = -\rho$, and $x = \rho$ and has no cosine terms.*

Proof: We have already remarked that f_o satisfies the conditions of Theorem 8.3.1, so we know its Fourier series converges to $f_o(x) = f(x)$ for x restricted to $(0, \rho)$. It converges to 0 at $0, -\rho, \rho$. (Why?) Thus, all we need to do is show that $a_n = 0$. We have

$$\begin{aligned} a_n &= \frac{1}{\rho}\int_{-\rho}^{\rho} f_o(x)\cos\frac{n\pi}{\rho}x\,dx \\ &= -\frac{1}{\rho}\int_{-\rho}^{0} f(-x)\cos\frac{n\pi}{\rho}x\,dx + \frac{1}{\rho}\int_{0}^{\rho} f(x)\cos\frac{n\pi}{\rho}x\,dx \end{aligned}$$

Now set $s = -x$ in the first integral. This leads to

$$\begin{aligned} -\frac{1}{\rho}\int_{-\rho}^{0} f(-x)\cos\frac{n\pi}{\rho}x\,dx &= \frac{1}{\rho}\int_{\rho}^{0} f(s)\cos\frac{n\pi}{\rho}s\,ds \\ &= -\frac{1}{\rho}\int_{0}^{\rho} f(s)\cos\frac{n\pi}{\rho}s\,ds = -\frac{1}{\rho}\int_{0}^{\rho} f(x)\cos\frac{n\pi}{\rho}x\,dx \end{aligned}$$

[4]Polynomials in odd powers of x and $\sin nx$ are two common collections of odd functions.

where we have replaced the dummy variable s with x to obtain the last integral. So in the expression for a_n, we see that the first integral is the negative of the second, and hence, $a_n = 0$ for all $n \geq 0$. □

Corollary 8.3.1. *Under the hypotheses of Theorem 8.3.2,*

$$f(x) = \sum_{n=1}^{\infty} b_n \sin \frac{n\pi}{\rho} x \tag{8.3.6}$$

where

$$b_n = \frac{2}{\rho} \int_0^{\rho} f(x) \sin \frac{n\pi}{\rho} x \, dx \tag{8.3.7}$$

Proof: As part of the proof of the theorem we showed that $a_n = 0$. In a similar manner, we derive

$$\begin{aligned} b_n &= \frac{1}{\rho} \int_{-\rho}^{\rho} f_o(x) \sin \frac{n\pi}{\rho} x \, dx \\ &= -\frac{1}{\rho} \int_{-\rho}^{0} f(-x) \sin \frac{n\pi}{\rho} x \, dx + \frac{1}{\rho} \int_0^{\rho} f(x) \sin \frac{n\pi}{\rho} x \, dx \end{aligned}$$

But,

$$\begin{aligned} -\frac{1}{\rho} \int_{-\rho}^{0} f(-x) \sin \frac{n\pi}{\rho} x \, dx &= -\frac{1}{\rho} \int_{\rho}^{0} f(s) \sin \frac{n\pi}{\rho} s \, ds \\ &= \frac{1}{\rho} \int_0^{\rho} f(s) \sin \frac{n\pi}{\rho} s \, ds = \frac{1}{\rho} \int_0^{\rho} f(x) \sin \frac{n\pi}{\rho} x \, dx \end{aligned}$$

so that

$$b_n = \frac{2}{\rho} \int_0^{\rho} f(x) \sin \frac{n\pi}{\rho} x \, dx$$

□

If $f(x)$ is sectionally continuous in $[0, \rho]$ and b_n is defined as in Corollary 8.3.1, then (8.3.6) is called the *Fourier sine* series expansion of $f(x)$ relevant to the interval $[0, \rho]$. We illustrate these ideas with an example.

Example 8.3.4. *Find the Fourier sine series expansion of*

$$f(x) = \begin{cases} x & 0 \le x \le 1 \\ 2 - x & 1 \le x \le 2 \end{cases}$$

Solution: We use the corollary to find b_n:

$$b_n = \int_0^1 x \sin \frac{n\pi}{2} x \, dx + \int_1^2 (2 - x) \sin \frac{n\pi}{2} x \, dx$$

The evaluation of these integrals can be effected by integration by parts or by table lookup.[5] We find that

$$b_n = 2 \left(\frac{2}{n\pi} \right)^2 \sin \frac{n\pi}{2}$$

So, by the corollary,

$$\begin{aligned} f(x) &= 2 \sum_{n=1}^{\infty} \left(\frac{2}{n\pi} \right)^2 \sin \frac{n\pi}{2} \sin \frac{n\pi}{2} x \\ &= \frac{8}{\pi^2} \sum_{n=0}^{\infty} \frac{(-1)^n}{(2n+1)^2} \sin(2n+1)\pi x \end{aligned}$$

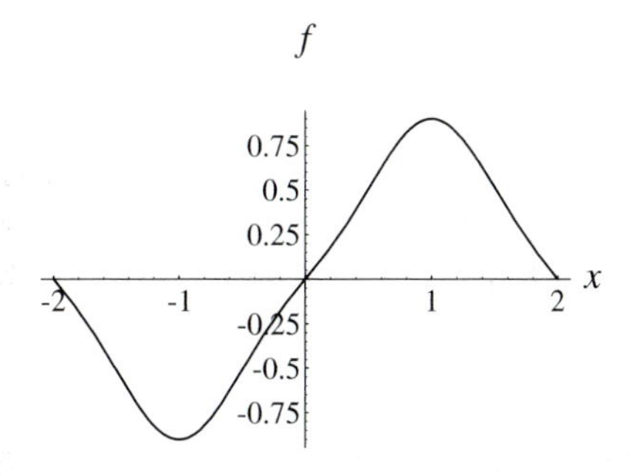

Figure 8.6: Two Terms of the Sine Series for the Function in Example 8.3.4

[5]The computer algebra systems Maple and Mathematica will also solve this type of problem; MATLAB cannot by itself handle these integrals, because the parameter n is symbolic, not numerical.

•EXERCISES

In Problems 1–5, $f(x)$ is defined in $[0, \rho]$, for some $\rho > 0$. (a) Define and sketch the function f_o, and compare its values at $x = 0$ and $x = \rho$ with $f(0)$ and $f(\rho)$. (b) Find the Fourier sine series expansion of each $f(x)$. (c) To what value does its Fourier sine series converge at $x = 0$? at $x = \rho$? Explain.

1. $f(x) = 1 - x, \qquad 0 \le x \le 1.$
2. $f(x) = x, \qquad 0 \le x \le 1.$
3. $f(x) = \cos x, \qquad 0 \le x \le \pi.$
4. $f(x) = x^2, \qquad 0 \le x \le 1.$
5. $f(x) = \begin{cases} x & 0 \le x \le 1/2 \\ 1 - x & 1/2 \le x \le 1 \end{cases}$
6. Show that Theorem 8.3.2 applied to the series deduced in Example 8.3.4 leads to

$$\pi^2 = 8 \sum_{n=0}^{\infty} \left(\frac{1}{2n+1} \right)^2$$

8.3.2 Fourier Cosine Series

Sometimes we need Fourier series expansions containing only cosine terms (including, possibly, a_0). When these circumstances arise, we often have f defined in $[0, \rho]$, satisfying the hypothesis of our Fourier theorem, Theorem 8.3.1. To obtain a pure cosine expansion, we simply adapt the technique of Section 8.3.1, only now we use the *even function* $f_e(x)$ defined as follows:

1. $f_e(x) = f(x), \qquad \text{for} \quad 0 \le x \le \rho$
2. $f_e(x) = f(-x), \qquad \text{for} \quad -\rho \le x \le 0$

Note that unlike the definition of f_o, f_e is necessarily continuous at $x = 0$. We can also verify that f_e is an even function.[6] Here are the "even" analogues of Theorem 8.3.2 and its corollary.

Theorem 8.3.3. *If $f(x)$ is sectionally continuous on $0 \le x \le \rho$, and $f_e(x)$ is as just defined, then the Fourier expansion of $f_e(x)$ converges to $f(x)$ on $0 < x < \rho$ to $f(0^+)$ and to $f(\rho^-)$ and has no sine terms.*

---

[6]Even functions have the property that $f(-x) = f(x)$. Polynomials in even powers of x and the functions $\cos nx$ are even.

Proof: The function $f_e(x)$ satisfies the hypothesis of Theorem 8.3.1, and so the Fourier series expansion of f_e converges to $f_e(x)$ in $[-\rho, \rho]$. In view of the fact that $f_e(x) = f(x)$ for all x in $(0, \rho)$, it follows that this same series converges to $f(x)$ in $(0, \rho)$. The cases $x = 0$ and $x = \rho$ need a bit more discussion, which we leave to the reader. We need only establish that $b_n = 0$ for all n. This is not hard, and the argument proceeds along the lines used in the proof of Theorem 8.3.3:

$$\begin{aligned} b_n &= \frac{1}{\rho}\int_{-\rho}^{\rho} f_e(x)\sin\frac{n\pi}{\rho}x\,dx \\ &= \frac{1}{\rho}\int_{-\rho}^{0} f(-x)\sin\frac{n\pi}{\rho}x\,dx + \frac{1}{\rho}\int_{0}^{\rho} f(x)\sin\frac{n\pi}{\rho}x\,dx \end{aligned}$$

Now set $s = -x$ in the first integral. This leads to

$$\begin{aligned} \frac{1}{\rho}\int_{-\rho}^{0} f(-x)\sin\frac{n\pi}{\rho}x\,dx &= \frac{1}{\rho}\int_{\rho}^{0} f(s)\sin\frac{n\pi}{\rho}s\,ds \\ &= -\frac{1}{\rho}\int_{0}^{\rho} f(s)\sin\frac{n\pi}{\rho}s\,ds = -\frac{1}{\rho}\int_{0}^{\rho} f(x)\sin\frac{n\pi}{\rho}x\,dx \end{aligned}$$

Since the first integral is the negative of the second, $b_n = 0$ for all $n \geq 0$. □

Corollary 8.3.2. *Under the hypotheses of Theorem 8.3.2,*

$$f(x) = \frac{1}{2}a_0 + \sum_{n=1}^{\infty} a_n \cos\frac{n\pi}{\rho}x \tag{8.3.8}$$

where

$$a_n = \frac{2}{\rho}\int_{0}^{\rho} f(x)\cos\frac{n\pi}{\rho}x\,dx \tag{8.3.9}$$

Proof: We have shown that $b_n = 0$. Now we derive an expression for a_n:

$$\begin{aligned} a_n &= \frac{1}{\rho}\int_{-\rho}^{\rho} f_e(x)\cos\frac{n\pi}{\rho}x\,dx \\ &= \frac{1}{\rho}\int_{-\rho}^{0} f(-x)\cos\frac{n\pi}{\rho}x\,dx + \frac{1}{\rho}\int_{0}^{\rho} f(x)\cos\frac{n\pi}{\rho}x\,dx \end{aligned}$$

But,

$$\begin{aligned}\frac{1}{\rho}\int_{-\rho}^{0} f(-x)\cos\frac{n\pi}{\rho}x\,dx &= -\frac{1}{\rho}\int_{\rho}^{0} f(s)\cos\frac{n\pi}{\rho}s\,ds\\ &= \frac{1}{\rho}\int_{0}^{\rho} f(s)\cos\frac{n\pi}{\rho}s\,ds\\ &= \frac{1}{\rho}\int_{0}^{\rho} f(x)\cos\frac{n\pi}{\rho}x\,dx\end{aligned}$$

so that

$$a_n = \frac{2}{\rho}\int_0^{\rho} f(x)\cos\frac{n\pi}{\rho}x\,dx$$

These formulas hold for all $n \geq 0$. However, it is necessary to multiply a_0 by $1/2$ if we use this formula for $n = 0$, as we shall always do! □

If $f(x)$ satisfies the hypothesis of our Fourier theorem and a_n is defined as in the corollary, then the series (8.3.8) is called the *Fourier cosine* series expansion of $f(x)$ relative to $[0, \rho]$.

Example 8.3.5. *Find the Fourier cosine series expansion of* $f(x) = \sin x$ *relevant to the interval* $0 \leq x \leq \pi$.

Solution: We begin by noting that $f_{\mathrm{e}}(x)$ is the periodic extension of $|\sin x|$. This observation does not change our computation! We have

$$a_0 = \frac{2}{\pi}\int_0^{\pi} \sin x\,dx = -\frac{2}{\pi}\cos x\Big|_0^{\pi} = \frac{4}{\pi}$$

For all $n > 1$,

$$a_n = \frac{2}{\pi}\int_0^{\pi} \sin x\cos nx\,dx = -\frac{2}{\pi}\left(\frac{1+(-1)^n}{n^2-1}\right)$$

which shows why we must consider $n = 1$ separately:

$$a_1 = \frac{2}{\pi}\int_0^{\pi} \sin x\cos x\,dx = 0$$

Hence,

$$f(x) = \frac{2}{\pi} - \frac{2}{\pi}\sum_{n=2}^{\infty}\frac{1+(-1)^n}{n^2-1}\cos nx$$

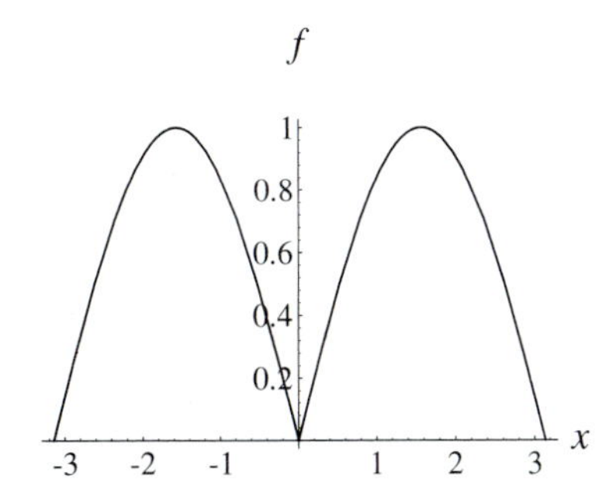

Figure 8.7: The Fourier Cosine Series of $|\sin x|$

•EXERCISES

Derive the Fourier cosine series expansions for the functions given in Problems 1–5. Sketch the corresponding $f_{\text{e}}(x)$. To what values does each series converge at $x = 0$?

1. $f(x) = x, \qquad 0 \le x \le 1.$
2. $f(x) = 1 - x, \qquad 0 \le x \le 1.$
3. $f(x) = \begin{cases} x & 0 \le x \le 1/2 \\ 1 - x & 1/2 \le x \le 1 \end{cases}$
4. $f(x) = \begin{cases} \sin x & 0 \le x \le \frac{\pi}{2} \\ 0 & \frac{\pi}{2} \le x \le \pi \end{cases}$
5. $f(x) = \sin x/2, \qquad 0 \le x \le \pi.$
6. Derive (8.3.8).

8.4 The Wave Equation

We now return to the problem of the vibrating string. We assume that a taut string of length L is held fixed at $x = 0$ and $x = L$ with an initial displacement $f(x)$ and a zero initial velocity. Suppose that $u(x, t)$ measures the displacement of the string for each x in $[0, L]$ and for all time $t > 0$.

Then $u(x,t)$ must satisfy the boundary-value problem

$$\begin{aligned} u_{tt}(x,t) &= a^2 u_{xx}(x,t) && (8.4.1)\\ u(0,t) &= u(L,t) = 0 && (8.4.2)\\ u_t(x,0) &= 0 && (8.4.3)\\ u(x,0) &= f(x) && (8.4.4) \end{aligned}$$

We use separation of variables on (8.4.1) and exploit (8.4.2) to write

$$u(x,t) = \sum_{n=1}^{\infty} \left(\alpha_n \cos \frac{na\pi}{L} t + \beta_n \sin \frac{na\pi}{L} t \right) \sin \frac{n\pi}{L} x \qquad (8.4.5)$$

Becaue $u_t(x,0) = 0$, we differentiate (8.4.5) and set $t = 0$. When this is done, we get

$$u_t(x,0) = 0 = \sum_{n=1}^{\infty} \beta_n \frac{na\pi}{L} \sin \frac{n\pi}{L} x \qquad (8.4.6)$$

Since (8.4.6) must hold for all x, we deduce that $\beta_n = 0$ for all n and that (8.4.5) reduces to

$$u(x,t) = \sum_{n=1}^{\infty} \alpha_n \cos \frac{na\pi}{L} t \sin \frac{n\pi}{L} x \qquad (8.4.7)$$

Finally, (8.4.4) leads to

$$u(x,0) = f(x) = \sum_{n=1}^{\infty} \alpha_n \sin \frac{n\pi}{L} x \qquad (8.4.8)$$

Equation (8.4.8) ties together the theory of Fourier series and the wave equation by telling us that the coefficients needed in the solution (or, more precisely, the Fourier sine series expansion of the solution) are the Fourier coefficients α_n in the expansion of the initial displacement function $f(x)$. Of course, the interval relevant to this expansion is $[0, L]$ and the expansion should have no cosine terms.

In order to conform to the notation in the previous section, let us write $b_n = \alpha_n$. An example follows.

Example 8.4.1. *Determine the motion of a plucked string of length* 2 m *which is initially at rest with both ends fixed. The initial displacement of the string is*

$$f(x) = \begin{cases} 0.05x & 0 \le x \le 1 \\ 0.1 - 0.05x & 1 < x \le 2 \end{cases}$$

Solution: The motion of the string is given by

$$u(x,t) = \sum_{n=1}^{\infty} b_n \cos \frac{na\pi}{L} t \sin \frac{n\pi}{L} x$$

where $L = 2$ and $b_n = \alpha_n$ is given by (8.3.7). The work in Example 8.3.4 yielded

$$b_n = 2\left(\frac{2}{n\pi}\right)^2 \sin \frac{n\pi}{2} \qquad \text{for all } n \geq 1$$

and since our function is just 0.05 of the value of the function in the example, we have

$$u(x,t) = \frac{1}{10} \sum_{n=1}^{\infty} \left(\frac{2}{n\pi}\right)^2 \sin \frac{n\pi}{2} \cos \frac{na\pi}{2} t \sin \frac{n\pi}{2} x$$

Now we treat the case where $f(x) = 0$ (i.e., no initial displacement) and $u_t(x,0) = g(x)$. We begin by setting $t = 0$ in (8.4.5). Then,

$$u(x,0) = 0 = \sum_{n=1}^{\infty} \alpha_n \sin n\pi Lx \tag{8.4.9}$$

and hence, $\alpha_n = 0$ for all $n \geq 0$. Thus (8.4.5) reduces to

$$u(x,t) = \sum_{n=1}^{\infty} \beta_n \sin \frac{na\pi}{L} t \sin \frac{n\pi}{L} x \tag{8.4.10}$$

We differentiate (8.4.10) with respect to t and then set $t = 0$ to get

$$u_t(x,0) = g(x) = \sum_{n=1}^{\infty} \frac{an\pi}{L} \beta_n \sin \frac{n\pi}{L} x \tag{8.4.11}$$

So $an\pi\beta_n/L$ are the coefficients of the Fourier sine series for $g(x)$. (Again we replace β_n by b_n to conform to the "standard" notation for the Fourier coefficients.) We conclude, therefore, that

$$b_n = \frac{2}{an\pi} \int_0^L g(x) \sin \frac{n\pi}{L} x \, dx \tag{8.4.12}$$

Example 8.4.2. *A string $L = \pi$ m long and fixed at both ends is given an initial velocity of $g(x) = 1$ m/s from equilibrium. The string has no initial displacement. Find an expression for the displacement function $u(x,t)$ valid for all x in $0 \leq x \leq \pi$ and all $t \geq 0$.*

Solution: As the discussion previous to this example has shown, we need the Fourier sine series expansion of $g(x)$. Equation(8.4.12) is relevant here and we find

$$\begin{aligned} b_n &= \frac{2}{an\pi}\int_0^{\pi} \sin nx\,dx \\ &= -\frac{2}{an^2\pi}\cos nx\Big|_0^{\pi} = \frac{2}{an^2\pi}(1-(-1)^n) \end{aligned} \tag{8.4.13}$$

Hence, from (8.4.10),

$$u(x,t) = \frac{2}{a\pi}\sum_{n=1}^{\infty}\frac{1-(-1)^n}{n^2}\sin nat \sin nx$$

Note that $u(x,0) = 0, u(0,t) = u(\pi,t) = 0$, and for all x, $0 < x < \pi$,

$$u_t(x,0) = \frac{2}{\pi}\sum_{n=1}^{\infty}\frac{1-(-1)^n}{n}\sin nx = 1$$

Suppose we denote the solution of the vibrating string with initial displacement $f(x)$ and with zero initial velocity by $u_1(x,t)$ and the solution of the vibrating string with zero initial displacement and initial velocity $g(x)$ by $u_2(x,t)$. Let

$$u(x,t) = u_1(x,t) + u_2(x,t) \tag{8.4.14}$$

In Problem 5 in the forthcoming exercise set, the reader is asked to provide the proof that $u(x,t)$ satisfies the boundary-value problem (8.4.1)–(8.4.4) , but with $u_t(x,0) = g(x)$.

The general nonhomogeneous problem

$$u_{tt}(x,t) = a^2u_{xx}(x,t) + y(x) \tag{8.4.15}$$

$$u(0,t) = u(L,t) = 0 \tag{8.4.16}$$

$$u_t(x,0) = g(x) \tag{8.4.17}$$

$$u(x,0) = f(x) \tag{8.4.18}$$

can be reduced to a homogeneous problem by a simple change of variables. The details and verifications are outlined in the Projects at the end of this chapter.

•EXERCISES

In Problems 1–3 a string of length π m with wave speed 40 m/s is fixed at both ends with zero initial velocity and is stretched from its equilibrium position to approximate $u(x,0) = f(x)$ in $0 \le x \le \pi$ as nearly as possible. Find the first three terms in the Fourier sine expansions of the solutions of these problems.

1. $f(x) = 1, \qquad 0 < x < \pi.$

2. $f(x) = \begin{cases} 0 & 0 \le x \le \pi/2 \\ 1 & \pi/2 \le x \le \pi \end{cases}$

3. $f(x) = \begin{cases} x & 0 \le x \le \pi/2 \\ 1 & \pi/2 \le x \le \pi \end{cases}$

4. A string of length π m is started in motion by giving the section $\pi/4 < x < 3\pi/4$ an initial velocity of 20 m/s. The wave speed is 60 m/s. Determine the resulting displacement $u(x,t)$.

5. Verify that the superposition of $u_1(x,t)$ with $u_2(x,t)$ as given in (8.4.13) does indeed satisfy (8.4.1) – (8.4.4).

8.5 The One-Dimensional Heat Equation

The one-dimensional heat equation

$$T_t(x,t) = kT_{xx}(x,t) \tag{8.5.1}$$

describes the temperature distribution $T(x,t)$ of an idealized, long heated rod with insulated lateral surfaces.[7] We consider a number of boundary conditions for which the method of separation of variables is applicable. Specifically, let the rod have length L, and suppose its ends are kept at 0° C. The initial temperature of the rod is given by $f(x)$, that is, $T(x,0) = f(x)$. Then the one-dimensional diffusion problem may be stated this way:

Find $T(x,t)$ satisfying

$$\begin{aligned} T_t(x,t) &= kT_{xx}(x,t) && (8.5.2)\\ T(0,t) &= T(L,t) = 0 && (8.5.3)\\ T(x,0) &= f(x) && (8.5.4) \end{aligned}$$

[7]Insulating the lateral surfaces means that heat cannot flow out of the cylindrical surface that bounds the rod, but can flow only along its axis.

We can separate x from t by assuming $T(x,t) = Y(t)X(x)$. Then, as in the wave equation, it is easy to see that this assumption is true if and only if the functions $Y(t)$ and $X(x)$ satisfy

$$Y'(t) - \lambda k Y(t) = 0 \tag{8.5.5}$$
$$X''(x) - \lambda X(x) = 0 \tag{8.5.6}$$

respectively. In these equations λ is negative.[8] We write $-b^2$ for λ. (The constant k depends on the specific material properties of the rod.) Equations (8.5.5) and (8.5.6) have the general solutions

$$Y(t) = Ae^{-b^2kt} \tag{8.5.7}$$
$$X(x) = B\sin bx + C\cos bx \tag{8.5.8}$$

respectively. We can expect from our experience with the wave equation that $T(x,t)$ will be represented by a series of the form

$$T(x,t) = \sum_{n=1}^{\infty} e^{-b_n^2 kt}\,(B_n \sin b_n x + C_n \cos b_n x) \tag{8.5.9}$$

where we have absorbed the arbitrary constant A in with the constants B_n and C_n. At this point, we note that $T(0,t) = 0$ used in (8.5.9) forces $C_n = 0$ for all n, otherwise $T(x,t)$ would be identically zero. Similarly, the condition $T(L,t) = 0$ implies that

$$\sum_{n=1}^{\infty} e^{-b_n^2 kt} B_n \sin b_n L = 0 \tag{8.5.10}$$

Here we choose $b_n L = \pi n$ for all $n \geq 1$ to effect (8.5.10). (The choice $n \leq 0$ either leads to a contradiction in the case $n = 0$ or leads to the same summation in the case $n < 0$.) Using what we have learned so far, we have

$$T(x,t) = \sum_{n=1}^{\infty} b_n e^{-(n^2\pi^2/L^2)kt} \sin \frac{n\pi}{L} x \tag{8.5.11}$$

The remaining constants b_n are determined by requiring (8.5.11) to reduce to $f(x)$ when $t = 0$. That is, (8.5.3) and (8.5.11) lead to

$$T(x,0) = f(x) = \sum_{n=1}^{\infty} b_n \sin \frac{n\pi}{L} x \tag{8.5.12}$$

We conclude from this that we need the Fourier sine series expansion to evaluate the coefficients b_n. Let us see how this works in an example.

---

[8]If $\lambda \geq 0$, then $Y(t)$ and $X(x)$ are increasing functions. But then, so is $T(x,t)$ and this is physically impossible.

Example 8.5.1. *A copper rod of length π m with insulated lateral surface and ends kept at $0°$ C has an initial temperature distribution given by $f(x) = 100°$ C in $0 < x < \pi$ where $k = 10^{-4}\text{m}^2/\text{s}$. Determine an expression for the temperature of the rod $T(x, t)$.*

Solution: This example fits the conditions set in the preceding analysis. So we only have to find the Fourier sine expansion of the constant functions $f(x) = 100$. We have

$$b_n = \frac{2}{\pi}\int_0^\pi 100 \sin nx\, dx = -\frac{200}{n\pi}\cos nx\bigg|_0^\pi = \frac{200}{n\pi}(1 - (-1)^n)$$

With these values in hand, we have as our solution

$$T(x, t) = \frac{200}{\pi}\sum_{n=1}^{\infty}\frac{1 - (-1)^n}{n} e^{-n^2 10^{-4} t} \sin nx$$

Instead of assuming that the temperature at the ends of the rod is held fixed at $t = 0°$ C, we may suppose that at $x = 0$, the temperature is T_0, and at $x = L$, the temperature is T_L. The solution in this case requires that we resurrect the solution obtained by setting $\lambda = 0$ in (8.5.5) and (8.5.6). Then the temperature function takes the form

$$T(x, t) = cx + d + \sum_{n=1}^{\infty} b_n e^{-n^2\pi^2 kt/L^2} \sin\frac{n\pi}{L}x \tag{8.5.13}$$

Nonzero values of c and d permit nonzero end conditions. (See the exercise set that follows.

A problem that requires a variation of the method used so far arises when we make the assumption that the ends of the rod are insulated. The insulation of the ends are translated into the conditions

$$T_x(0, t) = T_x(L, t) = 0 \tag{8.5.14}$$

and these conditions replace (8.5.3). To solve the diffusion problem subject to these constraints, we return to the product of (8.5.7) and (8.5.8):

$$\frac{\partial}{\partial x}\left[e^{-b^2 kt/L^2}\left(B \sin bx + C \cos bx\right)\right] = be^{-b^2 kt}\left(B \cos bx - C \sin bx\right) \tag{8.5.15}$$

We see that $T_x(0,t) = 0$ requires $B = 0$ and $T_x(L,t) = 0$ requires that

$$b_n L = n\pi, \qquad \text{for all } n \geq 0 \tag{8.5.16}$$

When the temperature at the ends of the rod was zero, the term in (8.5.11) arising from setting $n = 0$ was omitted because $\sin \frac{n\pi}{L} x = 0$ when $n = 0$. In contrast, the conditions in this problem require the $n = 0$ term. Using these expressions we get

$$T(x,t) = \frac{1}{2}a_0 + \sum_{n=1}^{\infty} a_n e^{-n^2\pi^2 kt/L^2} \cos \frac{n\pi}{L} x \tag{8.5.17}$$

The condition $T(x,0) = f(x)$ applied to (8.5.17) leads to

$$f(x) = \frac{1}{2}a_0 + \sum_{n=1}^{\infty} a_n \cos \frac{n\pi}{L} x \tag{8.5.18}$$

Summarizing, the constants in (8.5.18) are nothing more than the coefficients in the Fourier cosine series expansion of $f(x)$ in the interval $0 \leq x \leq L$.

Example 8.5.2. *Find the temperature $T(x,t)$ of a laterally insulated rod of length 2π m insulated at the ends, given the following initial temperature distribution:*

$$f(x) = \begin{cases} 0, & 0 \leq x < \pi/2 \\ 100, & \pi/2 < x < 3\pi/2 \\ 0, & 3\pi/2 < x \leq 2\pi \end{cases}$$

Solution: The first step is to determine the Fourier cosine series for the initial temperature distribution. For $n \geq 1$,

$$\begin{aligned} a_n &= \frac{1}{\pi}\int_0^{2\pi} f(x) \cos \frac{n}{2} x \, dx = \frac{1}{\pi}\int_{\pi/2}^{3\pi/2} 100 \cos \frac{n}{2} x \, dx \\ &= \frac{200}{n\pi} \sin \frac{n}{2} x \Big|_{\pi/2}^{3\pi/2} \\ &= \frac{200}{n\pi}\left(\sin \frac{3}{4} n\pi - \sin \frac{1}{4} n\pi\right) \end{aligned}$$

and

$$a_0 = \frac{1}{\pi}\int_{\pi/2}^{3\pi/2} 100 dx = 100$$

From these computations, we deduce

$$T(x,t) = 50 + \frac{200}{\pi}\sum_{n=1}^{\infty}\frac{1}{n}\left(\sin\frac{3n\pi}{4} - \sin\frac{n\pi}{4}\right)\cos\frac{n}{2}xe^{-n^2kt/4}$$

•EXERCISES

In Problems 1–9, let $T(x,t)$ be the temperature distribution of a 2-m rod, where $k = 10^{-4}\text{m}^2/\text{s}$. Find $T(x,t)$ subject to the boundary conditions given in each problem.

1. $T(x,0) = 100\sin\pi x$; $T(0,t) = 0, \quad T(2,t) = 0.$
2. $T(x,0) = 100\sin\frac{\pi}{2}x$; $T(0,t) = 0, \quad T(2,t) = 0.$
3. $T(x,0) = 80\sin\frac{\pi}{2}x$; $T(0,t) = 0, \quad T(2,t) = 80.$
4. $T(x,0) = 200(1+\sin\pi x)$; $T(0,t) = 200, \quad T(2,t) = 200.$
5. $T(x,0) = 100$; $T(0,t) = 0, \quad T(2,t) = 0.$
6. $T(x,0) = 100$; $T(0,t) = 0, \quad T(2,t) = 100.$
7. $T(x,0) = \begin{cases} 100x, & 0 < x \le 1 \\ 200 - 100x, & 1 < x < 2 \end{cases}$

with $T(0,t) = T(2,t) = 0$.

8. $T(x,0) = 100(2x - x^2)$; $T(0,t) = 8, \quad T(2,t) = 0.$
9. $T(x,0) = 50x^2$; $T(0,t) = 0, \quad T(2,t) = 200.$

In Problems 10–15, find the temperature at the middle of the rod for these specified times.

10. Problem 1 at $t = 1000$ s.
11. Problem 1 as $t \to \infty$.
12. Problem 2 as $t \to \infty$.
13. Problem 3 at $t = 2$ hr.
14. Problem 5 at $t = 1$ hr.
15. Problem 2 at $t = 1$ hr.

In Problems 16–20 find the times for the center of the rod to reach the temperature 50° C.

16. Problem 2.
17. Problem 3.
18. Problem 5.
19. Problem 6.
20. Problem 7.

In Problems 21–25, find $T(x,t)$ for a laterally insulated π-m rod if $k = 10^{-4}\text{m}^2/\text{s}$ for each of the given conditions.

21. $T(x,0) = 100\cos x$; $T_x(0,t) = 0$, $T_x(\pi,t) = 0$.
22. $T(x,0) = 100$; $T_x(0,t) = 0$, $T_x(\pi,t) = 0$.
23. $T(x,0) = 100\sin x/2$; $T(0,t) = 0$, $T_x(\pi,t) = 0$.
24. $T(x,0) = 100\sin x$; $T_x(0,t) = 0$, $T(\pi,t) = 0$.
25. $T(x,0) = \begin{cases} 100, & 0 < x \le \pi/2 \\ 0 & \pi/2 < x < \pi \end{cases}$
with $T(0,t) = T_x(\pi,t) = 0$.

26. Show that $T_t = kT_{xx}$, $T(0,t) = T_1$, $T(L,t) = T_2$, $T(x,0) = f(x)$ has the solution

$$T(x,t) = T_1 - \frac{1}{L}(T_2 - T_1)x + \sum_{n=1}^{\infty} b_n e^{-n^2\pi^2 kt/L^2} \sin\frac{n\pi}{L}x$$

where b_n are the coefficients in the Fourier sine series expansion of $f(x)$ relevant to the interval $0 \le x \le L$.

MATLAB

Sketch the temperature distribution at $t = 0$, 1000 s and 10 hr for the following problems.

27. Problem 1.
28. Problem 2.
29. Problem 3.
30. Problem 5.
31. Problem 7.

8.6 The Laplace Equation

A thin rectangular plate is positioned as in Figure 8.8 with a vertex at the origin and the two edges emanating from this vertex lying on the x- and y-axes. The far edge parallel to the y-axis is maintained at a temperature of $f(y)$, while the other three sides are held at 0° C. The top and bottom faces of this plate are insulated so there is no heat exchanged through these surfaces. The temperature $T(t,x,y)$ satisfies the diffusion equation

$$T_t(t,x,y) = k\left(T_{xx}(t,x,y) + T_{yy}(t,x,y)\right) \tag{8.6.1}$$

If we assume that there is a solution of (8.6.1) that is independent of time, then $T_t(x,y) = 0$, and (8.6.1) takes the simpler form

$$T_{xx}(x,y) + T_{yy}(x,y) = 0 \tag{8.6.2}$$

known as *Laplace's equation.* The function $T(x,y)$ is called a *steady-state* solution because the function representing the temperature does not vary with

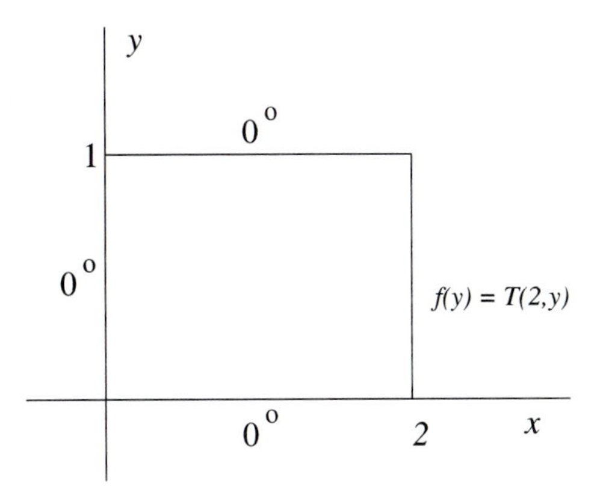

Figure 8.8: A Thin Plate with Insulated Flat Surfaces

time. A steady-state solution that provides the temperature distribution for the plate just described, must satisfy the boundary-value problem:

$$T_{xx}(x,y) + T_{yy}(x,y) = 0 \tag{8.6.3}$$
$$T(x,0) = T(x,1) = 0 \tag{8.6.4}$$
$$T(0,y) = 0 \tag{8.6.5}$$
$$T(2,y) = f(y) \tag{8.6.6}$$

We solve this boundary-value problem by the method of separation of variables. We write

$$T(x,y) = Y(y)X(x) \tag{8.6.7}$$

Using the now familiar argument, we find that

$$\frac{X''}{X} = -\frac{Y''}{Y} = b^2 \tag{8.6.8}$$

Hence,

$$X'' - b^2 X = 0 \tag{8.6.9}$$
$$Y'' + b^2 Y = 0 \tag{8.6.10}$$

From this pair of equations and our experience in earlier sections we expect to find the value of the constants b_n, A_n, B_n, C_n, D_n so that

$$T(x,y) = \sum_{n=1}^{\infty} \left(A_n e^{b_n x} + B_n e^{-b_n x}\right)(C_n \cos b_n y + D \sin b_n y) \tag{8.6.11}$$

Using the boundary conditions (8.6.4) and (8.6.5), we deduce

$$A_n + B_n = C_n = \sin b_n = 0 \tag{8.6.12}$$

from which we learn that $b_n = n\pi$ for all $n \geq 1$. Also, because $A_n = -B_n$, we can write $e^{b_n x} - e^{-b_n x} = e^{n\pi x} - e^{-n\pi x} = 2\sinh n\pi x$, and then write the solution in the far simpler form

$$T(x, y) = \sum_{n=1}^{\infty} b_n \sinh n\pi x \sin n\pi y \tag{8.6.13}$$

The last boundary condition, $f(2, y) = f(y)$, will determine the coefficients b_n. Indeed,

$$T(2, y) = f(y) = \sum_{n=1}^{\infty} b_n \sinh 2n\pi \sin n\pi y \tag{8.6.14}$$

Assuming that f satisfies the hypothesis of our Fourier theorem, we have

$$b_n \sinh 2n\pi = \int_0^1 \sin n\pi y \, dy = \frac{1 - (-1)^n}{n\pi} \tag{8.6.15}$$

After we substitute this value for b_n into (8.6.13), we get

$$T(x, y) = \sum_{n=1}^{\infty} \frac{1 - (-1)^n}{n\pi \sinh 2n\pi} \sinh n\pi x \sin n\pi y \tag{8.6.16}$$

We have assumed that the separation constant is positive by writing it as b^2. This was unavoidable, for otherwise $f(y) = T(x, y)$ could have never been satisfied. Had the nonzero temperature been given at the edge $y = 1$ rather than at $x = 2$, we would have had to choose $-b^2$ as the separation constant.

•EXERCISES

Find the steady-state temperature distribution in a 1 m × 1 m slab if the flat surfaces are insulated and the edge conditions are as follows:

1. $T(0, y) = T(x, 0) = T(1, y) = 0,$ $\quad T(x, 1) = 100 \sin \pi x.$
2. $T(0, y) = T(x, 0) = T(x, 1) = 0,$ $\quad T(1, y) = 100 \sin \pi y.$
3. $T(0, y) = T(x, 0) = T_x(1, y) = 0,$ $\quad T(x, 1) = 100.$
4. $T(0, y) = T_y(x, 0) = T_x(1, y) = 0,$ $\quad T(x, 1) = 100.$
5. $T(0, y) = T(x, 0) = T(x, 1) = 100,$ $\quad T(1, y) = 200.$

8.7 A Potential about a Spherical Surface

We conclude this chapter by studying the electric potential $V = V(r, \varphi)$ internal and external to a sphere of radius r_0, where φ is the polar angle. (See Figure 8.9.)

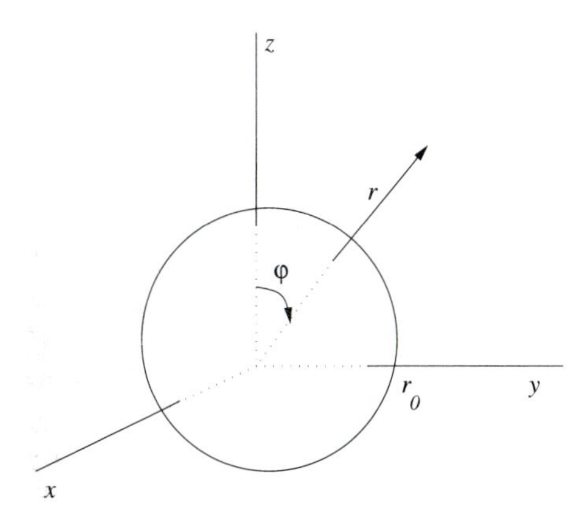

Figure 8.9: A Spherical Surface with Potential $f(\varphi)$

We assume that at the surface of the sphere,

$$V(r_0, \varphi) = f(\varphi) \tag{8.7.1}$$

The equation that describes V is Laplace's equation, which we give in spherical coordinates, the coordinate system of choice for spherically symmetric problems:

$$(r^2 V_r)_r + \frac{1}{\sin\varphi}(\sin\varphi V_\varphi)_\varphi = 0 \tag{8.7.2}$$

Physical considerations suggest that if an electric potential exists on the spherical surface then there is a zero potential at infinity. We add this to our list of conditions:

$$V(\infty, \varphi) = 0 \tag{8.7.3}$$

Once again, we assume that the solution to this boundary-value problem

$$\begin{gathered}(r^2 V_r)_r + \frac{1}{\sin\varphi}(\sin\varphi V_\varphi)_\varphi = 0 \\ V(r_0, \varphi) = f(\varphi), \qquad V(\infty, \varphi) = 0\end{gathered} \tag{8.7.4}$$

is a sum of terms of the form

$$V(r,\varphi) = R(r)\Phi(\varphi) \tag{8.7.5}$$

Letting $r \to \infty$ in (8.7.5), we see that (8.7.4) implies that

$$\lim_{r\to\infty} R(r) = 0 \tag{8.7.6}$$

When used in (8.7.2), the separation hypothesis (8.7.5) leads to

$$\frac{1}{R}(r^2R')' = \mu = -\frac{1}{\sin\varphi}\frac{1}{\Phi}(\varphi'\sin\varphi)' \tag{8.7.7}$$

So R satisfies the Cauchy-Euler equation

$$r^2R'' + 2rR' - \mu R = 0 \tag{8.7.8}$$

A general solution of (8.7.8) is given by

$$R(r) = c_1r^a + c_2r^{-a-1} \tag{8.7.9}$$

where

$$a = -\tfrac{1}{2} + \sqrt{\tfrac{1}{4} + \mu}$$

It is convenient to set $\mu = n(n+1)$ for reasons that will become apparent a bit later. On the other hand, Φ satisfies the much more difficult equation

$$\frac{1}{\sin\varphi}(\Phi'\sin\varphi)' + n(n+1)\Phi = 0 \tag{8.7.10}$$

Equation (8.7.10) is a disguised form of Legendre's equation, as we can see by making the change of variable

$$u(x) = u(\cos\varphi) = \Phi(\varphi) \tag{8.7.11}$$

(This change of variables requires a clear understanding of the chain rule. Rather than digress here by giving all these computations, we set the task as a project at the end of the chapter. The series solution of Legendre's equation was presented in Section 6.5).

We begin by considering the electric potential interior to the sphere. Assuming that $n \geq 0$ and that $V(0,\varphi)$ exists, we must choose $c_2 = 0$ in (8.7.9), since the term whose coefficient is c_2 would be infinite at $V(0,\varphi)$. Moreover, in order that the solution to (8.7.10) be continuous and have a continuous derivative at $\varphi = 0$ (i.e., at $x = 1$), we choose n a negative integer.[9] Then the solutions of (8.7.10) are the Legendre polynomials $P_n(x)$. Putting all this together, we conclude that (convergence questions aside)

$$V(r,\varphi) = \sum_{n=0}^{\infty} b_n r^n P_n(\cos\varphi) \tag{8.7.12}$$

for $r \leq r_0$. Finally, from the boundary condition (8.7.1), and from (8.7.12), we have

$$f(\varphi) = \sum_{n=0}^{\infty} b_n r_0^n P_n(\cos\varphi) \tag{8.7.13}$$

The coefficients a_n can be determined from (8.7.13) by exploiting a property the Legrendre polynomials share with the trigonometric functions:

$$\int_{-1}^{1} P_m(x)P_n(x)\,dx = 0, \qquad \text{if} \quad m \neq n \tag{8.7.14}$$

$$\int_{-1}^{1} P_n^2(x)\,dx = \frac{2}{2n+1}, \qquad \text{for} \quad n \geq 0 \tag{8.7.15}$$

Here is how these "orthogonality" conditions (8.7.14) and (8.7.15) can be used. If

$$f(x) = \sum_{n=0}^{\infty} a_n P_n(x) \tag{8.7.16}$$

then multiplying this equation through by $P_m(x)$ and integrating between -1 and 1 results in

$$\int_{-1}^{1} P_m(x)f(x)\,dx = \sum_{n=0}^{\infty} a_n \int_{-1}^{1} P_n(x)P_m(x)\,dx \tag{8.7.17}$$

But because of (8.7.14) and (8.7.15), the right-hand side of (8.7.17) reduces to a single term:

$$\int_{-1}^{1} P_m(x)f(x)\,dx = \frac{2a_m}{2m+1} \tag{8.7.18}$$

Hence, for each $n \geq 0$,

$$a_n = \frac{2n+1}{2}\int_{-1}^{1} P_n(x)f(x)\,dx \tag{8.7.19}$$

[9]This is the reason for the apparently "unreasonable" definition of μ.

So, from (8.7.12), $b_n = a_n/r_0^n$

$$V(r,\varphi) = \sum_{n=0}^{\infty} a_n (r/r_0)^n P_n(\cos\varphi) \tag{8.7.20}$$

where a_n is given by (8.7.19). Equation (8.7.20) represents the electric potential interior to the sphere $r \leq r_0$.

Now we turn to the electric potential exterior to this sphere. In order that the potential not grow infinitely large as $r \to \infty$, we take $c_1 = 0$ in (8.7.9) so that

$$R(r) = c_2 r^{-n-1} \tag{8.7.21}$$

Equation (8.7.5) now takes the form

$$V(r,\varphi) = \sum_{n=0}^{\infty} b_n r^{-n-1} P_n(\cos\varphi) \tag{8.7.22}$$

We set $r = r_0$ in (8.7.22) to obtain

$$f(\varphi) = \sum_{n=0}^{\infty} b_n r_0^{-n-1} P_n(\cos\varphi) \tag{8.7.23}$$

From the orthogonality conditions, we deduce that $b_n = a_n r_0^{n+1}$, and therefore, in $r \geq r_0$,

$$V(r,\varphi) = \sum_{n=0}^{\infty} a_n (r/r_0)^{-n-1} P_n(\cos\varphi) \tag{8.7.24}$$

Example 8.7.1. *Find the electric potential inside a spherical surface of radius* 1 *if the hemispherical surface* $\frac{1}{2}\pi < \varphi < \pi$ *is maintained at the constant potential* 1 *and the hemispherical surface* $0 < \varphi < \frac{1}{2}\pi$ *is kept at the constant potential zero.*

Solution: Inside the sphere, the solution is given by (8.7.20)

$$V(r,\varphi) = \sum_{n=0}^{\infty} a_n (r/r_0)^n P_n(\cos\varphi)$$

where

$$a_n = \frac{2n+1}{2} \int_{-1}^{0} P_n(x)\, dx$$

The integration to determine a_n is simply a polynomial integration – refer to (6.5.10)– and results in $a_0 = \frac{1}{2}$, $a_1 = -\frac{3}{4}$, $a_2 = 0$, $a_3 = \frac{7}{16}$, $a_4 = 0$, $a_5 = -\frac{11}{32}$. Therefore,

$$V(r, \varphi) = \frac{1}{2} - \frac{3}{4} r \cos \varphi + \frac{7}{16} r^3 P_3(\cos \varphi) - \frac{11}{32} r^5 P_5(\cos \varphi) + \ldots$$

•EXERCISES

1. Interior to a large spherically shaped mass is a spherical surface of radius 0.1 m that is maintained at a temperature of 250° C. Find the steady-state temperature distribution interior and exterior to this surface if $T \to 0°$ as $r \to \infty$.
2. The temperature on the spherical surface of a sphere with diameter 1m is $100 \cos \varphi$ degrees C. What is the steady-state temperature distribution interior to the surface?
3. Find the potential between two concentric spheres if the potential of the outer sphere is maintained at $V = 100$ and the potential of the inner sphere is maintained at zero. The radii of the spheres are 2 m and 1 m, respectively.

PROJECTS

1. Parseval's Theorem

Consider the problem of finding the constants α_k and β_k, $k = 1, 2 \ldots, N$ such that

$$S_N(x) = \frac{1}{2}\alpha_0 + \sum_{n=1}^{\infty}\left(\alpha_n \cos\frac{n\pi}{p}x + \beta_n \sin\frac{n\pi}{p}x\right)$$

is the "best" approximation to $f(x)$, N fixed. We measure the "goodness" of an approximation to $f(x)$ by the error term E_N:

$$E_N^2 = \frac{1}{p}\int_{-p}^{p}(f(x) - S_N(x))^2\, dx$$

In this expression, $f(x)$ and N are fixed, and E_N depends on $2n+1$ parameters $\alpha_0, \alpha_1, \ldots, \alpha_N$ and $\beta_0, \beta_1, \ldots, \beta_N$. We have a "best" approximation if E_N is minimal. Here's how we proceed.

(a) Show that

$$E_N^2 = \frac{1}{p}\int_{-p}^{p} f^2(x)\, dx - \frac{2}{p}\int_{-p}^{p} f(x)S_N(x)\, dx + \frac{1}{p}\int_{-p}^{p} S_N^2(x)\, dx$$

(b) Show that

$$\frac{2}{p}\int_{-p}^{p} f(x)S_N(x)\, dx = \alpha_0 a_0 + 2\left(\sum_{n=1}^{N}(\alpha_n a_n + \beta_n b_n)^2\right)$$

(c) Show that

$$\frac{1}{p}\int_{-p}^{p} S_N^2(x)\, dx = \frac{1}{2}\alpha_0^2 + \sum_{n=1}^{N}\alpha_n^2 + \sum_{n=1}^{N}\beta_n^2$$

(d) Using the results in (a)–(c) establish

$$E_N^2 = \frac{1}{p}\int_{-p}^{p} f^2(x)\, dx - \alpha_0 a_0 - 2\left(\sum_{n=1}^{N}(\alpha_n a_n + \beta_n b_n)^2\right)$$
$$+\frac{1}{2}\alpha_0^2 + \sum_{n=1}^{N}\alpha_n^2 + \sum_{n=1}^{N}\beta_n^2$$

(e) By completing the squares in (d), deduce

$$\begin{aligned} E_N^2 &= \frac{1}{p}\int_{-p}^{p} f^2(x)\,dx + \tfrac{1}{2}(a_0-\alpha_0)^2 - \tfrac{1}{2}a_0^2 \\ &\quad + \sum_{n=1}^{N}(a_n-\alpha_n)^2 + \sum_{n=1}^{N}(b_n-\beta_n)^2 \\ &\quad - \sum_{n=1}^{N}a_n^2 - \sum_{n=1}^{N}b_n^2 \end{aligned}$$

(f) Use (e) to prove that for every choice of α_n and β_n,

$$E_N^2 \geq \frac{1}{p}\int_{-p}^{p} f^2(x)\,dx - \frac{1}{2}a_0^2 - \sum_{n=1}^{N}a_n^2 - \sum_{n=1}^{N}b_n^2$$

with equality when $a_n = \alpha_n$ and $b_n = \beta_n$.

(g) Assume that $a_n = \alpha_n$ and $b_n = \beta_n$ for all n. Show that

$$E_1^2 \geq E_2^2 \geq \cdots E_n^2 \geq \ldots$$

In fact, it can be shown that under suitable hypotheses $\lim_{n\to\infty} E_n^2 = 0$. This leads to *Parseval's theorem:*

If $f(x)$ is sectionally continuous in $-p \leq x \leq p$, then

$$\frac{1}{p}\int_{-p}^{p} f^2(x)\,dx = \frac{1}{2}a_0^2 + \sum_{n=1}^{N}a_n^2 + \sum_{n=1}^{N}b_n^2$$

(h) Use Parseval's theorem for the function

$$g(x) = \begin{cases} -1 & -\pi < x < 0 \\ 1 & 0 < x < \pi \end{cases}$$

to deduce

$$\pi^2 = 8\sum_{n=1}^{\infty}\left(\frac{2}{2n-1}\right)^2$$

(i) Use Parseval's theorem for the function $h(x) = \begin{cases} 0 & -\pi < x < 0 \\ \sin x & 0 < x < \pi \end{cases}$

to deduce

$$\pi^2 = 8 + \sum_{n=1}^{\infty}\left(\frac{4}{4n^2-1}\right)^2$$

2. Legendre's Equation in Spherical Coordinates

The equation

$$\frac{1}{\sin\varphi}(\Phi'\sin\varphi)' + n(n+1)\Phi = 0$$

is Legendre's equation in spherical coordinates. It arises from the spherical coordinate form of Laplace's partial differential equation.

(a) Set $x = \cos\varphi$ and $u(x) = \Phi(\arccos x) = \Phi(\varphi)$. Use the chain rule to show that

$$\frac{d}{d\varphi}\left(\sin\varphi\frac{d\Phi}{d\varphi}\right) = \cos\varphi\frac{d\Phi}{dx}\frac{dx}{d\varphi} + \sin\varphi\frac{d}{d\Phi}\left(\frac{d\varphi}{dx}\frac{dx}{d\varphi}\right)$$

(b) Simplify (a) further to show that

$$\frac{d}{d\varphi}\left(\sin\varphi\frac{d\Phi}{d\varphi}\right) = -\cos\varphi\sin\varphi\frac{d\Phi}{dx} - \sin\varphi\frac{d}{d\varphi}\left(\sin\varphi\frac{d\Phi}{dx}\right)$$

(c) Now simplify (b) to show that

$$\frac{1}{\sin\varphi}\frac{d}{d\varphi}\left(\sin\varphi\frac{d\Phi}{d\varphi}\right) = -2\cos\varphi\frac{d\Phi}{dx} + \sin^2\varphi\frac{d^2\Phi}{dx^2}$$

(d) Complete the transformation by showing that

$$\frac{1}{\sin\varphi}\frac{d}{d\varphi}\left(\sin\varphi\frac{d\Phi}{d\varphi}\right) = (1-x^2)u'' - 2xu'$$

3. Solving the Nonhomogeneous Vibrating String Problem

Equations 8.4.14—8.4.17,

$$\begin{aligned} u_{tt}(x,t) &= a^2 u_{xx}(x,t) + y(x) \\ u(0,t) &= u(L,t) = 0 \\ u_t(x,0) &= g(x) \\ u(x,0) &= f(x) \end{aligned}$$

define a general family of nonhomogeneous equations. The function $y(x)$ is the feature that makes these equations nonhomogeneous.

(a) Set $v(x,t) = u(x,t) + h(x)$ where $h(x)$ is to be determined (see Part (b) to follow). Replace $u(x,t)$ in the above equations by $v(x,t) - h(x)$ thus

obtaining a boundary-value problem in $v(x,t)$ and the function $h(x)$.
(b) Choose $h(x)$ so that the $v(x,t)$ satisfies the homogeneous problem

$$\begin{aligned} v_{tt}(x,t) &= a^2 v_{xx}(x,t) \\ v(0,t) &= v(L,t) = 0 \\ v_t(x,0) &= G(x) \\ v(x,0) &= F(x) \end{aligned}$$

where $G(x)$ and $F(x)$ are functions related to $f(x)$ and $g(x)$.
(c) Use this idea to solve the nonhomogeneous problem $u_{tt} = u_{xx} + K$, given the conditions $u(0,t) = u(\pi,t) = 0$ and $f(x) = g(x) = 0$.
(d) Solve the nonhomogeneous problem $u_{tt} = u_{xx} + Kx$, given the conditions $u(0,t) = u(\pi,t) = 0$ and $f(x) = g(x) = 0$.

ANSWERS TO SELECTED ODD-NUMBERED PROBLEMS

Chapter 0

Section 0.1

3(a). $3 + 6i$.

3(c). $-13 + 11i$.

3(e). $11 - 10i$.

5. $\frac{5}{169} + \frac{12}{169}i$.

Section 0.2

1(a). $5e^{-i\arctan(4/3)}$.

1(c). $5e^{-i\arctan(4/3)}$.

1(e). $5e^{i\arctan(4/3)}$.

1(g). $4\sqrt{2}e^{i\pi/4}$.

1(i). -10.

3(a). $-e^2$.

3(c). $e^2(\cos(\pi t) + i\sin(\pi t))$.

5(a). $2 + i$.

Section 0.3

1(a). $2 - 2x + x^2$.

1(c). $2 - x - 2x^2 + x^3$.

1(e). $a^2 + b^2 - 2ax + x^2$.

Section 0.4

1(a). $\left[\begin{array}{ccc|c} 1 & 0 & 0 & 0 \\ 0 & 1 & 0 & 1 \\ 0 & 0 & 1 & 1 \end{array}\right]$.

1(c). $\left[\begin{array}{ccc|c} 1 & 0 & 0 & -1 \\ 0 & 1 & 1 & -1 \end{array}\right]$.

3(e). $\left[\begin{array}{ccc|c} 1 & 0 & 0 & 1 \\ 0 & 1 & 0 & 1 \\ 0 & 0 & 1 & 1 \end{array}\right]$.

Section 0.5

7. $x_1 = 5/3$, $x_2 = -1/3$.

9. $x_1 = 4$, $x_2 = 0$.

11. $x_1 = -2$, $x_2 = t/3$, $x_3 = t$.

13. $x_1 = 9$, $x_2 = -6\, x_3 = 1$.

Section 0.6

11(a). $\begin{bmatrix} 3 & 2 & 1 \\ 1 & -1 & -2 \\ 6 & 3 & -3 \end{bmatrix}$.

11(c). $\begin{bmatrix} 5 & 5 & 0 \\ 1 & 1 & 0 \\ 5 & 5 & -4 \end{bmatrix}$.

11(e). $\begin{bmatrix} 12 & 8 & 4 \\ 4 & -4 & -8 \\ 24 & 12 & -12 \end{bmatrix}$.

11(g). $\begin{bmatrix} 2 & 1 & 4 \\ 1 & -1 & 2 \\ 0 & -2 & 0 \end{bmatrix}$.

Section 0.7

1(a). $\begin{bmatrix} 0 & 0 & 0 \\ -2 & -4 & 1 \\ 0 & 0 & 0 \end{bmatrix}$.

1(c). $\begin{bmatrix} 1 & 0 & 4 \\ 0 & 2 & 4 \\ 0 & 2 & 3 \end{bmatrix}$.

1(e). $\begin{bmatrix} 1 & 0 \\ 0 & 2 \end{bmatrix}$.

1(g). $\begin{bmatrix} -1 & 0 & 2 \\ 1 & 2 & 1 \\ 2 & -1 & -1 \end{bmatrix}$.

3(a). $\begin{bmatrix} 0 & 0 & 0 \\ -2 & -4 & 1 \\ 0 & 0 & 0 \end{bmatrix}$.

3(c). $\begin{bmatrix} -2 \\ 0 \\ 0 \end{bmatrix}$.

3(e). $\begin{bmatrix} 0 & 9 & 9 \end{bmatrix}$.

7(a). $\begin{bmatrix} 0 & -6 \\ 0 & 0 \end{bmatrix}$.

7(c). $\begin{bmatrix} 0 & -6 \\ 0 & 0 \end{bmatrix}$.

9(a). $x^2 - 2xy + y^2 + 2xyz - z^2$.

15. $\begin{bmatrix} 0 & 0 & 0 \\ 0 & 0 & 0 \\ 0 & 0 & 0 \end{bmatrix}$.

Section 0.9

1. $1/2\begin{bmatrix} 1 & 1 \\ -1 & 1 \end{bmatrix}$.

3. $1/2\begin{bmatrix} 1 & 0 \\ 0 & 2 \end{bmatrix}$.

5. $1/5\begin{bmatrix} 1 & -2 \\ -2 & 1 \end{bmatrix}$.

7. Singular.

9. $\begin{bmatrix} 1/2 & 0 & 0 \\ 2 & -1 & 0 \\ 2 & -1 & -1 \end{bmatrix}$.

13. Singular.

Section 0.10.1

3. -12.

5. -5.

7. 50.

9. -257.

Section 0.10.2

1. 36.

Section 0.11

1(a). 0.

1(c). 2.

1(e). Dependent.

1(g). 2.

1(i). 0.

3(a). $x_1 == x_2 = -t$, $x_3 = x_4 = t$.

Chapter 1

Section 1.2

1(a). Linear, 1st-order.

1(c). Nonlinear, 1st-order.

1(e). Linear, 2st-order.

3(a). $c_1 + 2t + \frac{1}{3}t^3$.

3(c). $c_1 + \frac{1}{2}(t^2 + 2\sin t)$.

3(e). $c_1 + c_2 t + c_3 t^2 + e^t$.

5. te^{-2t}.

Section 1.3.1

1. $1 + 5t^2$.

3. e^{-2+2t}.

5. $-t$.

7. $\frac{1}{2} - \log t$.

9. $\frac{1}{2} - \log(-t)$.

Section 1.3.2

1(a). $2 + c_1 e^{-t}$.

1(c) $e^{-t^2/2}(c_1 + 10\int e^{t^2/2}dt$.

1(e). $e^{-t}(c_1 + \frac{1}{2}t^2)$.

1(g). $t^2(c_1 + \int e^{t/t^2}dt$.

3(a). $(2 + t)e^{-2t}$.

3(c). $-1 - t + 2e^t$.

9. $e^{-2t}(c_1 + \int e^{2t}f(t)dt$.

11. Let $W(t)$ be the general solution of $W' + p(t)W = f(t)$. Then $\int W(t)dt$ is the solution.

Section 1.5.1

1. $(c_1 - \log(t))^{-1}$.

3. $(c_1 - 10t)^{-1}$.

7. te^{kt}.

9. $\frac{1}{4}(-5 - 2t - \sqrt{7}\tanh(\sqrt{7}t + c_1\sqrt{7}))$.

13. $2t\sqrt{t}$. **Section 1.5.2**

1(a). $c_1/(2+t^2)$.

1(c). $c_1 \csc 2t$.

1(e). $c_1 \csc t e^{-t}$.

3. $Q(t) = e^{\int p(t)dt}$

Chapter 2

Section 2.2

1. $\left(-1, \begin{bmatrix} -2 \\ 1 \end{bmatrix}\right), \left(5, \begin{bmatrix} 1 \\ 1 \end{bmatrix}\right)$.

3. $\left(-1, \begin{bmatrix} 0 \\ 1 \end{bmatrix}\right), \left(2, \begin{bmatrix} 1 \\ 0 \end{bmatrix}\right)$.

5. $\left(0, \begin{bmatrix} -1 \\ 1 \end{bmatrix}\right), \left(1, \begin{bmatrix} -2 \\ 1 \end{bmatrix}\right)$.

7. $\left(2-2i, \begin{bmatrix} -i \\ 1 \end{bmatrix}\right), \left(2+2i, \begin{bmatrix} i \\ 1 \end{bmatrix}\right)$.

9. $\left(a, \begin{bmatrix} -1/a \\ 1 \end{bmatrix}\right), \left(b, \begin{bmatrix} -1/b \\ 1 \end{bmatrix}\right)$.

11. $\left(0, \begin{bmatrix} 1 \\ -1 \\ 1 \end{bmatrix}\right), \left(1, \begin{bmatrix} -2 \\ 1 \\ 1 \end{bmatrix}\right). \left(4, \begin{bmatrix} 1 \\ 1 \\ 1 \end{bmatrix}\right)$.

13. $\left(1, \begin{bmatrix} 0 \\ 0 \\ 1 \end{bmatrix}\right), \left(1, \begin{bmatrix} 1 \\ 9 \\ 0 \end{bmatrix}\right)$.

15. $\left(-6, \begin{bmatrix} -1 \\ 2 \\ 0 \end{bmatrix}\right), \left(4, \begin{bmatrix} 0 \\ 0 \\ 1 \end{bmatrix}\right), \left(14, \begin{bmatrix} 2 \\ 1 \\ 0 \end{bmatrix}\right)$.

17. $\left(-2, \begin{bmatrix} -3 \\ 1 \\ 0 \end{bmatrix}\right), \left(6, \begin{bmatrix} 0 \\ 0 \\ 1 \end{bmatrix}\right), \left(8, \begin{bmatrix} 1 \\ 3 \\ 0 \end{bmatrix}\right)$.

19. $\left(0, \begin{bmatrix} 1 \\ 0 \\ 0 \end{bmatrix}\right), \left(0, \begin{bmatrix} 0 \\ 1 \\ 0 \end{bmatrix}\right), \left(0, \begin{bmatrix} 0 \\ 0 \\ 1 \end{bmatrix}\right)$.

21. $\left(\cos t - i\sin t, \begin{bmatrix} i \\ 1 \end{bmatrix}\right), \left(\cos t + i\sin t, \begin{bmatrix} -i \\ 1 \end{bmatrix}\right)$.

23. Consider $\det(\mathbf{A} - 0\mathbf{I}) = 0$.

25. Subtract μ from each eigenvalue.

27. Multiply each eigenvalue by μ.

29. Let $\mathbf{A} = \begin{bmatrix} 1 & 1 \\ 0 & 1 \end{bmatrix}$.

Section 2.3

1(a). $\begin{bmatrix} 1 \\ 1 \end{bmatrix}$.

1(c). $\frac{2}{3}\begin{bmatrix} 1-e^{3t} \\ 1+e^{3t} \end{bmatrix}$.

1(e). $e^{3t}\begin{bmatrix} 2 \\ -1 \end{bmatrix}$.

3(a). $e^{2t}\begin{bmatrix} 1 \\ 1 \end{bmatrix}$.

3(c). $\frac{1}{2}\begin{bmatrix} -1+e^{2t} \\ 1+e^{2t} \end{bmatrix}$.

3(e). $\frac{1}{2}\begin{bmatrix} 3+e^{2t} \\ -3+e^{2t} \end{bmatrix}$.

5(a). $\frac{1}{3}e^{t}\begin{bmatrix} 2e^{3t}+1 \\ 4e^{3t}-1 \end{bmatrix}$.

5(c). $\frac{1}{3}e^{t}\begin{bmatrix} e^{3t}-1 \\ 2e^{3t}+1 \end{bmatrix}$.

5(e). $\frac{1}{3}e^{t}\begin{bmatrix} e^{3t}+5 \\ 2e^{3t}-5 \end{bmatrix}$.

7(a). No solutions of the form $c_1\mathbf{u}_1e^{\lambda_1}+c_2\mathbf{u}_2e^{\lambda_2}$.

7(c). No solutions of the form $c_1\mathbf{u}_1e^{\lambda_1}+c_2\mathbf{u}_2e^{\lambda_2}$.

7(e). No solutions of the form $c_1\mathbf{u}_1e^{\lambda_1}+c_2\mathbf{u}_2e^{\lambda_2}$.

9(a). $\frac{1}{2}e^{2t}\begin{bmatrix} -1+e^{2t}+2e^{4t} \\ 2e^{2t}(-1+e^{4t}) \\ -3+e^{2t}+2e^{4t} \end{bmatrix}$.

9(c). $\frac{1}{2}e^{2t}\begin{bmatrix} 1+e^{2t}-2e^{4t} \\ 2(1-e^{4t}) \\ 3+e^{2t}-2e^{4t} \end{bmatrix}$.

9(e). $e^{6t}\begin{bmatrix} 1 \\ 1 \\ 1 \end{bmatrix}$.

11(a). $e^{t}\begin{bmatrix} 1 \\ 0 \\ 0 \end{bmatrix}$.

11(c). $\frac{1}{2}e^{-2t}\begin{bmatrix} 0 \\ e^{4t}-1 \\ e^{4t}+1 \end{bmatrix}$.

11(e). $e^{2t}\begin{bmatrix} e^{-t} \\ 1 \\ 1 \end{bmatrix}$.

21. $\mathbf{x}_1(t)=e^{2t}\begin{bmatrix} 1 \\ 0 \\ 0 \end{bmatrix}$.

23. $\mathbf{x}_2(t)=e^{2t}\begin{bmatrix} t \\ 1 \\ 0 \end{bmatrix}$.

Section 2.4

1. $\frac{1}{4}\begin{bmatrix} e^{-2t}+3e^{2t} & -3e^{-2t}+3e^{2t} \\ -e^{-2t}+e^{2t} & 3e^{-2t}+e^{2t} \end{bmatrix}$.

3. $\frac{1}{3}\begin{bmatrix} 2e^{t}+e^{4t} & -e^{t}+e^{4t} \\ -2e^{t}+2e^{4t} & e^{t}+2e^{4t} \end{bmatrix}$.

5. $\frac{1}{3}\begin{bmatrix} e^{-2t}+2e^{t} & -2e^{-2t}+2e^{t} \\ -e^{-2t}+e^{t} & 2e^{-2t}+e^{t} \end{bmatrix}$.

7. $\begin{bmatrix} \cos t\cosh t+\cos t\sinh t & \sin t\cosh t+\sin t\sinh t \\ -\sin t\cosh t-\sin t\sinh t & \cos t\cosh t+\cos t\sinh t \end{bmatrix}$.

9. $\begin{bmatrix} e^{2t} & e^{t}-e^{2t} \\ 0 & e^{t} \end{bmatrix}$.

11. $\frac{1}{4}\begin{bmatrix} 3e^{t}+e^{5t} & -e^{t}+e^{5t} \\ -3e^{t}+3e^{5t} & 1e^{t}+3e^{5t} \end{bmatrix}$.

13. $\begin{bmatrix} -e^{-t}+e^{t}+e^{2t} & e^{-t}-e^{2t} & e^{-t}-e^{t} \\ -e^{-t}+e^{t} & e^{-t} & e^{-t}-e^{t} \\ -e^{-t}+e^{2t} & e^{-t}-e^{2t} & e^{-t} \end{bmatrix}.$

15. $\frac{1}{2}\begin{bmatrix} 1+e^{2t} & -1+e^{2t} & -1+e^{2t} \\ -1+e^{2t} & 1+e^{2t} & -1+e^{2t} \\ 0 & 0 & 1 \end{bmatrix}.$

17. $\frac{1}{4}\begin{bmatrix} e^{-2t}+3e^{2t} \\ -e^{-2t}+e^{2t} \end{bmatrix}.$

19. $\frac{1}{2}\begin{bmatrix} -e^{-2t}+3e^{2t} \\ e^{-2t}+e^{2t} \end{bmatrix}.$

21. $\begin{bmatrix} (e^{-2t}-e^{t})/3 \\ -e^{-2t}+2e^{-t} \\ 2e^{-2t}-2e^{-t} \end{bmatrix}.$

23. $\begin{bmatrix} (e^{-2t}-e^{t})/3 \\ -e^{-2t}+e^{-t} \\ 2e^{-2t}-e^{-t} \end{bmatrix}.$

27. $\frac{1}{2}\begin{bmatrix} -ie^{-it} & e^{-it} \\ ie^{it}+e^{it} & \end{bmatrix}.$

Section 2.6.1

1. $\begin{bmatrix} 1 \\ 1 \end{bmatrix}+\frac{1}{2}e^{it}\begin{bmatrix} 1 \\ 1 \end{bmatrix}-\frac{1}{2}e^{it}\begin{bmatrix} 1 \\ 1 \end{bmatrix}.$

3. $e^{-t}\begin{bmatrix} 1 \\ 0 \end{bmatrix}+e^{t}\begin{bmatrix} 0 \\ 1 \end{bmatrix}.$

5. $\frac{1}{2}e^{3it}\begin{bmatrix} 1 \\ 0 \end{bmatrix}+\frac{1}{2}e^{-3it}\begin{bmatrix} 1 \\ 0 \end{bmatrix}+\frac{1}{2i}e^{2it}\begin{bmatrix} 0 \\ 1 \end{bmatrix}-\frac{1}{2i}e^{-2it}\begin{bmatrix} 0 \\ 1 \end{bmatrix}.$

7. $(\cos t+\sin t)\begin{bmatrix} -2 \\ -1 \end{bmatrix}.$

9. $\begin{bmatrix} -e^{-t} \\ -1/3-1/4e^{-t} \end{bmatrix}.$

11. $-e^{-2t}(\cos t+3\sin t)\begin{bmatrix} 1 \\ 1 \end{bmatrix}.$

13. $e^{-t}\cos t\begin{bmatrix} 1 \\ -1 \end{bmatrix}.$

17. $\frac{1}{3}\begin{bmatrix} -\cos t+2\sin t \\ -\sin t \end{bmatrix}.$

Section 2.6.2

1. $\frac{1}{2}e^{2t}\begin{bmatrix} 1 \\ -1 \end{bmatrix}.$

3. $-(2+\cos t-\sin t)\begin{bmatrix} 2 \\ 1 \end{bmatrix}.$

5. $\frac{1}{4}\begin{bmatrix} 1-4e^{t} \\ -2 \\ 0 \end{bmatrix}.$

7. $-e^{-2t}(\cos t+3\sin t)\begin{bmatrix} 1 \\ 1 \end{bmatrix}$

9. $\frac{1}{2}e^{-t}(\cos t+\sin t)\begin{bmatrix} -2 \\ 1 \end{bmatrix}.$

11. $\begin{bmatrix} -1-t \\ -2-t \end{bmatrix}=-\begin{bmatrix} 1 \\ 2 \end{bmatrix}-t\begin{bmatrix} 1 \\ 1 \end{bmatrix}.$

Section 2.7

1. $\begin{bmatrix} 1-e^{t}+2te^{t} \\ e^{t}+2te^{t} \end{bmatrix}.$

3. $\frac{1}{2}(e^{t}-e^{-t})\begin{bmatrix} 1 \\ 1 \end{bmatrix}.$

Chapter 3

Section 3.4

1(a). e^{2t}.

1(c). $e^{t^2/2}$.

Section 3.5

1. $1 + a(t-1) + b(t^2 - 2t + 2)$.

Section 3.6

1. $c_1 e^{-2t} + c_2 e^{3t}$.

3. $c_1 e^{-2t} + c_2 e^{6t}$.

5. $c_1 e^{-3t} + c_2 e^{3t}$.

7. $2e^{5t}$.

9. $\frac{5}{2}(5e^t - e^{5t})$.

11. $c_1 + c_2 t$.

13. $e^{-2t}(c_1 + c_2 t)$.

15. $2 + 4t$.

17. $4e^{4t}(1 + 4t)$.

19. $e^{-2t}(c_1 \cos t + c_2 \sin t)$.

21. $e^{-t}(c_1 \cos 3t + c_2 \sin 3t)$.

23. $e^{-t}(\cos 3t + \sin 3t)$.

25. $e^{-t/2}(\cos \frac{\sqrt{3}t}{2} - \frac{1}{\sqrt{3}} \sin \frac{\sqrt{3}t}{2})$.

Section 3.7.1

1. $\phi'' + \frac{g}{L} \sin \phi$. Nonlinear. $\phi(t) = c_1 \sin \omega t + c_2 \cos \omega t$.

3(b). 0.7854 sec

3(d). 1.405 sec.

Section 3.7.2

1(a). $2e^{-3t} - 1.5e^{-4t}$.

1(c). $0.5454e^{-t} - 0.454e^{-12t}$.

3(a). $20(e^{-2t} - e^{-8t})/3$

3(c). $40(e^{-t} - e^{-12t})/11$.

5(a). 3.15 m.

5(c). 2.654 m.

7(a). $2(1 + 4t)e^{-4t}$.

7(c). $(2 + 15t)e^{-7.5t}$.

9(a). 1.016 m.

9(c). 0.3479 m.

11(a). $\cos 3.742t + 1.603 \sin 3.742t)e^{-6t}$.

13(a). -2.891 m/s.

13(c). 20 m/s.

15(a). $\sqrt{2}/2$

15(c). $\sqrt{101}/101$.

17. 1.2403 sec.

Section 3.8

1(a). $-70.t \sin 707t$.

1(c). $100 \sin 100t$.

5(a) 6.766A at 0.001225sec.

5(c). -455A at 2.607$\times$ 10$-$4sec.

7(a). -403A at 0.01726sec.

7(c). -1.294A at 0.001081sec.

Section 3.9

1. t. 3. $\frac{1}{2}(e^{-t} - \cos t + \sin t)$.
5. $\frac{1}{18}(10\sin t - \sqrt{10}\sin\sqrt{10}t)$.
7. $e^{-2t}t^2/2$.
9. $\frac{4}{5}((2-2e^t)\cos t + (1+e^t)\sin t)$.
11. $\frac{1}{2}(-t\cos t + \sin t)$.
13. $\frac{1}{27}(e^{-2t}(2+3t) + e^t(-2+3t))$.
15. $\frac{1}{5}(e^{2t} - \cos t - 2\sin t)$.
17. $\frac{1}{81}(-2 + 9t^2 + 2\cos 3t)$.
19. $\frac{1}{8}(9 - 2t + 2t^2 - e^{-2t}(9+16t))$.
21. $\frac{1}{75}(15t\cos 2t + 12\sin 2t - 13sin3t)$.
23. $\frac{1}{290}(-20e^{-3t} + 29e^{-2t} + e^{2t}(-9\cos 2t + 8\sin 2t))$.
25. $\frac{1}{8}(3 - 4t + 2t^2 + e^{-2t}(-3+2t))$.
27. $8 + 8t + 5t^2 - 8e^t$.
29. $-3e^{-3t} + 5e^{-t} - 2\cos t - 4\sin t$.
31. $2e^t(2\cos t + \sin t) - 4\cos 2t - 3\sin 2t$.

Section 3.10.2

1. $c_1 + c_2e^{-Ct/M} - Mgt/C$. 3. 5.56 sec, 8.78 sec.
7(a). $0.001t\sin 5t$. 7(c) $-\frac{1}{36}(\cos 6t + \sin 6t - e^{-6t})$.
7(e). $0.0202\sin 7t - 0.02\sin 7.071t$.
8(a). $10t\sin 10t$. 8(c). $120t\cos 100t - 1.2\sin 100t$.
9. 0.0056 Hz, 2 m. 11. 0.796 Hz, 1218 A.

Section 3.10.3

3. 2.2 m, 1.616 rad/sec.

Section 3.10.3

5(a). 1.356 m. 5(c). 5 m.

Section 3.11

1(a). $c_1t^{-2} + c_2t^{-4}$. 1(c). $-2t + c_1t^{-3} + c_2t^4$.

Section 3.12

1. $\cos t\log\left(\dfrac{1-\sin t}{\cos t}\right)$. 3. $\frac{1}{6}t^3e^{-2t}$.
5. $t\log t\, e^t$. 7. -9.
9. $-\frac{1}{6}t$.

Chapter 4

Section 4.2.1

1(a). $u^{(2)} - u^{(1)} = 0.$

1(c). $u^{(3)} - u^{(2)} + u^{(1)} + u = 0.$

1(e). $u(4) - u = 0.$

1(g). $u(4) - 12u^{(3)} + 62u^{(2)} - 156u^{(1)} + 169u = 0.$

1(i). $u^{(2)} + 4u = 0.$

3(a). $c_1e^{-t} + c_2e^{t} + c_3e^{2t}.$

3(c). $c_1e^{-t} + c_2e^{2t/5} + c_3e^{t/2}.$

3(e). $c_1e^{-3t/2} + c_2e^{-t/2} + c_3e^{2t}.$

5(a). $c_1e^{t/2} + c_2\cos 4t + \sin 4t.$

5(c). $c_1 + c_2e^{-t} + c_3e^{t} + c_4e^{2t}.$

5(e). $c_1e^{-4t} + c_2e^{t} + (c_3 + c_4t)e^{2t}.$

Section 4.2.2

1(a). $\frac{1}{2}(1 + \cosh 2t).$

1(c). $-5e^{t} + 8e^{2t} - 3e^{3t}.$

1(e). $2 - t.$

3(a). $2(\cos 2t + \cosh 2t.$

3(c). $2 - (2 + t)e^{-t}.$

Section 4.3

5. $1 - \frac{1}{3}e^{-t} - e^{t} + \frac{1}{3}e^{2t}.$

7. $1 + t - e^{t}.$

9. $\frac{1}{4}(\cos t + \sin t + (3 + t^2)e^{t}).$

11. $\frac{1}{8}(8 - \cos t + \sin t + (\cos t + \sin t)e^{-2t}.$

15. $\frac{2}{5}\cos 2t - \frac{3}{10}\sin 2t - \frac{2}{5}e^{t} + te^{t}.$

Section 4.4

1. $$\begin{bmatrix} 1 & e^{-t} & e^{3t} \\ 0 & -e^{-t} & 3e^{3t} \\ 0 & e^{-t} & 9e^{3t} \end{bmatrix}.$$

3. $$\begin{bmatrix} 1 & e^{t} & e^{3t} \\ 0 & e^{t} & 6e^{6t} \\ 0 & e^{t} & 36e^{6t} \end{bmatrix}.$$

5. $$e^{t}\begin{bmatrix} 1 & t & t^2 \\ 1 & t+1 & 2t + t^2 \\ 1 & t+2 & 2 + 4t + t^2 \end{bmatrix}.$$

7. $$\begin{bmatrix} e^{-2t} & e^{-t} & e^{t} & e^{3t} \\ -2e^{-2t} & -e^{-t} & e^{t} & 3e^{3t} \\ 4e^{-2t} & e^{-t} & e^{t} & 9e^{3t} \\ -8e^{-2t} & -e^{-t} & e^{t} & 27e^{3t} \end{bmatrix}.$$

9. $$\begin{bmatrix} 1 & e^{-2t}\cos t & e^{-2t}\sin t \\ 0 & -2e^{-2t}(\cos 2t + \sin 2t) & 2e^{-2t}(\cos 2t - \sin 2t) \\ 0 & 8e^{-2t}\sin 2t & -8e^{-2t}\cos 2t \end{bmatrix}.$$

11. $e^t\begin{bmatrix} e^t & \cos 2t & \sin 2t \\ e^t & -2\sin 2t & 2\cos 2t \\ e^t & -4\cos 2t & -4\sin 2t \end{bmatrix}$.

13. $e^t\begin{bmatrix} e{-}t & e^t & \cos t & \sin t \\ -e^{-t} & e^t & -\sin t & \cos t \\ e^{-t} & e^t & -\cos t & -\sin t \\ -e^{-t} & e^t & \sin t & \cos t \end{bmatrix}$.

15. $e^{-t}\begin{bmatrix} 1 & t & t^2 \\ -1 & -t+1 & 2t-t^2 \\ 1 & t-2 & 2-4t+t^2 \end{bmatrix}$.

Section 4.5

1. $\frac{1}{4}(3e^{-t}+e^{3t})$.
3. $-\frac{7}{3}+\frac{17}{5}e^t-\frac{1}{15}e^{6t}$.
5. t^2e^t.
7. $-\frac{8}{15}e^{-2t}+\frac{7}{4}e^{-t}+\frac{5}{6}e^t-\frac{1}{20}e^{3t}$.
9. $\frac{1}{5}(1-(\cos t+2\sin t)e^{-2t})$.
11. $e^t+\sin 2t$.
13. $\frac{1}{4}(5+e^t-2\cos t)$.
15. $e^{-t}(1+t+t^2)$.

Section 4.6

1. $c_1\sin t+c_2\cos t+\log(\sec t+\tan t)$.
3. $\frac{1}{6}t^3e^{-2t}$.
5. $t\log te^t$.
7. $2\log te^{-t}$.
9. $\frac{1}{4}t^2e^{-t}(-3+\log t)$.

Chapter 5

Section 5.2

1. $\dfrac{2}{s^2}$.
3. $\dfrac{-1}{s^2}+\dfrac{2}{s}$.
5. $\dfrac{e^{-3}}{s-2}$.
7. $\dfrac{s}{s^2+16}$.
9. $\dfrac{4s}{(s^2+4)^2}$.
11. $\dfrac{16s^3}{(s^2+b^2)^3)}-\dfrac{12s}{(s^2+b^2)^2)}$.
13. $\dfrac{2}{s^2-4}$.
15. $\dfrac{s}{s^2-16}$.

Section 5.3

1. $\dfrac{3}{s}-\dfrac{1}{s-1}$.
3. $\dfrac{2}{s-2}-\dfrac{6}{s^2+4}$.
5. $\dfrac{2}{s}-\dfrac{15}{s^2+9}$.
7. $\dfrac{3}{(s-3)^2}$.
9. $\dfrac{6s^2}{(s^2+16)^2}-\dfrac{3}{s^2+16}$.
11. $\dfrac{48s^3}{(s^2+1)^4}-\dfrac{24s}{(s^2+1)^3}$.
13. $\dfrac{8s^3}{(s^2+1)^3}-\dfrac{6s}{(s^2+1)^2}$.
15. $\dfrac{5}{s+2}+\dfrac{4}{(s+2)^2}+\dfrac{2}{(s+2)^3}$.

17. $-\frac{2}{(s^2+1)^2} - \frac{-2s}{(s^2+1)^2} + \frac{8s^2}{(s^2+1)^3}$.

19. $\frac{24}{(s+2)^5}$.

21(a). $\frac{2}{4+(s-2)^2} + \frac{2}{4+(s+2)^2}$.

21(c). $\frac{3s-3}{1+(s-1)^2} - \frac{3s+3}{1+(s+1)^2}$.

21(e). $2(\frac{4}{(s-2)^2-16} - \frac{4}{(s+2)^2-16})$.

23(a). $2\frac{s+1}{((s+1)^2+1)^2}$.

23(c). $-\frac{8}{(s+2)^2+16} + \frac{s+2}{(s+2)^2+16}$.

25. $\frac{2b^2}{(s^2-2as+a^2+b^2)^2}$.

Section 5.4

1(a). $\frac{1}{s}$.

1(c). $-\frac{1}{s} + \frac{2}{s}e^{-s}$.

3(a). $\frac{1}{s}(e^{-s} - e^{-2s})$.

3(c). $(\frac{2}{s} + \frac{3}{s^2} + \frac{2}{s^3})e^{-2s} + (\frac{1}{s} + \frac{1}{s^2})e^{-s}$.

5(a). $-\frac{12}{s} - \frac{8}{s^2} - \frac{2}{s^3})e^{-4s} + (\frac{4}{s^2} + \frac{2}{s^3})e^{-2s}$.

5(c). $-\frac{1}{s^2+1}e^{-2\pi s} - \frac{1}{s^2+1}e^{-\pi s}$.

Section 5.5

1. $\frac{2}{s^2+4}\frac{e^{\pi s/2}}{e^{\pi s/2}-1}$.

3. $\frac{1}{s}\frac{e^s}{e^s+1}$.

5. $\frac{1}{s^2} - \frac{1}{2s}\frac{1}{e^s-1} - \frac{s+2}{2s^2}\frac{1}{e^s+1}$.

Section 5.6

1. t^2+t-2.
3. $e^{a-t}u_a(t-a)$.
5. $\frac{1}{3}(12\cos 3t - 5\sin 3t)e^{-2t}$.
7. $\frac{1}{3}e^{-t}(1+8e^{6t})$.
9. $\frac{1}{2} - \frac{1}{2}e^{-2t}$.
11. $\frac{1}{4}(1-\cos 2t)$.
13. $\frac{1}{4}(-1-2t+e^{2t})$.
15. $t - \frac{1}{2}\sin 2t$.
17. $-t + \frac{1}{3}\sinh 3t$.
19. $-1 + 2e^{-t}$.
21. $\frac{1}{3}(e^{2t} - e^{-t})$.
23. $\frac{2}{9}(e^{2t}(1+6t) - e^{-t})$.

Section 5.7.1

1. $\frac{1}{2}(1-e^{-2t})$.
3. $-1+2e^{-t}$.
5. $\frac{1}{3}(e^{2t}-e^{-t})$.
7. $(1-e^{2-t})u_2(t)$.
9. $(-3+t+e^{2-t})u_2(t)$.
11. $\frac{1}{8}(2t-\sin 2t)$.
13. $\frac{1}{5}(-4\cos t-3\sin t+4e^{2t})$.
15. $\frac{1}{3}(e^t-\cos(\sqrt{3}t/2)e^{t/2}-\frac{3}{\sqrt{3}}\sin(\sqrt{3}t/2)e^{t/2})$.
17. $\frac{1}{10}(4\cos 2t+3\sin 2t-4e^{-t})$.
19. $\frac{1}{3}(5e^{-4t}-2e^{-t})$.
21. $\frac{1}{3}(-3+2e^{2t}+e^{-t})$.
23. $\frac{1}{6}(e^{-t}-9e^{t}+8e^{2t})$.
25. $(a\sin at-b\sin bt)/(a^2-b^2)$.
27. $\frac{1}{2}(a^2e^{-at}-b^2e^{-bt}/(a^2-b^2)-a^2e^{at}-b^2e^{bt})/(a^2-b^2)$.

Section 5.7.2

1. $2(1-3t)e^{-3t}$.
3. $t-\frac{1}{2}\sin 2t$.
5. $\frac{1}{4}(-1-2t+e^{2t})$.
7. $\frac{1}{4}u_1(t)(1+(1-2t)e^{-2(t-1)})$.
9. $1-t-\cos t+\sin t$.
11. $\frac{1}{4}(e^{-3t}+(3+4t)e^{t})$.
13. $\frac{5}{9}(-9+5e^{4t}+(4-15t)e^{t})$.
17. $\frac{1}{8}(3t\cos 2t+8\sin t-\frac{11}{16}\sin 2t)$.
19. $\frac{1}{54}(3t(\cos 3t+6t\sin 3t)-\sin 3t)e^{2t}$.
21. $\cos t-\cos 2t-\frac{3}{4}t\sin 2t$.

Section 5.9

1. $5\sin 2t$.
3. $2\sin t$.
5. $\frac{1}{2}t\sin 2t$.
7. $e^{-2t}-e^{-3t}$.
9. $\frac{1}{3}(e^{4t}+2e^{-2t})$.
11. $-4+2t+(4+2t)e^{-t}$.
13. $\frac{1}{2}\cos 2t+\sin 2t-\frac{1}{2}(\cos 2t+3\sin 2t)e^{-2t}$.

Section 5.10

1. $e^t\begin{bmatrix}\cos t & \sin t\\ -\sin t & \cos t\end{bmatrix}$.
3. $\frac{1}{2}\begin{bmatrix}\cosh t & \sinh t\\ \sinh t & \cosh t\end{bmatrix}$.
5. $\frac{1}{4}\begin{bmatrix}3e^{3t}+e^{-t} & e^{3t}-e^{-t}\\ 3e^{3t}-3e^{-t} & e^{3t}+3e^{-t}\end{bmatrix}$.
7. $\frac{1}{3}\begin{bmatrix}e^{3t}+2 & 2e^{3t}-2\\ e^{3t}-1 & 2e^{3t}+1\end{bmatrix}$.

Chapter 6

Section 6.3

7. $t + t^2 + t^3 + t^4 + \cdots$.
9. $t^2 - \frac{1}{3}t^4 + \cdots$.
11. $-t - \frac{1}{2}t^2 - \frac{1}{3}t^3 - \frac{1}{4}t^4 + \cdots$.
13. $2t + \frac{4}{3}t^3 + \cdots$.
15. $-2t - 2t^3 + \cdots$.
17. $\frac{1}{2} - \frac{3}{4}t + \frac{7}{8}t^2 - \frac{15}{16}t^3 + \frac{31}{32}t^4 + \cdots$.
19. $e + 2et + 2et^2 + e\frac{4}{3}t + e\frac{2}{3}t^4 + \cdots$.
21. $t^2 + \cdots$.
23. $\log\frac{1}{2} + t - \frac{1}{2}t^2 + \frac{1}{3}t^3 - \frac{1}{4}t^4 + \cdots$.
25. $\log 2 - \frac{3}{2}t + \frac{3}{8}t^2 - \frac{3}{8}t^3 + \frac{15}{64}t^4 + \cdots$.
27. $t - t^2 - \frac{1}{3}t^3 + \cdots$.
29. $1 - \frac{1}{4}t^2 - \frac{1}{32}t^4 + \cdots$.
31. $\frac{1}{4} - \frac{5}{16}t + \frac{21}{64}t^2 - \frac{85}{256}t^3 + \frac{341}{1024}t^4 + \cdots$.
33. $t - \frac{1}{2}t^2 + \frac{1}{3}t^3 - \frac{1}{4}t^4 + \cdots$.
35. $\frac{1}{2}t^2 - \frac{1}{4}t^4 + \cdots$.
37. $\frac{1}{2}t^2 - \frac{1}{6}t^3 + \frac{1}{12}t^4 + \cdots$.
39. $-\frac{1}{2} + \frac{1}{2}t^2 - \frac{1}{6}t^4 + \cdots$.
41. $-\frac{1}{3} - \frac{2}{9}(t-1) - \frac{7}{27}(t-1)^2 - \frac{20}{81}(t-1)^3 - \frac{61}{243}(t-1)^4 + \cdots$.
43. $(t-1) - \frac{1}{2}(t-1)^2 + \frac{1}{3}(t-1)^3 - \frac{1}{4}(t-1)^4 + \cdots$.
45. $1 + \frac{1}{2}t^2 + \frac{5}{24}t^4 + \cdots$.

Section 6.4

7. $c_1 + c_2(t - \frac{1}{6}t^2$.
9. $c_1(-2t + \frac{4}{3}t^3 + c_2(1 - 2t^2 + \frac{2}{3}t^4)$.
11. $c_1(1 - \frac{1}{6}t^3 + c_2(t - \frac{1}{12}t^4)$.
13. $c_1(1 - 2t^2 + \frac{2}{3}t^3 + \frac{1}{2}t^4) + c_2(t - \frac{1}{2}t^2 - \frac{1}{2}t^3 + \frac{7}{24}t^4)$.
15. $c_1(1 - \frac{1}{12}t^4) + c_2 t$.
17. $c_1(1 - 2t^2 + \frac{1}{3}t^4) + c_2(t - \frac{2}{3}t^3)$.
19. $c_1(1 + t^2) + c_2(t + \frac{1}{3}t^3)$.
21. $1 - \frac{1}{2}t^2 + \frac{1}{18}t^4$.
23. $1 - \frac{t^2}{2} + \frac{t^3}{6} + \frac{t^4}{24}$.

Section 6.5

3(a). $1 + c_1P_2(t) + c_2Q_2(t)$.
3(c). $-\frac{1}{30} + \frac{1}{2}t^2 + c_1P_5(t) + c_2Q_5(t)$.

Section 6.6

1. $12 - 48t^2 + 16t^4$.
3. $2 + 10t + \frac{2}{3}t^3 + \frac{5}{3}t^4$.
5. $1 + \frac{5}{3}t^3 + \frac{1}{9}t^6$.

Section 6.7

3(a). $c_1J_2(t) + c_2Y_2(t)$.
3(c). $c_1J_4(t) + c_2Y_4(t)$.

Section 6.8

1. $-\frac{1}{4}t^2 - \frac{3}{128}t^4 + \frac{3}{512}t^6$.

Chapter 8

Section 8.2

1. $u(0,t) = u(L,t) = u(x,0) = 0, \qquad u_t(x,0) = g(x).$

3. $Ae^{\sqrt{\mu}at} + Be^{-\sqrt{\mu}at}, \qquad Ce^{\sqrt{\mu}x} + De^{-\sqrt{\mu}x}.$

5. $\sum_{n=1}^{\infty} \sin\frac{n\pi x}{L}(A_n \cos\frac{n\pi at}{L} + B_n \sin\frac{n\pi at}{L}).$

Section 8.3

1. Series converges to 1 at $x = \pm 1$.

$$\frac{1}{2} + \frac{2}{\pi^2}\sum_{n=1}^{\infty} \frac{(-1)^n - 1}{n^2} \cos n\pi x$$

3. Series converges to 1 at $x = \pm 1$.

$$\frac{1}{3} + \frac{4}{\pi^2}\sum_{n=1}^{\infty} \frac{(-1)^n}{n^2} \cos n\pi x$$

5. Series converges to $1/2$ at $x = 1$ and to 0 at $x = -1$.

$$\frac{1}{4} + \sum_{n=1}^{\infty} \frac{(-1)^n - 1}{\pi^2 n^2} \cos n\pi x - \frac{(-1)^n}{\pi n} \sin n\pi x$$

7. Series converges to 0 at $x = \pm\pi$.

$$\frac{1}{\pi}(1 - \sum_{n=1}^{\infty} \frac{(-1)^n + 1}{n^2 - 1} \cos nx$$

Section 8.3.1

1. Series converges to 0 at $x = 0, \pm 1$.

$$\frac{2}{\pi}\sum_{n=1}^{\infty} \frac{1}{n} \sin n\pi x$$

3. Series converges to 0 at $x = 0, \pm\pi$.

$$\frac{2}{\pi}\sum_{n=1}^{\infty} \frac{n}{n^2 - 1}(1 + (-1)^n) \sin n\pi x$$

5. Series converges to 0 at $x = \pm 1$ and to $\pm 1/2$ at $x = \pm 1/2$.

$$\frac{4}{\pi^2}\sum \frac{\sin(n\pi/2)}{n^2} \sin n\pi x.$$

Section 8.3.2

1. Series converges to 1 at $x = \pm 1$.

$$\frac{1}{2} + \frac{2}{\pi^2}\sum_{n=1}^{\infty} \frac{(-1)^n - 1}{n^2} \cos n\pi x$$

3. Series converges to 0 at $x = \pm 1$ and to 1/2 at $x = \pm 1/2$.
$\frac{1}{4} - \frac{2}{\pi^2} \sum \frac{1}{n^2}(1 + (-1)^n - 2\cos(n\pi/2)) \cos n\pi x.$
5. Series converges to 1 at $x = \pm\pi$ and to 0 at $x = 0$.
$\frac{2}{\pi}(1 + 2\sum_{n=1}^{\infty} \frac{1}{4n^2 - 1} \cos n\pi x.$

Section 8.4

1. $\pi \cos 40t \sin x + \frac{\pi}{3} \cos 120t \sin 3x + \frac{\pi}{5} \cos 200t \sin 5x + \cdots.$
1. $\pi \cos 40t \sin x - \frac{3\pi}{8} \cos 80t \sin 2x + \frac{\pi}{9} \cos 120t \sin 3x + \cdots.$

Section 8.5

1. $100e^{-\pi^2 kt} \sin \pi x.$
3. $80e^{-\pi^2 kt/4} \sin \pi x/2.$
5. $\sum_{n=1}^{\infty} b_n e^{-\pi^2 kt/4} \sin \frac{n\pi x}{2},$
$$b_n = 200(1 - \cos n\pi)/n\pi.$$
7. $\sum_{n=1}^{\infty} b_n e^{-\pi^2 kt/4} \sin \frac{n\pi x}{2},$
$$b_n = \frac{400}{n^2\pi^2}(2\sin\frac{n\pi}{2} - \sin n\pi).$$
9. $100x + \sum_{n=1}^{\infty} b_n e^{-\pi^2 kt/4} \sin \frac{n\pi x}{2},$
$$b_n = 50 \int_0^2 (x^2 - 2x) \sin \frac{n\pi x}{2} dx.$$

11. 0°.
13. 53.5^o.
15. 41.1^o.
17. 8428sec.
19. ∞.
21. $100e^{-kt} \cos x$.
23. $100e^{-kt/4} \sin x/2$.
25. $\frac{400}{(2n-1)\pi}\left[1 - \cos(\frac{2n-1}{4}\pi)\right].$

Section 8.6

1. $42.55(e^{\pi y} - e^{-\pi y}) \sin \pi x.$
3. $\sum_{n=1}^{\infty} b_n (e^{(2n-1)\pi y/2} - e^{-(2n-1)\pi y/2}) \sin \frac{(2n-1)\pi x}{2},$
$$b_n = \frac{400}{(2n-1)\pi}(e^{(2n-1)\pi/2} - e^{-(2n-1)\pi/2}).$$

5. $$100 + \sum_{n=1}^{\infty} b_n(e^{n\pi x} - e^{-n\pi x}) \sin n\pi y,$$
$$b_n = \frac{200}{n\pi}(1 - \cos n\pi)(e^{n\pi} - e^{-n\pi}).$$

Section 8.7

1. $250^o \quad (r < 0.1), \qquad 25/r \quad (r > 0.1).$
3. $200(1 - 1/r).$

Index